Counterterrorist Detection Techniques of Explosives

Counterterrorist Detection Techniques of Explosives

Edited by

Jehuda Yinon
Department of Environmental Science,
Weizmann Institute of Science,
Rehovot, Israel

ELSEVIER

Amsterdam • Boston • Heidelberg • London • New York • Oxford • Paris
San Diego • San Francisco • Singapore • Sydney • Tokyo

Elsevier
Radarweg 29, PO Box 211, 1000 AE Amsterdam, The Netherlands
Linacre House, Jordan Hill, Oxford OX2 8DP, UK

First edition 2007

Library of Congress Cataloging-in-Publication Data
A catalog record for this book is available from the Library of Congress

British Library Cataloguing in Publication Data
A catalogue record for this book is available from the British Library

ISBN: 978-0-444-52204-7

For information on all Elsevier publications
visit our website at books.elsevier.com

Printed and bound by CPI Group (UK) Ltd, Croydon, CR0 4YY

Transferred to Digital Print 2011

To Bracha, the love of my life

Contents

Introduction

In the battle against terrorism, the detection of hidden explosives has become an issue of upmost importance. While terrorism is not new to the international community, terrorist attacks during the last years, such as the attack on the twin towers in New York on September 11, 2001, the Bali bombing in 2002, the train bombing in Madrid 2004, and the London subway and bus bombings in 2005, have brought the issue of detection of explosives to a high priority and have generated a great demand for rapid, sensitive, and reliable methods for detecting hidden explosives.

In Israel, a series of suicide bombers' attacks in buses, nightclubs, markets, and restaurants has shown the need for detecting explosives from a distance.

However, several attempts, such as Richard Reid's shoe bomb in 2001 and the London plot involving liquid explosives in 2006, were prevented.

The various bombings as well as the bombing attempts show the ingenuity and versatility of the various terrorist organizations. These result in major challenges to the scientific community who has to design and apply security screening technologies including new sensors and detecting devices which will prevent the smuggling of bombs and explosives in airports, train and bus stations, federal buildings, mail storerooms, etc.

The purpose of this book is to assemble in one volume a series of detection technologies of explosives, written by a group of scientists who are experts in each one of these technologies. It contains techniques that are already in use as well as techniques that are in advanced stages of development. Each technique is described in detail, including its principles of operation, as well as its applications in the detection of explosives.

The book includes several chapters on vapor and trace detection: chemiluminescence, mass spectrometry, ion mobility spectrometry, electrochemical methods, and micro mechanical sensors, such as microcantilevers. Other chapters deal with bulk detection techniques: neutron techniques, nuclear quadrupole resonance, X-ray diffraction imaging, millimeter-wave imaging, terahertz imaging, and laser techniques. Special chapters are devoted to personnel portals and to biological detection.

The purpose of this book is to be a reference volume for scientists and engineers in universities, research institutes, and industry involved in research and development of explosives detection techniques, as well as a textbook for graduate students in the natural sciences and in engineering.

Finally, I thank the contributing authors who have made the publication of this book possible. Many thanks are due to authors and publishers for permission to reproduce copyrighted material.

Jehuda Yinon

About the Editor

Dr. Yinon received his B.Sc. and M.Sc. degrees in electrical engineering from the Technion, Israel Institute of Technology (Haifa, Israel) and his Ph.D. in chemistry from the Weizmann Institute of Science (Rehovot, Israel).

Dr. Yinon conducted research, as a senior research fellow, at the Weizmann Institute of Science for over 35 years. He was a research associate (1971–1973) and a senior research associate (1976–1977) at Caltech's Jet Propulsion Laboratory (Pasadena, California) and spent sabbatical leaves as a visiting scientist at the National Institute of Environmental Health Sciences (Research Triangle Park, North Carolina) (1980–1981), the EPA Environmental Monitoring Systems Laboratory (Las Vegas, Nevada) (1988–1989), and the University of Florida (Gainesville, Florida) (1993–1994). His last appointment in the U.S., before returning to Israel, was as visiting professor and assistant director at UCF's National Center for Forensic Science (Orlando, Florida) (2000–2003).

Dr. Yinon's main activities involve applications of novel analytical techniques for the detection and analysis of hidden explosives.

Dr. Yinon has authored, co-authored, and edited nine books and published over 100 papers in the scientific literature.

He is a member of the editorial boards of several scientific journals, is a member of the NATO Advisory Explosives Detection Working Group, and serves as chairman of the International Committee for Symposia on Analysis and Detection of Explosives.

Dr. Yinon is currently an independent consultant on detection and analysis of explosives. He lives in Rehovot, Israel, where his consulting office is located.

Contributors

Robert Barat
Otto York Department of Chemical Engineering
New Jersey Institute of Technology
Newark, NJ 07102, USA

Avi A. Cagan
Department of Chemistry and Biochemistry
New Mexico State University
Las Cruces, NM 88003, USA

Frank C. DeLucia, Jr.
US Army Research Laboratory
AMSRD-ARL-WM-BD
Aberdeen Proving Ground, MD 21005, USA

Gary A. Eiceman
Department of Chemistry and Biochemistry
New Mexico State University
Las Cruces, NM 88003, USA

John F. Federici
Department of Physics
New Jersey Institute of Technology
Newark, NJ 07102, USA

Kenneth G. Furton
Department of Chemistry and Biochemistry
International Forensic Research Institute
Florida International University
Miami, FL 33199, USA

Dale Gary
Department of Physics
New Jersey Institute of Technology
Newark, NJ 07102, USA

Jennifer L. Gottfried
US Army Research Laboratory
AMSRD-ARL-WM-BD
Aberdeen Proving Ground, MD 21005, USA

Thomas E. Hall
Pacific Northwest National Laboratory
Richland, WA 99354, USA

Geoffrey Harding
GE Security Germany
D-22453 Hamburg, Germany

Ross J. Harper
Nomadics Inc., ICx Nomadics Inc.
Stillwater, OK 74074, USA

Ana M. Jimenez
Department of Analytical Chemistry
Faculty of Pharmacy
University of Seville
41012 Seville, Spain

Richard C. Lanza
Department of Nuclear Science and Engineering
Massachusetts Institute of Technology
Cambridge, MA 02139, USA

Kevin L. Linker
Contraband Detection Technologies
Sandia National Laboratories
Albuquerque, NM 87185, USA

Douglas L. McMakin
Pacific Northwest National Laboratory
Richland, WA 99354, USA

Kevin L. McNesby
US Army Research Laboratory
AMSRD-ARL-WM-BD
Aberdeen Proving Ground,
MD 21005, USA

Zoi-Heleni Michalopoulou
Department of Mathematical Sciences
New Jersey Institute of Technology
Newark, NJ 07102, USA

Joel B. Miller
Chemistry Division, Code 6120
Naval Research Laboratory
Washington, DC 20375, USA

Andrzej W. Miziolek
US Army Research Laboratory
AMSRD-ARL-WM-BD
Aberdeen Proving Ground,
MD 21005, USA

Chase A. Munson
US Army Research Laboratory
AMSRD-ARL-WM-BD
Aberdeen Proving Ground,
MD 21005, USA

Maria J. Navas
Department of Analytical Chemistry
Faculty of Pharmacy
University of Seville
41012 Seville, Spain

Hartwig Schmidt
Department of Chemistry
and Biochemistry
New Mexico State University
Las Cruces, NM 88003, USA

Larry Senesac
Nanoscale Science and Devices Group
Oak Ridge National Laboratory
Oak Ridge, TN 37831, USA

David M. Sheen
Pacific Northwest National Laboratory
Richland, WA 99354, USA

Thomas Thundat
Nanoscale Science and Devices Group
Oak Ridge National Laboratory
Oak Ridge, TN 37831, USA

Joseph Wang
Biodesign Institute
Arizona State University
Tempe, AZ 85287, USA

Jehuda Yinon
Department of Environmental Science
Weizmann Institute of Science
Rehovot 76100, Israel

Chapter 1

Detection of Explosives by Chemiluminescence

Ana M. Jimenez and Maria J. Navas

Department of Analytical Chemistry, Faculty of Pharmacy, University of Seville, 41012 Seville, Spain

Counterterrorist Detection Techniques of Explosives
Jehuda Yinon (Editor)

Contents

1. Introduction

The aim of this chapter is to explore the real and potential advantages of chemiluminescence (CL)-based trace detection systems for the detection of a broad group of explosives containing nitrogen. The literature reviewed covers about 25 years of research and advances in the use of CL for detecting explosives, from the 1980s to May 2006. Because the earliest papers in the 1970s and 1980s comprehensively describe how these techniques operate, we have used them to provide concepts and general information. The literature has been summarized to provide a contemporary understanding not only of the potential of CL systems but also of their implementation in many laboratories for the detection of trace explosives.

The chapter has been structured into several sections: first, there is a brief introduction, including a description of the concept and general principles of CL and how it is used in the field of explosives. The second section describes CL applications in the field of explosives and focuses in particular on the thermal energy analyser (TEA) because of its important role in the trace detection of explosives. The recent applications of luminol CL and electrochemiluminescence (ECL) to explosive detection are also described. Finally, because much of the research into explosive detectors has been directed towards civilian safety, a third section describes how CL is used as a security measure to detect explosives.

1.1. Principles of chemiluminescence

Energy can be transferred into (and out of) matter in many different ways, as heat and light or by chemical reactions. When energy is released by matter in the form of light, it is referred to as luminescence. An exception is usually made for matter that has such a high temperature that it simply glows; this is called incandescence. When energy in the form of light is released from matter because of a chemical reaction, the process is called CL [1]. CL is the generation of electromagnetic radiation (ultraviolet, visible or infrared) as light by the production of an electronically excited species from a number of reactants which goes on to release light in order to revert to its ground state energy. While the light can, in principle, be emitted in the ultraviolet, visible or infrared region, reactions emitting visible light are the most common. They are also the most interesting and useful. When the chemiluminogenic reaction occurs within living organisms, the phenomenon is named bioluminescence. CL in analytical chemistry has numerous advantages, such as superior sensitivity, safety, rapid and simple assay and controllable emission rate. However, it also has such disadvantages as poor reproducibility and long observation times. The CL process takes place at the rate of a chemical reaction, and the factors that affect emission intensity are in fact a combination of chemical reaction rate and luminescence considerations. A CL reagent may yield significant emission not just for one unique analyte, CL emission intensities are sensitive to a variety of environmental factors such as temperature, solvent, ionic strength, pH and other species present in the system, and the emission intensity from a CL reaction varies with time [2]. Today, many chemiluminescent systems are known, both biological and non-biological. Numerous inorganic and organic chemical reactions produce light because one of the reaction products is formed in an electronically excited state and emits the radiation on falling to the ground state.

A general description of the reactions is as follows:

$$A + B \rightleftarrows C^* + D$$

$$C^* \rightleftarrows C + \text{light}$$

where '*' indicates an electronically excited state [3]. This would be a rather simple case and would be classified as direct CL. In many systems, the initially excited state molecule is used as a conduit of energy to excite a second or third molecule that is the actual emitting species. When energy is transferred to an effective fluorophore (F) added to the system, the luminescence intensity increases considerably.

$$C^* + F \rightleftarrows C + F^*$$

$$F^* \rightleftarrows F + \text{light}$$

This would be classified as indirect, sensitized or energy transfer CL.

Nowadays, CL methods are a real alternative in analytical fields, and the applications that determine a wide variety of compounds have been extensively discussed in the literature.

1.2. Using CL to reveal the presence of explosives

Explosive detection techniques can be broadly classified into two main categories: bulk detection and trace detection. CL is commonly regarded as a trace detection technology. Trace detection involves the chemical detection of explosives by collecting and analysing tiny amounts of explosive vapour or particles (a microscopic amount of explosives) [4] and looking for residue or contamination from handling or being in proximity to explosive materials. Microscopic particles of solid explosive materials can adhere to a wide variety of surfaces (Teflon, glass, metal, plastic, etc.), and they can be detected by wiping the surface. Vapour detectors examine the vapour emanating from a liquid or a solid explosive, and because some explosives have a low vapour pressure, these detection techniques would be very sensitive.

Nitrogen-explosive compounds usually analysed by CL may be classified under three structural categories: (i) nitroaromatic compounds, (ii) nitrate esters and (iii) nitramines. Examples of nitro-substituted hydrocarbons are nitromethane, trinitrobenzene (TNB), trinitrotoluene (TNT) and pentantiroaniline. Nitroglycerine (NG), ethylene glycol dinitrate (EGDN) and pentaerythritol tetranitrate (PETN) are nitrate esters [5]. The nitro-explosive compounds that are the result of the presence of nitro and nitrate groups can be detected by CL methods, as we shall describe below (Section 2). A list of the most widely used nitrogen-containing explosives commonly determined by CL is given in Table 1. As explosives are usually referred to with an abbreviation, the Table gives both the compound name and the abbreviation. Figure 1 gives the structures of nitroaromatic and nitramine explosives commonly determined by CL.

Reviews and other papers (e.g. official reports or guides about instrumentation designed for explosive detection) describe CL as a reliable trace detection system for nitro explosives [4,6–13]. In a recent review, Moore [12] pointed out that vapour or particle detection systems are best suited for personnel monitoring. He has also stated how difficult it is to detect explosives, among other reasons, because of the low vapour pressure of most common explosives, vanishingly small in the case of the highest priority explosives, the attenuation of vapour by packaging or the thermal degradation of the material when the temperature is raised to achieve higher vapour pressure. Therefore, methods that relay on sampling of air spaces need to either sample very large volumes or have exceedingly small detection limits. Fine et al. [14] have pointed out that an ideal technique for ultra-trace analysis of explosives would be both simple and rapid, require minimal sample clean-up, be sensitive to as little as 1–10 pg quantities of all the compounds of interest and would work equally well both on complex samples from the real world and on high-grade laboratory standards made up in pure solvents. CL systems have many of the aforementioned qualities, and so it is one of the more commonly used techniques for detecting explosives [7].

Nevertheless, the use of CL for detecting explosives does have some drawbacks although some of them can be overlooked. First, it is limited to nitrogen-containing explosives. This does not reduce its usefulness too much because most major explosives are nitrogen compounds although it clearly makes it impossible to use CL to detect some types of explosives, such as peroxide explosives, that have become particularly important in forensic investigations because of the emergence of terrorist threats and crimes in which these explosives have been used [15]. The second disadvantage has to do with selectivity: CL per se is not capable of identifying what type of explosive molecule is present because several compounds, including those commonly used as taggants in plastic explosives, also contain NO_2 groups. This lack of selectivity inherent to most CL methods can be overcome by coupling CL with separation techniques. CL explosive detection technologies are usually fitted with a front-end gas chromatograph column. High-performance liquid chromatography (HPLC) or supercritical fluid chromatography (SFC) has been less used. The use of separation methods is also recommended because explosive residues usually have to be determined in complex matrices.

CL technology is a real alternative in trace explosive detection. It has excellent sensitivity and selectivity when combined with high-speed gas chromatography (GC). A wide range of explosives is detectable, including EGDN, NG, ammonium nitrate fuel oil (ANFO), TNT, dinitrotoluene (DNT), cyclotrimethylene trinitramine (RDX) and PETN. The detector does not require a radioactive source, and this can reduce both time and paperwork when transporting the system [16]. Commercial systems, easily hand-held or portable, are available and being used as security systems in different kinds

Table 1. List of common nitrogen-containing explosives analysed by chemiluminescence

Compound name	Abbreviation	Chemical formula	Class	CAS no.	Vapor pressure (High, moderate or low)
Cyclotrimethylene trinitramine	RDX	$C_3H_6N_6O_6$	Nitramine	121-82-4	Low
Cyclotetramethylene tetranitramine	HMX	$C_4H_8N_8O_8$	Nitramine	2691-41-0	Low
2-Nitrotoluene	2-NT	$C_7H_7NO_2$	Nitraromatic	88-72-2	High
3-Nitrotoluene	3-NT	$C_7H_7NO_2$	Nitraromatic	99-08-1	High
4-Nitrotoluene	4-NT	$C_7H_7NO_2$	Nitraromatic	99-99-0	High
2,3-Dinitrotoluene	2,3-DNT	$C_7H_6N_2O_4$	Nitroaromatic	602-01-7	High
2,4-Dinitrotoluene	2,4-DNT	$C_7H_6N_2O_4$	Nitroaromatic	121-14-2	High
2,5-Dinitrotoluene	2,5-DNT	$C_7H_6N_2O_4$	Nitroaromatic	619-15-8	High
2,6-Dinitrotoluene	2,6-DNT	$C_7H_6N_2O_4$	Nitroaromatic	606-20-2	Low
2,4,6-Trinitrotoluene	TNT	$C_7H_5N_3O_6$	Nitroaromatic	118-96-7	Moderate–low
Nitroglycerine	NG	$C_3H_5N_3O_9$/ $CH_2(ONO_2)$-CH (ONO_2)-$CH_2(ONO_2)$	Nitrate ester	55-63-0	High
Pentaerythritol tetranitrate	PETN	$C_5H_8N_4O_{12}$/ $C(CH_2ONO_2)_4$	Nitrate ester	78-11-5	Low
Ethylene glycol dinitrate	EGDN	$C_2H_4N_2O_6$/ NO_2-OCH_2CH_2O-NO_2	Nitrate ester	628-96-6	High

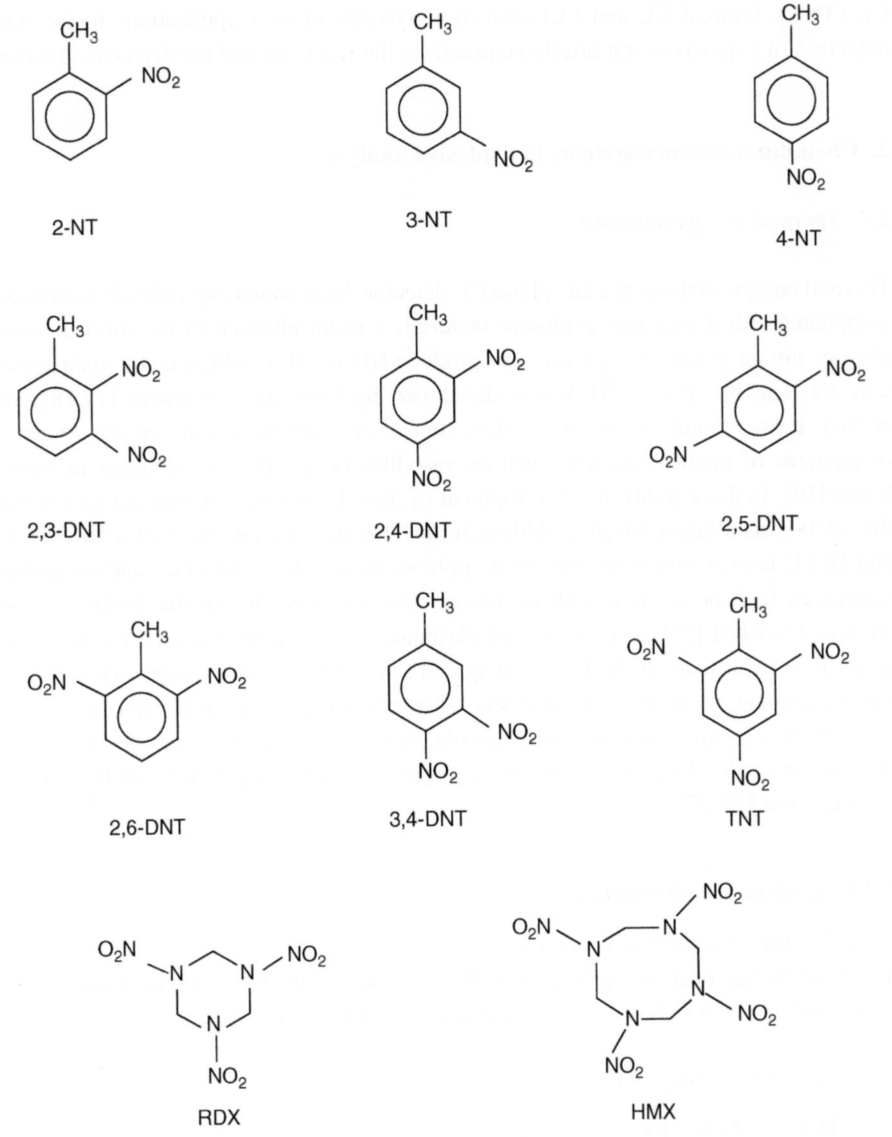

Fig. 1. Structures of nitroaromatic and nitramine explosives commonly determined by chemiluminescence. DNT, dinitrotoluene; HMX, cyclotertamethylene tetranitramine; NT, nitrotoluene; RDX, cyclotrimethylene trinitramine; TNT, trinitrotoluene.

of installations, such as in airports or forensic laboratories. A considerable number of papers have reported the application of luminol CL, anyway ECL and immunoassay (IM) techniques for explosive detection. Nevertheless, most of these papers deal with the CL reaction of the nitrosyl (NO) radical with ozone and the use of the TEA manufactured by Thermo Electron Corporation (Waltham, MA, USA). Section 2 deals with NO–ozone

CL (TEA), luminol CL and ECL detection systems in their applications to the trace detection of explosives and briefly summarizes the reactions and mechanisms involved.

2. Chemiluminescence systems in explosive analysis

2.1. Thermal energy analyser

Thermal energy analyser is a gas-phase CL detector that responds specifically to nitrogen compounds. Most common explosive materials contain nitrogen in the form of either nitro or nitrate groups that pyrolyse to produce NO or NOx, which is the main reason why CL can be applied. TEA was developed by Fine and coworkers [17,18], and at first, it was mainly used for the detection of *N*-nitrosamines in complex matrices or analytes of medical interest, such as vasodilators and their metabolites in human blood [19]. In these fields, the development of the CL-specific nitrogen detector during the 1970s was a major advance. Although the TEA detector can be used with both gas and HPLC in explosive detection, most applications couple TEA to GC, and the earliest references to it being used with explosives that appeared during the 1980s [20–24]. TEA is described [23] as a very useful chromatographic detector for the forensic trace analysis of explosives. Its high sensitivity and selectivity for nitro compounds allow the rapid screening of contaminated extracts for traces of explosives, the majority of which contain nitro groups, with the minimum likelihood of false-positive results. A detailed description of the detector and the physical principles of its operation can be found in the literature [25–27].

2.1.1. Analytical considerations

2.1.1.1. The chemiluminescence reaction
The fundamental operating principle of TEA is based on the CL reaction between nitric oxide and ozone, which can be schematized as follows [28]:

$$NO + O_3 \rightarrow NO_2^* + O_2$$

$$NO_2^* \rightarrow NO_2 + h\nu$$

The nitric oxide reacts with ozone to produce an electronically excited specie (NO_2^*) that decays back to its ground state emitting light in the infrared spectral region (600–2800 nm).

2.1.1.2. Mode of operation
The earliest references above mentioned (Subsection 2.1) describe the fundamental operating principle of the TEA, which can be briefly summarized as follows: the components of interest eluting from the chromatograph are introduced into a catalytic pyrolyser where NO_2 is released from organic nitro compounds and simultaneously converted into a

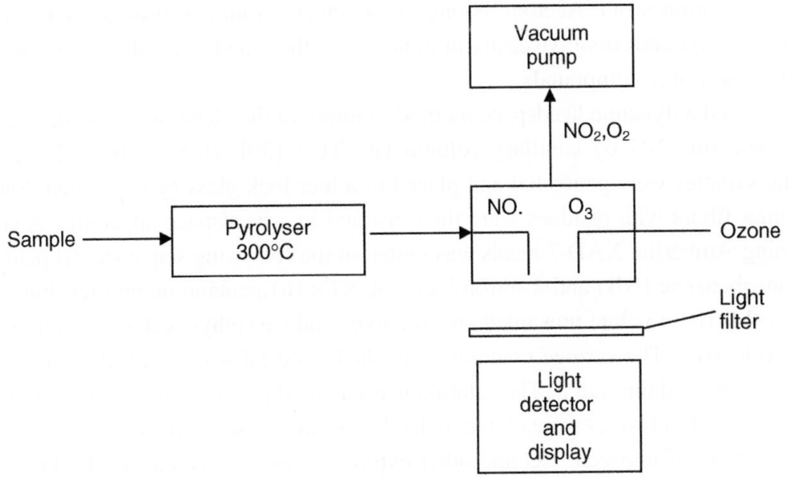

Fig. 2. Simplified schematic of the thermal energy analyser. Reprinted from Fine et al. [25]. Copyright (1975), with permission from Elsevier.

NO radical by the catalytic surface. The gaseous pyrolysis products are swept by a carrier gas from the pyrolyser through a cold trap (−100 or −150°C) or a solid sorbent device where most contaminants are condensed, but NO radicals survive. The ozone reaction generally takes place at reduced pressure in a reaction chamber where the NO radical is oxidized by ozone to form the electronically excited species NO_2^* that is detected by an appropriate photomultiplier tube (PMT). This operating principle produces peaks only for substances containing nitro or nitroso groups. Because the intensity of light from the chemiluminescent reaction can be increased at reduced pressure, a vacuum pump is always used. A red cut-off filter is used to eliminate light emission resulting from potential interferences. A schematic diagram of the TEA is shown in Fig. 2.

One nice feature of CL detectors is that they do not use a radioactive ionization source [29] and thus avoid some of the paperwork and regulatory oversight that may be associated with ion mobility spectrometer detectors.

2.1.1.3. *Sampling and preparation of the samples*

The papers reviewed show that the general procedure for collecting organic explosive residues from surfaces includes cotton swabs or vacuum onto a filter. Owing to the complexity of the samples (debris from the scene of an explosion, contaminated soil, etc.) and/or the need for pre-concentration, samples often require pre-treatment, which eliminates possible interfering substances. So the literature reviewed also contains reports on such sampling methodologies as solid-phase extraction (SPE) or supercritical fluid extraction (SFE), which describe how they are used to isolate explosives in combination with chromatographic techniques.

A common procedure for taking samples consists of using solvent-moistened cotton wool swabs. The same solvents are then often used to extract the explosive residues from the soaked cotton swab. Acetone is one of the most commonly used solvents because explosive compounds dissolve easily in it; however, the drawback is that it also dissolves other non-desirable compounds.

Douse used a dynamic headspace method to improve the clean-up of gunshot residues before detecting NG by capillary column GC–TEA [30]. Handswabs and vacuumed clothing samples were generated and placed in a luer lock glass syringe fitted. Syringes containing filters with residues were then inserted in a headspace apparatus. A column containing Amberlite XAD-7 beads was eluted in the following sequence: (i) pentane to elute nitrobenzene (NB) and 4-nitrotoluene (4-NT); (ii) pentane-methyl-tert-butyl ether (pentane-MTBE) to elute unwanted coextractives and (iii) ethyl acetate to elute NG and other explosives. The syringe containing residual material was clamped vertically, and MBTE was passed through it. The solution was passed through a glass column containing XAD-7 beads. Explosives were eluted with ethyl acetate, and this fraction was evaporated under a stream of nitrogen. NG and other explosives were detected by GC–TEA.

Thompson et al. [31] compared the aqueous recovery and acetone extraction from cotton swabs of organic explosive residue, which was generated by mixing standards in motor oil on aluminium foil, by detonating four different bombs hidden inside suitcases filled with clothing and by handling plastic explosives. The explosives were isolated by SPE, using a poly-*N*-vinylpyrrolidone-divinyl-benzene sorbent. Samples were screened by liquid chromatography-ultraviolet (LC-UV), and the presence of explosives was confirmed by LC– or GC–mass spectrometry (MS) and fast GC–TEA (EGIS). The authors concluded that the water extraction/SPE procedure was an effective process for treating organic explosive residue on cotton swabs for subsequent analysis by LC.

Kolla [32] used SPE and GC–TEA to prepare and analyse samples of explosives and found that SPE was the most useful way of reducing the contamination of the column by accompanying substances. The SPE columns were 500-mg RP18 (3 cm) columns. The original sample, whether dust or swabs, was extracted with acetone in a Soxhlet extractor. The acetone was then evaporated and the sample dissolved in acetone and diluted at 1:10 with water. After washing, the elution was performed with three volumes of methanol. After the methanol had been evaporated, the sample was dissolved in a small volume of acetone for GC–TEA.

The effectiveness of using SFE coupled to GC with a thermal desorption modulator (TDM) interface and TEA detection to analyse explosives has been evaluated by Francis et al. [33] in soil samples and standards.

2.1.1.4. Pyrolyser

In the analysis of nitroaromatic explosives, TEA is similar to the systems used for nitrosamines except for the fact that temperatures must be higher. Several papers have reported that the temperature of the TEA pyrolysis furnace can affect the selectivity of the detector. A study carried out by Douse [34] to detect traces of several explosives at the low nanogram level revealed that the chromatograms obtained by GC–TEA for both pure compounds and spiked handswabs clearly showed the superior selectivity of TEA in reducing the temperature from 800 to 700°C. Reduction from 700 to 550°C further demonstrated the selectivity of TEA, but this was offset by a partial loss of response for all explosives. It was found, however, that the selectivity of TEA using a pyrolysis of 550°C was such that as much as 13% of the sample could be analysed with little effect on the background. On the contrary, from the literature reviewed, it can be seen that, for nitroaromatic compounds, the release of nitric oxide was maximum between 800 and 950°C (though even 1000°C has been reported) [21]. Lafleur and Mills [20] studied the pyrolytic release of nitric oxide (produced from a series of NTs) as a function of temperature. The response was maximum at temperatures above 800°C and decreased by almost three orders of magnitude at temperatures below 500°C. They finally selected a temperature of 900 °C to ensure maximum thermal decomposition and thus maximum release of nitric oxide. The relative response was highly reproducible, and the release of nitric oxide was, therefore, also reproducible but not necessarily molar; that is, one nitro group did not necessarily yield one molecule of nitric oxide. Relative molar response ranged from 0.59 (2-NT) to 2.25 (2,4,6-TNT). A pyrolysis temperature of 950°C was chosen by Phillips et al. [35] to determine nitroaromatics, NB, 2,4-DNT and 2,6-DNT in biosludges by GC/TEA.

Selavka et al. [36] have reported that for LC, a pyrolysis chamber temperature greater than 550°C gives rise to substantial baseline noise, which severely limits the detector's sensitivity and selectivity. Thus, LC–TEA is always performed in the 'nitroso mode' (the pyrolyser is held at 550°C).

2.1.1.5. Selectivity

The use of a selective detection system is of primordial importance in reducing or eliminating interference in the signal of the analytes of interest. In this respect, TEA coupled to chromatographic techniques is one of the most selective devices. When TEA is used, the analytical signal is not affected by compounds without nitro functional groups, which may be in the sample. Fine et al. [14,26] have suggested that four factors are responsible for TEA's high degree of selectivity. First, only compounds that have NO_2 or NO functional groups can give a response. Second, the reactive species must survive the −100°C cold trap. For highly contaminated samples, this temperature can even be −160°C. At these temperatures, the vapour pressure of the NO radical is greater than 1 atm, whereas the vapour pressure of almost all organic compounds is substantially less. Third, the reactive species must react with ozone to produce a chemiluminescent light in a narrow range between 600 and 800 nm. Fourth, the reaction with ozone must be sufficiently rapid to take place during the residence time of the species in the reaction chamber. Phillips et al. [35] have compared the TEA results with those of other selective GC detectors (thermionic specific, electron capture and Hall electrolytic conductivity) to analyse nitroaromatics in biosludges. The chromatograms obtained from spiked extracts with the other detectors all contained many interfering compounds that eluted at retention times close to the nitroaromatics, which made it necessary to modify the sample workup so that the samples were amenable to the different selective detectors. They found that the selectivity of the TEA made sample clean-up unnecessary and solved all the problems of the solvent effects.

The Guide for the Identification of Intact Explosives from The Technical Working Group on Fire and Explosion (TWGFEX) of the National Center of Forensic Science [37] classified the analytical techniques into four categories, and GC–TEA and LC–TEA were in the category that provides a high degree of selectivity.

2.1.1.6. Sensitivity

Traces of explosives are commonly present in very low levels in samples that are analysed, so it is important to take sensitivity into account when designing detectors for explosive detection. As a rough 'rule of thumb', Nambayah and Quickenden [38] reported that a method suitable for direct explosive vapour detection should be able to detect explosive concentrations at less than 1 ng/L. They made an exhaustive study of the lowest experimental detection limits achieved with various analytical techniques reported in the literature on traces of explosive, and they informed that headspace GC-electron capture detector (ECD) followed by immunosensor techniques achieves the lowest detection limits (from 0.07 to 20 ng/L).

The literature reviewed shows that GC/TEA methods can be extremely sensitive and respond to a few picograms of common explosives. Fine et al. [14] using GC/TEA (capillary columns) found that the minimum detectable level at a signal-to-noise ratio of 3:1 is estimated to be 4 pg for TNT and RDX; 5 pg for EGDN, NG and DNT; and 25 pg for tetryl when they determined explosives in 'real world' samples without previous clean-up.

Although the reported sensitivity of TEA when it is used in combination with HPLC or packed column GC to analyse explosives is only in the low nanogram range, Douse [21] described a method that uses fused silica capillary column GC in conjunction with TEA detection for trace analysis of explosives in the low picogram range. He reported that silica capillary column GC can be used to overcome the problems of very polar explosives adsorbing on the packed columns used for GC analysis and in the transfer lines of the TEA. In this way, the detection limits were lowered. He also concluded that the minimum detectable levels (15,10 and <200 pg for NG, TNT and RDX, respectively) of the compounds studied were similar to those obtained when ECD was used.

2.1.2. Applications of TEA to the analysis of explosives

For complex matrices, a separation technique used to be necessary to obtain a clean sample for explosive analysis. Also, because of the lack of selectivity of CL, explosives are generally introduced into the TEA after separation. Chromatographic techniques coupled to sensitive detection systems are widely used to detect explosives. Most of the TEA methods for detecting and identifying nitrogen-containing explosives rely on GC separation techniques, as can be seen in most of the literature reviewed and in the commercial technologies available today. TEA in combination with certain chromatographic techniques can exhibit detection limits in the low picogram range. Its selectivity is good, and it has become the method used for detection and confirmation by a considerable number of laboratories. As far as we know, there has been little bibliography on HPLC and SFC, but some earlier papers suggest that they may be able to be used for explosive analysis. The selective and sensitive detection of explosive compounds by coupling HPLC with TEA has been limited to nitrate esters and nitramines, because when nitroaromatic compounds have been investigated under the same conditions, the molar response is relatively poor [20,39].

2.1.2.1. *Gas Chromatography–thermal energy analyser*

In GC, a sample containing the compounds of interest, which may be gas or liquid, is pushed through a column by an inert carrier gas (the mobile phase). The sample moves through a packed or capillary column at different speeds. The sample's components are separated and emerge at different 'retention times' because they are distributed differently between the mobile and the stationary phases. The detection system responds to the presence and concentration of the separated compounds, thus making it possible to identify and quantify them.

The potential of GC, HPLC and ion chromatography (IC) coupled with sensitive detection techniques, including TEA, for the trace detection of explosives has been investigated by Kolla [40]. GC is preferred for organic compounds, which can be vapourized without decomposition. Most of the explosives that can be detected by TEA belong to this group. TEA with GC is presented as one way of solving some of the problems associated with MS (i.e. the identification of nitrate esters or RDX in complex matrices). Kolla has pointed out that GC is the method of choice in general analytical operations, but the usual GC systems have to be applied with slight modifications because not every explosive is stable at higher temperatures. In order to be able to analyse all explosives, a compromise has to be found for GC conditions. An injection port with an insert that traps the non-vapourizable substances should be used (a split-splitless injector with a fritted glass insert). The optimum injection temperature was 170°C. The column used was a 10-m DB-5 capillary column. The column temperature was initially 50°C and was increased gradually to 250°C (10°C/min). A polymethylphenyl (5%) siloxane (DB-5) was used to enhance the selectivity of the column for the nitro compounds. Despite the specificity for nitro and nitroso compounds, in some samples, in addition to the explosive constituents, many unidentified peaks appeared in the chromatogram. The author pointed out the need for confirmation with at least one different method that must have a similar sensitivity and should have also a selective detector. A simple column change to a stationary phase of different polarity (that is to say, a stationary phase with a higher phenyl content or with additional cyanopropyl groups in polymethylsiloxane) in the GC–TEA system may make this confirmation possible. The author also reported that the TEA detector can be coupled to HPLC, but the greatest disadvantage of this approach is the lower sensitivity that results from broad peak shape and poor resolution. Confirmation is also possible by GC–MS or electron capture detector. The first, though well suited to the determination of nitroaromatics, presents some of the problems already mentioned, and the second, because of its lower selectivity, is only recommended if the matrix is relatively clean. Another paper published by Kolla and Sprunkel [41] deals with the identification of dynamite explosives in post-explosion residues. In the post-blast residues of dynamites that contain nitroaromatic compounds (i.e. European dynamites), the isomers of DNT and/or TNT can be detected, so it is possible to identify dynamite by determining the ratios of the nitroaromatic constituents. Six German dynamites were selected, and their compositions were analysed. The dynamites selected were detonated as free hanging charges, and the samples, taken from metal plates placed at different distances from the centre of the explosion, were analysed by GC–TEA.

The effectiveness of coupling SFE/GC with TDM and TEA detection was evaluated by Francis et al. [33] as a system for analysing explosive compounds [1-nitropyrene (1-NP), 2-nitronaphthalene (2-NN), 2,4-DNT, 2,6-DNT, 1,3,5-TNB, TNT, PETN, RDX and NG] in soil samples and standards. The system can be used for screening small samples in short periods of time. Citing other investigators, the authors claim that the system has the following advantages: it allows the extraction process to be monitored in real time, it provides very small injection bands for sample introduction and it is compatible with narrow-bore capillary columns. This study used neat or modified carbon dioxide (CO_2) as mobile phase and capillary columns (50 mm ID) with methylpolysiloxane and p,p'-cyanobiphenylmethylpolysiloxane stationary phases. Nitroaromatic compounds were easily extracted and eluted, but nitrate esters (PETN and NG) decomposed during thermal desorption in the modulator (a direct correlation between the temperature of the TDM and the extent of the decomposition was observed), which is one of the limitations of the method. Organic modifiers (15% acetone) had to be used to improve the extraction of some explosives, such as TNB or 1-NP. Nitramine or RDX could not be extracted under these conditions, and the use of methanol as a modifier for their extraction led to plugging of the 5-mm ID restrictor. The analysis of 200 mg of soil samples containing 24 ppb of 2,4-DNT (data obtained by GC/MS and SFC/MS) was complete in less than 10 min, and the minimum detectable quantity for 2,4-DNT was found to be 2.6 ppb (S/N = 3). The authors concluded that the proposed method can quickly analyse relatively volatile and thermally stable nitro compounds from solid matrices.

Trace explosive detection is a typical activity of forensic laboratories, and several laboratories involved with explosive investigations have adopted GG–TEA for this purpose. In several publications, Douse (from The Metropolitan Police Forensic Science Laboratory, London, UK) has described the application of capillary column GC–TEA to the trace analysis of explosives. Some of his preliminary papers [21,34] have shown that this combination leads to very broad peaks that impair resolution and thus reduce the power of the method for identifying and discriminating peaks due to unknown explosives. Attempts at improving the peak shape by optimizing the chromatographic conditions were unsuccessful. So, subsequent studies describe modifications in the TEA detector to improve selectivity [22]. Samples were cleaned up using columns of Amberlite XAD-7 beads and mixture of methyl tert-butyl ether and n-pentane as eluting solvent when necessary. TEA operating conditions included TEA pyrolysis oven temperature between 550 and 800°C, an ozone flow rate of 1.3 mL/min and a reaction chamber pressure of 0.29 mm Hg at an oven temperature of 60°C. One modification consisted of replacing the standard amplifier by a different amplifier and noise filtration system that gave better peak shape. Under the optimum conditions, 5 pg of NG, TNT and RDX were readily detectable. Moreover, selectivity was better when the pyrolyser catalyst tube normally fitted on the TEA detector was replaced with a length of silica capillary tubing. The selectivity of the system was such that vacuum extracts from clothing could be analysed for traces of organic firearms discharge residue without the need for sample clean-up. In addition, the cryogenic trap could be eliminated when TEA was used for trace analysis in contaminated extracts. Another advantage of the silica pyrolysis tubes reported by the author was the low price of replacement lengths of uncoated capillary, their long lifetime and the extreme ease of replacement of the pyrolysis column. As Douse has reported, these modifications considerably simplify routine operation and the maintenance of the TEA detector. Collins continued these studies [24] and in a subsequent paper identified the source of the peak broadening within the electronic circuitry. He also described a simple modification of the TEA [the original capacitor C2 (0.47 μF) was first replaced with a 0.1-μF component and later with a 0.022-μF capacitor], which results in much better capillary GC peak shape. Nevertheless, the above modifications resulted in more noise, which was obtrusive at higher sensitivities. To minimize this last problem, the author described the use of a supplementary electronic filter.

Subsequently, Douse [30] used the GC/TEA system to detect NG in gunshot residues. With pure standards, it was possible to detect down to about 5 pg of NG, but such sensitivity cannot be achieved with extracts from handswabs or vacuumed samples from clothing, for such extracts were substantially contaminated with lipids and other materials. The clean-up is adequate for many samples, but during experiments on clothing, it was frequently found that contamination was too great. This was studied, and a procedure was developed that exploits the volatility of NG to provide a primary clean-up by trapping the NG vapour on to XAD-7 beads, followed by solvent elution and GC–TEA detection. The dynamic headspace procedure described effectively separates volatile explosives such as NG from involatile impurities that reduce the selectivity. By trapping the volatilized explosives on to XAD-7 beads and effecting additional clean-up, detection limits of sub-nanogram quantities per swab or filter can be achieved. The results confirmed that the GC/TEA system is a robust combination that provides both sensitivity and selectivity.

Since 1989, GC–TEA has been used by the Forensic Explosive Laboratory (FEL; Kent, UK) as one of the techniques for explosive trace analysis. In this year, the FEL established a weekly quality assurance testing regime in its explosive trace analysis laboratory in an attempt to prevent the accumulation of explosives, which could result in contamination of samples and controls. GC–TEA was usually used for initial sample screening and identification. Confirmation was carried out when possible by GC–MS. The GC–TEA systems adopted were essentially as described in a previous paper by Douse [34]. Since 1989, several articles have described the results obtained by the FEL investigators during approximately a decade. The general procedures and conditions for organic explosives are summarized below. They use cotton wool swabs or vacuum onto a filter for sampling (with two different sampling solvents: a mixture of ethanol and water in equal volumes and methyl-tert-butyl ether), a standard TEA solution, generally containing 11 high common explosives (EGDN, 2-NT, 3-NT, 4-NT, NG, 2,4-DNT, 2,6-DNT, 3,4-DNT, 2,4,6-TNT, PETN and RDX) at low concentrations (between 0.1 and 0.75 ng/μL) analysed before and after the sample, three different types of columns – dimethylsiloxane (type BP1), 5% diphenyl-dimethylsiloxane (SGE type BP5) and 7% cyanopropyl-7% phenyl-1% vinyl-dimethylsiloxane (type CPSIL-19) – for GC and confirmation when possible by GC with MS. The markers that were commonly used to provide reference peaks in GC analyses were 2-fluoro-5-NT and the fragrance musk tibetine (2,6-dinitro-3,4,5-trimethyl-*tert*-butylbenzene). Hiley [42] used the GC–TEA system in these conditions to study chemical analysis methods for post-explosion investigation and in particular those methods developed by the FEL's Chemical and Electronic Sector and applied to explosive incidents on the UK mainland. The paper focuses on the importance of promptly collecting, analysing and identifying samples of material that are not visible to the naked eye, if the investigation is to be satisfactory. He reports that the systems will detect at worst 50 pg per 0.8 μL injection of the major explosives, which corresponds to about 6 ng in a 100-μL sample.

The presence of dinitrosopentamehylenetetramine (DNPMT), a potential interference in the detection of explosive traces, was revealed in samples taken from a soft suitcase using GC–TEA [43]. This compound is widely used as a chemical blowing agent in the manufacture of foamed polymers and, when analysed by GC–TEA, produces a strong response that might incorrectly be taken as indicative of the presence of PETN or RDX. For GC–TEA analysis, 'UNICEL 100' (DNPMT) was used and, for purposes of comparison, the mixed standard solution (TEA standard) containing a low concentration of the common high explosives was used. The pyrolysis furnace was held at 750°C and directly connected into the reaction chamber. In comparison with chromatogram of the TEA standard, the chromatograms of the DNPMT solution showed strong sharp peaks. The FEL standard method for explosive trace identification by GC/TEA required that the relative retention times of the suspect explosive peak and standard explosive peak agreed within ±0.5%. Under this criterion, DNPMT did not coincide sufficiently with PETN or RDX. Nevertheless, the author concluded that GC/TEA analysis with columns of a similar polarity to those used in the study would incorrectly consider DNPMT as indicative of the presence of PETN or RDX.

In another FEL study, Crowson et al. [44] carried out a survey to determine the background level of explosive traces in public places. Samples were taken at various transport sites (buses, taxis, underground trains, passenger aircraft, etc.), and police sites were also sampled to assess how likely it is that a suspect could be contaminated. A pyrolysis oven temperature of 750°C, an interface oven temperature of 250°C and a reaction chamber pressure reading of 0.5–2 mm Hg were the instrument settings. Although the most common high explosives can be detected by the technique in question, the explosive cyclotertamethylene tetranitramine (HMX) and the propellant ingredient nitrocellulose are not sufficiently volatile to be detected. Limits of detection for the complete procedure were estimated for the explosives NG, TNT, PETN and RDX in both clean and dirty samples using the three columns mentioned above. In the case of the BP-5 column and clean samples, the limits of detection (ng) were 1.0, 1.1, 2.2 and 0.6.

The results of this survey led the authors to conclude that traces of high explosives were rare within the general public environment. In fact, no traces of NG, TNT or PETN were detected at any of the public sites sampled during this project. Only four low level traces of RDX were detected. Therefore, they concluded that it was unlikely that a member of the public would be innocently contaminated with a significant quantity of explosives.

A second survey from FEL was presented in 2004 [45]. This study examined levels in four of the UK's major cities: Birmingham, Cardiff, Glasgow and Manchester. Samples were collected from taxis, buses, trains, airports, hotels, privately owned vehicles, and clothing purchased from charity shops. The samples were analysed as described in the study by Crowson et al. [44] with the exception of a modification to the temperature programme of the gas chromatographs to allow screening for NB. Traces of the high explosives NG, TNT, PETN and RDX were rare within the general public environment. The results of the survey indicate that it is unlikely that persons visiting public areas could become significantly contaminated with explosives.

Warren et al. [46] described a simple processing and analysis scheme for explosive trace swab samples which deals both with organic and inorganic materials. Swabs, wetted with ethanol or ethanol/water mixture, were extracted with ethanol/water mixture and applied directly through a simple column containing an acrylonitrile/styrene copolymer adsorbent. The adsorbent retained common organic explosives that were recovered with an efficiency of 30–50% and analysed using GC-CL or GC–MS detection. Figure 3

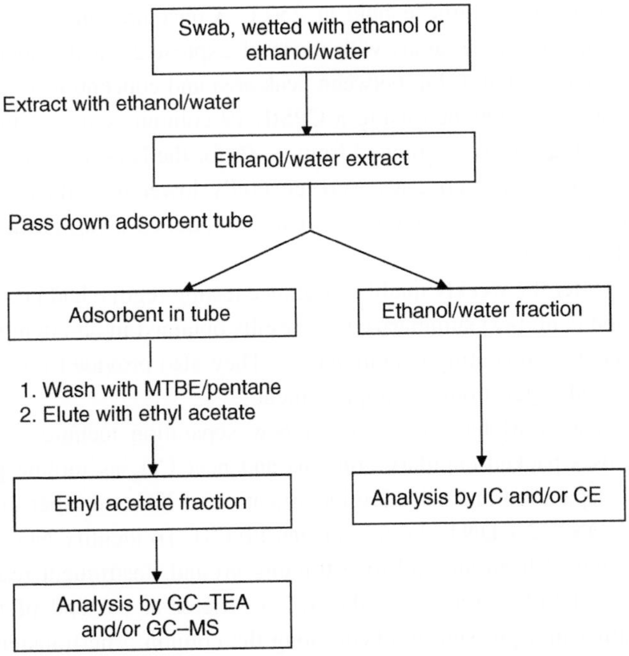

Fig. 3. Processing scheme. Reprinted with permission from Warren et al. [46]. Copyright (1999) Forensic Science Society. CE, capillary electrophoresis; GC, gas chromatography; TEA, thermal energy analyser; IC, ion chromatography.

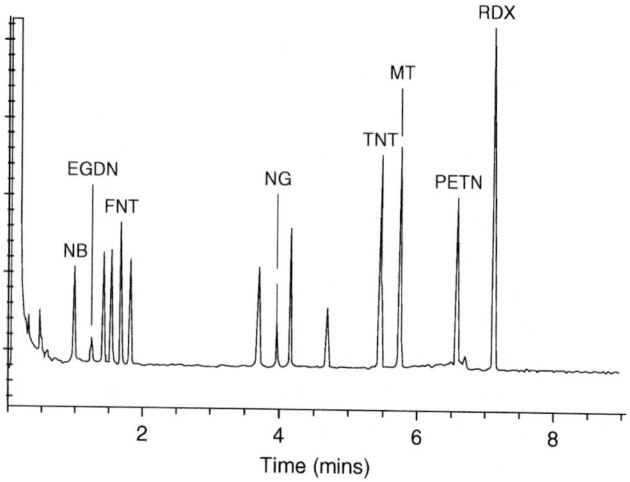

Fig. 4. Example of thermal energy analyser (TEA) standard analysis using CPSIL-19 column type. EGDN, Ethylene glycol dinitrate; FNT, 2-fluoro-5-nitrotoluene; MT, musk tibetine; NB, nitrobenzene; NG, nitroglycerine; PETN, pentaerythritol tetranitrate and RDX, cyclotrimethylene trinitramine; TNT, trinitrotoluene. Reprinted with permission from Warren et al. [46]. Copyright (1999) Forensic Science Society.

shows a processing scheme. For GC/TEA analysis, the three columns mentioned above are used with a pyrolysis furnace held at 750°C. Concentrations of explosives were estimated by comparing peak areas with those of explosives in the standard solutions and assuming a linear relationship between peak area and concentration. A typical TEA standard chromatogram obtained using a CPSIL-19 column is shown in Fig. 4. Mean recoveries of organic explosives ranged from ~4% for the NTs to over 50% for PETN. The authors reported that recoveries were generally lower than those obtained using XAD7 clean-up and that higher recoveries can be obtained using clean-up tubes with lower sample flow rates.

Twelve years after the weekly quality assurance testing regime had been implemented in FEL, Crowson et al. [47] summarized the results obtained in an attempt to determine the best methods for preventing contamination. They also provided information on the lessons learned and suggestions for improvement.

Bowerbank et al. [48] have described a new separation technique, solvating GC (SGC), which uses packed capillary columns and neat CO_2 as mobile phase coupled to TEA for the separation and selective identification of NG and other nitro-containing explosives (2,6-DNT, 2,4-DNT, 2,4,6-TNT and PETN). To identify NG, they used two soil samples obtained from an explosive training ground. Instrument requirements are similar to those of SFC, except that the SFC restrictor at the end of the column is removed, resulting in a pressure gradient along the column. The transfer line between the SGC column and the TEA pyrolyser unit consisted of a 50 cm length of large bore (530 μm ID × 700 μm OD) fused silica tubing. This was made necessary by the fixed distance between the outlet of the SGC column and the inlet to the pyrolyser. In this

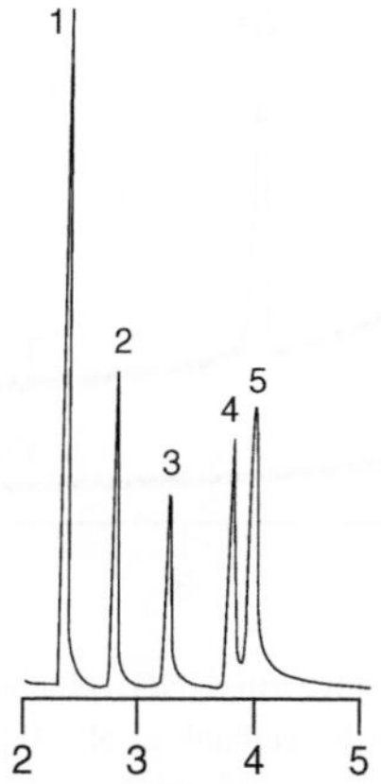

Fig. 5. Solvating gas chromatography (SGC) separation of a standard explosive mixture. Injected amounts: 0.33 μg nitroglycerine (NG), 3.3 μg all others. Peaks (1) nitroglycerine, (2) 2,6-dinitrotolene, (3) 2,4-dinitrotoluene, (4) 2,4,6-trinitrotoluene, (5) pentaerythritol tetranitrate (PETN). Reprinted from Bowerbank et al. [48]. Copyright (2000), with permission from Elsevier.

study, the authors pointed out the following benefits and advantages of coupling SGC to TEA: the CO_2 mobile phase does not give rise to added background noise; it does not increase the limit of detection of the TEA due to the absence of nitro or nitroso functional groupsand the selective nature of the TEA makes it possible to use certain organic modifiers in the CO_2 mobile phase to elute highly retentive compounds. Modifiers such as dichloromethane or methanol cannot be used with flame ionization detection (FID) because they significantly increase the baseline signal. The TEA calibration curve for NG showed linearity in the sub-micrograms per millilitre range. The authors reported an SGC separation of a standard mixture of nitroaromatic explosives in ~4 min using neat CO_2 and TEA as selective detector. Figure 5 shows a chromatogram obtained for nitroaromatic compounds (conditions: 65 cm × 320 μm ID column packed with octadecylsilane (ODS)-bonded porous particles, neat CO_2 mobile phase, 250 atm, 130°C isothermal oven, 740°C pyrolyser, 250°C transfer line, 200 nL injection and 12:1 split ratio) in which peak tailing is attributed by the authors to the use of the fused silica transfer mentioned above. When the same packed column was placed in an SGC-FID, no peak tailing was observed. The results obtained in a standard NG sample containing 0.40 μg/mL (0.2-μL injection), which corresponded to ~6 pg on-column after taking the split ratio (12:1) into account, were compared with those obtained with flame ionization detector (the most commonly used with SGC). As Fig. 6 shows, the TEA gives a strong NG peak, whereas FID failed to detect the compound (conditions as in Fig. 5). The authors concluded that SGC is capable of performing separations with half-height peak widths in the order of 200 ms, but a detector with greater sensitivity than FID (TEA or MS) is necessary for sensitive and selective detection of NG and nitroaromatics.

Numerous terrorist activities as well as many serious criminal offensives involve firearms. When a firearm is discharged, a variety of materials is emitted by the muzzle (accompanying the projectile), including primer and gunpowder (propellant) residues

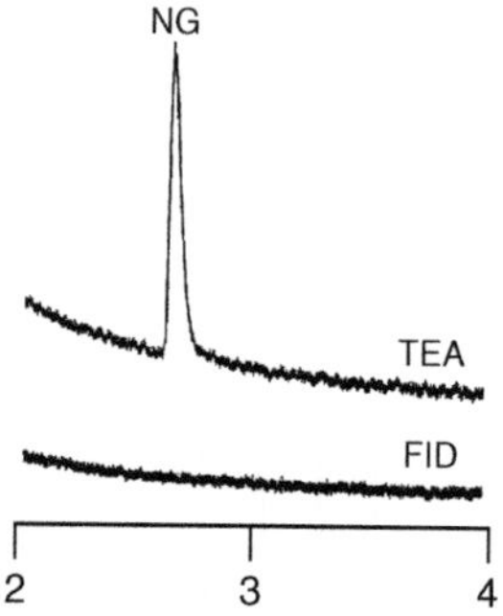

Fig. 6. Chromatograms of standard nitroglycerine (NG) obtained using thermal energy analyser (TEA) and FID (0.0004 mg/mL). Reprinted from Bowerbank et al. [48]. Copyright (2000), with permission from Elsevier.

from the projectile. Detection of gunpowder (propellant) residues on suspects or their clothing may link them to such an incident. Modern propellants for small arms ammunition almost exclusively contain nitrocellulose, and other explosive ingredients may also be present, e.g., NG and nitroguanidine, stabilizers such as diphenylamine and flash inhibitors such as 2,4-DNT, 2,6-DNT and plasticizers. Thus, analytical methods for detecting propellant or their residues are similar to those used for analysing explosives [49–51]. Numerous studies have been published on the analysis of propellant residues [49]. However, only a few have proposed operational methods for sampling, detecting and identifying these residues on shooters and/or their clothing.

Zeichner et al. [50] have conducted experiments to collect gunpowder (propellant) residues from shooters' clothing using a portable vacuum sampler and the Teflon filters supplied with the commercial IMS instrument. They analysed them by GC/TEA, IMS and GC/MS. The samples were centrifuged and/or filtered, concentrated by evaporation and analysed without any additional clean-up steps. The clothing is first sampled by double-side adhesive-coated aluminium stubs followed by vacuum collection for propellant residue examination. The detection limits (in ng) of the GC/TEA (15-m length column) are 0.2 for NG, 0.05 for 2,4-DNT and 0.05 for 2,6-DNT. The sensitivity of the system was lower when the authors used a 30-m long column in the GC/TEA (0.5–1.0 for NG, 0.1 for 2,4-DNT and 0.1 for 2,6-DNT) or IMS (0.3 for NG, 2,4-DNT and 2,6-DNT). The considerably lower sensitivity of GC/TEA for NG compared with DNT is due to the thermal decomposition of NG in GC columns, which is more pronounced when the 30-m column is used. With the 30-m column, the NG peak in most of the chromatograms was preceded by a smaller peak that could be identified as one of the possible thermal decomposition products of NG. The formation of two NG peaks in the GC/TEA chromatograms increases the probability of it being identified by this method.

On the basis of this last study, Zeichner and Eldar [51,52] developed an efficient method for extracting and analysing gunpowder residues on double-side adhesive-coated stubs. The method should be applied after the primer residues have been examined. The major problem of achieving this goal is minimizing the co-extraction of adhesive and skin debris components, which interfere in the analysis of propellant residues. The optimal procedure for the extraction consists of two stages: (i) extraction of the stubs with a mixture of 80% v/v aqueous solution of 0.1% w/v of sodium azide and 20% v/v of ethanol using sonication at 80°C for 15 min and (ii) further extraction of residues of the obtained extract with methylene chloride. The methylene chloride phase was concentrated by evaporation prior to analysis by GC/TEA and IMS. The optimal procedure was tested in extraction experiments of single gunpowder particles and of standard solutions of NG and 2,4-DNT deposited on stubs, as well as in additional shooting tests. The recovery of NG and DNT was quite variable and found to be in the range 30–90% for the adhesives tested. Neither the adhesives nor the skin was observed to interfere in any significant way in the analysis, but interferences were observed in the GC/TEA analysis of samples collected from hair. The applications of GC–TEA to explosive analysis are summarized in Table 2.

2.1.2.2. High-performance liquid chromatography–thermal energy analyser

High-performance liquid chromatography is a type of LC in which a liquid or an appropriately dissolved solid sample is forced through a column by liquid at high pressure. HPLC with UV detection, specially reversed phase methods, has been extensively used for the analysis of explosives [53–55]. HPLC-UV is standardized in the US Environmental Protection Agency (EPA) Method 8330 [56] for monitoring nitroaromatic explosives in soils, ground water and sediments, which provides HPLC conditions for detecting ppb levels of certain explosive residues. In our review of the literature, we have not found many papers about HPLC with TEA detection in recent years. However, at the beginning of TEA, a number of papers, which must be mentioned here, show that it can be useful for explosive detection. So, Lafleur and Morriseau [39] carried out a very detailed study using HPLC with TEA to identify and determine explosives and other related compounds [EGDN, NG, PETN, Petrin (2-hydroxymethyl-2-nitroximethyl-1,3-propanediol-1,3-dinitrate), RDX, 1-nitroguanidine (NGu) and HMX (octahydro-1,3,5,7-tetranitro-1,3,5,7-tetrazocine)] with thermally labile nitro or nitroxy groups. They used columns packed with Lichrosorb Si-60 or Lichrosorb NH_2, a pyrolyser temperature of 550°C and cold traps between −80 and −90°C. A mixture of isooctane and ethanol was used as mobile phase. Chromatographs obtained for the analysis of a mixture of four explosives (NG, PETN, RDX and HMX) showed increased retention times on the NH_2 column compared with silica. This was particularly evident in the case of PETN, which elutes after NG on the NH_2 column but before it on silica. According to the authors, this shift can be a valuable method for identifying components in certain mixtures of explosives. The HPLC–TEA analysis of an 80-ng sample of a standard dynamite with a nitrate ester composition of nine parts EGDN to one part NG showed that NG shifts to a longer retention time on an NH_2 column whereas EGDN does not. The authors pointed out that this behaviour could be very useful for identifying dynamite compositions. Retention data for 62 HPLC–TEA analyses of eight different explosives were reported. All of the components were clearly resolved on the silica column, but the retention times of PETN and NG overlapped significantly on the NH_2 column. Ancillary

Table 2. Applications of gas chromatography–thermal energy analyser (GC–TEA) to explosive analysis

Analyte	Observations	Pyrolysis temperature	Minimum working level	Reference
NB, 2,4-DNT and 2,6-DNT	A shakeout-centrifugation procedure is used to isolate the nitroaromatics from the biosludges	950°C	0.05 mg/L (LOD) NB was not detected (< 0.05 mg/L wet weight)	[35]
NG, TNT and RDX	Silica capillary column GC	1000°C	Low pg level	[21]
NG, DNT, TNT, RDX, EGND and tetryl	Parallel GC–TEA/HPLC–TEA. No need for sample clean-up before analysis	900°C	Low pg level. Sensitivity of 4–5 pg injected on-column	[14]
NG, EGND, TNT, RDX, tetryl BTT, NB, TNB and 2,4-DNT	Handswab extracts using columns of Amberlite XAD-7 porous polymer beads	500–900°C	Low ng level	[34]
NG, TNT and RDX	Sample clean-up using columns of Amberlite XAD-7 beads.	550–800°C	5 pg of NG, TNT and RDX	[22]
NG, EGND, BTN, TEGDN, TNT, RDX, NB, 4-NT, 2,4-DNT and musk tibetine	Dynamic headspace sampling. Columns of Amberlite XAD-7	625°C NB, 4-NT and 2,4-DNT required 750°C	Sub-ng level	[30]
NG, EGND, TNT, DNT, PETN and RDX	Sample pre-treatment: solid phase extraction	900°C	pg level	[32]
2-NN, 2,4-DNT, 2,6-DNT, 1-NP, 1,3,5-TNB, NG, PETN, TNT and RDX	SFE/TDM/GC	800°C	MDQ for 2,4-DNT 2.6 ppb	[33]

Table 2. (Continued)

Analyte	Observations	Pyrolysis temperature	Minimum working level	Reference
DNPMT (chemical blowing agent)	Extraction in methyl-tert-butyl ether	750°C	–	[43]
NG, TNT, PETN and RDX	Samples taken at various transport sites	750°C	LOD (BP-1 column) and relatively dirty samples NG (ng), 3.1 TNT(ng), 2.9 PETN (ng), 4.0 RDX (ng), 2.3	[44]
NG, 2,6-DNT, 2,4-DNT, 2,4,6-TNT and PETN	SGC using packed capillary columns and neat carbon dioxide as mobile phase	740°C	Low pg range	[48]
NG, 2,4-DNT and 2,6-DNT	Gunpowder residues collected by a portable vacuum sampler	850°C	LOD 0.2 ng (NG), 0.05 ng (2,4-DNT), and 0.05 ng (2,6-DNT)	[50]
NG and 2,4-DNT	Extraction and analysis of gunpowder residues on double-side adhesive-coated stubs	850°C	In the range of 0.1–1 ng	[51,52]

BTT and BTN, butane-1,2,4-triol-trinitrate; DNT, dinitrotoluene; EGDN, ethylene glycol dinitrate; HPLC, high-performance liquid chromatography; LOD, limits of detection; MDQ, minimum detectable quantities; NB, nitrobenzene; NG, nitroglycerine; NN, nitronaphthalene; NT, nitrotoluene; PETN, pentaerythritol tetranitrate; RDX, cyclotrimethylene trinitramine; SFE, supercritical fluid extraction; SGC, solvating gas chromatography; TDM, thermal desorption modulator; TNB, trinitrobenzene and TNT, trinitrotoluene.

components such as plasticizers and stabilizers presented no interferences even when standard explosive preparations were not purified. The 90% confidence interval for the quantification of RDX was ±1.6% at the 4-ng level and 3.2% for PETN at the 6-ng level. The authors conclude that the method proposed is useful for identifying explosive compounds at trace levels and for analysing explosive compositions.

Liquid chromatography with TEA detection or an ultraviolet HPLC detector has been evaluated as an analytical procedure that simultaneously determines EGDN and NG in air [57]. Samples were collected by drawing a known volume of air through large sampling tubes containing Tenax-GC resin. The samples were desorbed with methyl alcohol and analysed by LC with TEA or an UV HPLC detector. A Dupont Zorbax CN

(4.6 mm × 25 cm) column was used. Temperatures of 550 and −80°C were selected for the HPLC pyrolyser and the cold trap, respectively. The detection limit of the overall procedure was 186 ng/sample for EGDN and 280 ng/sample for NG. The equivalent air concentrations were 12 μg/m^3 (2.0 ppb) for EGDN and 19 μg/m^3 (2.0 ppb) for NG. These were considered to be reliable quantitation limits within the requirements of a recovery of at least 75%. The report mentions that one disadvantage is the high cost of the TEA detector.

Interfacing the TEA to both a gas and a HPLC has been shown to be selective to nitro-based explosives (NG, PETN, EGDN, 2,4-DNT, TNT, RDX and HMX) determined in real world samples, such as pieces of explosives, post-blast debris, post-blast air samples, hand swabs and human blood, at picogram level sensitivity [14]. The minimum detectable amount for most explosives reported was 4–5 pg injected into column. A pyrolyser temperature of 550°C for HPLC–TEA and 900°C for GC/TEA was selected. As the authors pointed out, GC uses differences in vapour pressure and solubility in the liquid phase of the column to separate compounds, whereas in HPLC polarity, physical size and shape characteristics determine the chromatographic selectivity. So, the authors reported that the use of parallel HPLC–TEA and GC–TEA techniques provides a novel self-confirmatory capability, and because of the selectivity of the technique, there was no need for sample clean-up before analysis. The detector proved to be linear over six orders of magnitude. In the determination of explosives dissolved in acetone and diluted in methanol to obtain a 10-ppm (weight/volume) solution, the authors reported that no extraneous peaks were observed even when the samples were not previously cleaned up. Neither were they observed in the analysis of post-blast debris. Controlled experiments with handswabs spiked with known amounts of explosives indicated a lower detection limit of about 10 pg injected into column.

Selavka et al. [36] have studied how the incorporation of on-line, post-column UV irradiation improves the HPLC–TEA detectability of mono-NT, DNT and TNT. This photolytically assisted TEA (PAT) detection approach included the use of an ultraviolet lamp and a knitted open tubular photochemical reactor design that significantly reduced the band broadening that resulted whenever long lengths of tubing (11.9 m long) were used to construct post-column reaction chamber. TEA was operated in the nitroso mode (pyrolyser temperature at 550°C). The mechanism was explored using batch irradiation followed by LC–TEA and GC/MS and was postulated to involve photochemically induced isomerization leading to homolytic cleavage of the C–NO_2 bond, followed by hydrogen abstraction. The authors concluded that PAT improved the TEA detectability of TNT and DNT by a factor of 30 and 16, respectively. LC–PAT has been used to confirm the presence of TNT in several bomb cases and makes the examination results more reliable.

2.1.2.3. *Supercritical fluid chromatography–thermal energy analyser*

Supercritical fluids are produced by heating a gas above its critical temperature or compressing a liquid above its critical pressure. In SFC, the sample is carried through a separating column by a supercritical fluid (typically CO_2) that is used as mobile phase. The solvating powers of supercritical fluids can be similar to those of organic solvents. Their diffusivities are higher, however, and their viscosity and surface tension lower. To our knowledge, SFC has not been extensively applied to the detection of explosives: only two earlier references describe the coupling of SFC to TEA and its application to the detection of explosives.

A representative group of compound explosives, including those compounds that are difficult to analyse by HPLC–TEA or GC–TEA because they are thermally unstable or non-volatile, have been satisfactorily analysed by capillary SFC with thermal energy analysis detection at the nanogram level [23]. Some of the explosives studied present the characteristics mentioned above: hexanitrobibenzyl, which has low vapour pressure, tetryl, which decomposes to *N*-methylpicramide, or PETN, which requires a clear GC system to prevent decomposition. Pure compounds were dissolved in ethyl acetate and analysed directly. The chromatographic column was 6.8 m × 0.05 mm ID SB Octyl 50 Superbond cross-linked flexible silica (Flexsil) capillary with a stationary phase film thickness of 0.25 μm. The restrictor penetrated 11.5 cm into the TEA pyrolyser interface block, and the pyrolysis temperature was 625°C. The SFC density and temperature programmes operated simultaneously and were described for specific applications. The operating conditions described by the author made it possible to analyse many nitramine, nitrate ester and nitroaromatic explosives in pure standards and hand swab extracts. The author reported that the minimum detectable levels (in pure standards) ranged from 20 to 60 pg injected for a representative group of compounds (NG, 23 pg; PETN, 40 pg; tetryl, 60 pg and hexanitrobibenzyl, 55 pg). Moreover, compounds containing very polar groups (amino and hydroxyl groups) could not be analysed. The peak shape of HMX was poor, probably because of its low solubility in supercritical CO_2. The effect of dirty extracts on column stability was also investigated, and the results obtained allowed the author to conclude that capillary SFC–TEA is more resistant than capillary GC–TEA to contamination by extractives.

Francis et al. [58] have reported the coupling of SFC with TEA for the analysis of such nitro- and nitroso-containing compounds as tobacco-specific nitrosamines and explosives (synthetic mixture: NG, 2,6-DNT, 2,4-DNT, PETN, 2,4,6-TNT, 1,3,5-TNB, tetryl, RDX and HMX). TEA was interfaced directly to the outlet restrictor of the SFC. Its potential use for thermally labile compounds and environmental samples has been evaluated by the authors. The sensitivity of TEA was good, and its linear dynamic range was broad; minimum detectable quantities (MDQ) were below 20 pg for PETN with a dynamic range of over four orders of magnitude. The analysis of HMX was accomplished by using modified CO_2 with a corresponding MDQ of 67 pg. Owing to the superior selectivity of the cyanobiphenyl stationary phases (medium polarity) for a variety of compounds, the authors used *m*,*p*-cyanobiphenyl and the *p*, *p*-cianobiphenyl phases. The selectivity of TEA for nitro and nitroso compounds can be adjusted by changing the pyrolysis temperature. Nitroso compounds, *O*-nitro (nitrate ester) and *N*-nitro compounds (nitramine), are more easily pyrolysed than the C-nitro compounds. The mixture of standard nitrated compounds was run at 400 and 800°C (although at this last pyrolysis temperature there was no selectivity towards the different types of nitrated groups). At 400°C only, the nitroso compounds, the nitrate esters and the nitramines were detected. The SFC/TEA system was used to study the environmental fate of propellant and explosive residues after military munitions had been disposed of. Air samples were obtained on quartz fibre filters when the compounds were burned or detonated in a 'bang box'. The major nitrated compound was TNT, there were smaller amounts of TNB and there was also a peak that eluted at the same retention time of RDX. Independent SFC/MS analysis of the sample confirmed the identification of the peak as RDX (contamination from an earlier test).

2.2. Luminol

Luminol is a relatively simple chemical containing only carbon, nitrogen, oxygen and hydrogen ($C_8H_7N_3O_2$). It is widely used as a chemiluminescent reagent, and it has quite a few other names: 5-amino-2,3-dihydro-1,4-phthalazine-dione, *o*-aminophthalyl hydrazide, 3-aminophthathic hydrazide and *o*-aminophthaloyl hydrazide, among others. It was discovered in the late nineteenth century and later, in 1928, Albrecht [59] discovered a specific chemical that, when placed in an aqueous alkaline solution, emitted a fairly intense blue-green light. This specific chemical was to be later called luminol. The chemiluminescent emitter is a 'direct descendent' of the oxidation of luminol by an oxidant in basic aqueous solution. Hydrogen peroxide is probably the most useful oxidant, however, other oxidants have been used. The presence of a catalyst is essential if this chemiluminescent method is to be used as an analytical tool. Many metal cations catalyse the reaction of luminol, H_2O_2 and OH^- in aqueous solution to increase light emission. The general CL reaction of luminol can be schematized as follows:

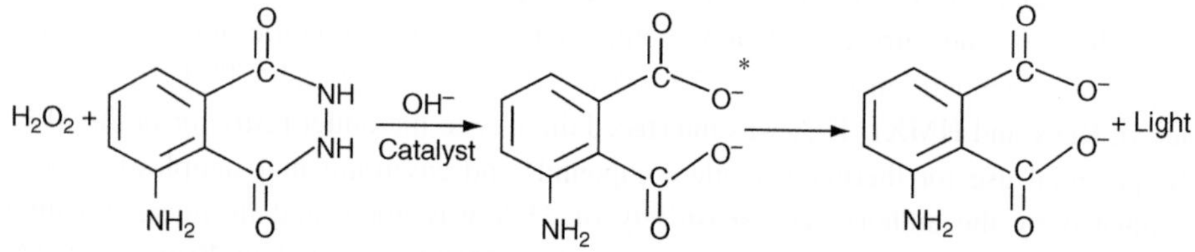

Analytical methods based on the CL reaction of luminol are famous for their inherent simplicity, speed, high sensitivity and wide dynamic range.

A highly selective and sensitive detector for measuring organic nitrates in atmospheric air samples has been developed by Hao et al. [60]. They describe the development of a chromatographic detector based on post-column thermal decomposition of organic nitrates (and other nitrogen-containing compounds) in a quartz tube pyrolyser to yield NO_2, which is subsequently detected using a highly sensitive and selective CL-based

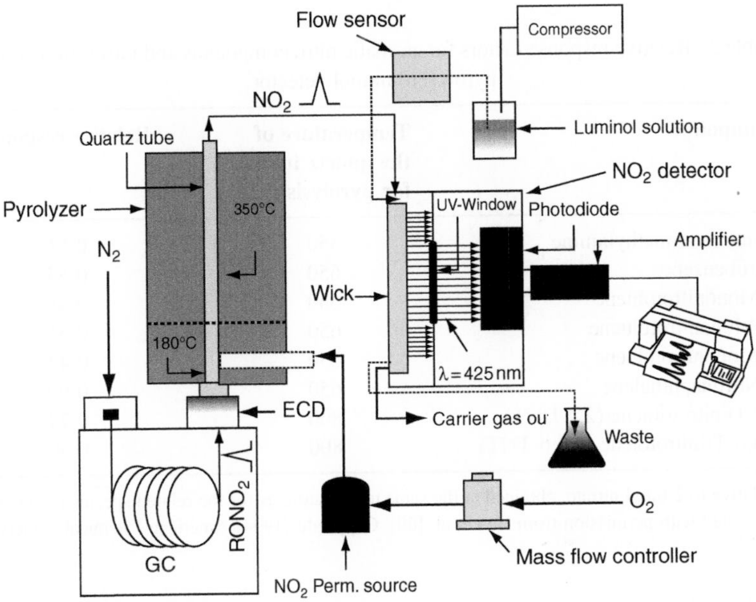

Fig. 7. Schematic diagram of the organic nitrate-specific detector. Reprinted with permission from Hao et al. [60]. Copyright (1994) American Chemical Society.

NO_2 detector, and luminol as the chemiluminescent reagent. Figure 7 shows a schematic diagram of the organic nitrate-specific detector.

In the case of organic nitrate pyrolysis, the reaction proceeds by O–N bond homolysis as follows:

$$RONO_2 \rightarrow RO + NO_2$$

The luminol detector is based on the sensitive chemiluminescent reaction between NO_2 and luminol in solution. The luminol in alkaline solution reacts with NO_2 to produce intensive CL centred on 425 nm. The detector cell itself is a 15 cm × 8 cm × 2 cm rectangular block, with inlets and outlets for the carrier gas and luminol flows. The reaction cell contains a fabric wick that is wetted with the luminol solution (1×10^{-4} M luminol, 0.2 M Na_2SO_3, 0.05 M NaOH, 1.5×10^{-4} ethylenediaminetetraacetic acid and 0.1% surfactant). The wick is viewed by a PMT through an acrylic window, which is transparent to the chemiluminescent light at 425 nm. When a NO_2 peak enters the cell, a fraction of the NO_2 dissolves in the solution on the surface of the wick, which then reacts with luminol to ultimately yield a strong CL. The results described indicate that this detector may be successfully used for the sensitive and selective detection of any nitrogen-containing species that may thermally decompose to yield NO_2, with a potential use for the detection of explosives (2,4-DNT and TNT).

Table 3. Relative response factors for aromatic nitro compounds and nitrosamine for the pyrolyser/luminol detector

Compound	**Temperature of the quartz tube for pyrolysis (°C)**	**Relative response[a]**
N-nitrosodimethylamine	350	0.51
nitrobenzene	650	0.53
o-Mononitrotoluene	650	0.48
m-Mononitrotoluene	650	0.51
p-Mononitrotoluene	650	0.44
1-Nitronaphthalene	650	0.99
2,4-Dinitrotoluene (2,4-DNT)	800	0.74
2,4,6-Trinitrotolene (2,4,6-TNT)	800	0.90

[a] Relative to 2-butyl nitrate, obtained at the same temperature as for the related species in the table. Reprinted with permission from Hao et al. [60]. Copyright (1994) American Chemical Society.

At sufficiently high temperatures, the detector also responds to a variety of nitroaromatics and to *N*-nitrosodimethylamine. NO_2 yields of 0.70 and 0.90 were obtained for 2,4-DNT and TNT, respectively. The detection system exhibits a linear response for all nitrogen-containing compounds studied, and the detection limit for the determination of organic nitrates was found to be ~0.05 pmol. The relative response factors for aromatic nitro compounds and nitrosamine for the pyrolyser/luminol detector are presented in Table 3.

The reversed micellar-mediated CL (RMM-CL) reaction of luminol with $[AuCl_4]^-$ was applied to the indirect CL determination at trace level of *p*-toluidine, 2-methyl-5-nitroaniline and 2,4-dinitroaniline in aqueous samples by Mohammadzai et al. [61]. These compounds are related to the degradation products of TNT but are not active in the CL reaction. Incorporating reversed micelles into the CL detection system has several advantages, including improved sensitivity and selectivity. In the reversed micellar solution, the surfactant molecules encompass tiny water droplets and are converted into homogeneously distributed micelles in the organic bulk. Acting as microreactors, the reverse micelles have unique properties, and it is believed that the RMM-CL reactions occur at water–surfactant interfaces of the reverse micelles. These microreactors have the ability to quantitatively transfer species of experimental interest into their water pools and to enhance sensitivity by several factors. The protonated analyte, $[HA]^+$, forms an ion-pair with the $[AuCl_4]^-$ ion in aqueous acidic medium, which is extracted into the organic phase of dichloromethane. Using a membrane phase separator, the on-line separated extracts are successively introduced into the mixing cell situated in a CL detector. In the cell, the extract containing the analyte as an ion-pair, $[HA]^+[AuCl_4]^-$, is mixed with the reversed micellar luminescent reagent (luminol in 0.3 M Na_2CO_3) prepared from cetyltrimethylammonium chloride in CH_2Cl_2-cyclohexane, pumped through the other tubing, and the CL signal produced from luminol oxidation by $[AuCl_4]^-$

is recorded. The detection limits (S/N > 3) reported were as follows: *p*-toluidine, 1.0×10^{-4} M; 2-methyl-5-nitroaniline, 1.0×10^{-7} M and 2,4-dinitroaniline, 1.0×10^{-7} M. The calibration curves were linear between 1.0×10^{-4} and 1.0×10^{-2} M for all the compounds. These data indicate that the proposed CL method could be a sensitive, fast and low-cost substitute for the existing procedures. With a simple experimental setup and low-cost instrumentation, the proposed method makes it possible to quantify *p*-toluidine, 2-methyl-5-nitroaniline and 2,4-dinitroaniline at trace levels in aqueous samples. The method has various advantages over the manual extraction procedure: simplicity, less human exposure to toxic reagents/solvents, throughput, sensitivity, precision, minimum risk of contamination, etc.

The great interest in detecting traces of explosives has originated several patents. Nguyen et al. [62] have provided a method and system for detecting small quantities of explosives and other contraband substances located on the surfaces of objects. The technique uses a chemical reaction between NO_2 gas and luminol in an aqueous, alkaline solution. Under suitable conditions, the vast majority of common types of explosives may be decomposed to produce NO_2 gas that is detected using the apparatus presented by the authors. Light is detected to signal the presence of the contraband substance. The system rapidly provides quantitative indications of the amount and location of a critical substance. It is especially well suited for use in applications that require throughput and accuracy, such as security screening associated with airlines and other forms of public transportation.

A portable advanced explosives detector (E 3500) based on luminol CL is manufactured by Scintrex Trace Corp. [63]. It offers both vapour and particulate sampling without the use of an external carrier gas or radioactive source. Vapours are sampled directly through the sampling nozzle. Particulates are sampled by swiping a suspected object with a cotton glove or other means and transferring any traces to the unit. This portable detector is capable of detecting plastic and high vapour pressure explosives rapidly, and it can detect minute traces (low nanogram level) of C-4, TNT, dynamite, PETN, semtex, EGDN, 2,3-dimethyl-2,3-dinitrobutane (DMNB), RDX, ANFO, ammonium nitrate, urea nitrate, NG and triacetone triperoxide (TATP), with low false alarm levels and few interferents.

2.3. Electrochemiluminescence

The application of ECL per se or in combination with enzyme IM to the detection of explosives has been reported and has extended the possibilities of CL in the field of explosives. ECL can be defined as the luminescence generated by the relaxation of excited state molecules that are produced during an electrochemically initiated reaction [64].

A wide range of reagents is known to produce these luminescence processes (organic, organic-metal ions including compounds of ruthenium, osmium, rhenium or other elements, and other kinds of molecules), but only a few of them are of interest for analytical purposes. Ruthenium trisbipyridine $(Rubpy)_3$ is probably the best known of the organic metal ion coordination complex ECL molecules. ECL combines the advantages of CL with the ability to control the time of the light-emitting reaction and is therefore a sensitive method for the analysis of some chemical species.

In explosive detection, Bruno and Cornette [65] have used ECL to determine several diaminotoluenes (DATs), degradation products of DNT. DNT is present in TNT, and DATs are related to other compounds that are TNT breakdown products. The determination is based on the low-level ECL reaction of these molecules with the group IB transition metals ions Au^{+} and Cu^{+2}. DAT isomers were screened for ECL against a battery of 32 metal ions, including Cu^{+2}, Eu^{+3}, Mg^{+2}, Ru^{+3} and Tb^{+3} associated with other known ECL complexes, at 1:3 added metal ion: ligand molar ratios. The 1:3 molar ratio presumed tris-bidentate octahedral metal coordination complex formation, which generally yielded optimal ECL intensity. An ORIGEN Analyzer (ECL sensor; IGEN Corp.) was used. DAT isomers (2,3-, 2,4-, 2,5-, 2,6- and 3,4-diaminotoluene), DNT isomers and 4-amino-3-NT were screened. Control solutions of metal ions without ligands, but with tripropylamine (TPA), an electron carrier, gave no detectable ECL, whereas some solitary DAT ligand solutions produced appreciable background ECL in the presence of TPA. The authors reported that the DNT and aminonitrotoluene metal ion data demonstrated that the presence of the nitro functional group disallowed or severely inhibited ECL (presumably because of steric hindrance), whereas amino functional groups appeared to interact with certain metal ions to produce noteworthy ECL reactions. Data from the preliminary screening showed that the most intense ECL response was obtained from Cu^{+2} for 3,4-DAT (light brown) and Au^{+1} for 2,4-DAT (dark brown). They suggested that the apparent specificity of Au^{+} for 2,4-DAT and Cu^{+2} for 3,4-DAT may be partly based on ionic size as Au^{+} has nearly twice the ionic diameter of Cu^{+2} and thus may form a coordination complex with the *meta* but not with the *orto* DAT. Other DAT isomers were screened and exhibited mildly enhanced ECL with various metal ions, but these enhancements were not statistically significant. An ECL intensity time dependence was observed in the Cu^{+2}-3,4-DAT system. A very linear ECL response of Cu^{+2} to 3,4-DAT from 1 to 200 ppm was obtained, and the detection limit was in the order of $\leq$1 ppm of 3,4-DAT. The Au^{+1}-2,4-DAT titration demonstrated no time dependence and did not appear to have a broad linear response range. Lower detection limits or sensitivities in the ppm range for both the metal ions and organic ligands involved can be achieved. Finally, the authors concluded that the ECL assays described are potentially useful for detecting and quantifying DATs or possibly triaminotoluenes as well as detection and quantitation of various transition metals in industrial wastewater streams and groundwater supplies.

A patent based on ECL reactions has been presented by these authors [66]. The invention is directed to novel ECL reactions between diaminoaromatic ligands and soluble metals ions and in particular to reactions between aminoaromatic ligands, such as 2,4-DAT, 3,4-DAT and 2,3-diaminonaphthalene, and metal ions, such as Au(I), Cu(II), Cr(VI), Fe(III), Ru(III), Se(IV) and V(V). The ECL assays are considered useful for carrying out field and laboratory analysis to detect TNT breakdown products and toxic metals in wastewater streams, soil and ground water supplies.

2.4. Immunoassay, chemiluminescence and electrochemiluminescence

Immunoassay makes use of the binding between an antigen and its homologous antibody to identify and quantify the specific antigen or antibody in a sample. IM methods cover quite an important field of analytical chemistry and are the method of choice for measuring analytes that are normally present at very low concentrations and which cannot be determined accurately by other less expensive tests. The sensitivity of any IM is a complex function of the underlying physicochemical basis of the technique and the size and source of 'experimental' errors [67]. Considerable effort has been made to improve the sensitivity (i.e. the introduction of time-resolved fluorescence IMs where femtomolar detection limits are achieved) but only at the cost of introducing more sophisticated and expensive equipment. High sensitivity can also be achieved by using chemiluminescent labels [68], which emit light when they are mixed with a chemical trigger. Although these labels produce a lower number of detectable events than fluorescent labels, there is no excitation source, and therefore the background emission is close to zero, and there is no interference due to scattering.

Competitive immunoanalytical methods are based on the competitive binding that occurs between a labelled and an unlabelled ligand for highly specific receptor sites on antibodies. Analysis is carried out by measuring a physical or chemical property associated with the label and by constructing a standard curve that represents a measured physical signal as a function of the concentration of the unlabelled ligand. Analyte concentrations are extracted from this calibration curve [2]. CL IM is now established as one of the best alternatives to conventional IM for quantifying low concentrations of analyte in complex samples. In recent years, ECL has been used for the detection step in IMs. The technology has evolved into analytical procedures that perform much better than IMs based on the use of radioactive labels. Several reports show the potential of IM combined with CL or ECL detection in the field of trace explosive detection.

Multianalyte detection (TNT, atrazine and fluorescein) with an affinity sensor array is presented by Schuetz et al. [69]. The normal assay procedure for an indirect enzyme-linked immunosorbent assay (ELISA) is used, with horseradish peroxidase (HRP)-labelled secondary antibodies in combination with a CL substrate. The authors have developed a parallel affinity sensor array (PASA) to overcome the most controversial disadvantage associated with ELISA: that is to say, cross reactivity. In the PASA system, parallelization is achieved by immobilizing different affinity molecules such as antibodies or haptens in distinct areas of a plain glass surface. If several antibodies are used for one analyte, substances should be able to be identified by pattern recognition. Multianalyte detection is made possible by using different antibodies for different analytes. The system is based on peroxidase-catalysed CL with luminol/hydrogen peroxide substrates (Pierce Supersignal, Pierce Supersignal ULTRA and DuPont Renaissance). The CL light is collected with a high-performance charge coupled device (CCD)-camera that allows high spatial resolution and a high density of different haptens. A standard liquid autosampler is used for standards and samples. The haptens were covalently coupled to a surface by the N-hydroxysuccinimide (NHS)-ester technique. Although results were best with the 'Super Signal ULTRA' substrate, the authors decided to use normal 'Supersignal' because the results were satisfactory, and it was cheaper than 'Super Signal ULTRA'. For the analyte TNT, the detection limit was about 0.2 μg/L.

Ciumasu et al. [70] have developed a portable and power-supply autonomous field immunosensor for environmental pollutants using TNT as the key target. Other explosives and pesticides (diuron and atrazine) have also been determined. Monoclonal antibodies (mAbs) were immobilized through adsorption on a gold surface with numerous pyramidal structures. The recognition reaction was enhanced in three ways: (i) through the enzymatic reaction, (ii) through the pyramidal structure with the gold surface cover and (iii) through the detection of the CL of the product through a very sensitive PMT. Immunoreagents (enzyme tracer and antibody) together with the environmental sample were located in a single-use chip that was replaced after each measurement.

This chip is the key to the versatility of the analytical system. In the final step, the competition between the enzyme tracer and the analyte for the antigen-binding sites of the antibodies yields a CL signal that is inversely proportional to the concentration of the analyte in the given range of detection. Reagents are transported with an automated, miniaturized flow-injection system. SuperSignal®ELISA Femto Maximun Sensitivity Substrate (Luminescence) and SuperFreeze™ Peroxidase conjugate Stabilizer were used. A battery of six enzyme tracers for nitroaromatic compounds was produced by hapten conjugation to HRP through *N*-hydroxysuccinimide active esters, and TNP–glycylglycine–HRP was established as the best enzyme tracer for mAb A1.1.1. After optimization on the microtiter plates, the ELISA procedure for each target analyte was adapted to the batch ELISA format, where the wells of the microtiter plate were replaced by the gold covered disk with numerous pyramid structures. The specific batch-ELISA conditions for TNT assays were as follows: the coating protein was protein A/G [alternatively, goat anti-mouse immunoglobulin G (IgG)] for mouse mAb A1.1.1, the coating concentration was 4 μg/mL and the recognition antibody was anti-TNT mAb A1.1.1, 2 μg/mL and 2 h at room temperature. After the blocking step, the TNT standard and the enzyme-tracer solution were incubated together on the structures for 10 min. When TNT was analysed with the optimized TNT-ELISA on microtiter plates, the goat anti-mouse IgG as coating antibody, mAb A1.1.1 and the mentioned enzyme tracer, the performance characteristics of the standard curve were as follows: detection limits (DL; IC_{20}) between 0.1 and 0.2 μg/L, linear dose–response region (IC_{20}–IC_{80}) from 0.1 to 10 μg/L and test midpoint (IC_{50}) between 0.3 and 1 μg/L. The pH range was from 4 to 10. The assay showed a high selectivity for TNT. The authors concluded that the sensitivities for all three analytes were very good, which was especially true for TNT. This platform had such important features as analytical flexibility for a wide range of analytes, sensitivity, automation, miniturization and low environmental hazardous waste.

Investigators from the Department of Chemistry of the University of Liverpool tried using ECL with enzyme IM for trace explosive detection (TNT and PETN). The detection step uses ECL and triggers light emission by an electrochemical reaction. The ECL step

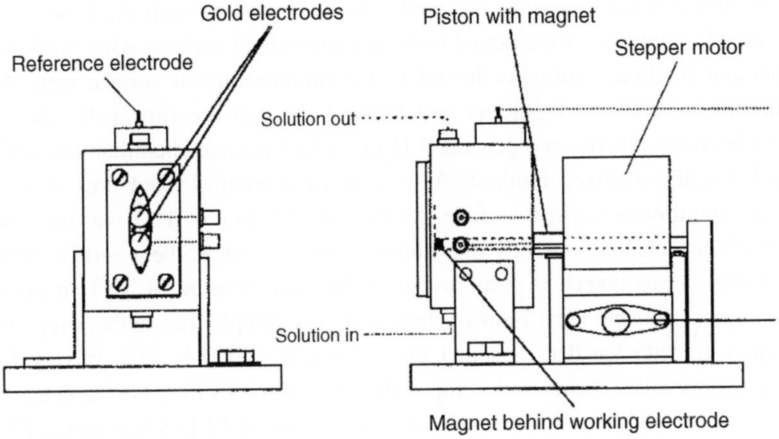

Fig. 8. Flow-cell with rare earth magnet positioned behind the working electrode. Reprinted from Wilson et al. [68]. Copyright (2003), with permission from Elsevier.

allows the time of the light-emitting reaction to be controlled, and so, the detection can be delayed until the antibody-binding reaction has taken place.

The enzymes glucose oxidase (Gox) and HRP were used in these assays. The authors designed and constructed a computer-controlled flow-injection chemiluminometer. Full details of the instrumentation used to carry out these IMs are given in the study by Wilson et al. [68,71,72]. The flow-injection electrochemiluminometer consists of a Pentium III computer, a flow injection analyser, a potentiostat and an electrochemiluminometer. The electrochemiluminometer consists of two enclosures, one outside the other (Fig. 8) surrounding a flow cell where ECL reaction takes place and a PMT that detects the light emitted. The enclosures are designed to exclude light from the PMT when the electrochemiluminometer is in use but allows access to the flow cell for routine maintenance. The inner enclosure contains a flow cell designed to facilitate the flow of solutions through the working compartment and to ensure that gas bubbles do not become trapped. It is made of PTFE (Teflon) to minimize the adsorption of molecules such as labelled antibodies. The main design features reported by the authors are the hydrodynamic shape of the working compartment through which solutions are pumped from the bottom to the top of the flow cell; the increasing diameter in the direction of flow; connection between the working and reference electrode compartments angled upwards, in the direction of the flow; and working and counter electrodes polished flush with the Teflon.

Wilson et al. [72] describe the ECL enzyme IM for TNT. The flow-injection electrochemiluminometer incorporates a three-electrode arrangement consisting of 3-mm diameter gold disk working and counter electrodes and a miniature Ag/AgCl reference electrode. The deposition of a re-usable immunosorbent dextran surface anchored to a gold surface in the flow cell by chemiadsorbed thiol groups is described. Antibodies were labelled with the enzyme glucose oxidase and used in competitive IM. In the ECL IMs, the antibodies were mixed with TNT. After TNT had occupied a fraction of the available antibody binding sites, the sample was pumped through the flow cell where unoccupied binding sites were bound to the immunsorbent surface. After washing away the unbound fraction, antibodies bound to the immunosorbent surface were detected by pumping a solution of glucose and luminol through the flow cell. The enzyme label oxidized the glucose and generated H_2O_2, which reacted chemiluminescently with electrochemically-oxidized luminol. A mixture of acetonitrile and urea was used to regenerate the immunosorbent surface. As the concentration of TNT was increased, the amount of the unbound antibody that bound to the immunosorbent surface decreased. Light intensity was inversely proportional to the concentration of TNT in the sample in the range 0–100 ppb. The limit of detection was 12 ppb. The authors reported that in comparison with the data obtained by Colorimetric ELISA, ECL is very fast: for example, colour takes 20 min to develop with Colorimetric ELISA, but the detection step in ECL takes only 100 s. However, the detection limit of ELISA was lower (2.3 ppb), and the authors mention that the limit of the proposed method must be improved for environmental analysis. Finally, they discuss the possibility of developing a portable instrument for use in the field.

A paramagnetic bead-based enzyme ECL IM for TNT has been described by Wilson et al. [68]. The paramagnetic beads are coated with haptenylated dextrans prepared by substituting biotin and analogues of TNT into aminodextrans. The IM protocol reported is shown in Fig. 9. The IMs are carried out in three stages: in the first stage, TNT and haptenylated dextrans compete for a limited number of antibody-binding sites; in the second stage, antibodies bound to the dextran are concentrated on the surface of paramagnetic beads; and in the third stage, the beads are magnetically concentrated on the surface of an electrode. In the presence of glucose, the glucose oxidase label conjugated to the antibodies generates H_2O_2, which reacts chemiluminescently with electrochemically oxidized luminol. The concentration of TNT in the sample is determined from the light intensity with reference to a calibration curve. The light intensity reaches a maximum at around +600 mV and then decreases as the potential increases. The limit of detection for TNT was 31 ppb. The flow cell can be used for at least 3 months without being cleaned manually. The Gox system described also has many shortcomings: comparatively low turnover, incompatible ECL and enzyme pH optima, a high CL background and susceptibility to interference from non-specifically bound label. The authors also point out that the detection step in the ECL assays described took 80 s compared with the 20 min for colour to develop in colorimetric IMs with similar sensitivity. The authors reveal that of the existing detection systems, the combination of HRP with electrochemically generated H_2O_2 comes closest to meeting the requirements, and so they carried out similar assays for TNT and PETN but based on HRP [71]. HRP catalyses the chemiluminescent oxidation of luminol in the presence of electrochemically generated H_2O_2. For enzyme IM, the haptens corresponding to the explosives were covalently attached to high-affinity dextra-coated paramagnetic beads, which were mixed with the corresponding Fab fragments and the sample. It is recommended to use Fab fragments rather than whole antibodies for maximum sensitivity. After adding a second HRP-labelled antispecies-specific antibody, the mixture was pumped into an electrochemiluminometer where

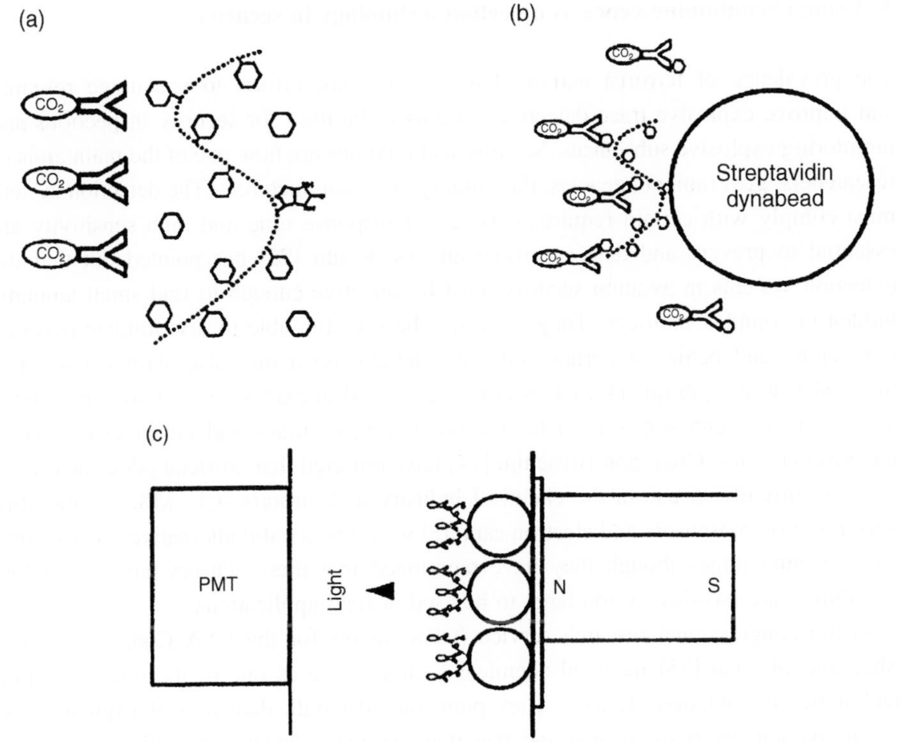

Fig. 9. Enzyme electrochemiluminescence (ECL) immunoassay protocol. (a) Trinitrotoluene (TNT) and dinitrophenyl of haptenylated dextran compete for antibody-binding sites. (b) Antibodies attached to dextran bound to streptavidin-coated paramagnetic beads. (c) ECL detection of enzyme-labelled antibodies magnetically concentrated on electrode. Reprinted from Wilson et al. [68]. Copyright (2003), with permission from Elsevier.

beads were concentrated on the working electrode magnetically. The amount of analyte in the sample was determined by measuring light emission when H_2O_2 was generated electrochemically in the presence of luminol and an enhancer, *p*-iodophenol. The HRP-ECL detection system is schematized by the authors as follows: (i) reducing dissolved oxygen electrochemically generates H_2O_2, (ii) H_2O_2 oxidizes HRP and (iii) HRP oxidizes *p*-iodophenol, which mediates the chemiluminescent oxidation of luminol.

The limit of detection of TNT was 0.11 ppb. Nevertheless, the limit of detection was 68 ppb when IMs were carried out with whole antibodies. The EPA has proposed a lifetime health advisory level of 2 ppb for TNT in drinking water. The detection limit for PETN was 19.8 ppb. The authors observed the relation between the limit of detection and the Ka (affinity of the antibodies) of the antibodies. The Ka value of the anti-TNT antibodies used in this work was 5.8×10^5, which corresponds to a maximum sensitivity of 0.03 ppb. Finally, the authors concluded that the simple protocol ensures that IMs for a previously undetected analyte can be up and running within a few weeks of the antibodies and corresponding haptens becoming available.

3. Using chemiluminescence as detection technology in security

The prevalence of terrorist activity has led to major efforts to encourage research and improve explosive trace detection systems to be used for security inspections and monitoring explosive substances. Security technologies are now one of the main aims of researchers, government agencies, the military and manufacturers. The detection system must comply with certain requirements: a fast response time and high sensitivity are essential to prevent and detect terrorist attacks. Rouhi [73] has pointed out that the detection systems in aviation security must be sensitive enough to find small amounts hidden in complex matrices. They must also be specific, able to discriminate between threatening and benign materials and not generate too many false alarms. Likewise, they must do the job quickly. CL is commonly cited in explosives' reports and guides as one of the technologies used for the security of civilians and the investigation of terrorist activities. Gross and Bruschini [74] have reported that artificial odour or vapour sensors currently in use in the chemical industry and airports (CL, MS, ion mobility spectrometry, biosensors and electron capture) would be a valid alternative for detecting antipersonnel mines though they have mentioned that these sensors have a too low sensibility, are too slow or too large to be used in field applications.

In the congressional research service (CRS) report for the USA Congress (2005), Shea and Morgan [75] included chemical luminescence as one of the trace detection techniques for aviation security. They point out that bulk detection of explosives on airline passengers is of interest and that these systems detect only bulk quantities of explosives, so they would not raise 'nuisance alarms' on passengers who have recently handled explosives for innocuous reasons. They also mention that trace detection techniques would also detect bulk quantities of explosives and may raise security concerns about passengers who have been in contact with explosives.

Hallowell [76] cites a patent from Thermedics Inc. [77] based on a high-speed GC-CL detector, as being in the category of trace detection technologies and states that it is a non-intrusive way of screening people for the presence of clandestinely concealed explosives, weapons or drugs for aviation security. Hallowell comments that this patent includes the development of a 'walk-in, walk-out' booth, containing suction vents that horizontally draw in a 'large volume' of air from around a human subject who enters the booth.

Gas chromatography–thermal energy analyser explosive detection systems (EGIS II and EGIS III systems manufactured by Thermo Electron Corporation) are now being used by several governments, institutions and laboratories, including the Federal Bureau of Investigation, the Bureau of Alcohol, Tobacco and Firearms (ATF), the Royal Canadian Mounted Police forensic laboratories, US and German government aviation security agencies and the British Military (to be deployed in Northern Ireland in support of the anti-terrorism mission) [78]. The specifications reported for EGIS II® and EGIS III® include a false alarm rate below 0.2% for typical luggage screening and lower detectable limits of 300 and 100 pg, respectively. The reader interested in more details and specifications can consult the last reference from the Thermo Electron Corporation. These systems have become the trace detection standard throughout Europe for the airport screening of bags. In 1995, the National Laboratory Center in Rockville, Maryland (from ATF's laboratory system), started using EGIS Explosives Detectors to examine explosives residue in post-blast explosive scenes.

The Explosion Investigation Section of the National Research Institute of Police Science of Japan [79] makes use of GC with a TEA and IC to analyse trace explosive residues from debris. An EGIS explosives detector was also used in the Oklahoma City bombing investigation [80].

In a collaborative police investigation involving the Indonesian National Police and the Australian Federal Police about the Bali bombings on 12 October 2002, which killed 202 people [81], the presence of the organic explosive TNT was detected using IMS, and it was confirmed by GC–TEA, among other systems.

References

[1] T.G. Chasteen. *Chemiluminescence Definitions and Primer.* Available at http://www.shsu.edu/~chm_tgc/chemilumdir/Define.html.

[2] W.R.G. Baeyens, S.G. Schulman, A.C. Calokerinos, Y. Zhao, A.M.G. Campaña, K. Nakashima and D. De Keukeleire. *J. Pharm. Biomed. Anal.*, 17 (1998) 941.

[3] A. Townshend. *Analyst*, 115 (1990) 495.

[4] L. Theisen, D.W. Hannum, D.W. Murray and J.E. Parmeter. *Survey of Commercially Available Explosives Detection Technologies and Equipment 2004*, Document No. 208861. Sandia National Laboratories, 2004. Available at http://www.ncjrs.gov/pdffiles1/nij/grants/208861.pdf.

[5] Environmental Health and Safety. University of Wisconsin. River Falls. *Environmental Safety Resources. Nitro explosive safety.* Available at http://www.uwrf.edu/ehs/2nitroexplosivesafety.htm.

[6] C.D. Bosco. *Hand Held Explosives Sensor System*, Report number 03306. University Transportation Center for Alabama (UTCA), Tuscaloosa, AL, 2003. Available at http://utca.eng.ua.edu/projects/final_reports/03306-Bosco-fin-rpt-16Sep03.pdf.

[7] E.B. Byall. *Explosives Report, Detection and Characterization of Explosives and Explosive Residue: A Review, 1998–2001.* 13th INTERPOL Forensic Science Symposium, Lyon, France, 16–19 October, 2001. Available at http://www.interpol.int/Public/Forensic/IFSS/meeting13/Reviews/Explosives.pdf.

[8] J.I. Steinfeld and J. Wormhoudt. *Annu. Rev. Phys. Chem.*, 49 (1998) 203.

[9] C. Bruschini. *"Commercial Systems for the Direct Detection of Explosives (for Explosive Ordnance Disposal Tasks" ExploStudy, Final Report* (2001). Available at http://www.eudem.vub.ac.be/publications/files/ExploStudy.pdf.

[10] J. Yinon. *Trac-trends. Anal. Chem.*, 21 (2002) 292.

[11] S. Singh and M. Singh. *Signal Processing*, 83 (2003) 31.

[12] D.S. Moore. *Rev. Sci. Instrum.*, 75 (2004) 2499.

[13] A.M. Jimenez and M.J. Navas. *J. Hazard. Mater.*, 106A (2004) 1.

[14] D.H. Fine, W.C. Yu, E.U. Goff, E.C. Bender and D.J. Reutter. *J. Forensic Sci.*, 29 (1984) 732.

[15] X. Xu, A.M. van de Craats, E.M. Kok and P.C.A.M. de Bruyn. *J. Forensic Sci.*, 49 (2004) 1.

[16] D.W. Hannum and J.E. Parmeter. *Survey of Commercially Available Explosives Detection Technologies and Equipment*, NCJ 171133. Sandia National Laboratories, 1998. Available at http://www.fortliberty.org/military-library/expsurvey.pdf.

[17] D.H. Fine, F. Rufeh and B. Gunther. *Analyt. Lett.*, 6 (1973) 731.

[18] D.H. Fine, F. Rufeh and D. Lieb. *Nature*, 247 (1974) 309.
[19] W.C. Yu and E.U. Goff. *Anal. Chem.*, 55 (1983) 29.
[20] A.L. Lafleur and K.M. Mills. *Anal. Chem.*, 53 (1981) 1202.
[21] J.M.F. Douse. *J. Chromatogr.*, 256 (1983) 359.
[22] J.M.F. Douse. *J. Chromatogr.*, 410 (1987) 181.
[23] J.M. Douse. *J. Chromatogr.*, 445 (1988) 244.
[24] D.A. Collins. *J. Chromatogr.*, 483 (1989) 379.
[25] D.H. Fine, D. Lieb and F. Rufeh. *J. Chromatogr.*, 107 (1975) 351.
[26] D.H. Fine, F. Rufeh, D. Lieb and D.P. Rounbehler. *Anal. Chem.*, 47 (1975) 1188.
[27] R.A. Scanlan. Chemilumnescence for measurements of N-nitrosamines in foods, in *Analysis of Food Contaminants* (Ed. by John Gilbert). Elsevier Applied Science Publishers, London, 321 (1984).
[28] U. Isacsson and G. Wetermark. *Anal. Chim. Acta*, 68 (1974) 339.
[29] C.L. Rhykerd, D.W. Hannum, D.W. Murray and J.E. Parmeter. *Guides for the Selection of Commercial Explosives Detection Systems for Law Enforcement Applications,* Guide 100-99. National Institute of Justice, Washington, DC, 1999, p. 18. Available at http://www.ncjrs.gov/pdffiles1/nij/178913-1.pdf.
[30] J.M.F. Douse. *J. Chromatogr.*, 464 (1989) 178.
[31] R.Q. Thompson, D.D. Fetterolf, M.L. Miller and R.F. Mothershead II. *J. Forensic Sci.*, 44 (1999) 795.
[32] P. Kolla. *J. Forensic Sci.*, 36 (1991) 1342.
[33] E.S. Francis, M. Wu, P.B. Farnsworth and M.L. Lee. *J. Microcolumn. Sep.*, 7 (1995) 23.
[34] J.M.F. Douse. *J. Chromatogr.*, 328 (1985) 155.
[35] J.H. Phillips, R.J. Coraor and S.R. Prescott. *Anal. Chem.*, 55 (1983) 889.
[36] C.M. Selavka, R.E. Tontarski Jr and R.A. Strobel. *J. Forensic Sci.*, 32 (1987) 941.
[37] TWGFEX Laboratory Explosion Group and Standards & Protocols Committe. *Recommended Guidelines for Forensic Identification of Intact Explosives.* Available at http://www.ncfs.org/twgfex/docs/Guide%20for%20identification%20of%20intact%20 explosives.pdf.
[38] M. Nambayah and T.I. Quickenden. *Talanta*, 63 (2004) 461.
[39] A.L. Lafleur and B.D. Morriseau. *Anal. Chem.*, 52 (1980) 1313.
[40] P. Kolla. *J. Chromatogr. A*, 674 (1994) 309.
[41] P. Kolla and A. Sprunkel. *J. Forensic Sci.*, 40 (1995) 406.
[42] R.W. Hiley. *J. Defence Sci.*, 1 (1996) 181.
[43] R.W. Hiley. *J. Forensic Sci.*, 41 (1996) 975.
[44] A. Crowson, H.E. Cullum, R.W. Hiley and A.M. Lowe. *J. Forensic Sci.*, 41 (1996) 980.
[45] H.E. Cullum, C. McGavigan, C.Z. Uttley, M.A.M. Stroud and D.C. Warren. *J. Forensic Sci.*, 49 (2004) 1.
[46] D. Warren, R.W. Hiley, S.A. Phillips and K. Ritchie. *Sci. Justice*, 39 (1999) 11.
[47] A. Crowson, R.W. Hiley and C.C. Todd. *J. Forensic Sci.*, 46 (2001) 53.
[48] C.R. Bowerbank, P.A. Smith, D.D. Fetterolf and M.L. Lee. *J. Chromatogr. A*, 902 (2000) 413.
[49] A. Zeichner. *Anal. Bioanal. Chem.*, 376 (2003) 1178.
[50] A. Zeichner, B. Eldar, B. Glattstein, A. Koffman, T. Tamiri and D. Muller. *J. Forensic Sci.*, 48 (2003) 1.
[51] A. Zeichner and B. Eldar. *J. Forensic Sci.*, 49 (2004) 1.
[52] A. Zeichner and B. Eldar. A novel method for extraction and analysis of gun powder residues on double-side adhesive coated stubs. The 8th Conference of the Israel Analytical Chemistry Society, 11–12 January 2005.
[53] T.F. Jenkins, D.C. Leggett, C.L. Grant and C.F. Bauer. *Anal. Chem.*, 58 (1986) 170.
[54] R.L. Marple and W.R. LaCourse. *Anal. Chem.*, 77 (2005) 6709.
[55] A.C. Schmidt, B. Niehus, F.M. Matysik and W. Engewald. *Chromatographia*, 63 (2006) 1.

[56] EPA Method 8330. *Nitroaromatics and Nitramines by High Performance Liquid Chromatography*. Available at http://www.epa.gov/SW-846/pdfs/8330.pdf.
[57] Ocupational Safety & Health Administration (OSHA), US Department of Labor. *Organic Methods Evaluation Branch OSHA Analytical Laboratory*, Method no. 43. 1983. Available at http://www.osha.gov/dts/sltc/methods/organic/org043/org043.html.
[58] E.S. Francis, D.J. Eatough and M.L. Lee. *J. Microcolumn. Sep.*, 6 (1994) 395.
[59] H.O. Albrecht. *Z. Phys. Chem.*, 135 (1928) 321.
[60] C. Hao, P.B. Shepson, J.W. Drummond and K. Muthuramu. *Anal. Chem.*, 66 (1994) 3737.
[61] I.U. Mohammadzai, T. Ashiuchi, S. Tsukahara, Y. Okamoto and T. Fujiwara. *J. Chin. Chem. Soc.*, 52 (2005) 1037.
[62] D.H. Nguyen, S. Berry, J.P. Geblewicz, G. Couture and P. Huynh. *Chemiluminescent Detection of Explosives, Narcotics, and Other Chemical Substances*, US patent no. 6 984 524. 2006. Available at http://www.freepatentsonline.com/6984524.html.
[63] Scintrex Trace Corp. Ottawa, Ontario, Canada. *Trace Detection Products and Systems for a Dangerous World*. Available at http://www.scintrextrace.com/.
[64] K.A. Fähnrich, M. Pravda and G.G. Guilbault. *Talanta*, 54 (2001) 531.
[65] J.G. Bruno and J.C. Cornette. *Microchem. J.*, 56 (1997) 305.
[66] J.G. Bruno and J.C. Cornette. *Electrochemiluminescence Assays Based on Interactions with Soluble Metal Ions and Diaminoaromatic Ligands*, US patent no. 5 976 887. 1999. Available at http://www.freepatentsonline.com/5976887.html.
[67] T.M. Jackson and R.P. Ekins. *J. Immunol. Methods*, 87 (1986) 13.
[68] R. Wilson, C. Clavering and A. Hutchinson. *J. Electroanal. Chem.*, 557 (2003) 109.
[69] A.J. Schuetz, M. Winklmair, M.G. Weller and R. Niessner. Multianalyte detection with an affinity sensor array, *Bioluminescence and Chemiluminescence – Perspectives for the 21st Century* (Ed. by A. Roda, M. Pazzagli, L.J. Kricka and P.E. Stanley, pp. 67–70). *Proc. 10th Intern. Symp.*, 1998, Bologna, John Wiley & Sons, Chichester, 1999. Available at http://www.ch.tum.de/ wasser/weller/BC99b.pdf.
[70] I.M. Ciumasu, P.M. Krämer, C.M. Weber, G. Kolb, D. Tiemann, S. Windisch, I. Frese and A.A. Kettrup. *Biosens. Bioelectron.*, 21 (2005) 354.
[71] R. Wilson, C. Clavering and A. Hutchinson. *Anal. Chem.*, 75 (2003) 4244.
[72] R. Wilson, C. Clavering and A. Hutchinson. *Analyst*, 128 (2003) 480.
[73] A.M. Rouhi. *Chem. Eng. News*, 73 (1995) 10.
[74] B. Gross and C. Bruschini. Sensor technologies for the detection of antipersonnel mines. A survey of current research and system developments: Paper presented at the International Symposium on Measurement and Control in Robotics (ISMCR'96), Brussels, May 1996, 564–569.
[75] D.A. Shea and D. Morgan. *Detection of Explosives on Airline Passengers: Recommendation of the 9/11 Commission and Related Issues*. Updated on 7 February 2005. http://fas.org/sgp/crs/homesec/RS21920.pdf.
[76] S.F. Hallowell. *Talanta*, 54 (2001) 447.
[77] M.D. Arney, G. Zaccai, E.K. Achter, E.J. Burke, G. Miskolczy and A.A. Sonin. *Air-sampling Apparatus with Easy Walk-in Access*, US patent no. 4896547. 1990. Availabele at http:/www.freepatentsonline.com/4896547.html.
[78] Thermo Electron Corporation. A line of portable, bench-top explosives detection systems. EGIS II and EGIS III. Available at http://www.thermoramsey.de/EGIS-englisch.pdf.
[79] National Research Institute of Police Science, Japan. Explosion Investigation Section. Available at http://www.nrips.go.jp/org/second/explosion/index-e.html.
[80] http://www.atf.gov/explarson/forensics-laboratory-support.htm.
[81] D. Royds, S.W. Lewis and A.M. Taylor. *Talanta*, 67 (2005) 262.

Chapter 2

Detection of Explosives by Mass Spectrometry

Jehuda Yinon

Department of Environmental Science, Weizmann Institute of Science, Rehovot 76100, Israel

Counterterrorist Detection Techniques of Explosives
Jehuda Yinon (Editor)

Contents

1. Introduction

The mass spectrometer meets the main performance requirements of an explosive detection system, which are sensitivity, selectivity, and speed of analysis. Additional requirements are mobility and cost.

During the last 10 years, mass spectrometers have become smaller and mobile and less expensive.

Various mass spectrometer configurations have been used for the detection of explosives, such as ion traps, quadrupoles, and time-of-flight (TOF) analyzers and tandem mass spectrometer (MS/MS) combinations. Also, various modes of ionization have been employed, depending on the specific application in the detection of explosives.

2. Mass spectrometry – principles of operation

Mass spectrometry is the field dealing with separation and analysis of substances according to the masses of the atoms and molecules of which the substance is composed.

The principle of mass analysis is that the parameters of time and space of the path of a charged particle in a force field in vacuum are dependent on the mass-to-charge (*m/e*) ratio.

The methods of separation and analysis can be divided in two:

(1) Methods based on time separation.
(2) Methods based on geometric separation.

The time separation method is based on the fact that ions having different *m/e* ratios have different times of flight and are thus collected one after the other. The TOF mass spectrometer is based on these principles of operation.

In the geometric separation method, ions having different *m/e* ratios are separated according to their geometric position at the collecting spot. The magnetic mass analyzer, the quadrupole mass analyzer, and the ion trap are based on these principles of operation. The magnetic mass spectrometer will not be discussed in this chapter as it has not been used for the detection of explosives.

2.1. Quadrupole mass analyzer

The quadrupole mass analyzer [1] consists of four parallel metal rods arranged as in Fig. 1. Two opposite rods are electrically connected and have an applied potential of $U + V \cos \omega t$, and the other two rods, also electrically connected, have a potential of $-(U + V \cos \omega t)$, where U is a DC voltage and $V \cos \omega t$ is an RF voltage, ω being the angular frequency ($\omega = 2\pi f$).

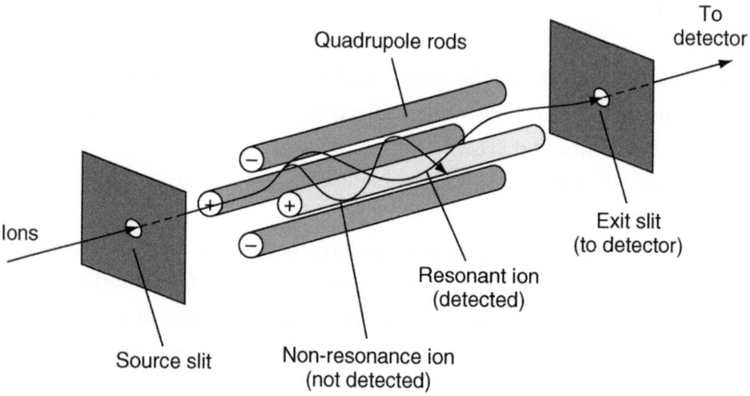

Fig. 1. Quadrupole mass analyzer. Reproduced from http://www.chm.bris.ac.uk/ms/images/quad-schematic.gif, with permission from Dr Paul Gates, School of Chemistry, University of Bristol, UK.

The motion of an ion injected into the field in the z direction can be described by the Mathieu differential equations:

$$m\frac{d^2x}{dt^2} + 2\frac{e}{r_o^2}(U + V_o \cos \omega t)\ x = 0$$

$$m\frac{d^2y}{dt^2} - 2\frac{e}{r_o^2}(U + V_o \cos \omega t)\ y = 0$$

$$m\frac{d^2z}{dt^2} = 0$$

where r_o is the radius of the four-rod system, m is the mass of the ion, and e is its charge.

The solution of these equations designate oscillations performed by an ion in the x and y directions, whereas the ion proceeds with constant velocity in the z direction.

The applied voltages affect the trajectory of ions traveling down the flight path centered between the four rods. For given DC and RF voltages, only ions of a certain m/e ratio pass through the quadrupole analyzer until they hit the detector, whereas all other ions are thrown out of their original path and hit the walls. A mass spectrum is obtained by monitoring the ions passing through the quadrupole analyzer as the voltages on the rods are varied. There are two methods: varying the frequency ω while holding U and V constant, or varying U and V but keeping U/V constant.

2.2. *Ion trap mass analyzer*

The ion trap mass analyzer [2] (Fig. 2) consists of two end-cap electrodes, held at ground potential, and an interposed ring electrode, to which DC and RF voltages are applied. The ring electrode is a single surface formed by a hyperboloid of rotation. The end-caps are complementary hyperboloids having the same conical asymptotes: z is an axis of

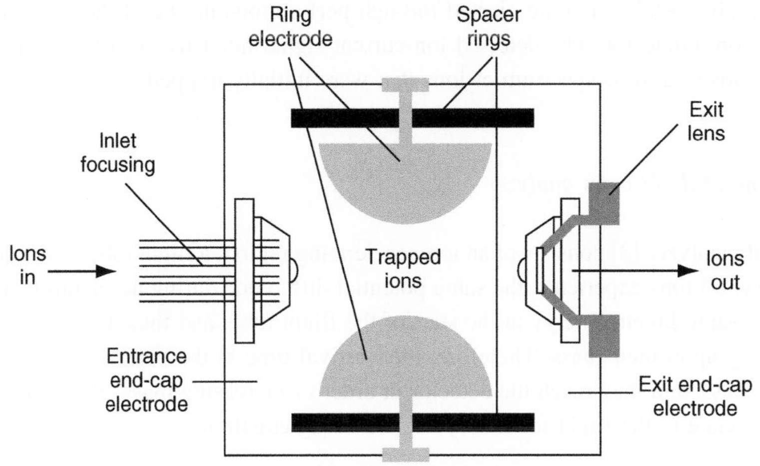

Fig. 2. Ion trap mass analyzer. Reproduced from http://www.chm.bris.ac.uk/ms/images/iontrap-schematic.gif, with permission from Dr Paul Gates, School of Chemistry, University of Bristol, UK.

cylindrical symmetry. These electrodes form a cavity in which it is possible to trap and analyze ions. Both end-cap electrodes have a small hole in their centers through which the ions can travel. The ring electrode is located halfway between the two end-cap electrodes.

Ions produced in the source enter the trap through the inlet focusing system and the entrance end-cap electrode.

As in the quadrupole mass analyzer, the motion of an ion in the trap having an m/e ratio can be described by the Mathieu differential equations:

$$m\frac{d^2 r}{dt^2} + 2\frac{e}{r_o{}^2}(U + V_o \cos \omega t)r = 0$$

$$m\frac{d^2 z}{dt^2} - 2\frac{e}{z_o{}^2}(U + V_o \cos \omega t)z = 0$$

where U is the DC voltage, V_o cos ωt is the RF voltage, ω is the angular frequency ($\omega = 2\pi f$), r_o is the radius of the cavity formed by the ring electrode (represents the trap size), z_o is the distance from the center of the trap to the end cap, and m and e are the mass and charge of the ion, respectively. For an ideal quadrupole field, $r_o^2 = 2z_o{}^2$.

Solution of these equations lead to two possibilities: r or z continues to increase with time giving unbounded and therefore unstable trajectories, and r and z are periodic with time, producing stable trajectories within the trap (assuming that the maximum distance of excursion from the center is less than r_o and z_o).

Mass analysis is based on mass selective instability. DC and RF voltages applied to the trap electrodes are such that ions over the entire *m/e* range of interest can be trapped within the field imposed by the electrodes. After this storage period, the parameters U and V_o are changed so that the trapped ions of consecutive values of *m/e* become successively unstable and are ejected through perforations in one of the electrodes and impinge on a detector. The detected ion current signal intensity, as a function of time, corresponds to a mass spectrum of ions that were initially trapped.

2.3. *Time-of-flight mass analyzer*

The TOF analyzer [3] consists of an ion-accelerating region, a flight tube, and a detector. In theory, all ions experience the same potential difference during acceleration and thus have the same kinetic energy at the start of the flight tube, and thus different velocities depending upon their mass. Therefore, their arrival time at the detector is proportional to their mass, and they reach the detector in order of increasing mass. The m/e ratios of the ions relate to the flight times t by the following equations:

$$\frac{mv^2}{2} = eV \qquad v = \sqrt{\frac{2eV}{m}}$$

$$t_f = \frac{l}{v} = l\sqrt{\frac{m}{2eV}} = l\sqrt{\frac{1}{2V}}\sqrt{\frac{m}{e}}$$

$$t_f = k\sqrt{\frac{m}{e}}$$

where t_f is the TOF, l is the length of the flight tube, v is the velocity of the ion of mass m and charge e, and V is the acceleration potential. The differences of flight times are, of course, the basis for resolving ions of different m/e ratios in the TOF analyzer, and the mass resolution depends on the flight time differences, which are proportional to $l(\sqrt{m_1} - \sqrt{m_2})$.

The major deficiency of a simple linear TOF instrument is its insufficient mass resolution, resulting from flight time variations of ions of the same m/e ratio. The ionization process adds a certain amount of initial kinetic energy to the molecules before acceleration. In addition, the different spatial positions in the source from where the ions are formed lead to varying values of l, and thus to flight time variations. As a result, modern TOF instruments commonly employ a technique called reflectron to enhance resolution. A reflectron TOF (RETOF) [4] (Fig. 3) is used to focus ions of the same

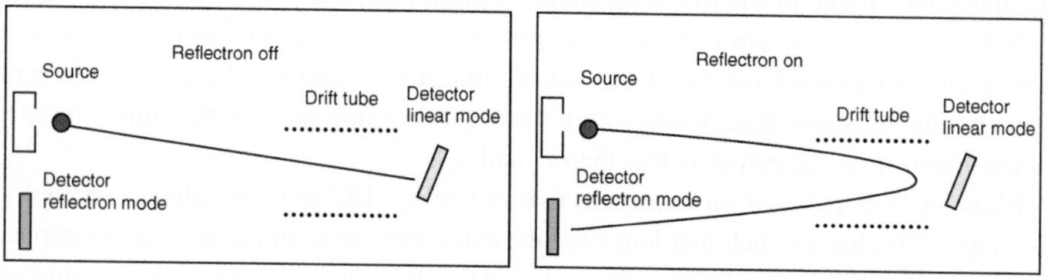

Fig. 3. Reflectron time-of-flight (TOF) mass analyzer.

m/e but of different kinetic energy. There are different designs of reflectrons, but often they consist of a series of rings or grids that act as an ion mirror. It is located after the drift tube, creating a retarding field that the ions penetrate. Depending upon their kinetic energy, they enter this field at different depths and then are reflected back into the flight tube, where they drift to the detector, which is placed close to the ion source. Consider the case of ions of the same *m/e* but significant energy spread, which would be detected as broad peaks in the linear TOF: the ion with a higher kinetic energy penetrates deeper and spends more time in the reflectron. The slow ion, on the other hand, returns to the flight tube faster, but with the same lower kinetic energy. The faster ion will catch up with the slow one at the detector, thereby improving mass resolution greatly.

2.4. *Tandem mass spectrometer (MS/MS)*

Tandem mass spectrometry, or MS/MS, is used for structure determination of molecular ions or fragments [5]. In MS/MS, the ion of interest, the precursor ion, is selected with the first analyzer (MS-1), collided with an inert gas, such as argon or helium, in a collision cell, and the fragments generated by the collision are separated by a second analyzer (MS-2) and detected (Fig. 4). This configuration of having two mass spectrometers in series is called tandem in space. Several combinations have been used, the most common one being the triple quadrupole. In this configuration, the second quadrupole, to which only an RF voltage is connected, serves as the collision cell.

In the ion trap, the experiments are carried out in one analyzer, and the various events are separated in time, not in space. This configuration is called tandem in time and is

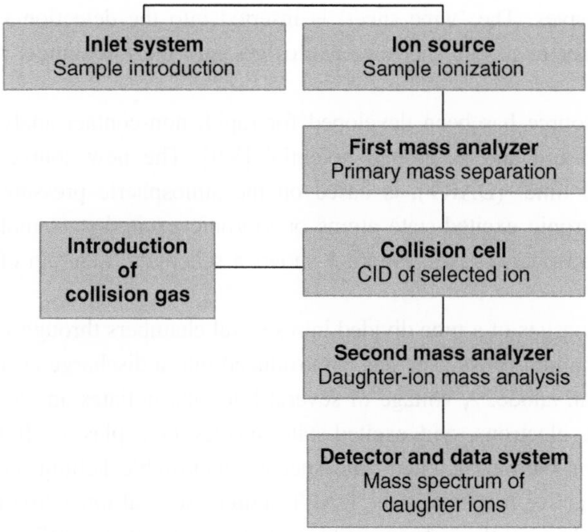

Fig. 4. Schematic MS/MS configuration.

not limited to simple MS/MS experiments but can achieve multiple MS/MS/MS.../MS measurements under optimal conditions.

The MS/MS mass spectrum provides a 'fingerprint' of the precursor ion and provides an additional dimension of selectivity and identification of the analyzed sample.

3. Detection of explosives by mass spectrometry methods

3.1. Explosive vapor and trace detection

An explosive detection system based on an atmospheric pressure chemical ionization (APCI) source with corona discharge ionization has been developed [6]. The direction of the sample gas flow introduced into the ion source is opposite to that of the ion flow produced by the ion source, a technique called 'counter-flow introduction' (CFI). In a conventional source, sample gas flows directly into the corona discharge region. Hence, the sample gas and primary ions produced by the corona discharge are moved toward the sampling aperture by the gas flow and by an electric field produced between the corona discharge electrode and the aperture electrode. In the present ion source, the sample gas flows in the opposite direction. The negative ions generated are extracted by the electric field in the direction opposite to the gas flow. The mass analyzer is an ion trap having a radius r_0 of 16 mm. The system, with a mass range of 300 amu, had a detection limit for trinitrotoluene (TNT) of 10–20 ppt. The detection limit was reduced to 0.3 ppt when using MS/MS.

A commercial explosive trace detection system, Hitachi DS-120E, based on a quadrupole mass spectrometer was built [7]. Ionization is carried out by corona discharge and APCI. Sampling is carried out by wiping surfaces on suspicious objects for traces of explosives. The 'wipe sheet' is inserted into the detection system, where it is heated in order to desorb the trace explosives into the ion source. Analysis time is 5–10 s.

A new ion source has been developed for rapid, non-contact analysis of materials at ambient pressure and at ground potential [8,9]. The new source, termed 'direct analysis in real time' (DART), is based on the atmospheric pressure interactions of long-lived electronic excited-state atoms or vibronic excited-state molecules with the sample and atmospheric gases. Figure 5 shows a schematic diagram of the DART ion source.

The source consists of a tube divided into several chambers through which a gas such as nitrogen or helium flows. The gas is introduced into a discharge chamber containing a cathode and an anode. A voltage of several kilovolts initiates an electrical discharge producing ions, electrons, and excited-state species in a plasma. It is believed that the electronic or vibronic excited-state species (metastable helium atoms or nitrogen molecules) are active reagents in the DART source. Several ionization mechanisms are possible, depending on the polarity and reaction gas, the proton affinity and ionization potential of the analyte, and the presence of additives or dopants. The simplest process

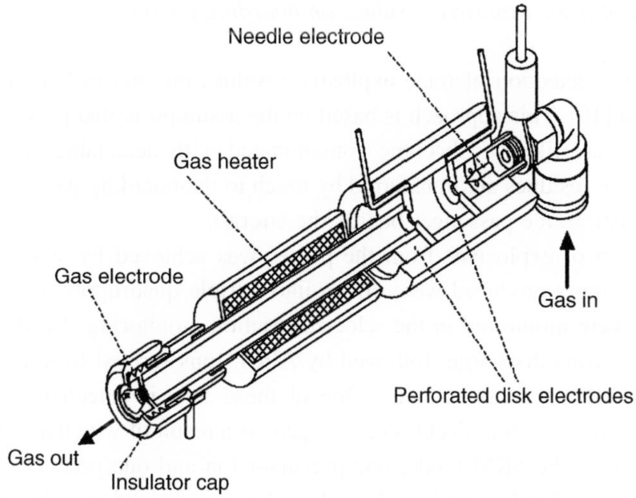

Fig. 5. Schematic diagram of direct analysis in real time (DART) ion source. Reproduced with permission from Cody et al. [9]. Copyright 2005 American Chemical Society.

is Penning ionization involving transfer of energy from the excited gas M^* to an analyte S having an ionization potential lower than the energy of M^*. This produces a radical molecular cation $S^{+\cdot}$ and an electron.

Penning ionization is a dominant reaction when nitrogen or neon is used in the DART source. Nitrogen or neon ions are effectively removed by electrostatic lenses and are not observed in the background mass spectrum. When helium is used, the dominant positive-ion formation mechanism involves the formation of ionized water clusters followed by proton transfer reactions. Negative-ion formation occurs by production of electrons by Penning ionization or by surface Penning ionization:

$$M^* + \text{surface} \rightarrow M + \text{surface} + e^-$$

These electrons, after being thermalized by collisions with atmospheric pressure gas, undergo electron capture by atmospheric oxygen to produce O_2^-, which reacts with the analyte to produce anions.

The DART source was installed on a high-resolution TOF mass spectrometer with a resolution of 6000. The system can directly detect compounds on surfaces without requiring sample preparation, such as wiping or solvent extraction. Negative molecular ions are observed for TNT. For other explosives, such as cyclotrimethylenetrinitramine (RDX), cyclotetramethylenetetranitrate (HMX), pentaerythritol tetranitrate (PETN), nitroglycerin (NG), and ethylene glycol dinitrate (EGDN), adduct ions like $[M + Cl]^-$ or $[M + \text{trifluoroacetate}]^-$ are observed, when a suitable additive is present.

3.2. Detection of trace explosive residues on boarding passes

A system for the detection of trace explosive residues on aircraft boarding passes has been developed [10]. This approach is based on the assumption that passengers involved with explosives are likely to become contaminated with detectable amounts of trace explosives. These residues are transferred by touch to the boarding pass and are detected by the system before the passenger boards the aircraft.

The desorption of explosives from the passes was achieved by short wave infrared radiation. The vapors produced were drawn into a triple quadrupole mass spectrometer (MS/MS) and were monitored in the selected reaction monitoring (SRM). Ionization is carried out by corona discharge, followed by APCI. Ions formed from most explosives are M^-, $[M-H]^-$, and adduct ions. One of these ions is selected to pass into the collision cell, to react with molecules of nitrogen, as a result of which a series of product ions are formed. In the SRM mode, one precursor ion and one product ion are chosen for each compound. The first and third quadrupoles are adjusted in order to enable SRM transition between these two ions.

When introducing an additive, such as dichloromethane, precursor adduct ions are observed for RDX, PETN, and NG. SRM transitions are as follows: for RDX, m/z 262 $[M+{}^{35}Cl]^- \rightarrow$ m/z 46 $[NO_2]^-$; for PETN, m/z 351 $[M+{}^{35}Cl]^- \rightarrow$ m/z 46 $[NO_2]^-$; and for TNT, m/z 227 $[M]^- \rightarrow$ m/z 210 $[M-OH]^-$.

Limits of detection were less than 100 pg for TNT, NG, PETN, and RDX. The system could handle 1000 boarding passes per hour.

A background study into the levels of explosive residues on used boarding passes was conducted by analyzing over 20 000 boarding passes from a number of airports in UK, USA, and Canada. Traces of explosives were detected on approximately 0.5% of passes analyzed. NG contributed to the majority of the positive signals observed, probably because of its use as a medication.

3.3. Personnel screening portal

An explosive detection personnel portal is a walk-through system for rapidly screening personnel for trace amounts of explosives at sites such as airports or Federal buildings. Such a portal, the Syagen Guardian MS-ETD Portal, has been built, using a mass spectrometer detector [11,12]. The basic functions of the portal are as follows: sample collection and concentration, selective detection of targeted compounds, and accurate measurement of the targeted compounds. When a person enters the portal, he feels a flow of air that removes and collects any particles (e.g., explosive materials) from the person's clothing and body. The air in the portal is filtered to collect particles and condensable vapors in a preconcentrator. Following this enrichment process, the collected sample is thermally desorbed in a glow discharge ion source coupled to a quadrupole ion trap TOF (QitTof) mass spectrometer system, having MS/MS capability.

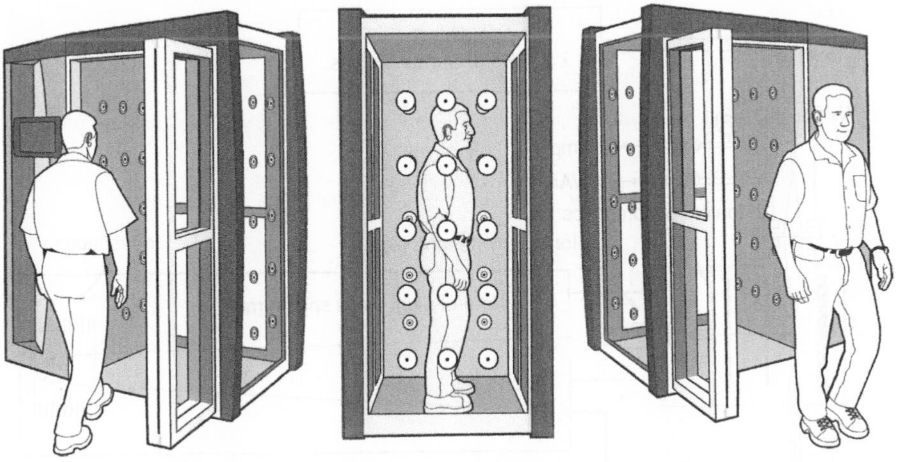

Fig. 6. Schematic diagram of how a portal works. Reproduced from http://www.syagen.com/datasheet/Syagen4pBroFin.pdf, with permission from Syagen Technology, Inc.

Explosives that were detectable included TNT, dinitrotoluene (DNT), RDX, HMX, PETN, EGDN, NG, Tetryl, ammonium nitrate fuel oil (ANFO), triacetone triperoxide (TATP), and hexamethylene triperoxide diamine (HMTD). Analysis time is less than 15 s. Figure 6 shows schematically how such a portal works:

(1) Front door opens and person enters the portal.
(2) Door closes and a puff of air dislodges trace materials from clothing.
(3) Portal air is concentrated to extract particles, residue, and condensable vapors.
(4) Enriched mixture is analyzed by mass spectrometer and results displayed.
(5) Rear door opens and person exits.

3.4. Laser photoionization mass spectrometry

Single-photon ionization (SPI) in combination with mass-selective detection by TOF mass spectrometry has been used for the detection of explosives [13].

The laser system consists of a Continuum Powerlite Precision 9010 Nd:YAG with a 5-ns pulse width and a repetition rate of 10 Hz. The 118.2 nm light, equivalent to a photon energy of 10.49 eV, is produced by frequency tripling the third harmonic output (354.6 nm) of the Nd:YAG laser. To accomplish this, the 354.6 nm laser beam is focused into a stainless steel tripling cell filled with Xe or Xe/Ar mixtures, which is attached directly to the ion source of the mass spectrometer. Figure 7 shows a schematic diagram of the SPI–TOF-MS instrument.

Ions produced by SPI are detected and mass analyzed by a reflectron TOF mass spectrometer having a resolution of $m/\Delta m = 1000$. The sample is introduced into the

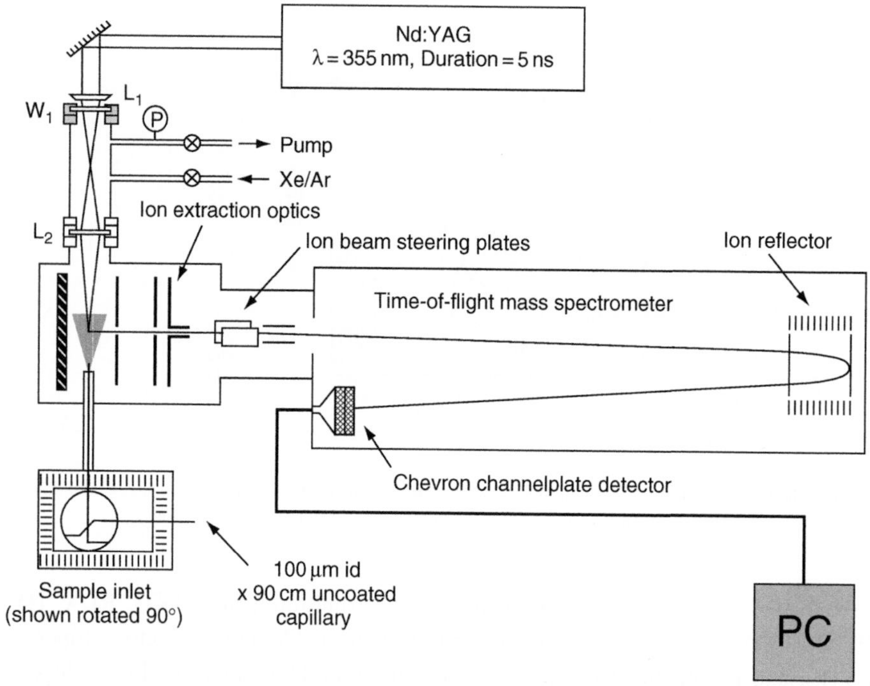

Fig. 7. Schematic diagram of the single-photon ionization–time-of-flight (SPI–TOF)-MS instrument. Reproduced with permission from Mullen et al. [13]. Copyright 2006 American Chemical Society.

ion source through a short deactivated capillary gas chromatography (GC) column that can be heated to 200 °C.

The SPI–TOF mass spectrum of TNT consists of the molecular ion at *m/z* 227 and the $[M - OH]^-$ fragment ion at m/z 210. The SPI–TOF mass spectrum of TATP consists of a molecular ion at *m/z* 222 and a series of fragment ions at m/z 43 (acetyl ion), *m/z* 58 (acetone ion), *m/z* 59 $[C_3H_7O]^+$, *m/z* 75 $[C_3H_7O_2]^+$, *m/z* 106 $[C_3H_6O_4]^+$, and m/z 122 (diacetone diperoxide $C_3H_6O_6{}^+$).

Limit of detection, measured for 2,4-DNT, was 40 ppb with a signal-to-noise ratio of $S/N \approx 2:1$.

The drawback of the 118.2-nm SPI method is its applicability to only those molecules with ionization energies below 10.49 eV.

3.5. Desorption electrospray ionization mass spectrometry

Desorption electrospray ionization (DESI) uses an aqueous spray directed at an analyte deposited on an insulating surface (Fig. 8) [14,15]. The sample is in the solid form at atmospheric pressure. The spray of charged droplets is produced, as in electrospray ionization (ESI), by passing an aqueous solution (i.e., methanol–water, containing some

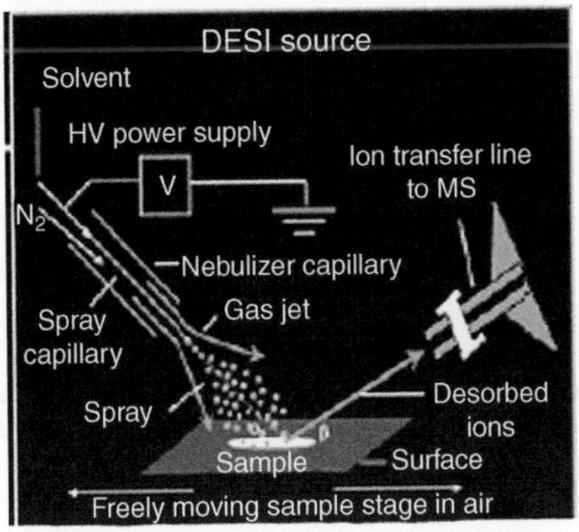

Fig. 8. Schematic diagram of the desorption electrospray ionization (DESI) source. Reproduced with permission from Cotte-Rodriguez et al. [14]. Copyright 2005 American Chemical Society.

additive) through a capillary held at high voltage. The impact of the charged micro-droplets with neutral molecules present on the surface produces gas-phase ions of the analyte. These ions are then sampled with an ion trap mass spectrometer, producing mass spectra that are similar to normal ESI mass spectra. Addition of reagents in the spray solution formed characteristic adduct ions that improved selectivity and identification capability. The type of ions observed depends on the nature of the sample, substrate, and reagent. For example, RDX is observed as the chloride adduct $[RDX+Cl]^-$ when electrosprayed with a dilute HCl solution but is observed as the protonated molecule $[RDX+H]^+$ when sprayed with pure water. Observed ions can be mass selected and examined by recording their MS/MS dissociation products mass spectrum.

Analyzed explosives included TNT, RDX, HMX, PETN, TATP, and the plastic explosives C-4, Semtex-H, and Detasheet. Limits of detection were in subnanogram to subpicogram range.

Figure 9(a) shows the negative ion DESI spectrum for a plastic explosive C-4 fingerprint on glass, using methanol–water–HCl (1:1:0.05%) as spray solvent. Identity of the ions was confirmed by MS/MS. Figure 9(b) shows the negative ion DESI spectrum of a C-4 fingerprint after five transfers onto a glass slide. The absolute abundance of the ion at *m/z* 257 $[RDX+{}^{35}Cl]^-$ for the five transfers was more than one order of magnitude lower than the one of the original fingerprint.

A DESI source was adapted to sample ions at a distance up to 3 m [16]. The mass spectrometer and the DESI source were connected by a long stainless steel flexible ion transport tube (1.8 mm ID, 3.18 mm OD), which allows the transport of ions through air to the mass spectrometer after they leave the surface. The transport tube allows movement of the DESI source across the checked surface, looking for explosives. Methanol–water (70:30) doped with 10 mM NaCl was used as spray solvent, resulting in typical chloride adduct ions.

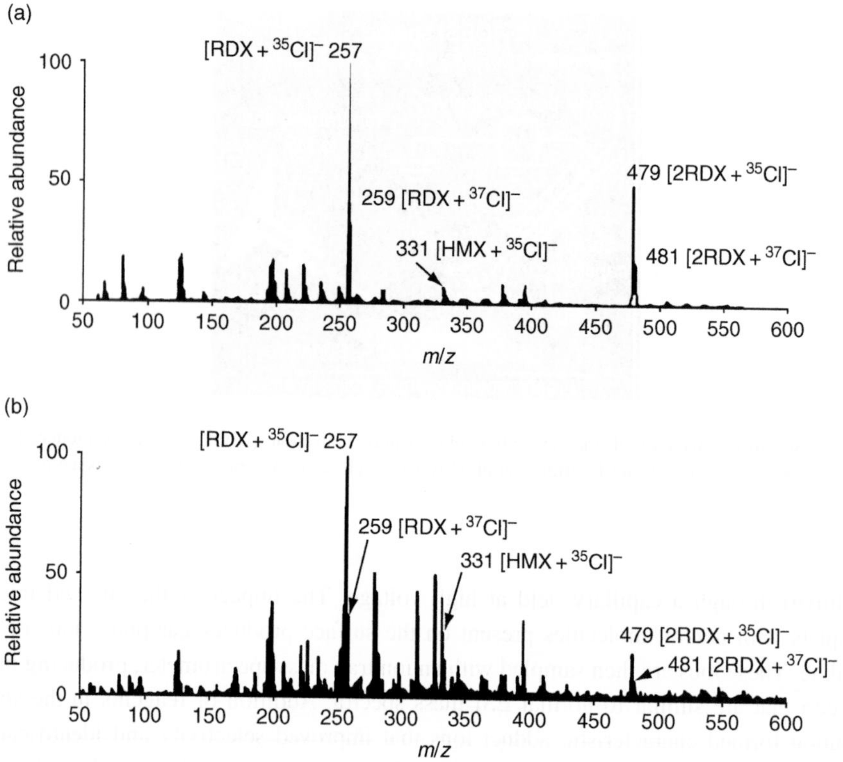

Fig. 9. (a) Negative ion desorption electrospray ionization (DESI) spectrum for a plastic explosive C-4 fingerprint on glass, using methanol–water–HCl (1:1:0.05%) as spray solvent. (b) Negative ion DESI spectrum of a C-4 fingerprint after five transfers onto a glass slide. Reproduced with permission from Cotte-Rodriguez et al. [14]. Copyright 2005 American Chemical Society.

Limits of detection at 1 m distance for TNT, RDX, HMX, TATP, and C-4, on different surfaces, were 0.5–1 ng (for PETN, 10 ng). At 3 m distance, the limits of detection were 5–10 ng (for PETN, 20 ng).

4. Miniature and mobile mass spectrometers

Mass spectrometers were always considered too complex, too large and heavy, and too power-consuming, in order to be used as mobile explosive detectors.

However, during the last 10 years, several groups have developed miniature and mobile mass spectrometers. Although these instruments were built for environmental applications, detection of chemical and biological weapons, and space exploration, most of them can easily be converted to detection of explosives.

Sections 4.1–4.5 describe some of these mass spectrometers.

4.1. Quadrupole ion trap TOF mass spectrometer

A 30-lb field-portable quadrupole ion trap TOF (QitTof) mass spectrometer with an atmospheric photoionization source was constructed [17]. The photoionization source has the ability to choose a narrow-band ionization energy that is sufficiently high to ionize and detect most compounds of interest but low enough to avoid ionization of most common air constituents, such as N_2, O_2, H_2O, CO_2, CO, Ar, etc. Fragmentation is minimal because ionization occurs just above the ionization thresholds, with very little excess energy. The ion trap stores ions from a continuous ionization source followed by pulsed extraction into the TOF mass analyzer.

The QitTof mass spectrometer has MS^n capability. Its mass resolution, $m/\Delta m$, for mass 181 amu, ranges from 100–200 for a 22-cm flight-path linear TOF analyzer to 400 for a 38-cm coaxial reflectron configuration.

Detection limits of 10–100 ppb for phosphonates and aromatic compounds were obtained.

4.2. Compact ion trap mass spectrometer

A compact ion trap ($r_o = 10\,\text{mm}$) mass spectrometer was developed for space-based applications [18]. The trap was made of titanium; its hyperboloid surfaces were machined to a tolerance of 0.02 mm. Electron ionization was used by generating electrons from a heated, spiral-wound tungsten wire and accelerated at 75 eV into the trap. The trap operates in an RF only mode without DC. No cooling gas, such as helium, is required. Mass range was 1–300 amu and resolution $m/\Delta m = 324$. The sensitivity of the trap was determined for N_2 and was found to be 2×10^{14} counts/torr.s.

4.3. Mass analyzer based on rectilinear geometry ion trap

A mass analyzer based on rectilinear geometry ion trap (RIT) was built [19]. The ion trap consists of two pairs (x and y) of rectangular electrodes supplied with RF voltages and a pair of z electrodes to which only a DC voltage is applied (Fig. 10). The RF signal is applied between the x and y electrode pairs and forms an RF trapping field in the xy plane, whereas the z electrode DC voltage generates a DC trapping potential well along the z axis. Slits in the RF electrodes and apertures in the z electrodes allow the electrons to be injected into the trap and ions to be ejected from the trap. As in the Paul trap, a buffer gas is used to facilitate trapping of ions through collisions. The half-distance between the x pair of electrodes (x_o) and y pair of electrodes (y_o) was 5.0 mm. The distance between the two z electrodes was 43.2 mm. The overall dimensional tolerance was ~0.1 mm.

Mass range was up to 650 amu and resolution in excess of 1000. Capabilities of the rectilinear trap included MS/MS.

(a)

(b)

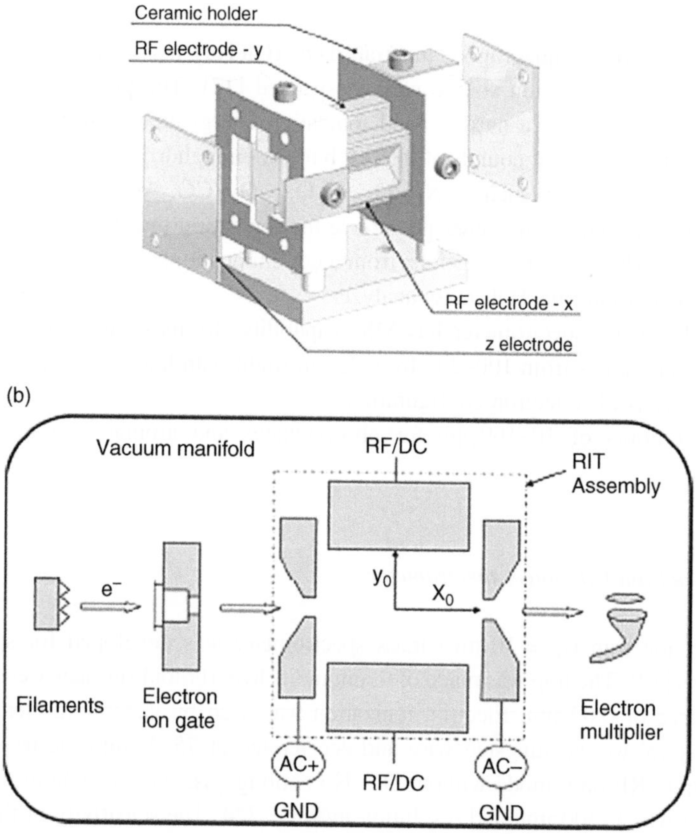

Fig. 10. Rectilinear geometry ion trap (RIT) assembly (a) and instrumentation (b) Reproduced with permission from Ouyang et al. [19]. Copyright 2004 American Chemical Society.

A handheld miniaturized mass spectrometer based on the RIT geometry was built [20]. It size is 13.5 inches (length) × 8.5 inches (width) × 7.5 inches (height) and weighs 10 kg, with a power consumption of less than 75 W.

4.4. *Mass spectrometer with APCI*

A miniaturized mass spectrometer with APCI was built [21]. The analyzer was a monopole with 54 mm rod length and 2 mm radius. A two-stage differentially pumping system and sampling nozzle of 80 μm enabled an inlet gas flow rate of 1 ml/s. The ion current generated by the corona discharge was 0.01–10 μA.

With a voltage of 20 V, $U/V = 0.12$ and a frequency of 5 MHz, a resolution of one mass unit in the range of 12–200 amu was obtained. The dimensions of the mass spectrometer were 185 × 100 × 70 mm, with a weight of 20 kg.

For phosphorous-containing substances, the limit of detection was in the ppt region.

4.5. Cylindrical ion trap mass spectrometer with APCI and ESI

A miniature cylindrical ion trap mass spectrometer with APCI and ESI capabilities was developed [22]. The system includes a three-stage, differentially pumped vacuum system and can be interfaced to many types of atmospheric pressure ionization sources.

The cylindrical ion trap, machined from stainless steel, has a radius (r_0) of 2.5 mm and a center to end-cap distance (z_0) of 2.7 mm. Planar end-cap electrodes allow for ion entrance and exit through 1-mm-radius apertures. The instrument measured ~46 × 50 × 38 cm, with a weight of 38 kg and 210 W power consumption while in operation.

Performance of the instrument was investigated by using a corona discharge source and an ESI source.

A mass range of 450 amu with unit mass resolution was obtained. Limits of detection for trace analysis in air of methyl salicylate (1.24 ppb) and for nitrobenzene (629 ppt) were achieved. Isolation and collision-induced dissociation efficiencies in MS/MS experiments were greater than 50%.

Desorption electrospray ionization was implemented on this portable mass spectrometer [23]. DESI experiments were carried out at ambient capillary temperature, at a spray voltage of 3 kV and a nebulizing N_2 gas flow of 80–120 psi. Detection of RDX from three different surfaces (paper, plastic, and metal) was demonstrated with this portable instrument in the positive-ion mode, with an analysis time of 5–10 s. The result obtained for 10 ng of RDX deposited on 1-cm^2 paper is shown in Fig 11(a). Figure 11(b) shows

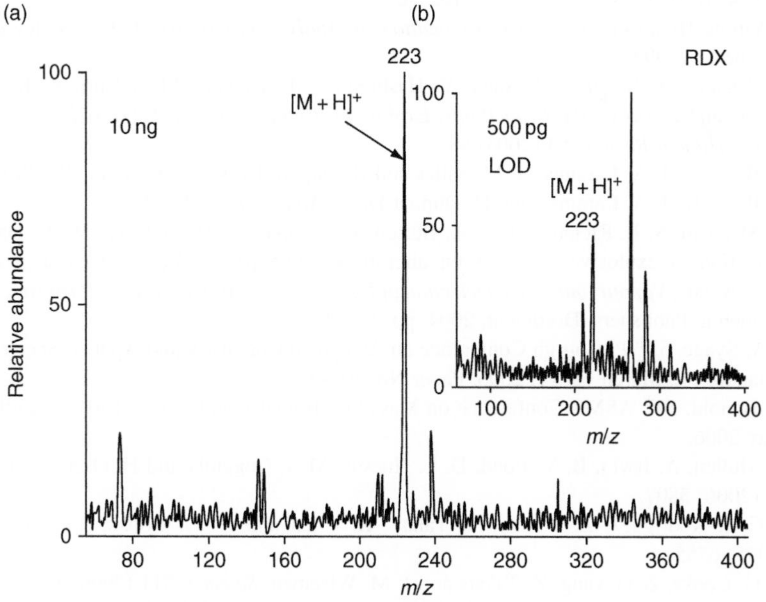

Fig. 11. Positive-ion desorption electrospray ionization (DESI) mass spectrum of (a) 10 ng of RDX and (b) 500 pg of RDX. Reproduced with permission from Mulligan et al. [23].

the mass spectrum obtained for 500 pg of RDX deposited on 1-cm^2 paper, with the actual area sampled by the nebulizing spray being significantly smaller.

5. Conclusions

Because of its high sensitivity and selectivity, mass spectrometry has always shown great potential for vapor and trace detection of explosives.

Issues of size, portability, and power requirements have been addressed during the last years, and miniature and mobile mass spectrometers have been developed, based on a variety of mass analyzers. Also, different ionization techniques, such as ESI and DESI, have been applied for the detection of explosives.

Integration of mass spectrometry with other technologies, i.e., bulk detection systems, would greatly enhance the detection capabilities of such integrated systems.

References

[1] P. H. Dawson, Principles of operation. In: P. H. Dawson, ed., *Quadrupole Mass Spectrometry and its Applications*, Elsevier, Amsterdam, 1976, pp. 9–64.

[2] R. E. March and J. F. J. Todd, eds., *Practical Aspects of Ion Trap Mass Spectrometry, Vol I: Fundamentals of Ion Trap Mass Spectrometry*, CRC Press, Inc., Boca Raton, 1995.

[3] M. Guilhaus, *J. Mass Spectrom.*, 30 (1995) 1519.

[4] R. Cotter, *Anal. Chem.*, 64 (1992) 1027A.

[5] J. Yinon, *Modern Methods and Applications in Analysis of Explosives*, John Wiley & Sons, Chichester, 1993.

[6] Y. Takada, H. Nagano, M. Suga, Y. Hashimoto, M. Yamada, M. Sakairi, K. Kusumoto, T. Ota and J. Nakamura, *Propellants, Explosives, Pyrotechnics*, 27 (2002) 224.

[7] Anon, *Hitachi Review*, 53 (2004) 88.

[8] R. B. Cody, J. A. Laramee, J. M. Nilles and H. Dupont Durst, *JEOL News*, 40 (2005) 8.

[9] R. B. Cody, J. A. Laramee and H. Dupont Durst, *Anal. Chem.*, 77 (2005) 2297.

[10] R. Sleeman, S. L. Richards, I. F. A. Burton, J. G. Luke, W. R. Stott and W. R. Davidson, Detection of explosives residues on aircraft boarding passes. In: M. Krausa and A. A. Reznev, eds, *Vapour and Trace Detection of Explosives for Anti-Terrorism Purposes*, Kluwer Academic Publishers, Dordrecht, 2004. pp. 133–142.

[11] J. A. Syage, 57th Pittsburgh Conference on Analytical Chemistry and Applied Spectroscopy, Orlando, FL, March 12–17, 2006. Paper No. 203–3.

[12] K. Hanold, 54th ASMS Conference on Mass Spectrometry and Allied Topics, Seattle, WA, May 2006.

[13] C. Mullen, A. Irwin, B. V. Pond, D. L. Huestis, M. J. Coggiola and H. Oser, *Anal. Chem.*, 78 (2006) 3807.

[14] I. Cotte-Rodriguez, Z. Takats, N. Talaty, H. Chen and R. G. Cooks, *Anal. Chem.*, 77 (2005) 6755.

[15] R. G. Cooks, Z. Ouyang, Z. Takats and J. M. Wiseman, *Science*, 311 (2006) 1566.

[16] I. Cotte-Rodriguez and R. G. Cooks, *Chem. Commun.*, 2968 (2006).

[17] J. A. Syage, M. A. Hanning-Lee and K. A. Hanold, *Field Anal. Chem. Technol.*, 4 (2000) 204.

[18] O. J. Orient and A. Chutjian, *Rev. Sci. Instr.*, 73 (2002) 2157.

[19] Z. Ouyang, G. Wu, Y. Song, H. Li, W. R. Plass and R. G. Cooks, *Anal. Chem.*, 76 (2004) 4595.
[20] Z. Ouyang, R. G. Cooks, L. Gao, S. Kothari and S. Qingyu, 57th Pittsburgh Conference on Analytical Chemistry and Applied Spectroscopy, Orlando, FL, March 12–17, 2006. Paper No. 2260–8.
[21] A. L. Makas, M. L. Troshkov, A. S. Kudryavtsev and V. M. Lunin, *J. Chromatogr. B*, 800 (2004) 63.
[22] B. C. Laughlin, C. C. Mulligan and R. G. Cooks, *Anal. Chem.*, 77 (2005) 2928.
[23] C. C. Mulligan, N. Talaty and R. G. Cooks, *Chem. Commun.*, 1709 (2006).

Chapter 3

Explosives Detection Using Differential Mobility Spectrometry

Gary A. Eiceman, Hartwig Schmidt and Avi A. Cagan

Department of Chemistry and Biochemistry, New Mexico State University, Las Cruces, NM 88003, USA

Counterterrorist Detection Techniques of Explosives
Jehuda Yinon (Editor)

Contents

1. Introduction

Ion mobility spectrometry (IMS) is an instrumental method where sample vapors are ionized and gaseous ions derived from a sample are characterized for speed of movement as a swarm in an electric field [1]. The steps for both ion formation and ion characterization occur in most analytical mobility spectrometers at ambient pressure in a purified air atmosphere, and one attraction of this method is the simplicity of instrumentation without vacuum systems as found in mass spectrometers. Another attraction with this method is the chemical information gleaned from an IMS measurement including quantitative information, often with low limits of detection [2–4], and structural information or classification by chemical family [5,6]. Much of the value with a mobility spectrometer is the selectivity of response that is associated with gas-phase chemical reactions in air at ambient pressure where substance can be preferentially ionized and detected while matrix interferences can be eliminated or suppressed. In 2004, over 20 000 IMS-based analyzers such as those shown in Fig. 1 are placed at airports and other sensitive locations worldwide as commercially available instruments for the determination of explosives at trace concentration [7].

In mobility spectrometers, including differential mobility spectrometers [8–10], ions are characterized for mobility as shown in Fig. 2, where an ion swarm or packet of ions develops a velocity in an electric field, established in a fixed distance between two potentials. This drift velocity (v_d), when normalized for field strength (v_d/E), is termed the mobility coefficient (K), and common magnitudes of these are 1–4 m/s for drift velocity and 2.5–1.5 cm^2/V s for mobility coefficients. Commonly, the mobility coefficient is adjusted for temperature and pressure yielding a reduced mobility coefficient, K_o, referenced to 760 torr and 273 K. The measurement of drift velocity occurs inside a drift tube (Fig. 2) where a set of conducting rings is used to establish the electric field and maintain a level of homogeneity in the field [11,12]. In an IMS measurement, ions

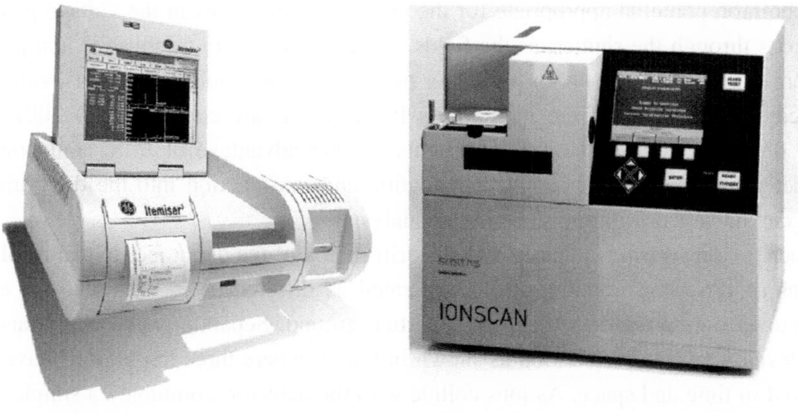

Fig. 1. Two commercial ion mobility spectrometry (IMS) explosive detectors used at airports worldwide: the Itemiser[3] (GE Ion Track, (GE Security, Wilmington, MA, USA)) and the IONSCAN (Smiths Detection (Mississauga, Ontario, Canada)).

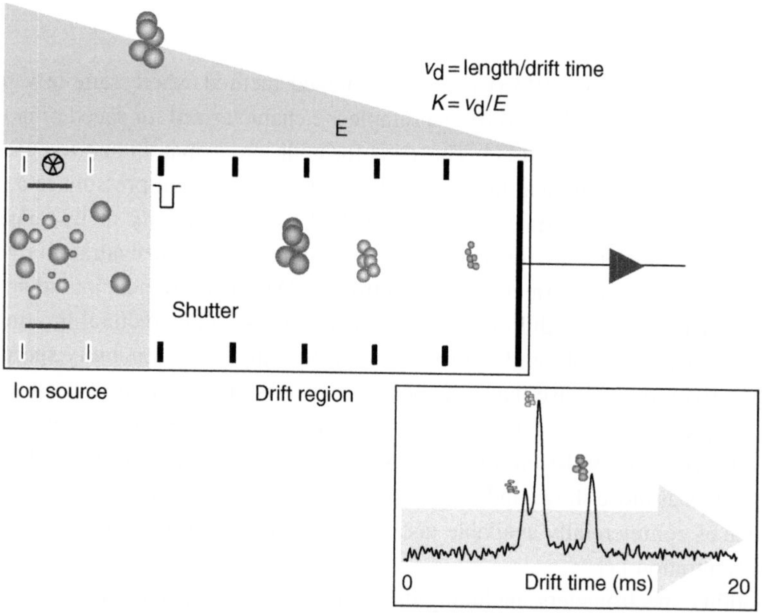

Fig. 2. Schematic representation of a mobility spectrometer. Ions created in the ion source are separated in the drift region based on their mobility. The ion swarms reach the detector where their drift times are recorded and plotted in the form of a mobility spectrum (Originally published in the article of Buryakov et al. [9]).

are introduced into the drift region using an ion shutter, which is commonly a set of parallelly interdigitated wires in a plane or as close to a plane as possible [12]. Electric fields are established between adjacent wires in the shutter so ion passage down the voltage gradient of the drift tube and through the shutter, which spans the cross-section of the drift tube, is stopped by ion annihilation on the wires. When the wires are brought to a common potential appropriate for the position of the wires in the voltage gradient, ions pass through the shutter; as the fields in the shutter are re-established, ion passage is again stopped forming an ion swarm or pulse of ions for mobility analysis,with width defined by the timing of the ion shutter. Pulse widths are commonly 50–300 μs and occur on a 10–30 Hz basis. This highlights another advantage of IMS, comparatively fast determinations, where sample preparation and introduction into the drift tube are often the slowest or limiting steps in the analysis.

When an ion swarm is injected into the drift region of the drift tube, spatial resolution of ions of differing mobility can be separated as differences in drift velocity as the ions move toward the detector, here at virtual ground. Separate packets or swarms of ions develop with the separation as shown in Fig. 2, where three ion swarms have been resolved in time and space. As ions collide with the detector, commonly a simple metal disc or Faraday plate, neutralization of ions is accompanied by electron flow in the detector plate; this is amplified and shown in the inset of Fig. 2. Thisplot of detector response(current or voltage) versus time (in ms) is called a mobility spectrum and is the

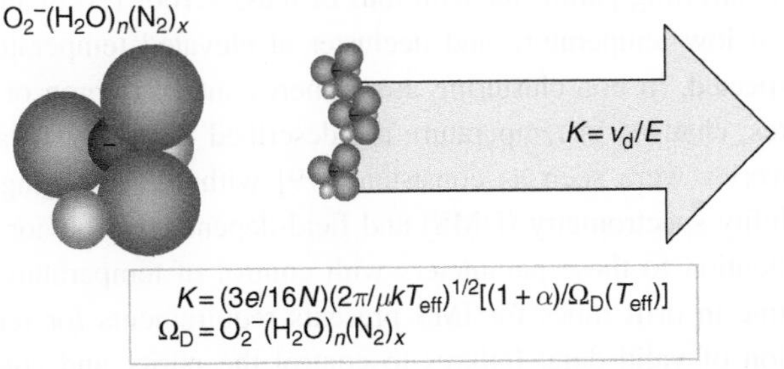

$O_2^-(H_2O)_n(N_2)_x$
$K = v_d/E$
$K = (3e/16N)(2\pi/\mu kT_{eff})^{1/2}[(1+\alpha)/\Omega_D(T_{eff})]$
$\Omega_D = O_2^-(H_2O)_n(N_2)_x$

Fig. 3. The mobility coefficient K describing the movement of an ion swarm in an electric field. The mobility coefficient depends on the cross-section (collision area) Ω, the reduced mass μ, and the effective temperature T_{eff} of the ion. Collision area Ω_D will depend on moisture, temperature, drift gas, and molecule.

source of drift time (hence drift velocity) and K_o values, quantitative response as peak height or area, and performance of the analyzer as peak width and shape.

The basis for mobility is shown in Fig. 3, where a link is made between the structure of individual ions and the drift velocity of the ion swarm. These are joined effectively by ion cross-section of collision (Ω_D), reduced mass (μ), and effective temperature of the ion (T_{eff}). The relationship is directly suitable for ions in a supporting atmosphere where ions do not form associations [13] with molecules that are strongly polarizable or contain permanent dipoles [14,15]. In this instance, increased temperature of the gas atmosphere leads to diminished mobility through increases in ion drag with increased collision frequency from increased speed of gas molecules. By contrast, negative ions formed in an IMS ion source in air with even low ppm levels of moisture will form cluster ions where O_2^-, the core ion at the center of a cluster, is surrounded by polar neutrals such as water as shown in Fig. 3. These associations occur through dipole charge or charge-induced dipole interactions, forming $O_2^-(H_2O)_n$ species, where n is controlled by temperature and moisture of the gas environment [16,17]. The interactions between water adducts and the core ion can be substantial leading to stable cluster identities; however, the addition or loss of adducts also can be highly dynamic increasing cluster size as association enthalpies decrease. Additionally, the buffer gas can be viewed as part of the cluster even though the associations may be very weak solely through the large number of collisions or high collision frequency. Clustered ions may be dissociated with heat, and the core ion of O_2^- or Cl^- could be regarded as thoroughly declustered with sufficient temperature or reduced moisture. The heat of association between H^+ and H_2O is so large that the terminal ion at elevated temperature and reduced moisture in positive polarity in IMS will be H_3O^+. When an ion is declustered extensively with increases in gas temperature, a decrease in ion size (seen as Ω_D) is observed, and mobility, as K or K_o, will be increased. This is apparently contradictory to the formula in Fig. 3, where K decreases with an increase in T_{eff}. Karpas highlighted the balance between Ω_D and T_{eff} where mobility with ions of mass >100–150 will be governed by T_{eff} even in an air atmosphere and elevated temperature [18]. However, the collision cross-section

Ω_D will be the controlling parameter with ions of mass <100–120 in an air atmosphere as ions cluster at low temperature and decluster at elevated temperature. This is not observed, as expected, in non-clustering atmospheres such as argon or helium, and in inert atmospheres, changes in temperature are described in the formula principally by T_{eff}. Suchdiscoveries were seen as consistent [19] with the emerging technology of differential mobility spectrometry (DMS) and field-dependent behavior as described in this chapter. Attention to these parameters with control of temperature, moisture, and ion residence time in drift tubes for IMS presents requirements for reliable operation and the collection of valid data; failures to control the purity and consistency of the supporting gas atmosphere in the ionization and drift regions will make the method unreliable and unpredictable. The fine engineering, seen with instruments such as those shown in Fig. 1, illustrates that such control is possible and commercially viable and that IMS analyzers have been entrusted with the security of commercial aviation.

Ion mobility spectrometry as a research tool and an analytical method has undergone a transformation since 1990 [1] and is today understood as a powerful supplement to mass spectrometry (MS) for biological studies [20] and as a robust, field instrument as a stand-alone analyzer.This transformation is due in large part to the successful utilization of IMS analyzers for explosives monitoring and for detection of chemical warfare agents [7]. Also, changes have occurred in understandings of the principles of IMS and the level of technology as illustrated in comparisons between the earliest summary of IMS (an edited monograph [21]) and the first integrated summary of IMS [22] and then the most recent edition of the same [1]. Although a progression of science and technology for IMS has occurred since 1970 when the first analytical IMS analyzer was described [23], there are still very few applications for IMS that eclipse military and security analyses and few prospects evident on the horizon of commercial activity. Still, no other advanced technology is used in such large numbers or in such mission-critical applications as mobility spectrometers. Recent reviews will provide a broadened perspective on these comments [24–27].

The first step in response in an IMS analyzer is the conversion of a sample neutral to a gas-phase ion through coulombic associations resembling displacement reactions as shown in Eqs 1 and 2 for positive and negative polarity, respectively.

$$\underset{\text{Sample neutral}}{\mathrm{M}} + \underset{\text{Positive reactant ion}}{\mathrm{H^+(H_2O)}_n} \rightarrow \underset{\text{Positive product ion}}{\mathrm{MH^+(H_2O)}_{n-x}} + x\mathrm{H_2O} \quad (1)$$

$$\underset{\text{Sample neutral}}{\mathrm{M}} + \underset{\text{Negative reactant ion}}{\mathrm{O_2^-(H_2O)}_n} \rightarrow \underset{\text{Negative product ion}}{\mathrm{MO_2^-(H_2O)}_{n-x}} + x\mathrm{H_2O} \quad (2)$$

These occur simultaneously in the ion source, and heretofore, the user chooses the polarity for analysis.Because properties of ionization of substances including explosives can sometimes be favored in one polarity over the other, a restriction to measuring a single polarity is not a minor limitation or simply an inconvenience. One of the key limitations with all mobility spectrometers, until recently, has been the use of either positive or negative charge singly in the mobility measurement providing chemical information for one ion polarity only while that for the other is lost. In the early 1990s, a solution to this was sought where two drift tubes shared a common ionization region and tubes were biased positive or negative with maximum potential at the detectors that were opto-isolated to an amplifier [28]. This design was termed the twin drift tube design, and a single sample, introduced to the analyzer, was simultaneously screened for positive and negative ions with two drift tubes of identical structure and a shared timing cycle for the ion shutters. This concept has now been commercialized in 2005, 15 years later, as the model IONSCAN 500DT from Smiths Detection [29].

Explosive analyzers based on IMS are popular as benchtop instruments that are transportable but not man-portable, and recent developments have been successful in making hand-held explosive analyzers [30]. Even so, these are still comparatively large and heavy due in large part to the size and mass of the drift tube and are expensive, though IMS analyzers are economical compared with mass spectrometers. In the late 1990s, efforts were made to create small drift tubes based on principles of mobility but unlike traditional drift tubes or methods. This new generation of analyzers was based on micro-fabrication techniques and was operated using field-dependent behavior of ions in gases. This was applied to the determination of explosives and has been extended from the early successes as described in this chapter.

2. Background and principles of DMS

Early in the first part of the twentieth century, studies with ions in gases revealed mobility concepts as described in Section 1 and also showed that mobility coefficients could be changed by E/p, where p is the gas pressure inside theanalyzer [31–33]. This is shown in Fig. 4, where the mobility coefficients of alkali ions are plotted against E/p: mobility is constant for most ions up to certain values for E/p, after which the ion is linearly dependent in mobility upon E/p. Both the point where the curve breaks from a plateau to a slope and the slope of the plot are characteristic of an ion (in a given gas environment). In the instance of Cs^+, there is no level portion for the plot, and a positive slope exists from the lowest measured value for E/p. The cause for this is understood as the heating of ions through the electric field and declustering of ions, a decrease in Ω_D with increased T_{eff}. Today, E/p has been replaced by E/N, where N is the number density for atoms in a gas, though interconversion between E/p and E/N is inconvenient [34]. In one obscure reference [35], a method for field-dependent ion characterization came very close to an analytical method though the subject of field dependence of ion mobility was apparently overlooked, forgotten, or ignored until the mid 1980s.

In the early to mid 1980s, the characterization of ions with field dependence of mobility was treated theoretically [36] and was based on the following concept: if ions had the same mobility but differing dependence of mobility with E/N, then the characteristic differences in mobility could be a basis for ion separation. Thus, a complete expression of mobility is shown in Eq. 3, where the mobility coefficient contains a non-linear

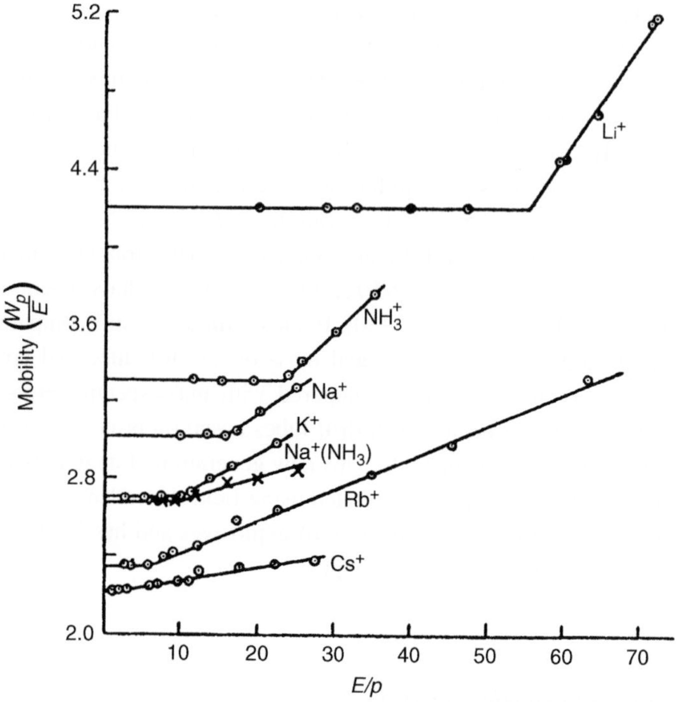

Fig. 4. Plots of mobility versus E/p for ions in helium (originally published in the article of Mitchell and Ridler [33]).

dependence on an electric field, and pressure (or E/N) is derived from plots such as those shown in Fig. 5:

$$K(E) = K(0)\left[1+\alpha_2\left(\frac{E}{N}\right)^2+\alpha_4\left(\frac{E}{N}\right)^4+\ldots\right] = K(0)\left[1+\alpha\left(\frac{E}{N}\right)\right] \tag{3}$$

where $K(0)$ is the mobility coefficient for electric field below 1000 V/cm, α_i parameters are functions that describe the dependence of K on the electric field, N is neutral density of drift gas in molecules/cm^3, and the electric field strengths at ambient pressure can be as large as 30 000 V/cm. The alpha parameter can be understood as shown in Eq. 4, where an alpha parameter describes the changes in the mobility coefficient with E for a constant N:

$$\alpha\frac{E}{N} = \frac{K(E)-K(0)}{K(0)} = \frac{\Delta K(E)}{K(0)} \tag{4}$$

Thus, alpha is associated with a difference in mobility of the ion may be positive or negative in sign, which discloses the direction of change in the coefficient of mobility (and compensation voltage) with E at constant N. In practice, ions are characterized by

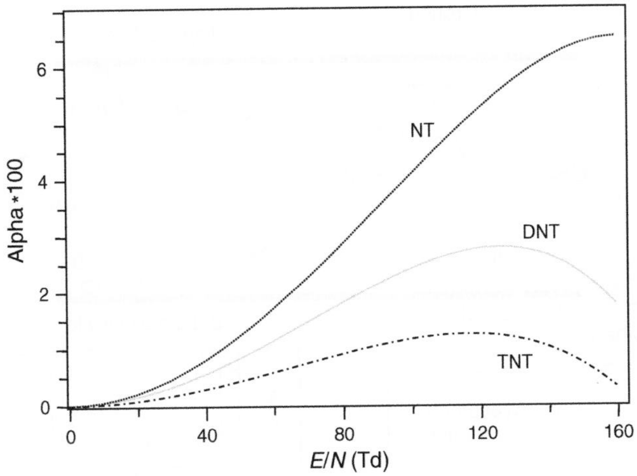

Fig. 5. Alpha functions of nitrotoluene (NT), dinitrotoluene (DNT), and trinitrotoluene (TNT).

repeatedly placing ions at high and low fields obtaining a differential mobility of ΔK, seen in Fig. 5 as different mobilities with E/N.

The practice of differential mobility determinations was demonstrated in 1993 by a team in Tashkent, Uzbekistan, who developed a method that was both practical and potentially scalable to miniaturized dimensions [8–10]. This approach is shown in Fig. 6, where ions are introduced with a gas flow between two plates separated by a distance from 0.5 mm to 0.1 cm. Ions entering the volume between the plates are exposed to an oscillating electric field shown in the inset of Fig. 6. This field, in the instance of small planar analyzers, is ~1.4 MHz and generated by voltages of 1000–100 V (fields of 1–30 kV/cm). During the high-field portion of the waveform, the ions move, under the formula $v_d = K(E/N) \times E$, in the direction of the ground plate (the top electrode). During the low-field portion, the ion is restored toward the bottom plate, now negative, according to formula $v_d = K(E/N) \times E$, where *K(E/N)* and E differ from the high-field portion. The waveform is arranged so that ions moving, under the formula $v_d = K(E/N) \times E$, will have no net displacement from the center of the gap if *K(E/N)* is not dependent on E/N. However, if the coefficient of mobility is dependent on E/N as shown in Fig. 5, a net displacement toward a plate will occur, and eventually, the ion will collide with the plate. The estimated maximum number of oscillations is ~10^3 for an ion spending ~1 ms in the analyzer region. Ions with large differences in *K(E/N)* or high ΔK will reach the wall after a lower number of oscillations compared with an ion with low ΔK between two values of E/N. Ions with negative dependence of mobility on *K(E/N)*, where the mobility decreases with increased field strength, will move in directions opposite to those with positive alpha.

Without intervention, all ions apart from those with no dependence of mobility on field will strike a wall, be neutralized, and be swept as a neutral from the analyzer. However, a DC voltage (the compensation voltage) can be added to the bottom electrode and superimposed on the megahertz waveform (Fig. 6). As this potential is applied, ion

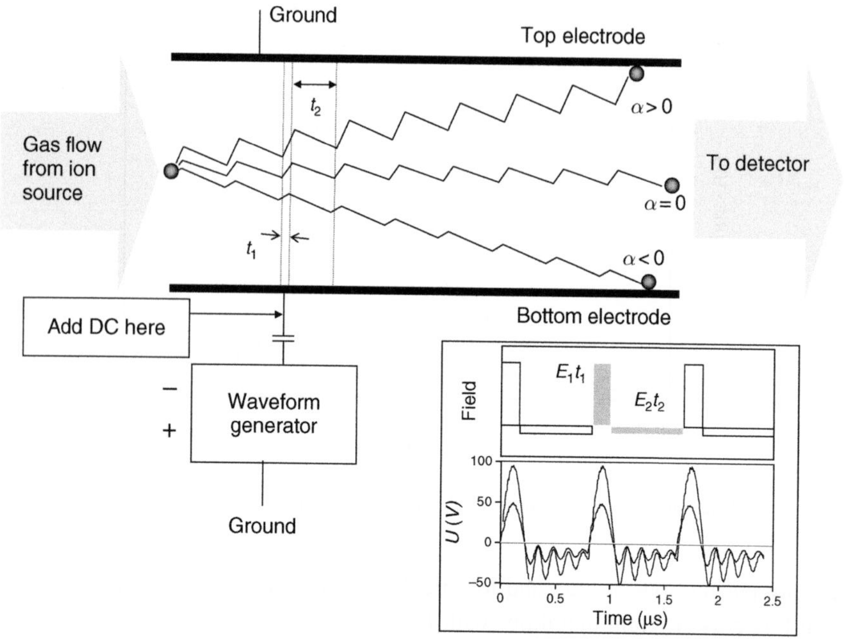

Fig. 6. Schematic of a differential mobility spectrometer showing the principles of ion separation in a differential mobility spectrometry (DMS) drift tube. Ion paths are governed by both the asymmetric electric field and field dependence of mobility for an ion. The inset displays the asymmetric waveform of separation electric field used in the DMS drift tube. The waveforms shown are theoretical (top part) and actual or experimental (bottom part) used in these experiments.

displacement between the plates can be affected and a certain potential may be found for each ion to be restored in displacement to the center of the channel. This DC voltage is also known as the compensation voltage, and a sweep of compensation voltage will, in increments, bring ions with characteristic dependences into the center of the gas flow and to the detector. Detectors are commonly Faraday plates, and a plot of detector current against compensation voltage yields a differential mobility spectrum. This is shown in Fig. 7, where the polarity of compensation voltage is associated with positive or negative dependences of K on E/N. The magnitude of compensation voltage is proportional to ΔK as shown in Fig. 7. As ions formed from sample differ in mass and in ΔK, in principle, each ion should have a characteristic compensation voltage for constant conditions or a unique plot of K versus E/N providing DMS with capabilities for ion separation.

The principle for alpha dependence is rooted in ion behavior through collisions and associations with molecules of the supporting atmosphere, and a comprehensive treatment has been given.

In summary, ions that exhibit a positive dependence of K on E/N can be understood to undergo clustering and declustering events as shown in Fig. 8. During a low-field portion of the waveform, ions are not electrically heated or are slightly over thermal energies, and associations with water and supporting atmosphere can occur with lifetimes proportional to the heats of association. These associations result in increases in Ω_D and decreases

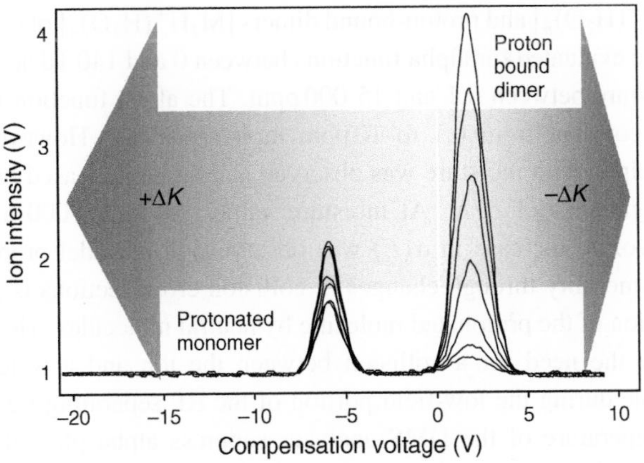

Fig. 7. Differential mobility spectra (intensity vs. compensation voltage plot). The relationship between ΔK and the compensation voltage is shown for ions with positive and negative alpha functions.

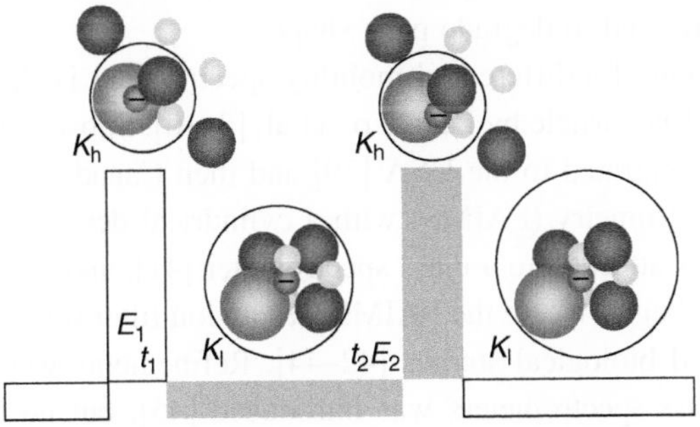

Fig. 8. Model for positive alpha with increasing electric field. During the high-field part of the duty cycle (E_1), ions are declustered, leading to an increase in mobility (K_h) due to a decrease in Ω. During the low-field part of the duty cycle (E_2), ions are re-clustered, leading to a decrease in mobility (K_l, $K_h > K_l$) due to an increase in $\Omega(\Omega_h < \Omega_l)$. With this model, a positive ΔK may be expected.

in K (Fig. 3). As an ion is exposed to the high-field portion of the waveform, the ion is no longer at or near thermal energies and the energy of the field causes an increase in T_{eff}. Although this will cause a small decrease in K due to drag, the greater effect is declustering with a decrease in Ω_D and an increase in K. Consequently, the ΔK and alpha function are positive. Ions that exhibit negative alpha and negative ΔK will be those where clustering and declustering exert a negligible effect and where ion drag through increased collision frequency dominates ion behavior as T_{eff} is increased. This is true of large ions and is seen principally for ions with masses over 150 amu, consistent with Karpas' early observation on the balance between T_{eff} and Ω_D in supporting atmospheres where clustering was possible [18].

The mechanism of clustering and declustering for ions with positive alphas was developed and supported by results from the study of the changes in alpha functions with moisture in the gas atmosphere in the DMS analyzer [37,38]. Gas-phase protonated monomers [$MH^+(H_2O)_n$] and proton-bound dimers [$M_2H^+(H_2O)_n$] of organophosphorus compounds were evaluated for alpha functions between 0 and 140 Td at ambient pressure in air with moisture between 0.1 and 15 000 ppm. The alpha function for $MH^+(H_2O)_n$ at 140 Td was constant from 0.1 to 10 ppm moisture in air. However, at 50 ppm, a dependence of alpha with moisture was observed and alpha increased twofold from 100 to 1000 ppm regardless of E/N. At moisture values between 1000 and 10 000 ppm, an additional twofold increase in $\alpha(E)$ was observed. In a model proposed here, field dependence for mobility through changes in collision cross-sections is governed by the degree of solvation of the protonated molecule by neutral molecules. The level of 50 ppm matched closely the need for a collision between the ion and a water molecule and the time available during the low-field portion of the RF separating field. Additionally, increases in temperature of the DMS analyzer suppress alpha plots illustrating that a declustered ion at high gas temperature will be weakly clustered, if any, during the low-field cycle. Consequently, control of moisture in a DMS experiment is as important as that in an IMS measurement; however, the direction of degraded response is somewhat different – DMS separations are improved with increased moisture. In an IMS analyzer, increases in moisture tend to degrade peak shape.

The first description of a differential mobility spectrometer is shown in Fig. 9 with a schematic from the 1993 article by Buryakov et al. [8–10]. Subsequently, the technology from this team was migrated to the USA [39] and then Canada [40] as field asymmetric ion mobility spectrometry (FAIMS) with a cylindrical design for the analyzer. The FAIMS analyzer was attached to a mass spectrometer [41], and a line of study on large instrumentation was begun where the FAIMS was an ion filter for the mass spectrometer in environmental and biological studies [42–44]. Refinements were made and a commercial inlet for mass spectrometers was introduced [45], but no determinations with

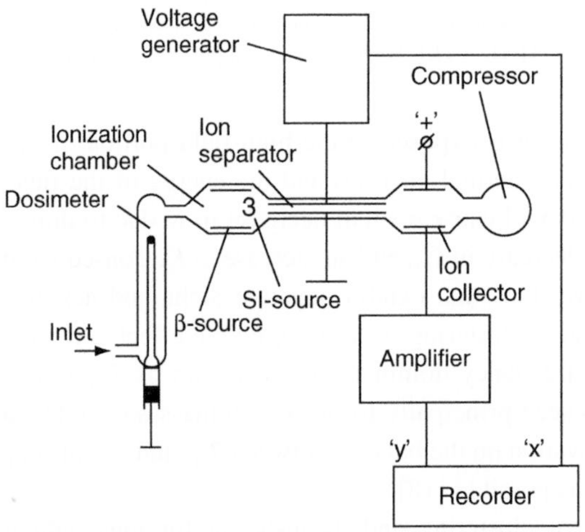

Fig. 9. Experimental arrangement used in one of the first differential mobility spectrometers (originally published in the article of Buryakov et al. [9]).

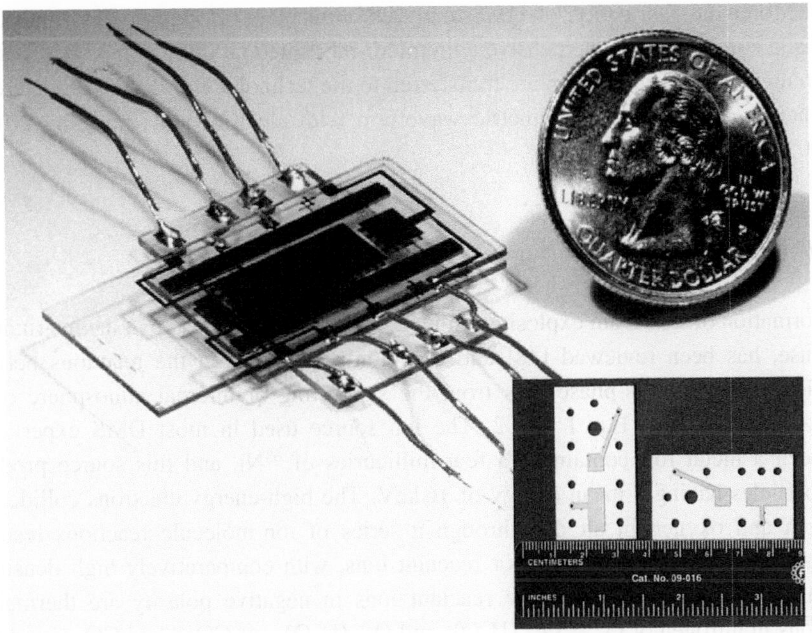

Fig. 10. Planar micro-fabricated differential mobility spectrometer. The photograph shows the device in comparison to a quarter (adapted from the article of Miller et al. [47]); the inset shows the size of the drift region (4 mm × 12 mm) and the size of the detectors (4 mm × 4 mm).

explosives were made by either team. However, a similar cylindrical design was refined or developed independently in Russia and applied for explosive detection (see pp. 76ff.).

The development of a micro-fabricated analyzer employing field-dependent mobility or differential mobility methods [46,47] was developed for simple motivations of small size, high portability, and low cost. In this approach, a planar configuration for the region of ion characterization was employed (similar to the first DMS manuscript [9]) to permit micro-fabrication methods where analyzers could be produced using mass production techniques. A photograph of the original glass-based drift tube is shown in Fig. 10, and planar units built using ceramic from designs at New Mexico State University (NMSU) are shown in Fig. 10 (inset). Here, the inside surfaces of the analyzer are shown. The detector is a 4 mm × 4 mm square metal plate, and the region for ion separation is a rectangular metal plate with dimensions of 4 mm × 12 mm.

3. Technology of DMS

Differential mobility spectrometers are attractive through the mechanical simplicity of an analyzer assembly without ion shutters, an aperture grid, or multiple components to establish a voltage gradient to move ions as found in conventional IMS drift tubes. Moreover, fabrication of DMS analyzers with planar designs is both uncomplicated and comparatively inexpensive with micro-fabricated production methods. Instead, requirements for sophistication are transferred to the technology or design of electronics to generate a ~1.2 MHz asymmetric waveform with electric fields reaching 20 000–30 000 V/cm.

3.1. Ion source

The formation of ions from explosives in IMS and DMS, the first step in the instrumental response, has been reviewed [48] and commonly is based on the reactions between sample vapors and gas-phase ions from the supporting or internal atmosphere of the analyzer as shown in Eqs 1 and 2. The ion source used in most DMS experiments has been a metal foil containing a few millicuries of ^{63}Ni, and this source produces beta particles having a mean energy of 16 keV. The high-energy electrons collide with nitrogen and oxygen in air and through a series of ion-molecule reactions result in the formation of long-lived ions, or reactant ions, with comparatively high density of 10^{10} ion/cm^3 [49]. Generally, the reactant ions in negative polarity are thermalized electrons in nitrogen or O_2^-, $O_2^-(H_2O)$, and $O_2^-(H_2O)_2$, or CO_2^- and CO_4^-, and their hydrates in air. Selectivity and sensitivity in response can be enhanced in IMS analyzers by the addition of a reagent gas into the gas atmosphere to form alternate reactant ions. The reagent gases, or dopant chemicals, are typically volatile organic compounds or substances that when heated release vapors. The reagent gases undergo reactions also in the ion source yielding alternate reactant ions such as Br^- or Cl^-. All response in an IMS or a DMS analyzer begins with the formation of an association between the substance and the reactant ion as shown in Eq. 2 for negative polarity. Subsequently, the product ion can undergo further reaction steps as shown for two possible pathways in Eqs 5 and 6 and summarized in whole in Fig. 11.
Charge exchange:

$$\underset{\text{Negative production}}{MCl^-(H_2O)_{n-x}} \rightarrow M^- + Cl + (n-x)H_2O \tag{5}$$

Proton abstraction:

$$\underset{\text{Negative production}}{MCl^-(H_2O)_{n-x}} \rightarrow (M-1)^- + HCl + (n-x)H_2O \tag{6}$$

All of these reactions occur in the ion source without user intervention or effort, an understood advantage of IMS and DMS technology. Although the chemistry in Fig. 11 is shown for a purified air atmosphere, the reagent gas and reactant ions (with O_2^- replaced by Cl^-) assist in suppressing response to background interferences and in simplifying response by directing ion chemistry through only one or two paths of Fig. 11. Thus, ion sources for DMS and IMS will be equipped with a reservoir of chemical, which normally should be thermostated or controlled so that vapor levels of the reagent gas are

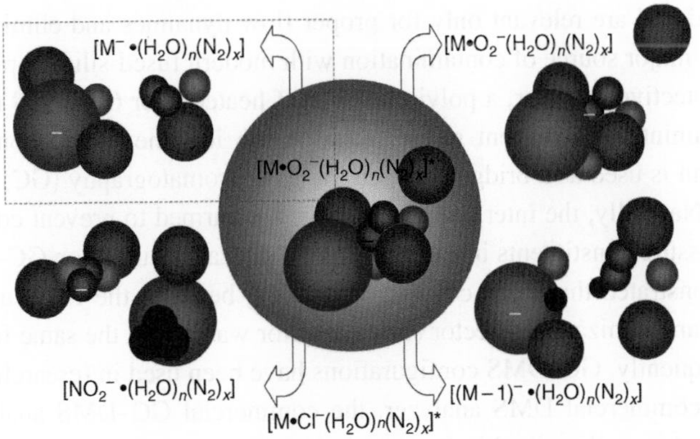

Fig. 11. Overview of different reaction pathways that an ion $[M \bullet O_2^-(H_2O)_n(N_2)_x]^*$ can take.

constant and reproducible. Recently, Hill et al. demonstrated the value of electrospray ionization with controlling ion chemistry with explosives determination by IMS [50,51]. The benefits of this method are difficult to imagine for on-site use with routine screenings owing to the operation of an electrospray ionization (ESI) source with condensed phase, i.e., aqueous solution.

3.2. Analyzer

The analyzer in Fig. 10 was the first micro-fabricated DMS analyzer and is shown independent of the housing, ion source, flow and power utilities, and data system. The plates of the units built at NMSU are held between two aluminum (Al) supports with Teflon cushions and insulators between the Al housing and the DMS plates and a 0.5-mm-thick gasket to separate plates and direct gas flow and ion through the analyzer.

3.3. Utilities

Flows of purified gases, scrubbed over molecule sieves, are passed into the DMS analyzer at flows of 0.3–1.2 L/min usually with a mass flow controller. Pressure is near ambient; however, DMS analyzers can operate at reduced pressures with scaled values of E as expected with E/N. The electronics of the DMS analyzer are based on a fly back transformer to give the RF waveform.

3.4. Combination to gas chromatograph

Because differential mobility spectrometers like IMS analyzers function at ambient pressure in air or nitrogen, combinations with gas chromatograph inlets are technically uncomplicated and are relevant only for proper flow dynamics and elimination of contamination. A major source of contamination with modern fused-silica capillary columns is the thin protective polymer, a polyimide, that if heated over 60°C will release vapors that become unintended reagent gases and alter the ion chemistry. Consequently, an Al-clad column is used as a bridge between the gas chromatography (GC) and the DMS flow system. Naturally, the interface region must be warmed to prevent condensation of low vapor pressure constituents including explosives. Early studies on GC–DMS configurations demonstrated that extra-column broadening between the DMS analyzer and a commercial flame ionization detector (FID) detector was nearly the same for a prototype design. Subsequently, GC–DMS configurations have been used in research and form the basis for the commercial DMS analyzer, the commercial GC–DMS analyzer, and the EGIS™ Defender, as described in Section 4.

4. Studies of explosives by field asymmetric IMS

Differential mobility spectrometry has been known by several names, including drift spectrometry and field asymmetric IMS, though the principles are shared in common with DMS and these other methods; various analyzers have common features though they differ in geometries and sizes. Unsurprisingly, explosives were examined by these other methods, and Buryakov et al. [52] found that ions for dinitrotoluene (DNT), trinitrotoluene (TNT), and pentaerythritol tetranitrate (PETN) could be separated (Fig. 12(c) and (d)) using a cylindrical configuration for the analyzer (Fig. 12(a)) in contrast to parallel plates [9]. The separation of negative ions was made in a supporting atmosphere of air with ~4% relative humidity yielding baseline separation of peaks for ions formed at 120 °C and passed through the characterization region with a gas temperature that was undisclosed but possibly nearly room temperature. The ions showed large dependence of V_c on V_{rf} and differences in dependence with separating fields up to ~18 kV/cm and a frequency of 160 kHz. In this first study, the response of the analyzer with respect to separating field and temperature of the ion source was characterized and optimized. The effect of water concentration in the gas atmosphere on the response was also noted or observed without reference to the effects of moisture on the position of peaks (i.e., the compensation voltages). The level of moisture will be seen as critical in the operation of field-dependent mobility instruments.

In a second report on FAIMS response to explosives [53], Buryakov extended the studies with a cylindrical tube to nitrobenzenes and nitrotoluenes, and plots of alpha versus E/N plots were determined (Fig. 12) and found that the alpha followed the relationship TNT < trinitrobenzene < DNT < dinitrobenzene < mononitrotoluene (MNT) < iodine. Ions for all these showed positive alpha coefficients for the negative ions (presumably M^-, $M-1^-$, or $M^*X(H_2O)_n{}^-$ species) though the TNT began to exhibit a negative dependence at the highest separating fields. This was consistent with a condition where the effects from ion cluster–decluster were greater than the effects of ion drag on mobility. These plots would be confirmed with a micro-fabricated DMS analyzer though

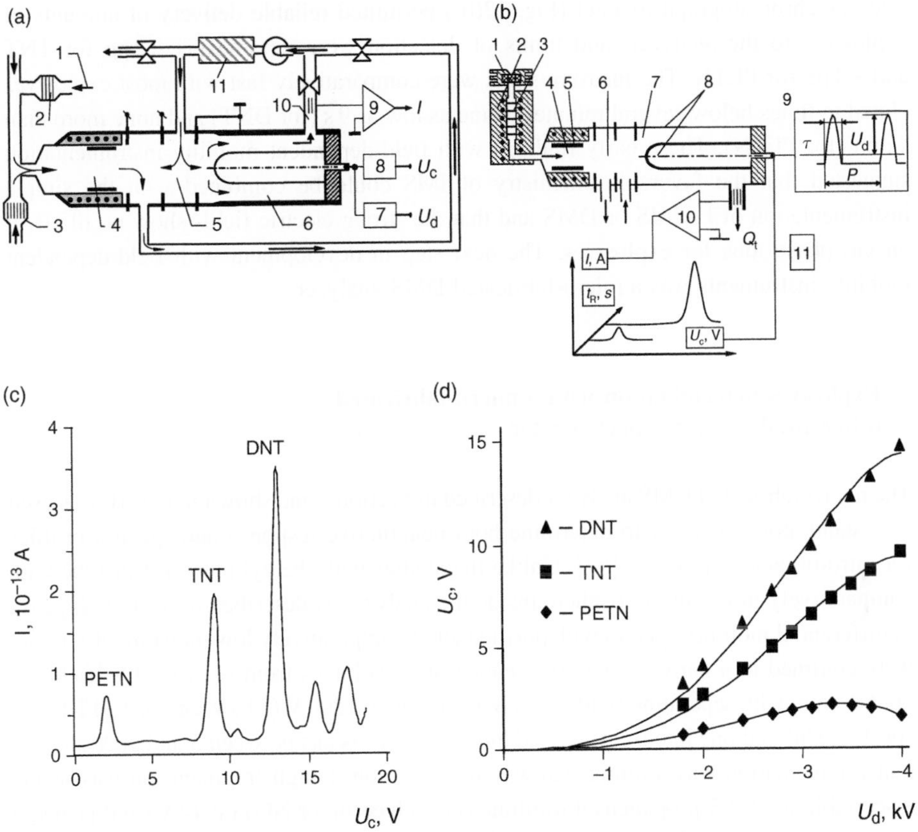

Fig. 12. (a) Schematic diagram of the differential mobility spectrometer (top left): 1, laboratory air; 2, water; 3, explosive; 4, ionization chamber; 5, field inlet; 6, separation chamber; 7, dividing voltage generator; 8, compensating voltage generator; 9, electrometer; 10, ion collector; and 11, gas filter (originally published in the article of Buryakov et al. [52], translated from *Zhurnal Analiticheskoi Khimii*, 56(4) (2001) 381). (b) Flow chart of the gas chromatography–differential mobility spectrometry (GC–DMS) system (top right): 1, sample injector; 2, thermostat; 3, multi-capillary column; 4, ionization chamber; 5, ^{63}Ni source of beta radiation; 6, field inlet; 7, separation chamber; 8, electrodes; 9, drift voltage generator; 10, electrometer; and 11, compensation voltage source (originally published in the article of Buryakov and Kolomiets [54], translated from *Zhurnal Analiticheskoi Khimii*, 58(10) (2003) 1057). (c) Differential mobility spectrum of ambient air containing dinitrotoluene (DNT), trinitrotoluene (TNT), and pentaerythritol tetranitrate (PETN) vapors (bottom left). (d) Experimental plots of compensating voltage against the amplitude of separation voltage for DNT, TNT, and PETN (bottom right, both originally published in the article of Buryakov et al. [52], translated from *Zhurnal Analiticheskoi Khimii*, 56(4) (2001) 381; in the original reference, the instrument is termed ion drift non-linearity spectrometer).

the effects here were exaggerated owing to the high levels of moisture in the supporting atmosphere.

In an article on drift tubes of comparatively large size and a cylindrical geometry [54], Buryakov and Kolomiets combined a multi-capillary column [55,56] with the analyzer to obtain measurements and quantitative response curves with a GC–DMS configuration.

The gas chromatograph as inlet (Fig. 12(b)) permitted reliable delivery of amounts of explosives to the analyzer, and limits of detection were reported as <2 pg for TNT and <4 pg for PETN. The measurements were comparatively fast with most explosives eluted in times below several minutes, some as low as 9 s (for DNT) and none more than 4 min (for PETN). These early findings with field-dependent mobility instrumentation suggested that the favorable chemistry of IMS could be combined with the simple instrumentation of FAIMS or DMS and that the strong electric fields show no ill effect on gas-phase ions for explosives. The next step in development with field-dependent mobility instruments was a micro-fabricated DMS analyzer.

5. Explosives determination with a micro-fabricated differential mobility spectrometer

The micro-fabricated DMS analyzer described in Section 2 and shown in Fig. 10 was used as a stand-alone analyzer to determine both quantitative response and spectral profiles for nitro-organic explosives [57]. Unlike the studies with the cylindrical drift tube with comparatively moist air atmosphere inside the analyzer as described in Section 4, peaks in differential mobility spectra with purified air at comparatively low moisture of ~1 ppm were confined to a narrow range of compensation voltages from −1 to +3 V [fields of 20–60 V/cm with separating fields at E/N values up to 1200 V (24 kV/cm) or 120 Td]. All spectra exhibited one intense peak, and the absence of peaks between 5 and 20 V showed that fragmentation was a minor pathway in ionization though fragment ion was barely discernible for 1,2,3-propanetriol trinitrate (nitroglycerin, or NG) (at 14 V) and peaks of low intensity for PETN (also at 14 V). The ions with the product ions were identified with a DMS analyzer fitted to a mass spectrometer, and the product ions observed with a ^{63}Ni ion source at ambient pressure in purified air at 100 °C were M^-, $M^*NO_2^-$, or $(M-1)^-$, as expected for these chemicals from previous decades of experience with IMS/MS investigations. Product ion peaks were concentration dependent, but ion identity and compensation voltages were independent of analyte concentration.

The compression of peaks for explosives with the small planar DMS in dried air to a narrow band of compensation voltages was analytically limiting and was consistent with the model of cluster–decluster for ions with positive alphas. The level of moisture was too low to permit ion hydration during the low-field portion of the waveform; however, the alpha function was understood to be adjustable through the controlled addition of a polar or polarizable neutral into the supporting atmosphere. This had been seen earlier for organophosphorous compounds and could be associated quantitatively with collision frequencies between ions and water molecules to form water clusters during the 280-ns period of low electric field strength. During the high-field portion of the waveform, the ions would be desolvated providing a difference in mobility constant with decreased collision area. Various dopants affected compensation voltages for DNT, and methylene chloride was shown to favorably enhance alpha for all explosives tested as shown in Fig. 13, where alpha functions were increased to values of 0.08–0.24 (at 100 Td), from

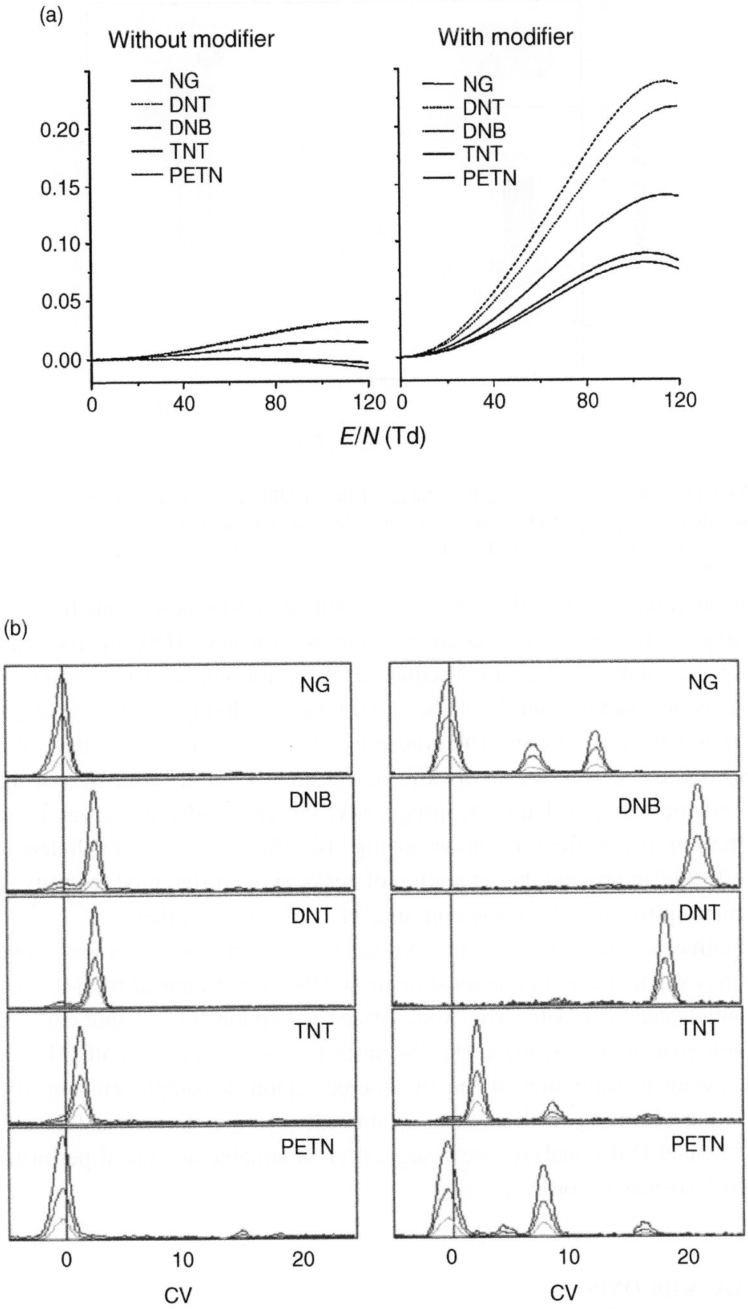

Fig. 13. (a) Alpha functions [alpha vs. *E* (Td)] of the explosives ions in pure air (top left) and a 1000 ppm mixture of methylene chloride/air (top right) [57]. (b) Differential mobility spectra of the explosives in purified air (bottom left) and 1000 ppm of methylene chloride in air (bottom right). Explosives are from top: 1,2,3-propanetriol trinitrate (NG); 1,3-dinitrobenzene (DNB); 2,6-dinitrotoluene (DNT); 2,4,6-trinitrotoluene (TNT); and pentaerythritol tetranitrate (PETN) [57].

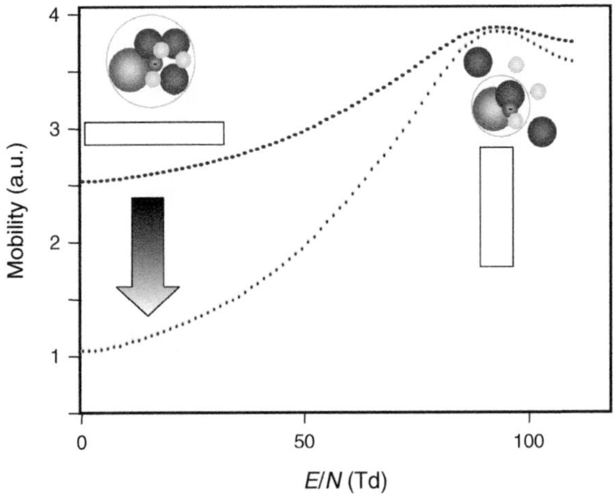

Fig. 14. Mobility versus E/N, showing the change of the mobility coefficient K as the ions are clustered through the addition of a dopant (e.g., methylene chloride). Under high-field conditions, K stays practically unaltered, whereas under low-field conditions, K is decreased, which leads to an increase in ΔK.

the clean gas values of 0.0–0.01, through the addition of 1000 ppm of methylene chloride to the drift gas. This improved separation of ions as shown in differential mobility spectra of Fig. 13 was understood as a consequence of the formation of ion clusters between product ions and vapor neutrals of the modifying gas during the low-field portion of the separation field waveform. The core ion when heated and unclustered during the high-field portion of waveform was affected in mobility to a lesser degree increasing ΔK and compensation voltages. Consequently, the effect of the reagent is to wet or solvate the ion at low field as shown in Fig. 14. The addition of methylene chloride had the effect of improving the separation of peaks in the differential mobility spectrum without altering the identity of the core ion, M^-, $M-1^-$, or other.

Quantitative response curves for the nitrobenzenes and nitrotoluenes were obtained for a continuous vapor stream of chemical in air and the plots extended from 0.1 to 100 ppb with an anticipated systematic error from surface adsorption at low vapor concentrations. At the high concentrations, the curves begin to flatten as seen with all IMS and DMS response owing to saturation of the ion source region, a complication in all uses of gas-phase ion chemistry with sources at ambient pressure. Detection limits from the micro-fabricated DMS analyzer were suggestive of suitable analytical performance for a trace explosives detector [57].

6. Fast GC with DMS

Traditionally, the majority of explosives monitors rely on the detection exclusively of negative ions due to the strong Coulomb interactions with negative ions (expressed as high electron affinity) of many, but not all, of the target molecules and comparatively low associations of matrix interferences with the negative reactant ion. One of the benefits of

DMS, not previously developed for explosives detection, is the simultaneous detection of ions of both polarities. This is seen in Fig. 15(a)–(d) for triacetone triperoxide (TATP), hexamethylene triperoxide diamine (HMTD), 4-MNT, and TNT in which product ions can be seen for TATP and 4-MNT only in positive polarity and for TNT only in negative polarity. Thus, a DMS analyzer operated would see ions for each of these compounds. The changes in intensity for reactant ion peaks (RIPs), however, suggest that reactions may occur in both polarities, but the product ions are unstable on the time scale of observation. For example, the negative RIP for 4-MNT (Fig. 15(c)) is decreased by nearly 50% with exposure to chemical vapor suggesting that O_2^- ions are attaching to the 4-MNT molecules. However, the absence of product ions suggests that the adduct $M^*O_2^-$ is undergoing simple dissociation without proton abstraction or charge transfer. This can also be seen slightly with TNT (cf. Fig. 15(d)), where the positive RIP is depressed slightly with increased concentration, but no positive ions appear in the differential mobility spectrum. Thus, $MH^+(H_2O)_n$ apparently was formed but did not survive until reaching the detector. Such observations have not been made previously. In some instances, a chemical will form stable product ions with positive and negative reactant ions, and these can be seen simultaneously in a planar DMS scan as shown for HMTD in Fig. 15(b). In this instance, a positive product ion can be seen at 5.5 V and a negative ion at 7.8 V. The selectivity of ionization and the benefit of detection of dual ions may not fully protect analytical response from interferences, and in a step to even more rugged response, pre-fractionation through chromatographic methods was applied.

The analytical result from a GC–DMS analysis of a mixture of explosives is shown in Fig. 16, where seven explosives were separated and detected within 3 min. The results from the measurement are displayed as topographic plots of retention time (x-axis), compensation voltage (y-axis), and ion intensity (z-axis), and each substance exhibited specific retention times and compensation voltages with ions in one or the other polarity, occasionally both polarities. Thus, the gas chromatograph provided an additional dimension of information in the retention time, e.g., ~15 s for HMTD, ~25 s for ethylene glycol dinitrate (EGDN), or ~2.5 min for TNT and ion separation in the DMS added to this. For example, isomers such as 2-MNT and 4-MNT can be distinguished using their slightly differing compensation voltages ($\Delta CV \approx 0.5$ V) and different retention times ($\Delta t_r \approx 10$ s). For the reliable detection of these substances, a concentration effect needs to be considered: both MNTs form two different ions of specific compensation voltages, and for both pairs a concentration-dependent intensity ratio can be explained by the formation of monomer and dimer ions. As the substance molecules elute from the GC column into the DMS analyzer in a chromatographic profile, the initially small analyte concentration favors the formation of monomer ions ($MH^+(H_2O)_n$), detected at ~2 V (Figs 15(c) and 16). At the center of the chromatographic profile, the analyte concentration may be high enough that dimer ions of the analyte, ($M_2H^+(H_2O)_n$), can be formed and provide response at ~0 V in the differential mobility spectra. At the end of the elution profile, the equilibrium returns to the preferred formation of monomer ions. Not surprisingly, at decreased sample amounts, only monomer ions are formed throughout the entire elution process. Because concentrations are often unknown, both compensation

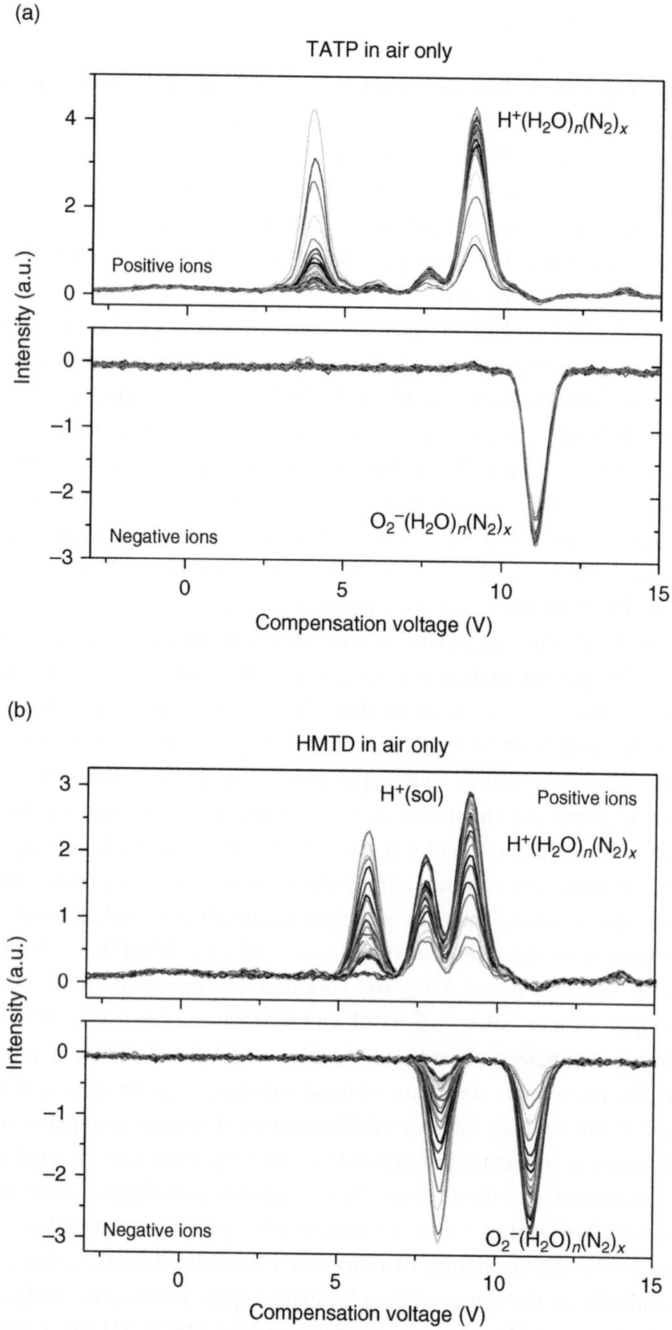

Fig. 15. (a) Differential mobility spectra of triacetone triperoxide (TATP) in air at different concentrations. Product ions can only be detected in the positive mode, negative product ions cannot be observed. (b) Differential mobility spectra of hexamethylene triperoxide diamine (HMTD) in air at different concentrations. Product ions can be detected both in positive and negative polarity. The additional peak in the positive mode could be attributed to a co-eluting solvent, acetonitrile (ACN).

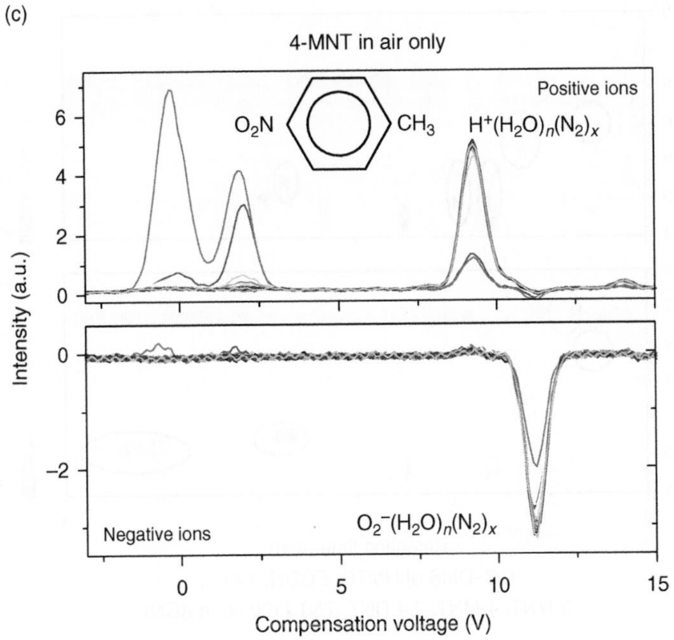

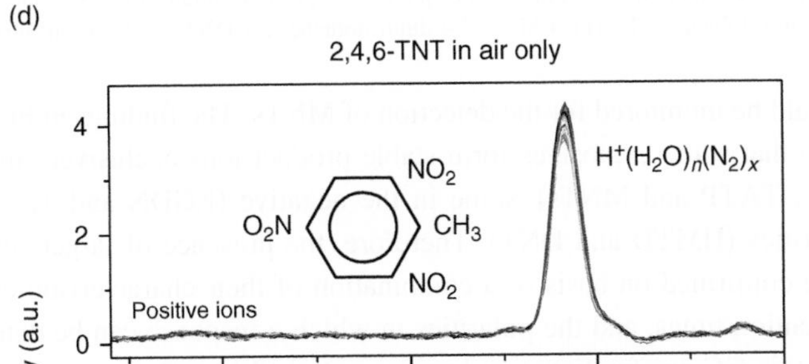

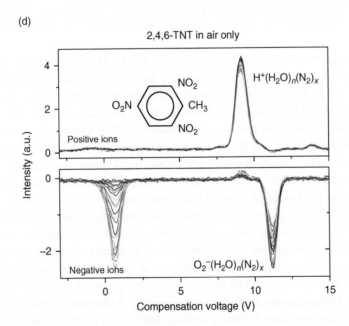

Fig. 15. (Continued) (c) Differential mobility spectra of 4-mononitrotoluene (4-MNT) in air at different concentrations. Product ions can only be detected in the positive mode, negative product ions cannot be observed. (d) Differential mobility spectra of 2,4,6-trinitrotoluene (TNT) in air at different concentrations. Product ions can only be detected in the negative mode, positive product ions cannot be observed. *Source*: A.A. Cagan, H. Schmidt, G.A. Eiceman, NMSU (unpublished results, 2006).

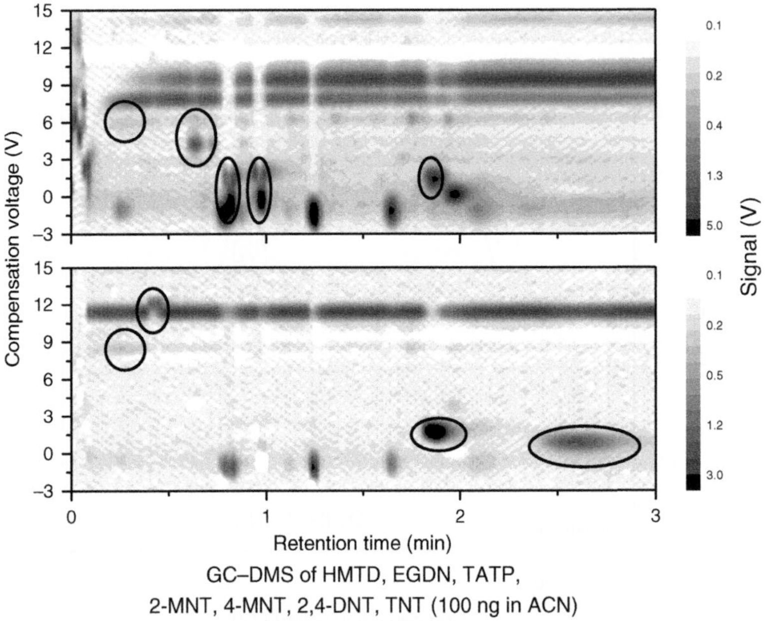

Fig. 16. Gas chromatography–differential mobility spectrometry (GC–(DMS) topographic plot [retention time (x-axis), compensation voltage (y-axis), intensity (z-axis)] of a mixture of seven explosives (100 ng in acetonitrile): hexamethylene triperoxide diamine (HMTD), ethylene glycol dinitrate (EGDN), triacetone triperoxide (TATP), 2-mononitrotoluene (2-MNT), 4-MNT, 2,4-dinitrotoluene (2,4-DNT), and 2,4,6-trinitrotoluene (TNT).

voltages should be monitored for the detection of MNTs. The findings in Fig. 16 amplify the condition that some explosives form stable product ions exclusively in the positive polarity (e.g., TATP and MNTs), some in the negative (EGDN and TNT), and some in both polarities (HMTD and DNT). Therefore, the presence of target substances can altogether be confirmed on basis of a combination of their characteristic compensation voltages, retention times, and the polarities in which a response can be detected.

However, this GC pre-fractionation has an analytical benefit much greater than only retention time. The chemistry of ionization is simplified as substances are provided individually to the ionization source of the DMS analyzer. As shown earlier, reactant ions are created in the ionization region and subsequently form analyte ions with sample molecules. However, as the whole sample is presented to the reactant ions at the same time, the competition between sample molecules of different proton affinities in the positive mode, or electron affinities in the negative mode, to form analyte ions can result in the suppression of response of certain molecules of lower proton or electron affinities. This may lead to a false-negative response to some analytes. A chromatographic pre-separation can improve the reliability regarding both qualitative and quantitiative response compared with the presentation of a whole sample vapor to the DMS. One disadvantage of the use of a chromatographic column is the extended total analysis time, because a fast response is a critical requirement in explosives detection. This problem can be solved through the use of short high-speed columns in combination with very fast temperature ramps as practiced with true high-speed GC. The results in Fig. 16 were collected using a conventional large benchtop gas chromatograph with no refinements or upgraded components needed to obtain good quality with high-speed GC. These limitations were ameliorated with the commercial GC–DMS analyzer as described in the next section.

7. The EGIS Defender: a commercial high-speed GC–DMS analyzer

A commercial version of a GC–DMS instrument [58,59] has been produced for detecting explosives and is known as the EGIS Defender from Thermo Scientific Corporation (Fig. 17). Much of this instrument had technical heritage with the EGIS analyzers such as the Thermedics EGIS 3000, which was a high-speed gas chromatograph with a chemiluminescence detector. Sample is collected as with most trace detectors for explosives by wiping a surface with a specially designed filter. The filter is then placed into an inlet where explosives are thermally desorbed for an 18-s analysis to determine the types and levels of explosives in the sample. A box diagram for the architecture of this instrument is also shown in Fig. 17, where vapors desorbed from a sample are passed to a gas chromatograph and then a DMS analyzer (sensor assembly). The gas chromatograph is based on high-speed control of temperature, which was re-developed and released as the EZ Flash upgrade kit designed to improve the speed of a conventional gas chromatograph. In this gas chromatograph, column temperatures can be increased at a rate up to 1200 °C/min with upper temperatures of 280–365 °C, depending on the column length.

The contours of the GC separation with the DMS analyzer as selective detector are visible in the graphic display of Fig. 17. The sample inlet is seen in the front of the analyzer at bottom right. A measure of the performance of the EGIS Defender is shown in Fig. 17 (bottom) where quantitative response is shown in part and illustrates that DMS and IMS analyzers are quantitative devices so long as sources are not saturated or overloaded. The analytical response of this unit is only partially disclosed here as response curves for two substances, TNT and EGDN. The plots are normal in appearance for IMS or DMS quantitative response, with a range, as shown here, in nanograms. As with all DMS and IMS instruments, response can become saturated at levels $\times 100$ or more than these making this and other DMS analyzers trace detectors with quantitative and proportional response across a limited range.

8. Next developments in DMS and IMS determinations of explosives

Although reliable and capable of operation by non-specialists with trace detection capabilities useful for explosives screening, IMS analyzers are comparatively low in resolution; although specificity of response is strong owing to the method of ionization of sample and the additional evaluation of product ions with mobility, increased specificity and expanded scope of response to an enlarged list of explosives is a reasonable

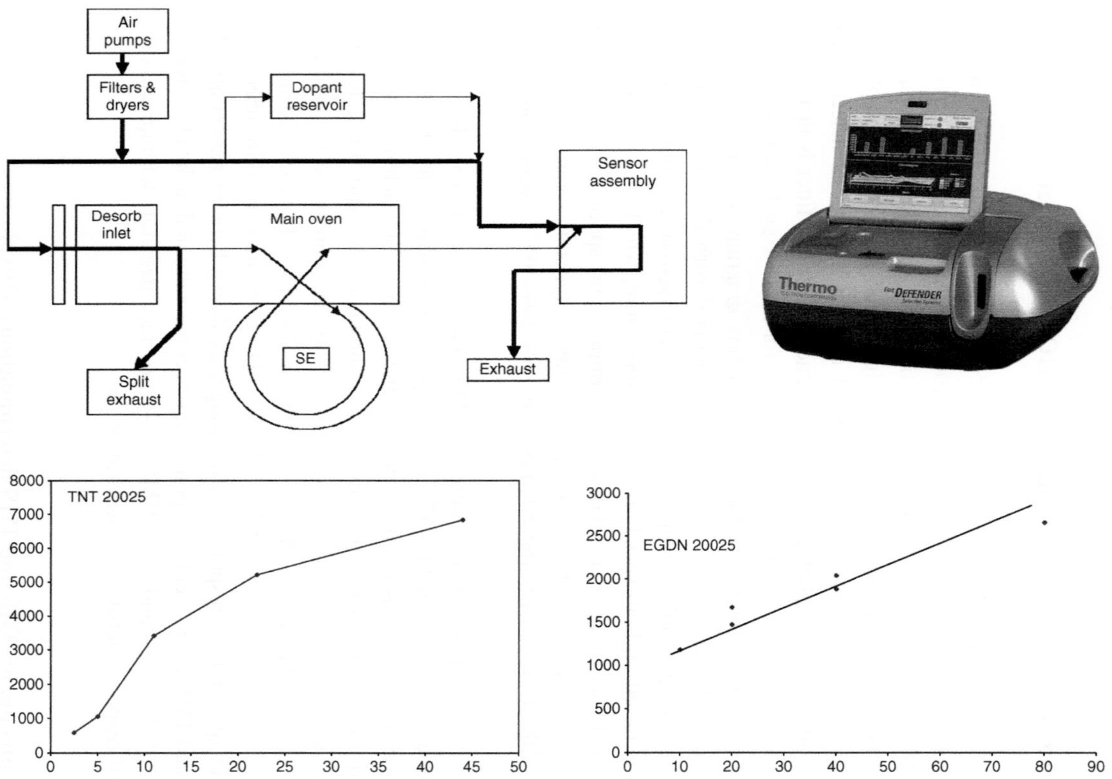

Fig. 17. A commercial configuration of high-speed gas chromatography–differential mobility spectrometry (GC–DMS) is the Defender (flow schematic top left, instrument top right), which is a successor of the EGIS and EGIS II explosives analyzers. Calibration curve of trinitrotoluene (TNT) (bottom left) and ethylene glycol dinitrate (EGDN) (bottom right) (signal vs. mass in nanogram) from the EGIS Defender.

expectation for maturation of technology. In the moment, the low resolution found with IMS is offset or compensated with practical benefits of simplicity of instrumentation sans vacuum pumps, the very low levels of detection, and the simplicity of use and maintenance. The emergence of DMS and GC/DMS as described in Sections 2–7 will aid some of these concerns as will the twin drift tube of the IONSCAN 500DT. Nonetheless, specificity of response remains a factor in false alarms, positive or negative, and the use of a single technology with stable parameters for a more comprehensive list of explosives would be a welcome enhancement with trace detectors.

The method of DMS–IMS^2 is based on characterizing ions using differences in the mobility (ΔK) followed by ion separation based on K in an IMS analyzer [60,61]. Whereas characterization of ions in IMS is based on cross-section of collision (as a first approximation), ions in DMS are characterized on the susceptibility of an ion to hydration or solvation and the mass of the ion. Although these concepts are not purely orthogonal, enough difference exists between ion characterizations by IMS and DMS so a combined or tandem analyzer of DMS–IMS is three to five times improved over either alone. When ions of both polarities are characterized simultaneously, two IMS drift tubes may be used to interrogate ions after DMS analysis. What is known for DMS determination of explosives and for IMS determination of explosives is neither lost nor compromised with the DMS–IMS^2 characterization of explosives.

Findings are shown in Fig. 18, where the response to different nitrotoluene isomers (2-MNT, 3-MNT, 2,4-DNT, 3,4-DNT, and 2,4,6-TNT) is shown in three-dimensional

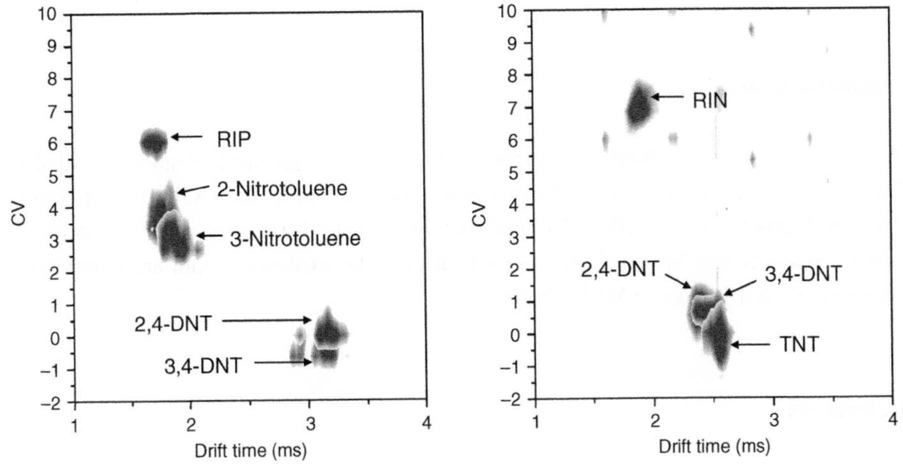

Fig. 18. Overlay of plots from DMS–IMS^2 response to several nitrotoluenes [2-mononitrotoluene (2-MNT), 3-MNT, 2,4-dinitrotoluene (2,4-DNT), 3,4-DNT, 2,4,6-trinitrotoluene (TNT)] in positive polarity (left). A reactant ion peak (RIP) is seen at 6.5 V and 1.9 ms, whereas protonated monomers are seen near 3 V and 2.2 ms, and proton-bound dimers are at 0 V and 3.5 ms. TNT does not show a response in positive polarity. Overlay of plots from DMS–IMS^2 response to several nitrotoluenes (cf. list in positive polarity) in negative polarity (right). A RIN is seen at 7 V and 1.8 ms, whereas molecular adducts or charge-exchanged ions are seen at 0.5 V and 2.5 ms. MNTs do not show a response in negative polarity. *Source*: Neil D. Paz, NMSU (unpublished results, 2006).

plots of compensation voltage (for the DMS) versus drift time (for the IMS) and ion intensity in both polarities. Emphasis should be given to the benefit of collecting positive and negative ion data simultaneously in the DMS–IMS^2 analyzer design. Also, the drift region in the IMS analyzers here was the smallest drift tube ever built at NMSU (1 cm drift length) and constituted the smallest considered viable for measurements. Doubling this length or even 3 cm would enhance analytical performance seen in Fig. 18 (the DMS analyzer was of normal size with 15-mm-long analyzer path). Nonetheless, the plots show that response to nitrotoluene isomers will provide enhanced resolution in the DMS scale over IMS, and larger ions (such as DNT or TNT) exhibit enhanced resolution in the IMS scale. The combination provides response to both ion polarities simultaneously and resolution across a range of explosives.

9. Conclusions

The benefits of characterizing ions in air at ambient pressure bring some compelling strengths for chemical measurements with explosives, and these have been expanded with the advancement toward micro-fabricated DMS analyzers. The principal advantage of small size and low cost with agreeable analytical performance can be enhanced with sample preparation through fast chromatography, as seen now in the EGIS Defender, and in additional ion evaluation with tandem DMS and IMS configurations with simultaneous characterization of ions of both polarities for mobility and differential mobility.

Acknowledgments

We acknowledge with thanks the supporting information on the Defender provided by Dan Dussault, ThermoScientific Corporation, and Raanan Miller, Sionex Corporation. The development of the DMS–IMS^2 analyzer was a joint effort with Hamilton Sundstrand Sensor Systems and Sionex Corporation [60], and the explosives data are unpublished courtesy of Neil D. Paz, NMSU, 2006.

References

[1] G.A. Eiceman and Z. Karpas, *Ion Mobility Spectrometry*, 2nd edition, CRC Press Inc., Boca Raton, FL (2005).
[2] D.D. Fetterolf and T.D. Clark, *J. Forensic Sci.*, 38 (1993) 28.
[3] M.J. Cohen, R.F. Wernlund and R.M. Stimac, *Nucl. Mater. Manage.*, 13 (1984) 220.
[4] R.F. Wernlund, M.J. Cohen and R.C. Kindel, The ion mobility spectrometer as an explosive or taggant vapor detector. *Proceedings of the New Concept Symposium and Workshop on Detection Identification of Explosives*, Reston, VA, USA, October/November 1978, pp. 185–189.

[5] S. Bell, E. Nazarov, Y.F. Wang, J.E. Rodriguez, and G.A. Eiceman, *Anal. Chem.*, 72 (2000) 1192.
[6] G.A. Eiceman, E.G. Nazarov and J.E. Rodriguez, *Anal. Chem. Acta*, 433 (2001) 53.
[7] G.A. Eiceman and J.A. Stone, *Anal. Chem.*, 76 (2004) 390A.
[8] I.A. Buryakov, E.V. Krylov, A.L. Makas, E.G. Nazarov, V.V. Pervukhin and U.K. Rasulev, *Pisma Zh. Tekh. Fiz.*, 17 (1991) 60.
[9] I.A. Buryakov, E.V. Krylov, E.G. Nazarov and U.K. Rasulev, *Int. J. Mass Spectrom. Ion Process.*, 128 (1993) 143.
[10] I.A. Buryakov, E.V. Krylov, A.L. Makas, E.G. Nazarov, V.V. Pervukhin and U.K. Rasulev, *J. Anal. Chem. USSR* (*Zhurnal Analiticheskoi Khimii*), 48 (1993) 156.
[11] G.A. Eiceman, V.J. Vandiver, T. Chen and G. Rico-Martinez, *Anal. Instrum.*, 18 (1989) 227.
[12] G.A. Eiceman, E.G. Nazarov, J.A. Stone and J.E. Rodriguez, *Rev. Sci. Instrum.*, 72 (2001) 3610.
[13] A.J. Midey and A.A. Viggiano, *J. Chem. Phys.*, 114 (2001) 6072.
[14] S.L. Gong and R.E. Jervis, *J. Chem. Phys.*, 103 (1995) 7081.
[15] M. Meot-Ner, *J. Am. Chem. Soc.*, 114 (1992) 3312.
[16] J.D. Payzant and P. Kebarle, *J. Chem. Phys.*, 56 (1972) 3482.
[17] M. Arshadi and P. Kebarle, *J. Phys. Chem.*, 74 (1970) 1483.
[18] Z. Karpas, Z. Berant and O. Shahal, *J. Am. Chem. Soc.*, 111 (1989) 6015.
[19] Z. Karpas, G.A. Eiceman, E.V. Krylov and N. Krylova, *Int. J. Ion Mobility Spectrom.*, 7(1) (2004) C8.
[20] C.S. Creaser, J.R. Griffiths, C.J. Bramwell, S. Noreen, C.A. Hill and C.L.P. Thomas, *Analyst*, 129 (2004) 984.
[21] T.W. Carr (ed.), *Plasma Chromatography*, Plenum, New York, 1984.
[22] G.A. Eiceman and Z. Karpas, *Ion Mobility Spectrometry*, CRC Press Inc., Boca Raton, FL, (1993).
[23] M.J. Cohen, *Plasma Chromatography. A New Dimension for Gas Chromatography and Mass Spectrometry*, presented at the Pittsburgh Conference on Analytical Chemistry and Applied Spectroscopy, March 1969.
[24] J.I. Baumbach, *Anal. Bioanal. Chem.*, 384 (2006) 1059.
[25] D.C. Collins and M.L. Lee, *Anal. Bioanal. Chem.*, 372 (2002) 66.
[26] G.A. Eiceman, *Trends Anal. Chem.*, 21 (2002) 259.
[27] H. Borsdorf and G.A. Eiceman, *Appl. Spectrosc. Rev.*, 41 (2006) 323.
[28] United States Patent 5,227,628 Ion mobility detector, July 13, 1993 Turner; Brian R. (Chesham, GB) Assignee: Graseby Dynamics Limited (Cambridge, GB2) Appl. No.: 741,472 Filed: August 5, 1991PCT Filed: February 07, 1990 PCT NO: PCT/GB90/00182 371 Date: August 05, 1991 102(e) Date: August 05, 1991 PCT PUB.NO.: WO90/09583.
[29] Smiths Detection, *IONSCAN 500DT, Simultaneously Detect Explosives and Narcotics Using Dual IMS Detectors*, available at http://trace.smithsdetection.com/products/Default.asp?Product=55 (2006).
[30] R. Wilson and A. Brittain, *Explosives Detection by Ion Mobility Spectrometry*, Special Publication - Royal Society of Chemistry 203 (Explosives in the Service of Man), Graseby Dynamics Ltd., Hertfordshire, p. 92 (1997).
[31] R.J. Munson, *Proc. R. Soc. Lond.*, A172 (1939) 51.
[32] A.M. Tyndall and C.F. Powell, *Proc. R. Soc. Lond.*, A134 (1931) 125.
[33] J.H. Mitchell and K.E.W. Ridler, *Proc. R. Soc. Lond.*, A146 (1934) 911.
[34] L.G.H. Huxley, R.W. Crompton, and M.T. Elford, *Br. J. Appl. Phys.*, 17 (1966) 1237.
[35] I.I. Balog, *Fizika*, 74 (1944) 123.
[36] M.P. Gorshkov, *Analysis of Impurities in Gases*, Inventor's Certificate of USSR No 966583, G01N27/62 (1982).

[37] E. Krylov, E.G. Nazarov, R.A. Miller, B. Tadjikov and G.A. Eiceman, *J. Phys. Chem. A*, 106 (2002) 5437.
[38] N. Krylova, E. Krylov, J.A. Stone and G.A. Eiceman, *J. Phys. Chem. A*, 107 (2003) 3648.
[39] B. Carnahan, S. Day, V. Kouznetsov and A. Tarassov, *Development and Applications of a Transverse Field Compensation Ion Mobility Spectrometer*, presented at the 4th International Workshop on Ion Mobility Spectrometry, Cambridge, UK, 6–9 August 1995.
[40] R. Guevremont, *Can. J. Anal. Sci. Spectrosc.*, 49 (2004) 105.
[41] R.W. Purves, R. Guevremont, S. Day, C.W. Pipich and M.S. Matyjaszczyk, *Rev. Sci. Instrum.*, 69 (1998) 4094.
[42] R.W. Purves, D.A. Barnett and R. Guevremont, *Int. J. Mass Spectrom.*, 197 (2000) 163.
[43] R.W. Purves, D.A. Barnett, B. Ells and R. Guevremont, *J. Am. Soc. Mass Spectrom.*, 11 (2000) 738.
[44] D.A. Barnett, B. Ells, R. Guevremont and R.W. Purves, *J. Am. Soc. Mass Spectrom.*, 13 (2002) 1282.
[45] Thermo Scientific, *FAIMS: The First Step in LC-MS Method Development Problem Solving*, Waltham, MA (2007) available at http://www.thermo.com/com/cda/resources/resources_detail/ 1,2166,112852,00.html.
[46] R.A. Miller, G.A. Eiceman, E.G. Nazarov and A.T. King, *Sens. Actuators B*, 67 (2000) 300.
[47] R.A. Miller, E.G. Nazarov, G.A. Eiceman and A.T. King, *Sens. Actuators A*, 91 (2001) 307.
[48] R.E. Ewing, G.J. Ewing, D.A. Atkinson and G.A. Eiceman, *Talanta*, 54 (2001) 515.
[49] M.W. Siegel, Atmospheric pressure ionization, in T.W. Carr (ed.), *Plasma Chromatography*, Plenum Press, New York, (1984).
[50] G.R. Asbury, J. Klasmeier and H.H. Hill, Jr., *Talanta*, 50 (2000) 1291.
[51] M. Tam and H.H. Hill, Jr., *Anal. Chem.*, 76 (2004) 2741.
[52] I.A. Buryakov, Yu.N. Kolomiets and B.V. Luppu, *J. Anal. Chem.*, 56 (2001) 336.
[53] I.A. Buryakov, *Talanta*, 61 (2003) 369.
[54] I.A. Buryakov and Yu.N. Kolomiets, *J. Anal. Chem.*, 58 (2003) 944.
[55] V.P. Zhdanov, V.N. Sidelnikov and A.A. Vlasov, *J. Chromatogr. A*, 928 (2001) 201.
[56] J.I. Baumbach, G.A. Eiceman, D. Klockow, S. Sielemann and A. von Irmer, *Int. J. Environ. Anal. Chem.*, 66 (1997) 225.
[57] G.A. Eiceman, E.V. Krylov, N.S. Krylova, E.G. Nazarov and R.A. Miller, *Anal. Chem.*, 76 (2004) 4937.
[58] Thermo Scientific Corporation, *EGIS Defender - Explosive Detector*, Waltham, MA, USA (2007) available at http://www.thermo.com/com/cda/product/detail/1,10119656,00.html.
[59] Thermo Scientific Corporation, *EGIS™ Defender, Portable Lightweight Desktop Explosives Trace Detection System,* Waltham, MA, USA (2007), available at http://www.thermo.com/eThermo/CMA/PDFs/Product/productPDF_30772.pdf.
[60] C.R. White, *Characterization of Tandem DMS-IMS² and Determination of Orthogonality Between the Mobility Coefficient (K) and the Differential Mobility Coefficient* (ΔK), MS Thesis, New Mexico State University, Las Cruces, NM, (2006).
[61] G.A. Eiceman, H. Schmidt, J.E. Rodriguez, C.R. White, E.G. Nazarov, E.V. Krylov, R.A. Miller, M. Bowers, D. Burchfield, B. Niu, E. Smith and N. Leigh, *Characterization of Positive and Negative Ions Simultaneously Through Measures of K and* ΔK *by a Tandem DMS-IMS²*, presented at the 14th International Conference for Ion Mobility Spectrometry, Maffliers, France, 25 July 2005.

Chapter 4

Electrochemical Sensing of Explosives

Joseph Wang

Biodesign Institute, Arizona State University, Tempe, AZ 85287, USA

Counterterrorist Detection Techniques of Explosives
Jehuda Yinon (Editor)

Contents

1. Introduction

The detection of explosive compounds has received considerable attention for national security and environmental applications [1]. Such security and environmental needs have generated major demands for effective and innovative field-deployable devices for detecting organic explosives in a sensitive, fast, simple, reliable, and cost-effective manner. New devices and innovative protocols providing effective on-site or in situ real-time monitoring capability are particularly needed for addressing escalating threats of terrorist activity and related explosive detection challenges.

This chapter highlights recent advances, primarily from our laboratory, aimed at developing electrochemical sensors and detectors for explosive compounds. Electrochemical sensors offer unique opportunities for addressing the need for on-site detection of various explosives. The advantages of electrochemical systems include high sensitivity and selectivity, a wide linear range, minimal space and power requirements, and low-cost instrumentation. Both the sensor and the controlled instrumentation can be readily miniaturized to yield hand-held meters. The coupling of modern electrochemical detection principles with recent advances in microelectronics and microfabrication has led to powerful, compact, and user-friendly analytical devices for decentralized (on-site) testing. Common examples are pocket-size glucose meters (based on amperometric enzyme electrodes) that are widely used for self-testing of diabetes.

Electroanalytical techniques are concerned with the interplay between electricity and chemistry, namely, the measurements of electrical quantities, such as current, potential, or charge, and their relationship to chemical parameters [2]. Controlled-potential (potentiostatic) techniques deal with the study of charge transfer processes at the electrode/solution interface and are based on dynamic situations. Here, the electrode potential is being used to derive an electron transfer reaction, and the resulting current is measured. The role of the potential is analogous to that of the wavelength in optical measurements. Such controllable parameter can be viewed as 'electron pressure', which forces the chemical species to gain or lose an electron (reduction or oxidation, respectively). In 1980s and 1990s have seen major advances in electroanalyticalchemistry, including the design of tailored interfaces, the development of ultramicroelectrodes, the coupling of biological components or nanoscale materials with electrical transducers, the microfabrication of molecular devices, and the introduction of 'smart' sensors and sensor arrays. The readers are referred to a recent book and a review for comprehensive information on electrochemical systems [2,3].

The inherent redox activity of nitroaromatic explosives [4], namely, the presence of easily reducible nitro groups, makes them ideal candidates for electrochemical (voltammetric) monitoring. Particularly suited for the rapid testing of military explosives is square-wave voltammetry (SWV), owing to its high sensitivity, fast scan rates, and compact low-power instrumentation [5]. SWV is a large-amplitude pulse-voltammetric technique in which a waveform composed of a symmetrical square wave superimposed on a base staircase potential (Fig. 1) is applied to the working electrode [2]. The current is sampled twice during each square-wave cycle, and the current difference offers

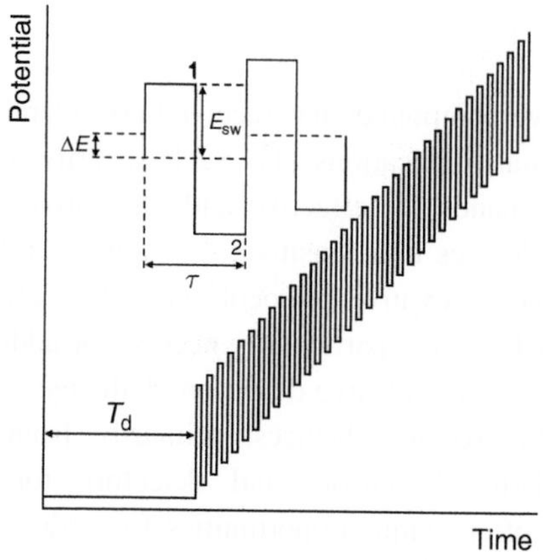

Fig. 1. Waveform used in square-wave voltammetry (SWV) showing the amplitude, E_{SW}; step height, ΔE; square-wave period, T; delay time, T_d; and current measurement times, 1 and 2. Reproduced with permission from Wang [2].

a substantial increase in the ratio of the analytical to the background current. The resulting peak-shaped voltammogram is symmetrical, with the peak potential providing the qualitative identification and the peak current being proportional to the analyte concentration.

Sections 3–5 will describe several recently developedpowerful electrical devices for on-site monitoring of explosive materials.

2. Electrochemistry of explosive materials

The nitro group is an excellent electron acceptor. The reduction of polynitroaromatic compounds is complicated by multi-step processes. These usually occur in two steps to form an amine via a hydroxylamine. Such processes are strongly dependent on the pH, the number of nitro groups per molecule, their relative position on the ring, and the nature and position of other substituents present on the aromatic system [4]. In general, trinitroaromatic compounds, such as TNT or picric acid, are more easily reduced than dinitro- and mononitroaromatic compounds. Nitramines and polynitrate esters are more difficult to reduce, and their reduction peaks appear at more negative potentials. Such redox activity of TNT and RDX can be exploited not only for sensing applications (described in Sections 3–5) but also for electrochemical 'cleanup' in connection with flow-throughporous electrodes [6].

Detection tags with high vapor pressure, particularly 2,3-dimethyl-2,3-dinitrobutane (DMNB), are commonly added to commercial explosives to facilitate the vapor detection of explosives with relatively low vapor pressure such as plastic explosives. Evans group [7] investigated the mechanism of the DMNB reduction at mercury electrodes in a dimethylformamide (DMF) medium, whereas Wang's team [8] recently exploited the electroreduction of DMNB at the carbon-fiber surface for its fast and sensitive SWV detection.

3. Easy-to-use disposable electrode strips for explosives

The performance of the voltammetric procedure is strongly influenced by the working electrode material. The working electrode is the one at which the reaction of interest occurs. The selection of the working electrode depends primarily on the redox behavior of the target explosive and the background current over the applied potential region. A range of materials have found application as working electrodes for reductive (cathodic) measurements of nitroaromatic and nitramine explosive compounds, including mercury, carbon, or gold/amalgam. The deliberate modification of conventional electrodes can lead to an enhanced explosive detection. Modified electrodes based on carbon nanotube [9] or mesoporous SiO_2-MCM-41 [10] coatings have been recently shown to be useful for enhancing the sensitivity through adsorptive accumulation of the target explosive.

Three-electrode cells are commonly used in controlled-potential experiments. The cell is usually a covered beaker of 5–20 ml volume and contains the three electrodes (working, reference, and auxiliary) that are immersed in the sample solution. To address the needs of field sensing of explosives, it is necessary to move away from such traditional electrodes and bulky cells. The exploitation of advanced microfabrication techniques allows the replacement of conventional 'beaker-type' electrochemical cells and electrodes with easy-to-use sensor strips. Both thick-film (screen-printing) and thin-film (lithographic) fabrication processes have thus been used for high-volume production of highly reproducible, effective, and inexpensive electrochemical sensor strips. Such strips rely on planar working, reference, and counter electrodes on a plastic, ceramic, or silicon substrate (Fig. 2). These strips can thus be considered as self-contained electrochemical cells onto which the sample droplet is placed. The thin-film fabrication route also facilitates the development ofcross-reactive electrode arrays [11], which are promising for multi-explosive electronic nose (e-nose)-type gas-phase detection (in connection with advanced signal-processing algorithms).

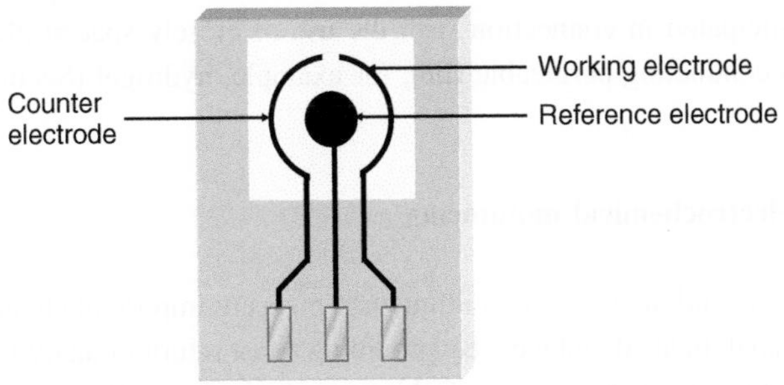

Fig. 2. A disposable thick-film voltammetric sensor for detecting organic explosives.

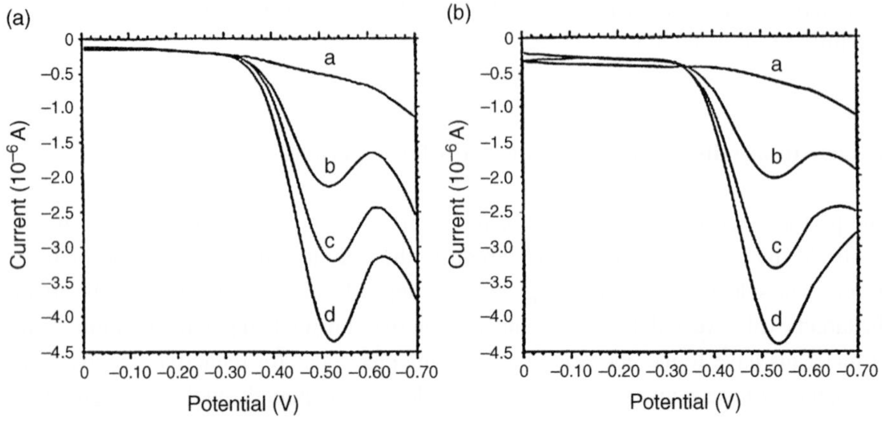

Fig. 3. Explosive detection at single-use screen-printed electrodes. Assays of river water (A) and groundwater (B) samples. Response to the sample (a) as well as for subsequent concentration increments of 3 ppm TNT (b–d). Reproduced with permission from Wang et al. [12].

It was demonstrated that a thick-film TNT sensor, based on the coupling of fast SWV and the screen-printing fabrication process [12]. Screen-printing is a well-established technology used for the mass production of low-cost, disposable electrochemical sensors [2,13]. This process involves printing patterns of conductors and insulators on surfaces of planner substrates by 'forcing' the corresponding ink through the openings of a patterned stencil. The screen-printed TNT sensor developed in our laboratory offers an attractive analytical performance, including high sensitivity (down to a detection limit of 200 ppb TNT), a wide linear range, fast response, and high selectivity [12]. Such high sensitivity and selectivity were demonstrated in connection with direct TNT measurements in ground and river water samples (Fig. 3). Hart's group [14] demonstrated the utility of similar screen-printed carbon electrodes for stripping voltammetric measurements of trace 2,6-dinitrotoluene. Reproducible measurements down to 160 ppb were reported in connection with 100 μl samples. Zen and coworkers [15] used the preanodization of screen-printed electrodes for obtaining sharper peaks for nitroaromaticcompounds. This allowed a useful approach to identifying substituents and facilitated single-run (internal standard) approach that obviated the need for standard additions. Applicability to vapor detection is anticipated in connection with the use of closely spaced planar electrodes covered with a conducting permeable film, for example, hydrogel (Section 4.5).

4. Real-time electrochemical monitoring

The use of sensors and detectors to continuously measure important chemical properties has significant analytical advantages. By providing a fast return of analytical information in a timely, safe, and cost-effective fashion, such devices offer direct and reliable monitoring of explosive compounds while greatly reducing the huge analytical costs and minimizing errors associated with the sampling process. Such real-time electrochemical monitoring of explosive compounds has been accomplished in our laboratory through on-line [16] and submersible [17] operations.

4.1. Flow analysis of TNT: toward on-line monitoring of explosives

Electrochemical measurements can be readily adapted for on-line monitoring. An electrochemical detector uses the electrochemical properties of target analytes for their determination in a flowing stream. An electrochemical flow system, based on an SWV operation at a carbon-fiber-based detector, for use in the on-line continuous monitoring of trace TNT in marine environments was developed [16]. Such flow detector offers selective measurements of sub-part-per-million concentrations of TNT in untreated natural water samples with a detection limit of 25 ppb. It responds rapidly to sudden changes in the TNT concentration with no apparent carryover. About 600 runs can be made every hour with high reproducibility and stability (e.g., relative standard deviation (RSD) = 2.3%, $n = 40$). The system lends itself to full automation and to possible deployment onto various stationary mobile platforms (e.g., buoys and underwater vehicles).

An electrochemical flow detector, based on four working electrodes coated with different permselective film types, for use in flow-injection measurements of multiple nitroaromatic explosives was recently developed [18]. The resulting array response (Fig. 4) offers unique 'fingerprints' of such explosive compounds. Electrochemical devices are extremely attractive for designing such e-nose-type multi-electrode arrays that combine several partially selective electrodes and lead to a distinct response pattern (signature) for mixtures of organic vapors without prior separation [2].

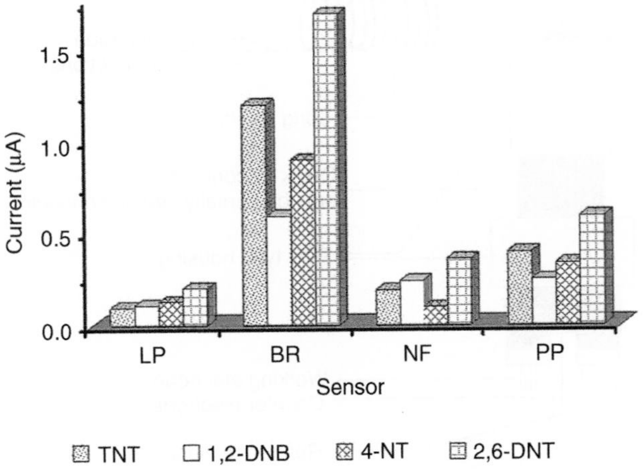

Fig. 4. Amperometric array response to various nitroaromatic explosives at a flow detector with four electrodes coated with different films. LP, lipid; BR, bare; NF, Nafion; PP, polyphenol.

4.2. *Remote monitoring of TNT*

Remotely deployable submersible sensors capable of monitoring contaminants both in time and in location have a variety of applications [19]. Such ability to perform in situ measurements of explosives is important for improving the efficiency of characterization and remediation of sites that are former military munitions manufacturing, storage, and demilitarized zones and are contaminated with high levels of explosive compounds.

A submersible electrode assembly, connected to a 50-ft-long shielded cable via environmentally sealed rubber connectors, for the real-time in situ monitoring of the TNT explosive in natural water was developed [17]. As illustrated in Fig. 5, this sensor assembly consists of the carbon-fiber, silver, and platinum working, reference, and counter electrodes, respectively, and operates in the rapid SWV mode. This remote/submersible probe circumvents the need for solution pumping and offers greater simplification and miniaturization. The facile reduction of the nitro moiety allowed convenient and rapid SWV measurements of sub-part-per-million levels of TNT. Lower (ppb) concentrations have been detected using a background subtraction operation. Such high sensitivity has been coupled with good selectivity and stability and the absence of carryover effects. The latter reflects the absence of recognition/binding events. These capabilities were illustrated in various natural water environments. For example, Fig. 6 displays voltammograms for seawater samples containing increasing levels of TNT in 250-ppb steps. Prolonged direct operations of such in situ electrochemical sensors in organic-rich ocean environments may be complicated by surface fouling by co-existing substances (e.g., surfactants, suspended particulates, and bacterial fauna). Prevention of electrode deactivation due to such surfactant adsorption has been accomplished by covering the working electrode with a protective (anti-fouling) polyphenol coating [20].

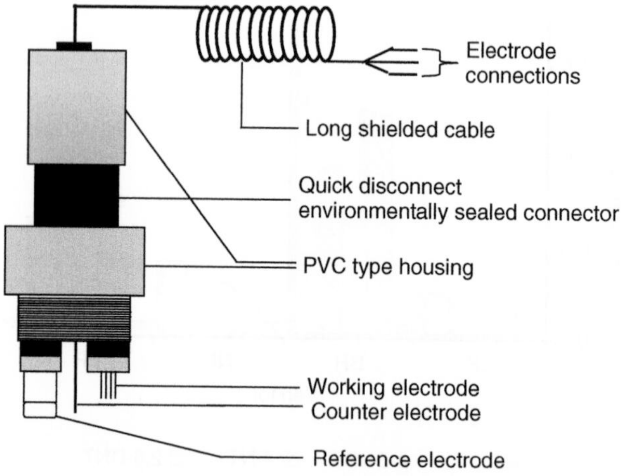

Fig. 5. A submersible electrochemical sensor for remote monitoring of explosive substances in natural water samples. Based on Wang [19].

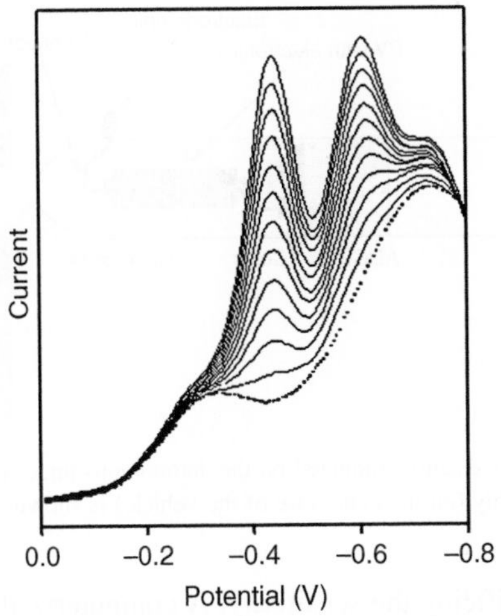

Fig. 6. Square-wave voltammograms recorded with remote/submersible electrochemical probe for seawater samples containing increasing levels of TNT in 250-ppb steps (scale: 1 MA per cm).

Such size-exclusion film is prepared by electropolymerizing polyphenol in the presence of the phenol monomer.

A self-contained system, which integrated the remote carbon-fiber electrochemical sensor with a voltammetric analyzer and a wireless communication system, was described by Fu et al. [21]. The mobile remote underwater systems were applied for field measurements of explosive residues in marine environments.

4.3. In situ electrochemical monitoring of TNT using underwater vehicle platforms

The voltammetric sensor was integrated onto an unmanned underwater vehicle (UUV). This integration was a part of the Office Naval Research (ONR) Chemical Sensing in Marine Environments (CSME) Program, whose goal was to demonstrate the effectiveness of advanced sensor technology for detecting explosives (leaking out of mines) in coastal regions of the ocean. Such chemical detection of explosives in a marine environment is an extremely challenging problem, as the explosive chemical signature emanating from a source can rapidly disperse in the environment, diluting the signature to trace levels that are difficult to detect. Highly sensitive and rapid sensors are thus desired for detecting and prosecuting plumes containing trace amounts of explosives that leak into the seawater, and this detection is used for identifying and marking mines.

The electrochemical TNT sensor was mounted onto the UUV as a separate module, and it serially sent data values to the REMUS UUV. The electrode assembly was mounted on the cone nose of the vehicle and connected to the internal microanalyzer (Fig. 7). Major attention was given to the optimization of variables of the square-wave waveform (including the frequency, amplitude, and potential step) essential for attaining high speed

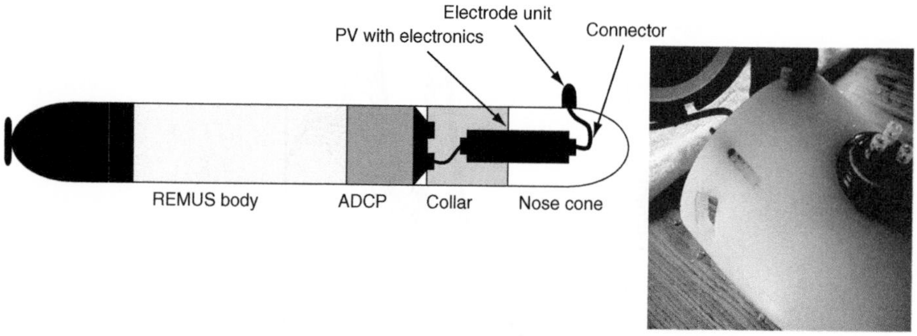

Fig. 7. Electrochemical explosive sensor mounted on the autonomous underwater vehicle (AUV). The three-electrode assembly (on the cone nose of the vehicle) is shown on the right side.

(1–4 s runs) without sacrificing the sensitivity. A computerized baseline subtraction was developed to compensate for the oxygen background contribution and hence to facilitate the detection of low part-per-billion levels of TNT (Fig. 8). The optimal device offered high sensitivity and selectivity, fast response, excellent precision and stability, and no matrix effect, hence meeting the demands of underwater sensing of TNT. A typical set of stability data involving 100 repetitive runs is displayed in Fig. 9. The UUV-deployed sensor was tested in tracking TNT plumes in several field missions lasting between 2 and 3 h.

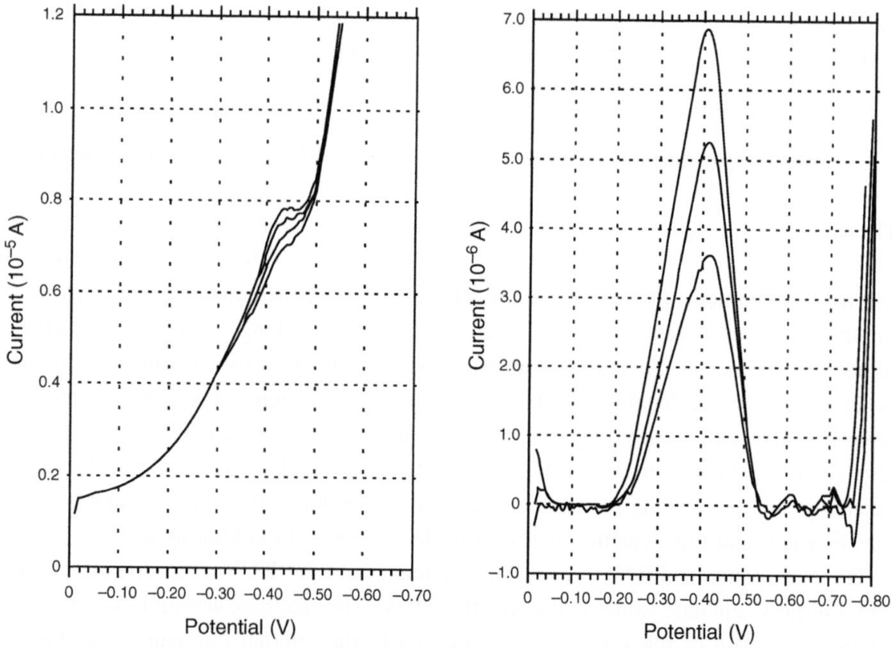

Fig. 8. Square-wave voltammograms for seawater samples containing increasing levels of TNT in 20-ppb steps (right and left, with and without background correction, respectively).

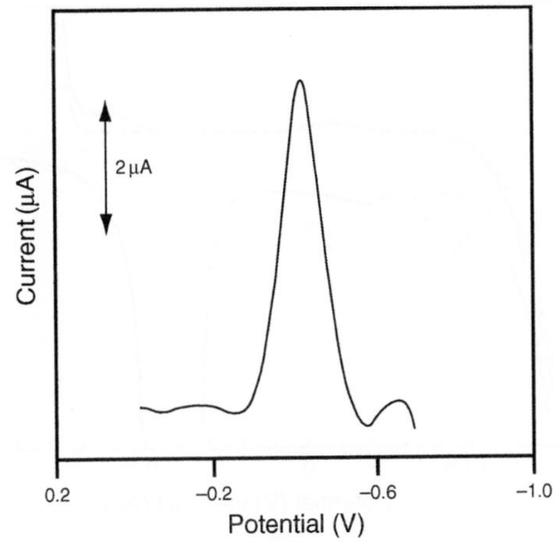

Fig. 9. Stability of 100 repetitive measurements of 3 ppm TNT by the submersible carbon-fiber electrode assembly.

4.4. Diver-held TNT sensing unit

The carbon-fiber-based voltammetric sensor was also integrated into a diver-held unit. This integration relied on the incorporation of a commercial compact (the PalmSense) hand-held voltammetric analyzer within a pressure vessel, with the electrode assembly sticking out of the vessel surface. This configuration allows the diver to observe in real time the corresponding voltammogram. Repetitive square-wave voltammograms are thus performed at 2–4 s intervals. Particular attention has been given to the adaptation of the PalmSense software to meet the requirements of continuous TNT detection. The new software was successfully tested, and the unit was deployed during a field test. Preliminary underwater testing has been very promising, with the current signal increasing and decreasing as the diver enters and departs from the explosive plume.

4.5. Gas-phase voltammetric detection

Real-time reliable detection of explosive vapors is extremely important for practical applications ranging from roadside checkpoints to baggage screening. Commercial amperometric devices are widely used for real-time environmental emission monitoring of toxic gases such as CO, SO_2, NO_x, O_3, CO, formaldehyde, or ethanol [2,22]. For amperometric gas sensors, the main challenge lies in the creation of a working electrode that is accessible for the sample gas while still being in contact with a liquid electrolyte solution (containing the reference and counter electrodes) [22]. Alternately, it is possible to use a solid polymer electrolyte membrane, such as Nafion, covering a planar micro-fabricated electrode arrangement, as solid-state sensors.

Krausa and Schorb [23] described such a solid-state TNT gas sensor based on 25-μm gold disk working electrode, surrounded by a gold ring reference/counter electrode,

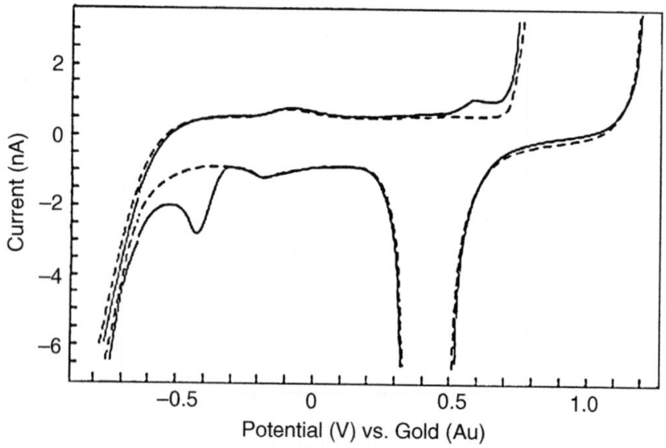

Fig. 10. Gas-phase detection of TNT: cyclic voltammograms for TNT (solid line) and control (dotted line) placed 10 cm above the TNT sample. Scan rate, 400 mV/s. Reproduced with permission from Krausa and Schorb [23].

arranged in a single plane, and coated with a sulfuric acid film. Figure 10 (solid line) displays a cyclic voltammogram recorded by placing the probe 10 cm away from the solid TNT sample. The shape of this voltammogram, with the characteristic TNT reduction peak at−0.4 V, is identical with that recorded upon immersing the electrode in a TNT solution (not shown). The gas-phase reduction peak increased with the exposure time up to 20 s and leveled off for longer periods. Different configurations of gas-phase amperometric TNT sensors are currently being developed in several laboratories, including ours.

5. Lab-on-a-chip electrochemical detection of explosives

Microfluidic analytical devices, which combine multiple sample-handling processes with the actual measurement step on a microchip platform, have received a considerable recent interest [24,25]. Complete assays, involving sample pretreatment (e.g., preconcentration/extraction), chemical/biochemical derivatization reactions, electrophoretic separations, and detection, can thus be accomplished on a single microchip platform. Because of this high degree of integration (of multiple functional elements and processes), these microfabricated devices are referred to as 'lab-on-a-chip' devices. These analytical microsystems rely on electrokinetic fluid 'pumping' and obviate the need for pumps or valves. Highly effective separations combined with short assay times have been achieved by combining long separation channels and high electric fields. Owing to their versatility, efficiency, speed, high degree of integration, miniaturization, and ability to handle nanoliter volumes, microchip platforms have proven themselves as attractive vehicles for 'counter-terrorism' assays, in general, and for fast and effective separation of explosive substances, in particular.

Electrochemistry offers great promise for such microsystems, with features that include high sensitivity (approaching that of fluorescence), inherent miniaturization and integration of both detector and control (potentiostatic) instrumentation, compatibility with advanced microfabrication and micromachining technologies, low power and cost requirements, and independence of optical path length or sample turbidity [26–28]. These properties make electrochemical detection extremely attractive for creating truly portable (and possibly disposable) standalone field-deployable microsystems. Various detector configurations, based on different capillary/working electrode arrangements and positions of the electrode relative to the flow direction, have been proposed to meet these requirements. The common characteristic of most of these configurations is the alignment of the detector at the exit of the separation channel. Electrochemical detection for microchip devices can work on different principles. Most commonly, it is performed by controlling the potential of the working electrode at a fixed value and monitoring the current as a function of time. Such fixed-potential amperometric measurements have the advantages of ease of operation and freedom of background current contributions. Negative potentials are particularly useful for monitoring reducible (nitro-containing) explosive compounds.

Our group and Hilmi and Luong have developed several effective capillary electrophoresis (CE)/amperometric microchip protocols for detecting nitroaromatic explosives down to the part-per-billion level [29,30]. Such amperometric detection relies on the application of a fixed negative potential at the working electrode and on subsequently monitoring the reduction current as a function of time. The current response generated in this manner reflects the concentration profiles of these explosives as they pass over the detector. For example, Fig. 11 displays microchip-based electropherograms for mixtures containing increasing levels of TNT and DNB in 200-ppb steps. A surfactant, such as sodium dodecyl sulfate (SDS), is commonly added to the run buffer to facilitate the separation of the neutral nitroaromatic explosives. An end-channel thick-film carbon electrode detector was used for this task. Hilmi and Luong [29] employed a gold working electrode, formed by electroless deposition onto the chip capillary outlet, for highly sensitive amperometric detection of nitroaromatic explosives (with a detection limit of 24 ppb TNT). Analysis of a mixture of four explosives (TNT, 2,4-DNT, 2,6-DNT, and 2,3-DNT) was accomplished within 2 min through the use of a borate/SDS run buffer at a detection potential of −0.8 V. This microchip system was successfully applied for the analysis of explosive compounds in groundwater and soil samples [30]. Such microchip system is particularly useful for monitoring explosive contamination in military sites, explosives-producing plants, and ammunition factories.

Convenient distinction between ‘total’ and ‘individual’ explosive compounds has been accomplished in connection with chip-based ‘flow-injection’ (fast screening) and ‘separation’ (fingerprint identification) operation modes [31]. The realization of this dual screening/identification mode protocol using a single microchannel chip manifold involved the rapid switching from a run buffer that did not contain SDS to an SDS-containing buffer (Fig. 12, right). As micellar electrokinetic chromatography (MEKC) separation of neutral (nitroaromatic) compounds requires the addition of a surfactant,

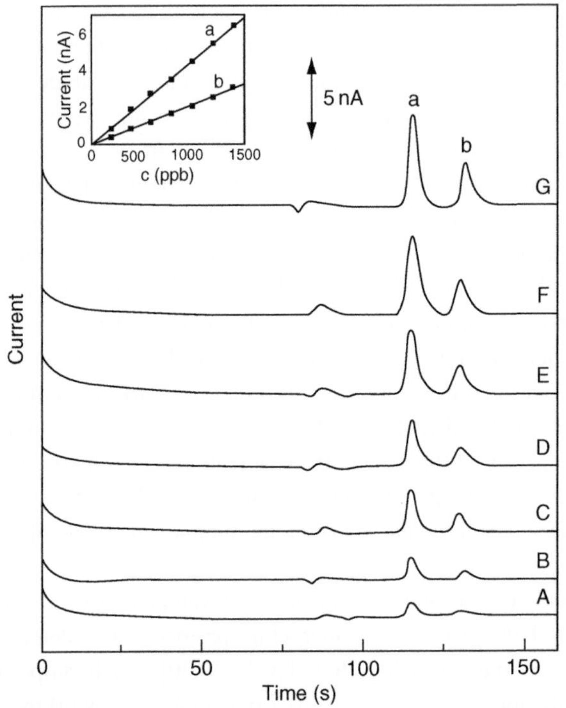

Fig. 11. Microchip electropherograms for mixtures containing increasing levels of DNB and TNT in 200-ppb steps, along with the resulting calibration plots. Thick-film amperometric carbon detector was held at −0.7 V; borate buffer (15 mM, pH 8.7) contained 25 mM sodium dodecyl sulfate (SDS).

such as SDS, an operation without SDS leads to high-speed ('flow-injection') measurements of the 'total' content of these hazardous compounds. As desired for various security scenarios, this allowed for the generation of repetitive fast screening assays of the 'total' content of explosive compounds and for a switch to detailed fingerprint identification once such substances were detected. Figure 12 (left) illustrates such 'total' and 'individual' measurements for a mixture of nitroaromatic organic explosives.

In addition to their effectiveness in detecting high-energy organic explosives, microchip devices offer great promise for the separation and detection of ionic components of low explosives [32]. A contactless-conductivity detection system has been particularly useful for this task. This detector can sense all ionic species with conductivity values different from that of the background electrolyte. The low electroosmotic flow (EOF) of the poly(methylmethacrylate) (PMMA) chip material facilitated the rapid switching between analyses of explosive-related cations and anions using the same microchannel and run buffer; this led to rapid (<1 min) measurements of seven explosive-related cations and anions down to the low micromolar level. Addition of an 18-crown-6 ether modifier has been used to separate the peaks of co-migrating potassium and ammonium ions. In addition to sequential injection of anionic and cationic

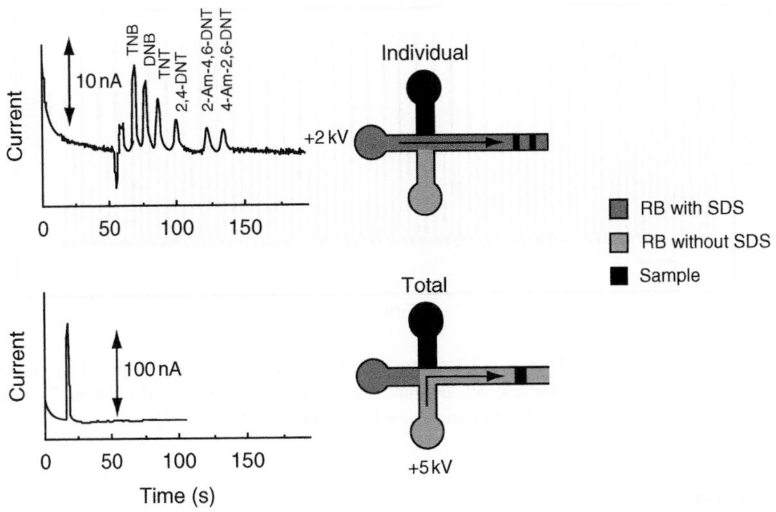

Fig. 12. 'Total' and 'individual' capillary electrophoresis (CE) microchip measurements of nitroaromatic organic explosives (based on rapid switching between flow-injection and separation modes, respectively). Reproduced with permission from Wang et al. [31].

residues, it is possible to use a special chip-based dual-end opposite injection protocol for simultaneous measurements of explosive-related cations and anions [33]. For this purpose, mixtures of cations and anions were injected simultaneously from both sides of the chip to the separation channel. The cations and anions thus migrated in opposite directions and were detected in the center of the separation channel by a movable contactless-conductivity detector. Simultaneous measurements of explosive-related ions and nerve-agent degradation products were also demonstrated.

Lab-on-a-chip devices have been used for a wide variety of explosive analyses. However, integrating on chip sampling into these explosive measurements is still in the preliminary stages. A new 'macro-to-micro' chip interface based on a sharp sample inlet that allowed rapid, convenient, and reproducible introduction of continuously flowing explosive sample stream into narrow chip microchannels was developed [34,35]. Such simple and yeteffective interface facilitates the use of hydrodynamically pumped large-volume samples without perturbing the CE separation and hence the realization of on-line chip-based explosive monitoring. For example, Fig. 13 displays the response for 80 alternate flow-injection measurements of 10 and 5 ppm TNT solutions at ahigh sample throughput of approximately 100 samples per hour.

The power and utility of such microfluidic explosive assays will be greatly enhanced by integrating additional sample processing functions (e.g., preconcentration) into their protocol. On-going collaborative efforts by several laboratories (including ours) are aimed at developing a self-contained completely functional multi-channel 'counter-terrorism' field-portable (hand-held) microanalyzer for providing early and timely simultaneous detection of different classes of explosives and chemical warfare agents.

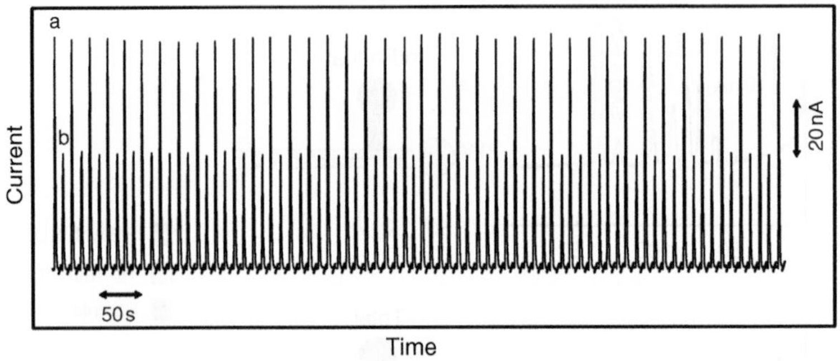

Fig. 13. Rapid capillary electrophoresis (CE) microchip measurements of TNT obtained by direct sample introduction into the separation channel: 80 alternate injections of (a) 10 and (b) 5 ppm TNT solutions. Reproduced with permission from Chen and Wang [34].

6. Conclusions

This chapter has illustrated the power and versatility of modern electrochemical devices for detecting explosives. Electrochemical devices offer attractive opportunities for addressing the growing needs for sensing of explosives. Such devices would allow field testing for major explosives to be performed more rapidly, sensitively, inexpensively, and reliably and should greatly facilitate the realization of in situ detection of explosive compounds. The resulting real-time monitoring capability should thus have a major impact on the way explosive materials are monitored and on the prevention of terrorist activity. Although early efforts have focused on liquid-phase electrochemical measurements of nitroaromatic explosive compounds, current activity focuses on vapor-phase explosive detection.

Acknowledgments

The author thanks the coworkers and collaborators who have contributed to the research on electrochemical sensors, detectors, and microsystems. This work was supported by grants from the ONR, DHS, DOE, and EPA.

References

[1] J. Yinon, *Trends Anal. Chem.*, 21 (2002) 292.
[2] J. Wang, *Analytical Electrochemistry* (3rd Ed.), John Wiley, New York, 2006.
[3] J. Wang, *Acc. Chem. Res.*, 35 (2002) 811.
[4] K. Bratin, P. Kissinger, R. Briner, C. Bruntlett, *Anal. Chim. Acta*, 130 (1981) 295.
[5] J. Wang, E. Ouziel, C. Yarnitzky, M. Ariel, *Anal. Chim. Acta*, 102 (1978) 99.
[6] D.M. Gilbert, T.C. Gale, *Environ. Sci. Technol.*, 39 (2005) 9270.
[7] W.J. Bowyer, D.H. Evans, *J. Org. Chem.*, 53 (1978) 5234.

[8] J. Wang, S. Thongngamdee, D. Lu, *Electroanalysis*, 18 (2006) 971.
[9] J. Wang, S.B. Hocevar, B. Ogorev, *Electrochem. Commun.*, 6 (2004) 176.
[10] H. Zhang, A. Cao, J. Hu, L. Wan, S. Lee, *Anal. Chem.*, 78 (2006) 1967.
[11] K. Albert, N.S. Lewis, C. Schaurer, G. Soltzing, S. Stitzel, T. Vaid, D.R. Walt, *Chem. Rev.*, 100 (2000) 2595.
[12] J. Wang, F. Lu, D. MacDonald, J. Lu, M.E.S. Ozsoz, K.M. Rogers, *Talanta*, 46 (1998) 1405.
[13] S. Wring, J. Hart, *Analyst*, 117 (1992) 1281.
[14] K.C. Honeychurch, J.P. Hart, P.R. Pritchard, S.J. Hawkins, N.M. Ratcliffe, *Biosens. Bioelectron.*, 19 (2003) 305.
[15] J.C. Chen, J.L. Shih, C.H. Liu, M.Y. Kuo, J.M. Zen, *Anal. Chem.*, 78 (2006) 3752.
[16] J. Wang, S. Thongngamdee, *Anal. Chim. Acta*, 485 (2003) 139.
[17] J. Wang, R.K. Bhada, J. Lu, D. MacDonald, *Anal. Chim. Acta*, 361 (1998) 85.
[18] J. Wang, S. Thongngamdee, J. Zima, in preparation.
[19] J. Wang, *Trends Anal. Chem.*, 16 (1997) 84.
[20] J. Wang, S. Thongngamdee, A. Kumar, *Electroanalysis*, 16 (2004) 1232.
[21] X. Fu, R. Benson, J. Wang, D. Fries, *Sens. Actuators B Chem*, 106 (2005) 296.
[22] R. Knake, P. Jacquinot, A.W. Hodgson, P. Hauser, *Anal. Chim. Acta*, 549 (2005) 1.
[23] M. Krausa, K. Schorb, *J. Electroanal. Chem.*, 461 (1999) 10.
[24] D.R. Reyes, D. Iossidis, P.A. Aurox, A. Manz, *Anal. Chem.*, 74 (2002) 2623.
[25] P.A. Aurox, D.R. Reyes, D. Iossidis, A. Manz, *Anal. Chem.*, 74 (2002) 2637.
[26] J. Wang, *Talanta*, 56 (2002) 223.
[27] J. Wang, *Electroanalysis*, 17 (2005) 1133.
[28] J. Wang, B. Tian, E. Sahlin, *Anal. Chem.*, 71 (1999) 5436.
[29] A. Hilmi, J.H.T. Luong, *Anal. Chem.*, 72 (2007) 4677.
[30] A. Hilmi, J.H.T. Luong, *Environ. Sci. Technol.*, 34 (2000) 3046.
[31] J. Wang, M. Pumera, M.P. Chatrathi, A. Escarpa, M. Musameh, *Anal. Chem.*, 74 (2002) 1187.
[32] J. Wang, M. Pumera, G.E. Collins, F. Opekar, I. Jelinek, *Analyst*, 127 (2002) 719.
[33] J. Wang, G. Chen, A. Muck Jr., G.E. Collins, *Electrophoresis*, 24 (2003) 3728.
[34] G. Chen, J. Wang, *Analyst*, 129 (2004) 507.
[35] J. Wang, W. Siangproh, S. Thongngamdee, O. Chailapakul, *Analyst*, 130 (2005) 1390.

Chapter 5

Explosive Vapor Detection Using Microcantilever Sensors

Larry Senesac and Thomas Thundat

Nanoscale Science and Devices Group, Oak Ridge National Laboratory, Oak Ridge, TN 37831, USA

Counterterrorist Detection Techniques of Explosives
Jehuda Yinon (Editor)

Contents

1. Introduction

Explosive-based terrorism is an eminent threat to a civilized and free society. Accurate and cost-effective explosive sensors are, therefore, essential for combating the terrorist threat. Some of the main performance characteristics for explosive sensors include sensitivity, selectivity, and real-time fast operation. As the vapor pressures of commonly used explosives are extremely small, highly sensitive sensors are essential for detecting trace quantities of explosives. Moreover, the sensors should have high selectivity to have an acceptable rate of false positives. Also, these sensors should have the capability of mass deployment because of the breadth of terrorist threats involving explosives [1]. Finally, these sensors should have fast detection and regeneration time for fast operation. Currently available sensors are unable to satisfy these requirements.

Silicon microcantilever sensors that can be mass-produced using currently available microfabrication techniques, however, have the potential to satisfy the conditions of sensitivity, miniature size, low power consumption, and real-time operation [2]. Microcantilevers are generally micromachined from silicon wafers using conventional techniques. Typical dimensions of a micromachined cantilever are 100 μm in length, 40 μm in width, and 1 μm in thickness. The primary advantage of a cantilever beam originates from its ability to sensitively measure displacements with sub-nanometer precision. Sensitive detection of displacement leads to sensitive detection of forces and stresses.

Microcantilever sensor technology has been demonstrated in sensitive detection of chemical, physical, and biological analytes [3,4]. Microcantilever sensors can be operated in dynamic mode where mass loading due to molecular adsorption is monitored as a variation in resonance frequency of the cantilever or in static mode where adsorption-induced surface stress is monitored as bending of a cantilever. In the static mode, adsorption of molecules on one side of the cantilever makes the cantilever bend [5,6]. The bending mode of operation is unique to thin structures where adsorption-induced forces cause the cantilever to bend if the adsorption is confined to a single side. Differential adsorption on the surfaces results in differential stress. Low-frequency, low-spring-constant cantilevers result in large bending due to adsorption-induced forces. In addition to bending, the resonance frequency of the cantilever varies sensitively as a function of mass loading. Resonance frequency mode is very similar to the operation of other gravimetric sensors, such as quartz crystal microbalance (QCM) and surface acoustic wave (SAW) transducers. The sensitivity of the dynamic mode of operation is directly related to the frequency of the cantilever; the higher the frequency, the higher the sensitivity. For the cantilevers used in our experiments, the bending mode sensitivity is much higher than that of the dynamic mode, and therefore, we will discuss only bending mode of operation of the cantilever in this chapter. Figure 1 shows a schematic diagram of cantilever bending due to differential molecular adsorption.

Microcantilever-based sensing satisfies many requirements for an ideal explosive sensor. Microcantilever sensors have extremely high sensitivity and are compatible with array arrangement for simultaneous detection of multiple analytes. They have the advantages of low power consumption and miniature size. However, microcantilever

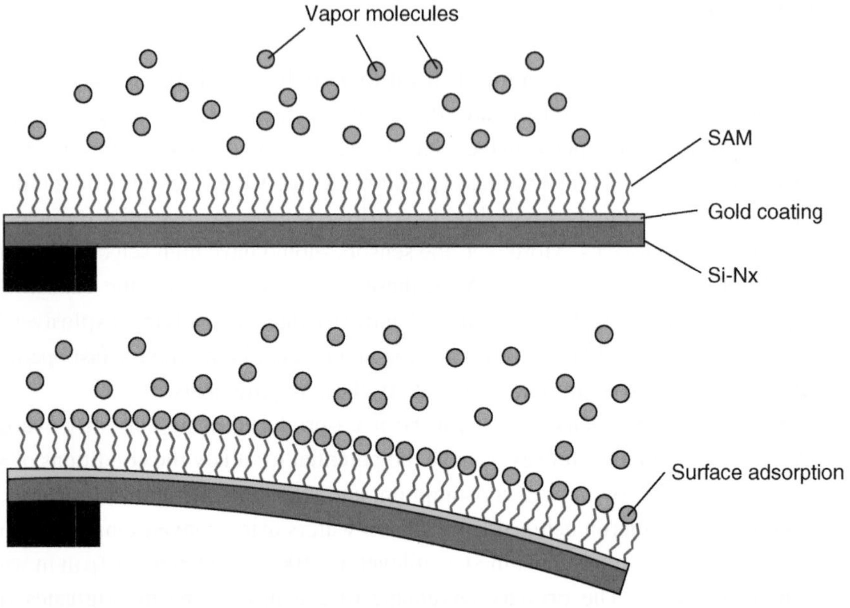

Fig. 1. Schematic diagram of a cantilever bending due to differential adsorption of analyte molecules.

sensors lack chemical selectivity. Chemical selectivity is accomplished using selective layers based on receptor-based detection.

2. Theory

2.1. Cantilever bending

The microcantilever is an ideal displacement sensor. The ability to detect the motion of a cantilever beam with nanometer precision makes the cantilever ideal for measuring bending. Cantilever bending can be related to adsorption/desorption of molecules through adsorption forces. As molecular reactions on a surface are ultimately driven by free energy reduction of the surface, the free energy reduction leads to a change in surface stress. Although they would produce no observable macroscopic change on the surface of a bulk solid, the adsorption-induced surface stresses are sufficient to bend a cantilever if the adsorption is confined to one surface. Adsorption-induced forces, however, should not be confused with bending due to dimensional changes such as swelling of thicker polymer films on cantilevers. The sensitivity of adsorption-induced stress sensors can be three orders of magnitude higher than those of frequency variation mass sensors (for resonance frequencies in the range of tens of kilohertz) [3].

Microcantilever deflection changes as a function of adsorbate coverage when adsorption is confined to a single side of a cantilever (or when there is differential adsorption on opposite sides of a cantilever). As we do not know the absolute value of the initial surface stress, we can only measure its variation. A relation can be derived between cantilever bending and changes in surface stress from Stoney's formula and equations that describe cantilever bending [7]. Specifically, a relation can be derived between the radius of curvature of the cantilever beam and the differential surface stress:

$$\frac{1}{R} = \frac{6(1-\nu)}{Et^2}\delta s, \tag{1}$$

where R is the radius of curvature of the cantilever, ν and E are Poisson's ratio and Young's modulus for the substrate, respectively, t is the thickness of the cantilever, and $\delta s = \Delta\sigma_1 - \Delta\sigma_2$ is the differential surface stress.

Surface stress, σ, and surface free energy, γ, can be related using the Shuttleworth equation [8]:

$$\sigma = \gamma + \frac{\partial \gamma}{\partial \varepsilon}. \tag{2}$$

The surface strain, $\partial\varepsilon$, is defined as the ratio of change in surface area, to original surface area $\partial\varepsilon = (\mathrm{d}A/A)$. As the bending of the cantilever is very small compared with the length of the cantilever, the strain contribution is only in the part-per-million (10^{-6}) range. Therefore, one can possibly neglect the contribution from surface strain effects and equate the free energy change to surface stress variation [9].

Using Eq. 2, a relationship between the cantilever deflection, z, and the differential surface stress, δs, is obtained as

$$z = \frac{3L^2(1-\upsilon)}{Et^2}\delta s, \tag{3}$$

where L is the cantilever length. Therefore, the deflection of the cantilever is directly proportional to the adsorption-induced differential surface stress. Surface stress has units of N/m or $\mathrm{J/m^2}$. Equation 3 shows a linear relation between cantilever bending and differential surface stress.

Adsorption-induced forces are applicable only to monolayer films and, as mentioned earlier, should not be confused with bending due to dimensional changes such as swelling of thicker polymer films. It should also not be confused with deflection due to the weight of the adsorbed molecules. The deflection due to weight is extremely small—for example, for a cantilever with a spring constant of 0.1 N/m, the bending due to weight of 1 ng of adsorbed material will be 0.1 nm.

The minimum detectable signal for cantilever bending depends on the geometry and the material properties of the cantilever. For a silicon nitride cantilever that is 200 μm long and 0.5 μm thick, with $E = 8.5 \times 10^{10}\,\mathrm{N/m^2}$ and $\nu = 0.27$, a surface stress of $0.2\,\mathrm{mJ/m^2}$ will result in a deflection of 1 nm at the end. Because a cantilever deflection strongly depends on geometry, the surface stress change, which is directly related to molecular adsorption on the cantilever surface, is a more convenient quantity of the reactions for comparison of various measurements. Changes in free energy density in biomolecular reactions are usually in the range of 1–50 mJ/m^2 but can be as high as 900 mJ/m^2.

2.2. *Thermal motions of a cantilever*

As the cantilevers are very small structures, they execute thermal (Brownian) motion. The longer the cantilever, the more sensitive it is for measuring surface stresses. However, increasing the length also increases the thermal vibrational noise of the cantilever, which from statistical physics is [10,11]

$$\delta_n = \sqrt{\frac{2k_B TB}{\pi k f_0 Q}}. \tag{4}$$

Here, k_B is the Boltzmann constant (1.38×10^{-23} J/K), T is the absolute temperature (300K at room temperature), B is the bandwidth of measurement (typically ~1000 Hz for dc measurement), k is the cantilever spring constant, f_0 is the resonant frequency of the cantilever, and Q is the quality factor of the resonance, which is related to damping. It is clear from Eq. 4 that lower spring stiffness produces higher thermal noise. Higher thermal noise leads to poor signal-to-noise ratio. This thermal motion, however, can be used as an excitation technique for resonance frequency mode operation.

3. Apparatus

3.1. *Cantilevers*

Cantilevers are usually microfabricated from silicon by using conventional photomasking and etching techniques. Typical dimensions of a cantilever are 100 μm (length), 40 μm (width), and 1 μm (thickness). Silicon and silicon nitride cantilevers and cantilever arrays that utilize optical beam deflection for signal transduction are commercially available. Piezoresistive cantilever arrays are also commercially available. Piezoresistive cantilevers are 120 μm in length, 40 μm in width, and 1 μm in thickness.

3.2. *Readout techniques*

There exist a number of readout techniques based on optical beam deflection, variation in capacitance, piezoresistance, and piezoelectricity. Piezoelectricity is more suited for a detection method based on resonance frequency than the method based on cantilever bending. The capacitive method is not suitable for liquid-based applications. The piezoresistive readout has many advantages, and it is ideally suited for handheld devices.

The piezoresistive readout method has been gaining attention recently. Doped silicon exhibits a strong piezoresistive effect. The resistance of a doped region on a cantilever can change reliably when the cantilever is stressed with deflection [12]. Boisen et al. [13] developed piezoresistive cantilever sensors with integrated differential readout. Each cantilever had a thin resistor element made of doped silicon [13]. The silicon is fully encapsulated in silicon nitride for electrical insulation. The thickness of silicon nitride coating on the doped silicon is adjusted in such a way that the doped silicon resistor is asymmetric with respect to the neutral axis of the cantilever. As the cantilever bends, the resistance of the doped silicon varies linearly because of its asymmetric location with respect to the neutral axis of the entire cantilever. Each sensor was composed of a measurement cantilever and a built-in reference cantilever, which enabled differential signal readout. The two cantilevers were connected in a Wheatstone bridge, and the surface stress change on the measurement cantilever was detected as the output voltage from the Wheatstone bridge. The typical signal-to-noise ratio of the resistance measurement was 26 during the experiments. For cantilevers that were 120 μm long, 40 μm wide, and 1.3 μm thick, the sensor was determined to have a minimum detectable surface stress change of approximately $5\,mJ/m^2$. Figure 2 shows an array of piezoresistive cantilevers [14].

The piezoresistive readout technique has several advantages over commonly used optical beam deflection methods. For example, optical beam deflection probes the bending of the free end of the cantilever. It is assumed that the bending is uniform along the length of the cantilever. The piezoresistive method, however, measures the integrated bending of the cantilever. Piezoresistive cantilevers can be encapsulated in silicon nitride

Fig. 2. Scanning electron micrograph of a piezoresistive cantilever array.

for operation under solution, thus avoiding the long-standing problems associated with optical path lengths and variations in refractive index. In addition, because no external optical components are required, the electronic readout is more amenable to miniaturization and is ideal for portable devices. An electronic readout is compatible with array arrangements because both cantilevers and readout circuits can be fabricated simultaneously on the same chip. However, currently available piezoresistive cantilever sensors are an order of magnitude less sensitive than those using optical readout techniques. This discrepancy in sensitivity, however, is vanishing because of recent progress in piezoresistive cantilever development.

3.3. Vapor generators

Producing highly calibrated, trace quantities of explosive vapor is a challenging task. As the vapor pressures of most explosives are extremely small at room temperature, their vapors are often produced by maintaining the sources at higher temperatures. This leads to the condensation of vapor at cold spots, which should be avoided to deliver highly calibrated quantities of explosive vapors to the cantilever sensor.

We have used a custom-built vapor generator that can deliver precise concentrations of explosive vapors in these experiments. The explosive vapors were produced from the headspace vapors from granules of explosives deposited on thickly packed glass wool kept at a constant temperature. These explosive vapors were then mixed as needed with vapors of interferents such as water, alcohol, and acetone, prior to exposure to cantilevers. As the vapors were created at temperatures higher than room temperature, the delivery lines were heated to avoid condensation in these lines. We have also used a vapor generator developed at Idaho National Laboratory (INL) to generate the PETN, RDX, and TNT vapor streams. Flowing ambient air through a reservoir containing explosives (PETN, TNT, or RDX) kept at a constant temperature generated the vapor stream. The reservoirs of the vapor generators consisted of 0.1 g of explosives dissolved in acetone and deposited on glass wool contained in a stainless steel block. The temperature of the reservoir was controlled by two thermoelectric elements that cooled or heated the reservoir, generating a saturation vapor pressure within it. All the experiments were conducted at a constant flow rate of 100 standard cubic centimeters per minute (sccm) to eliminate the parasitic cantilever deflection that would have been induced by variations in flow. In some experiments involving RDX and PETN, the vapor generators were operated at 50°C to produce enough vapor to be delivered to the cantilever chamber. The vapor concentrations for PETN and RDX were 1.4 and 0.3 ppb, respectively, at 50°C [15].

3.4. Sensitivity and false positives

The selectivity of the sensor is determined by the efficiency of the coating material. Developing selective coatings is an ongoing research area. Figure 3 shows a useful metric for sensor performance called the receiver operating characteristic (ROC) curve [16].

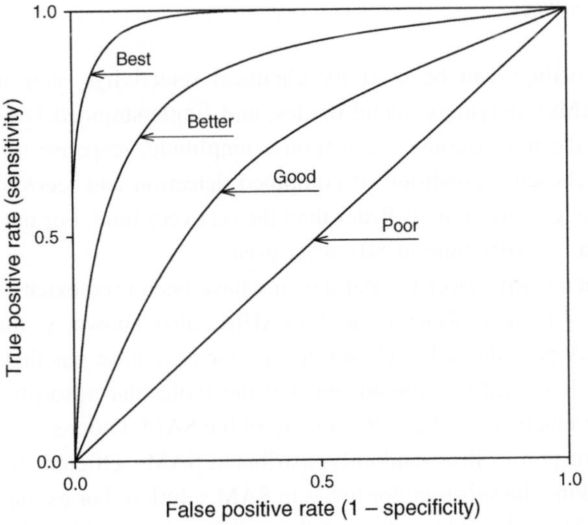

Fig. 3. Receiver operating characteristic (ROC) curve.

The y-axis of the ROC curve shows the rate of true positives (related to sensitivity), whereas the x-axis shows the rate of false positives (1 – specificity). A poor sensor response will lie along the diagonal. An ideal sensor performance curve lies on the y-axis, that is, zero rate of false positives and high sensitivity. However, the performance of most current sensors lies between the y-axis and the $y = x$ diagonal line as shown in Fig. 3.

The key to achieving chemical selectivity lies in using coatings that can take the response curve away from the $y = x$ line toward the y-axis. In the case of microcantilevers, the ability to functionalize one surface of the microcantilever with receptive molecules in such a way that the explosive molecules preferentially bind to the treated surface results in high selectivity. Choosing receptor molecules that can provide highest affinity, therefore, can control the selectivity of detection. Another important requirement for a sensor system is fast regeneration (recovery), so that the sensor can be used repetitively. Satisfying these two conditions requires developing coatings that interact with explosives with interaction energies that are close to the thermal energy, $k_B T$, where k_B is the Boltzmann constant and T is the room temperature. This condition, however, severely restricts the types of interactions that one can choose in designing specific coatings. This also results in high false positives as there exists a large number of interfering molecules that can interact with the coating materials.

This problem may be solved by using cantilever arrays where each cantilever element produces a unique response due to the chemically specific coatings used. The higher the uniqueness of an individual cantilever, the higher the specificity and the lower the number of cantilevers needed in an array. The design goal for the array is to move away from the $y = x$ diagonal in the ROC curve toward the y-axis.

3.5. Selective coatings

A number of coatings can be used for chemical selectivity, such as self-assembled monolayers (SAMs), polymers, metal oxides, and single-stranded DNA. The choice of coatings must take into account the response amplitude, response time, and recovery time. We have chosen a condition of combined detection and recovery time less than 1 min. Often, the response time is faster than the recovery time. For our case, a response time of 10 and a recovery time of 50 s were used.

One of the chemically selective coatings that have been used extensively in our study was a SAM of 4-mercaptobenzoic acid (4-MBA; also known as thiosalicylic acid). SAMs of 4-MBA provide carboxyl end groups for acid–base reactions with explosive molecules [17]. It should be pointed out that the molecular adsorption-induced stress generation is extremely sensitive to the quality of the SAM. Ordered, well-formed SAMs produce higher responses than randomly distributed SAMs. Often, it is necessary to age the SAM by soaking the substrate for hours in SAM solution. For example, the formation of a 4-MBA SAM on the gold surface of the cantilever was achieved by immersing the cantilever into a 6×10^{-3} M solution of 4-MBA (97% from Aldrich Chemical Company) in absolute ethanol for 2 days. Prior to SAM immobilization, the cantilevers were thoroughly cleaned and coated on one side with 3 nm of titanium followed by 30 nm of gold. Upon removal from the solution, the cantilever was rinsed with ethanol and then dried before use in the experiments. The monolayer coating was shown to be quite stable for several months under normal operating conditions. We have also found that electrochemical deposition of SAMs reduces the aging time of the SAMs to minutes. The 4-MBA SAM on gold-coated cantilever has a pK_a in the range of 5–7 and binds with strongly basic groups such as nitro-substituted molecules of explosive vapors.

For the pattern recognition (PARC) approach, we have coated the piezoresistive cantilevers with different selective layers. Each piezoresistive cantilever had four cantilever elements. Two of these cantilevers were coated with gold, whereas the other two served as reference cantilevers. We have used four separated chips arranged into an array in a single vapor chamber. Each cantilever chip was coated with a different selective agent. The four coatings used in our study include 4-MBA, Au (evaporated), $CH_3(CH_2)_{11}$-SH, and a complex of β-cyclodextrin and alkane.

4. PARC for analyte identification

4.1. Background

Coatings that can identify explosive molecules with no interference do not exist at present time. This requires the use of partially selective coatings on arrays that can provide responses that are orthogonal or close to orthogonal. For each analyte vapor, an ideal detector array produces a unique chemical response signature. A catalog of these signatures may be produced for a large number of analytes of interest, and this catalog may then be used for the identification of unknown vapors. For single-component analytes, this task is relatively straightforward, but for mixtures of several different analytes, a sophisticated computational algorithm is required.

Initially, PARC algorithms for this purpose were based on linear system theory. Algorithms such as linear discriminant analysis (LDA), principle component analysis (PCA), and cluster analysis (CA) were used because the action of these linear algorithms on the data is easier to visualize and understand than for non-linear techniques. These linear algorithms map the response data from the individual analytes onto clusters of points in a multidimensional space, where each cluster corresponds to a different analyte. As the number of analytes increases, this invariably leads to an overlap of some of the clusters, resulting in a region where the mapping of a point no longer produces a unique classification. Similarly, as the concentration of the analytes decreases, more of the clusters overlap until at some low concentration they all overlap and no analyte identification is possible. Although algorithms based on these linear techniques have been successful at recognizing fairly large numbers of individual analytes and many binary mixtures [18], they have never been successful for the identification of mixtures of more than two components. For explosive detection, it will be necessary to identify not only single compounds such as TNT, PETN, and RDX individually but also when they are in the presence of associated compounds and common environmental interferents. For these multicomponent mixtures, an approach utilizing non-linear algorithms will be required.

The latest PARC algorithms [19,20] include non-linear systems that mimic natural systems. Artificial neural networks (ANNs) were created to mimic the neuronal networks of the human brain. ANN algorithms [21] vary in structure, in learning methodology, and in their performance strengths and weaknesses. The most popular and versatile ANN is the multi-layer perceptron (MLP) with back-propagation error correction (BP-EC), but others include probabilistic neural networks (PNNs), neuro-fuzzy systems (NFSs), learning vector quantization (LVQ), radial basis transfer function (RBF), and the newest algorithm, support vector machines (SVMs). Several reviews comparing the characteristics of the different PARC algorithms conclude that the best choice of algorithm is application specific and that hybrid approaches (algorithms combining two or more of the above techniques) show the greatest potential [18,22–24]. It is therefore necessary to develop a PARC algorithm that can meet the specific project goals.

4.2. Analyte signature pattern

For the purpose of analyte identification, the response of the detector array to exposure to the analyte vapor must be recorded, and a unique pattern or chemical signature for each analyte must be extracted from the data. A good signature pattern highlights the characteristic features of the data that yield high differentiation between patterns and eliminates those features that provide little or no differentiation information.

This process of feature selection [20] begins by preprocessing the raw data to remove artifacts such as outliers and noise that do not reproduce from one data run to another, making the patterns inconsistent. An example of an outlier would be a typical data spike where one data point may be significantly outside the average data value. The most common noise in the data may come from the electronics used to read and record the data. For microcantilever sensors, the relevant response signal takes place over a period of a few seconds, where by contrast the noise has a relatively high-frequency characteristic. A Fourier high-frequency filter such as a Fourier–Gaussian convolution filter may be used to remove most or all of the noise from the data while causing little or no distortion of the underlying low-frequency response. As rapid changes in signal amplitude require higher frequency Fourier components, caution must be taken to reduce the unwanted noise without severely attenuating (or rounding off) sharp data transitions. An added bonus of this filtering is that the aforementioned outlier spikes in data are of very high frequency and are therefore severely attenuated from the signal.

After preprocessing the data, the search begins to find the features in the data that show the largest differences between patterns. This process is very much dependent on the type of detectors in the detection system but may involve comparing the relative amplitudes of the different detectors in the array, the derivative of the response, or even mathematical transforms of the data to select which features show the most differentiation between the patterns of different analytes. For our examples described in Section 5.3., we chose to use 120-data point analyte signature patterns sampled from the entire shape of the signal from a detector array of four cantilevers with coatings described in Section 3.5.

4.3. PARC – linear algorithm – CA

For the linear system approach to PARC, we will examine a simple form of CA [20]. We wish to plot the feature vectors of length N (our 120-data point patterns, $N = 120$) in an N-dimensional vector space. We started with three data sets (three signature patterns) for each of five known analytes. This plotting produces five clusters of three points each in the 120-dimensional vector space. The coordinates of the centers of the five clusters are calculated and stored in the CA algorithm. Once 'trained' in this manner, the CA algorithm can map new signature patterns in the 120-dimensional vector space and calculate each pattern's of RMS distance from the five stored cluster points. If the new point falls within a predetermined distance from one of the five cluster points, the pattern is identified with that cluster point analyte. For this type of analysis, the amplitude of the pattern is important. For our array of four detectors, the response to changes in analyte vapor concentration is fairly linear, so the shape of the response for a given analyte is unchanged by differences in analyte concentration, but the amplitude of the response can vary considerably. If uncorrected, this would lead the CA algorithm to map these patterns at large distances from any of the known analyte clusters. Therefore, for this type of algorithm, all the patterns are normalized between 0 and 1 to yield uniform mapping of the analyte signatures regardless of analyte concentration.

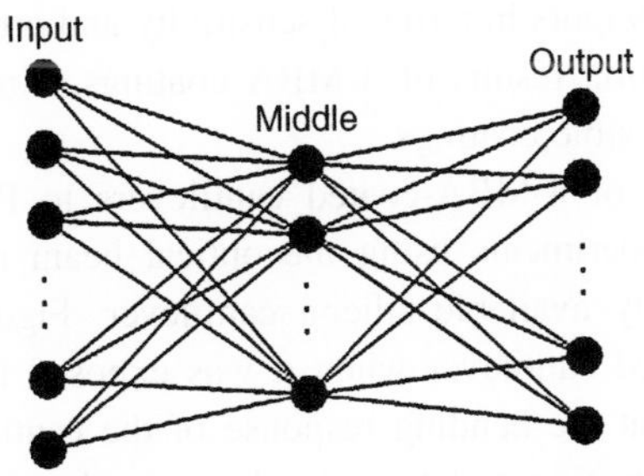

Fig. 4. Simple three-layer artificial neural network (ANN).

4.4. *PARC – non-linear algorithm – ANNs*

Our choice for the non-linear system approach to PARC is the ANN. The ANN is composed of many neurons configured in layers such that data pass from an input layer through any number of middle layers and finally exit the system through a final layer called the output layer. In Fig. 4 is shown a diagram of a simple three-layer ANN. The input layer is composed of numeric scalar data values, whereas the middle and output layers are composed of artificial neurons. These artificial neurons are essentially weighted transfer functions that convert their inputs into a single desired output. The individual layer components are referred to as nodes. Every input node is connected to every middle node, and every middle node is connected to every output node.

The ANN is trained using special sets of data called training sets. Each training set consists of an input set of data for the ANN and an output target set. The target set is simply the desired output of the ANN in response to the input set. The goal of training is for the ANN output to match the target set. In our case, the input set is the analyte signature pattern described in Section 4.2, and the target is a set of five numbers corresponding to the five analytes we wish to identify. The training procedure is as follows: the input data set is placed into the input layer of the ANN and is fed forward through the ANN to its output layer. The output is then compared with the desired target set, and the error is calculated. The source of these errors is then back-propagated (traced back through the neurons), and the weight factors are corrected for every neuron connection. This training process is repeated until the ANN output for every training input set converges in agreement with the corresponding target set. An ANN that incorporates this learning technique is referred to as a feed-forward BP-EC ANN [21].

5. Results and discussion

5.1. *Single coating*

To be used in an array, the coatings will have to be selected by testing out many different candidate coatings. We have tested each selective coating individually to characterize its response to explosive vapors in terms of sensitivity and orthogonality. Here, we will describe the characterization results of 4-MBA coatings. Similar studies were carried out on many different possible coatings.

To test the sensitivity of 4-MBA-coated cantilevers to PETN vapors, we carried out some preliminary experiments using the optical beam readout. In this case, we have used a commercially available silicon cantilever. Figure 5 shows the bending response of a SAM-coated cantilever when it was exposed to a PETN vapor stream. It is clear from Fig. 5 that the bending response of the cantilever to the PETN vapor exposure is extremely sensitive and fast. As the noise level of the bending response in these experiments is approximately 2 nm (3× standard deviation of the noise level), the detection resolution corresponding to Fig. 5 is approximately 0.1 ppb. Maximum bending of the cantilever is achieved within 20 s. The amount of PETN delivered by the generator in 20 s was approximately 700 pg. The adsorbed mass calculated from cantilever resonance frequency variation was approximately 100 pg. As the bending due to the weight of the adsorbed mass is around 4 pm, almost all of the bending is due to surface stress. The 200-nm deflection for 100 pg of adsorbed PETN corresponds to a limit of detection (LOD) of a few picograms [16].

Additionally, we carried out temporal response studies of the cantilever as a function of repeated exposure to PETN vapor. One of the important characteristics of a sensor,

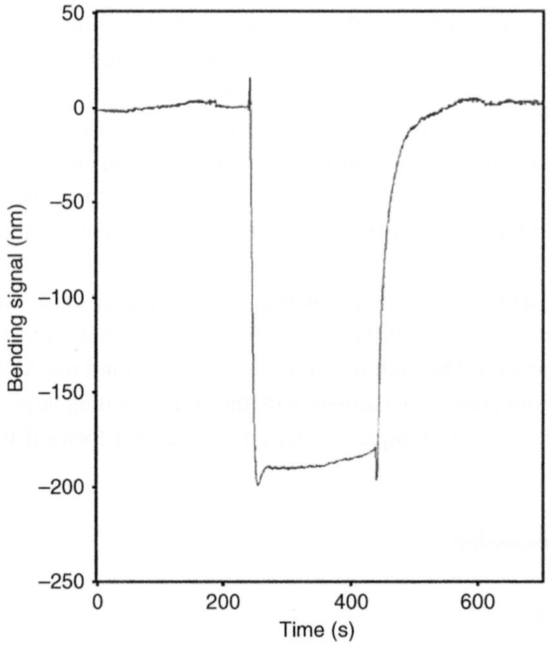

Fig. 5. The response of a 4-mercaptobenzoic acid (4-MBA)-coated silicon cantilever to a PETN vapors of 1.4 ppb concentration in ambient air. This experiment was performed using the optical readout.

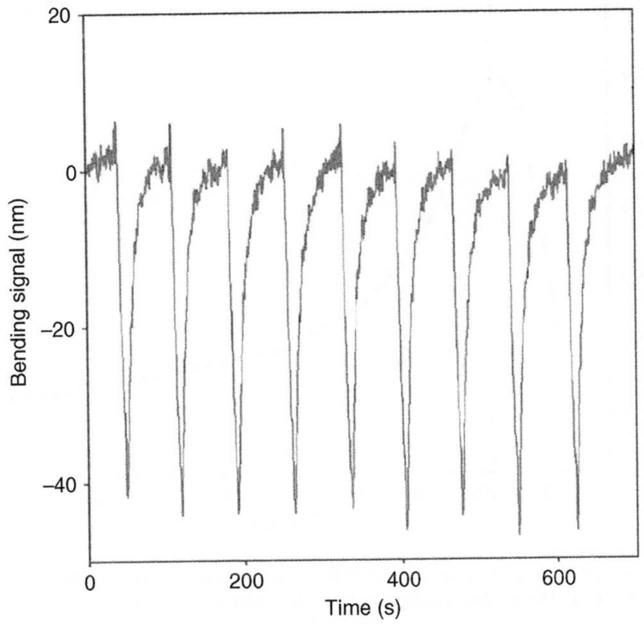

Fig. 6. Temporal response of a 4-mercaptobenzoic acid (4-MBA)-coated cantilever as a function of repeated exposure to PETN vapor.

in addition to selectivity and sensitivity, is its ability for regeneration. For useful applications as an explosive sensor, the cantilever should be able to regenerate itself within seconds for continued operation. Therefore, techniques based on receptors should utilize weak bonding that can be broken at room temperature. Figure 6 shows the temporal response of the cantilever when exposed to a stream of PETN vapor. It is obvious from the Figure that the cantilever responds within 10 s to a stream of PETN vapor at a sub-part-per-billion vapor concentration. When the PETN vapor stream is turned off, the cantilever returns to the original position within 60 s. Figure 5 clearly shows that the cantilever can be operated repeatedly without losing its sensitivity.

A test was also performed using a piezoresistive cantilever with the associated Wheatstone bridge readout. The cantilever was prepared with a coating of 4-MBA as described in Section 3.5. Figure 7 shows the bending response of the 4-MBA-coated piezoresistive cantilever as it was exposed to a 10 s pulse of PETN vapor. The Figure also shows the first 40 s of analyte desorption following the exposure. Again, we see the rapid response followed by a return of the cantilever to its original position.

The mechanism for cantilever bending is assumed to be adsorption-induced stress. The adsorption decreases the surface free energy, and surface free energy density is surface stress. Hydrogen bonding between the nitro groups of the explosive molecules and the hydroxyl group of 4-MBA may be responsible for the easily reversible adsorption of explosive vapors on the SAM-coated top surface of the cantilever. This hydrogen

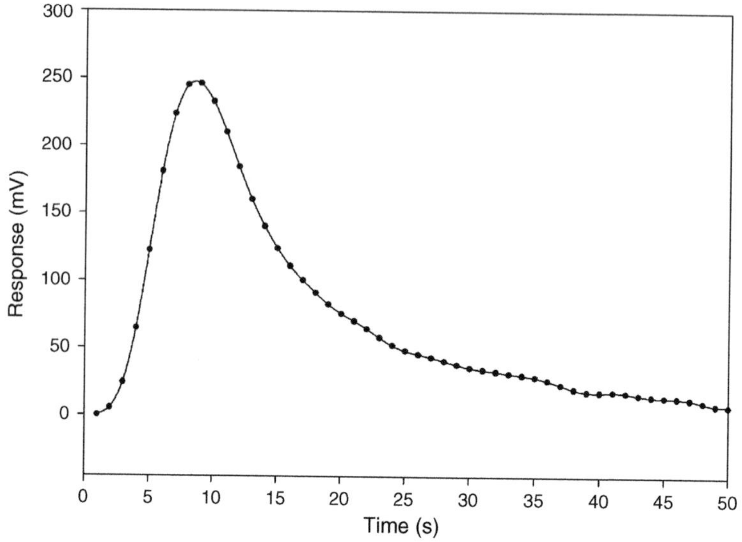

Fig. 7. Response of a 4-mercaptobenzoic acid (4-MBA)-coated piezoresistive cantilever as a function to the exposure to PETN vapor. The readout electronics were designed to produce a positive signal in response to cantilever bending.

bonding creates a differential surface stress because hydrogen bonding is confined to only one of the surfaces.

5.2. Four-coating array

Using just a 4-MBA receptor layer for selective detection of explosives will bring many false-positive signals because the fundamental mechanism is an acid–base reaction. However, there are many other SAMs with different head groups that could be used to provide orthogonal signals. A group of coatings that can provide partially selective binding are metals and oxides of metals deposited on cantilever surfaces. Other coatings we have experimented with include $CH_3(CH_2)_{11}$-SH and a complex of β-cyclodextrin and alkane. Once a number of these layers are identified, an array of cantilevers can be modified with different SAMs to create unique responses. The response from an orthogonal array can be analyzed with PARC techniques to identify the explosive molecules.

5.3. PARC: analyte signature patterns

In Fig. 8 are shown the chemical signature patterns for five known analytes and one unknown analyte. The patterns were extracted from the response of an array of four coated microcantilevers to exposure to each analyte vapor. The four cantilever coatings

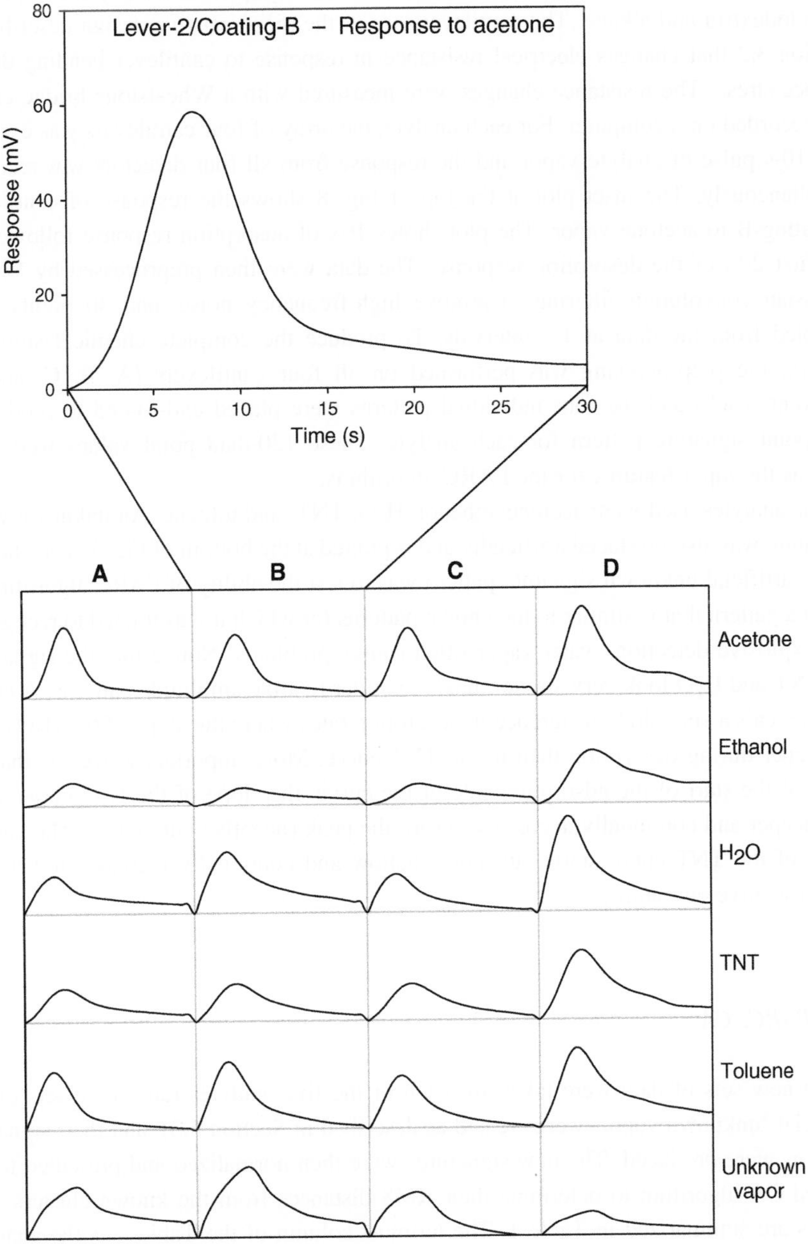

Fig. 8. Analyte chemical signature patterns. The inset at the top of the Figure shows the temporal response of one of the four cantilevers in the detector array. The responses for all four detectors were recorded simultaneously, and the signature pattern for each analyte was created by combining the four responses sequentially.

were (A) 4-MBA, (B) Au (evaporated), (C) $CH_3(CH_2)_{11}$-SH, and (D) a complex of β-cyclodextrin and alkane. The cantilevers are of the piezoresistive design described in Section 3.2 that changes electrical resistance in response to cantilever bending due to surface stress. The resistance changes were measured with a Wheatstone bridge circuit and recorded on a computer. For each analyte, the array of four cantilevers was exposed to a 10-s pulse of analyte vapor and the response from all four detectors was recorded simultaneously. The inset plot at the top of Fig. 8 shows the response of cantilever-2/coating-B to acetone vapor. The plot shows 10 s of adsorption response followed by the first 20 s of the desorption response. The data were then preprocessed by Fourier Gaussian convolution filtering to remove high-frequency noise, and 30 points were sampled from the data at 1-s intervals. To produce the complete chemical signature pattern, the preprocessing was performed on all four cantilevers (A, B, C, and D, 30 points each) and the four individual patterns were placed end-to-end to produce a 120-point signature pattern for each analyte. These 120-data point values were then used as the input features for the PARC algorithms.

The analytes used were acetone, ethanol, H_2O, TNT, and toluene. An unknown vapor signature was also produced artificially and is plotted at the bottom of Fig. 8. The purpose of the artificial unknown signature pattern was to test the ability of PARC algorithms to reject a pattern that is similar to the known patterns for which it was trained to recognize. For explosive detection, water vapor often causes problems. Notice that the signatures for TNT and H_2O look very similar, having similar relative amplitude patterns. A closer look reveals a very slight difference in desorption rates where the slope of the H_2O curve is steeper during desorption than in the TNT curve. More importantly, we see that for most of the start of the adsorption side of the curve, the slope of the H_2O curve starts out steeper and continually decreases toward the peak (negative curvature), whereas the slope of the TNT curve starts out more shallow and continually increases toward the peak (positive curvature).

5.4. PARC: CA

Three new sets of data were taken for each of the five analytes (and three sets of the artificial 'unknown vapor' were created as described in Section 5.3), and their signature patterns were produced. The new signatures were then normalized and presented to the trained CA algorithm to determine their RMS distances from the known clusters. The results are summarized in Table 1. The leftmost column of the Table lists the analytes (three rows each for the three data sets) to be identified by the CA algorithm, and the other columns list the distances of the analyte from the known clusters. If we assign a value of between 6.0 and 7.0 to be the threshold between positive identification and negative identification, we see that the algorithm correctly identifies all the analytes (the shaded cells in the Table) with no false positives. Also note that the unknown analyte was rejected, not mapping close to any of the known clusters. For explosive detection,

Table 1. Pattern recognition of analytes by cluster analysis

Analytes	Root mean square distance from center of each cluster				
	Acetone	Ethanol	H_2O	TNT	Toluene
Acetone	5.79	21.44	16.53	14.15	12.06
	5.70	12.17	15.93	7.32	12.31
	1.32	17.13	15.60	10.35	10.78
Ethanol	18.83	3.77	20.50	11.85	19.89
	14.48	3.45	16.09	7.09	15.27
	17.34	2.30	18.67	9.75	17.98
H_2O	15.12	19.83	5.02	12.47	10.02
	15.19	17.43	1.70	10.29	8.53
	16.86	18.46	3.58	11.83	9.72
TNT	10.42	8.04	12.13	1.54	11.15
	9.63	9.68	10.42	1.27	9.50
	9.90	10.09	10.51	1.43	9.49
Toluene	13.01	17.90	8.80	10.95	3.22
	11.45	17.63	8.62	10.30	1.73
	8.78	17.93	10.35	10.23	4.34
Unknown	18.95	22.76	19.47	18.13	19.18
	16.72	18.07	16.87	14.14	15.89
	17.43	19.13	17.89	15.12	16.77

it is particularly interesting to note that the TNT and H_2O signatures do not map close to each other, making them easy to differentiate.

5.5. PARC: ANN

For our ANN training, we used the same set of signature patterns used in training the CA algorithm above (three runs each for the five analytes of interest), and the patterns were preprocessed as described in Section 5.3. One important note here: whereas, in the case of the CA algorithm, it was necessary to normalize the signature patterns to prevent misidentification of analytes at different concentrations, ANNs can be trained to recognize both the species and the concentration of the analytes and have been used to recognize the species and the concentrations of analytes individually as well as the components of binary mixtures [25]. For consistency and comparability with the CA algorithm, we chose to normalize the data here as well. The target output of the ANN is a set of five numbers corresponding to the five analytes we wish to identify. For each analyte training pattern, the output target has a value of 1.0 in the position corresponding to the correct analyte and a value of 0.0 in the other four positions. The ANN is then

Table 2. Pattern recognition of analytes by artificial neural network (ANN)

Analytes	**ANN recognition of analyte components**				
	Acetone	**Ethanol**	**H_2O**	**TNT**	**Toluene**
Acetone	0.993	0.005	0.000	0.000	0.000
Ethanol	0.010	0.993	0.000	0.000	0.016
H_2O	0.004	0.000	0.965	0.006	0.033
TNT	0.004	0.000	0.004	0.968	0.017
Toluene	0.001	0.005	0.004	0.005	0.956
Unknown	0.034	0.001	0.009	0.022	0.011

trained to produce a value near 1 for the output corresponding to the correct analyte and a value near 0 for incorrect analytes. The training for this ANN took less than 1 min on a desktop computer.

A new set of data was taken for each of the five analytes (and one set of the artificial 'unknown vapor' was created), and their signature patterns were produced as described in Section 5.3. The new signatures were then normalized and presented to the trained ANN algorithm, and the results are summarized in Table 2. The leftmost column of the Table lists the analytes to be identified by the ANN, and the other columns list the five ANN output components. The output values can be thought of as representing the relative probability of detection for each of the five output analytes. The small differences of the outputs from the desired 0 or 1 probabilities are due to the fact that test set data are not identical to any of the training sets used to train the ANN. Allowing for these random fluctuations and assigning a threshold value of 5% for positive detection, the ANN correctly identifies all five analytes (the shaded cells in the Table) with no false positives. The ANN also rejects the 'unknown' signature pattern. For explosive detection and the discrimination between the TNT and H_2O signatures, we see that the output values show no significant recognition of TNT in the H_2O signature and no significant recognition of H_2O in the TNT signature.

6. Conclusions

Adsorption-induced bending of microcantilevers in an array where each cantilever element is modified with different receptor layers can be used for the detection of explosive vapors. The detector response is fast, reversible, and repeatable. Receptive layers such as SAMs provide selectivity to cantilever sensing. As no chemically selective coatings are absolutely specific, arrays of cantilevers coated with partially selective coatings were used, and data were analyzed with PARC algorithms. Comparing the two PARC algorithms used here, we saw that both algorithms successfully identified the target analytes. A closer look at the CA output shows that the range of positive detection output is 1.27–5.79 and for negative detection is 7.09–22.76, giving a separation of 1.3 between positive and negative ranges. Therefore, the separation between positive and negative detection is only 0.29 times the range of positive detection. By contrast, the ANN output shows that the range of positive detection output is 0.956–0.993 and for negative detection is 0.000–0.034, giving a separation of 0.922 between positive and negative ranges. This separation between positive and negative detection is 24.9 times the range of positive detection, giving the ANN algorithm almost two orders of magnitude greater separation between positive and negative detection than the CA algorithm. This makes the ANN much less likely to produce a false-positive analyte identification.

The array of four microcantilever detectors proved adequate for differentiating the five individual analytes used here, but as we strive to distinguish hundreds or thousands of analytes and mixtures of analytes, larger arrays of detectors and coatings will be necessary to produce the orthogonal responses needed to produce response signature patterns that are unique to each analyte. Our current research involves using arrays of 32 cantilever sensors to detect explosives such as TNT, RDX, and PETN individually and in the presence of associated compounds and common environmental interferents. As the complexity of the mixtures increases, we find it necessary to modify and combine multiple PARC algorithms to properly detect and identify the mixture components.

Acknowledgments

This work was supported by the Department of Homeland Security, Alcohol, Tobacco, and Firearms (ATF), and Oak Ridge National Laboratory (ORNL). Initial experiments were carried out with help from E. Hawak, V. Boidjev, Lal Pinnaduwage, T. Ghel, and D. Hedden. ORNL is managed by UT-Battelle, LLC, for the U. S. Department of Energy under contract DE-AC05-00OR22725.

References

[1] L.A. Pinnaduwage, Hai-Feng Ji, and T. Thundat, *IEEE Sensors Journal*, 5, 775 (2005).
[2] T. Thundat, P.I. Oden, and R.J. Warmack, *Microscale Thermophysical Engineering*, 1, 185 (1997).
[3] T. Thundat and A. Majumdar, Microcantilevers for physical, chemical, biological sensing, in *Sensors and Sensing in Biology and Engineering*, F.G. Barth, J.A.C. Humphry, and T.W. Secomb (eds.) pp 338–355, SpringerWein, New York (2003).
[4] N.V. Larvick, M.J. Sepaniak, and P.G. Datskos, *The Review of Scientific Instruments*, 75, 2229 (2004).
[5] T. Thundat, R.J. Warmack, and D.P. Allison, *Applied Physics Letters*, 77, 4061 (2000).
[6] R. Berger, Ch. Berger, H.P. Lang, and J.K. Gimzewski, *Science*, 276, 2021 (1997).
[7] G.G. Stoney, *Proceedings of the Royal Society of London. Series B. Biological Sciences*, 82, 172 (1909).
[8] R. Shuttleworth, *Proceedings of the Physical Society of London*, 63, 444 (1950).
[9] H.-J. Butt, *Advances in Colloid and Interface Science*, 180, 251 (1996).
[10] D. Sarid, *Scanning Force Microscopy*, Oxford University Press, New York (1991).

[11] M.V. Salapaka, S. Bergh, J. Lai, A. Majumdar, and E. McFarland, *Journal of Applied Physiology*, 81, 2480 (1997).
[12] M. Tortonese, R.C. Barrett, and C.F. Quate, *Applied Physics Letters*, 62, 834 (1993).
[13] A. Boisen, J. Thaysen, H. Jensenius, and O. Hansen, *Ultramicroscopy*, 62, 834 (2000).
[14] Available from Cantion Inc. A/S, c/o NanoNord A/S, Skjernvej 4A, DK9220 Aalborg 0, Denmark. www.cantion.com.
[15] B.C. Dionne, D.P. Rounbehler, E.K. Achter, J.R. Hobbs, and D.H. Fine, *Energetic Material*, 4, 447 (1986).
[16] S. Srinath, PhD Dissertation, Department of Mechanical Engineering, University of California, Berkeley, CA (2006).
[17] L.A. Pinnaduwage, V. Boiadjiev, J.E. Hawk, and T. Thundat, *Applied Physics Letters*, 83, 1471 (2003).
[18] M. Hsieh, E.T. Zellers, *Analytical Chemistry*, 76, 1885 (2004).
[19] R.O. Duda, P.E. Hart, and D.G. Stork, *Pattern Classification – 2nd Edition*, John Wiley and Sons, New York (2001).
[20] S. Theodoridis and K. Koutroumbas, *Pattern Recognition – 3rd Edition*, Academic Press, Amsterdam (2006).
[21] M.T. Hagan, H.B. Demuth, and M. Beale, *Neural Network Design*, PWS Publications Co., Boston, MA (1995).
[22] T.P. Vaid, M.C. Burl, and N.S. Lewis, *Analytical Chemistry*, 73, 321 (2001).
[23] R.E. Shaffer, S.L. Rose-Pehrsson, and R.A. McGill, *Analytica Chimica Acta*, 384, 305 (1999).
[24] P.C. Jurs, G.A. Bakken, and H.E. McClelland, *Chemical Reviews*, 100, 2649 (2000).
[25] L.R. Senesac, P. Dutta, P.G. Datskos, and M.J. Sepaniak, *Analytica Chimica Acta*, 558, 94 (2006).

Chapter 6

Neutron Techniques for Detection of Explosives

Richard C. Lanza

Department of Nuclear Science and Engineering, Massachusetts Institute of Technology, Cambridge, MA 02139, USA

Counterterrorist Detection Techniques of Explosives
Jehuda Yinon (Editor)

Contents

1. Introduction

The problem under consideration is the detection of explosive materials that may be made of a variety of materials and generally are hidden so as to make the detection of their presence difficult, especially to a casual inspection. The goal of any inspection technique is to non-intrusively determine the presence of such materials in a manner that is consistent with not interrupting the normal scheme of commerce and that, at the same time, exhibits a high probability of detection and a low probability of false alarms. A great deal of work has been reported in the literature on neutron-based techniques for the detection of explosives, with by far the largest impetus coming from the requirements of the commercial aviation industry for the inspection of luggage and, to a lesser extent, cargo. There, the major alternative techniques are either X-ray based or chemical trace detection methods that look for small traces of explosive residues. The limitations of the X-ray and trace methods are well known, but as of the date of this writing (2006), it is safe to say that despite extensive development, no neutron- or nuclear-based technique is in use in aviation security on a routine basis. More recent events (September 11, 2001) have focused interest on other places for which surveillance is required, but unfortunately, these new places lack much of the control over inspected objects, which is characteristic of the aviation industry; materials must of necessity pass through a defined location before being loaded on the aircraft. In light of these changes, it is now useful to re-examine neutron methodologies.

1.1. Characteristics of explosive materials

Explosive materials vary widely in their chemical composition, and in some sense, one could argue that the common characteristic of these materials is that they explode. Although it is certainly true that many commercial and military explosives have high concentrations of nitrogen, recent experience has revealed a new set of non-nitrogenous explosives that are often fabricated by terrorists from readily available and fairly ordinary materials that may be purchased without arousing suspicion. It is essential that the designer of detection equipment recognize the vast variety of potential threats and thus understand clearly what the limitations of a particular technique may be. It should be emphasized here that the implications of the widespread use of these improvised explosives imply that the detection of nitrogen alone will often be inadequate, and indeed, the use of assumed elemental ratios for explosive composition may lead to difficulties in detection [1].

A list of potential explosives was summarized by Makky [2] and Gozani [3]. Data from Krauss are shown in Figs 1–3. Note that the Figures are summaries of explosives of concern that have high concentrations of nitrogen and oxygen, and in fact, these elements Figure prominently in most of the current literature.

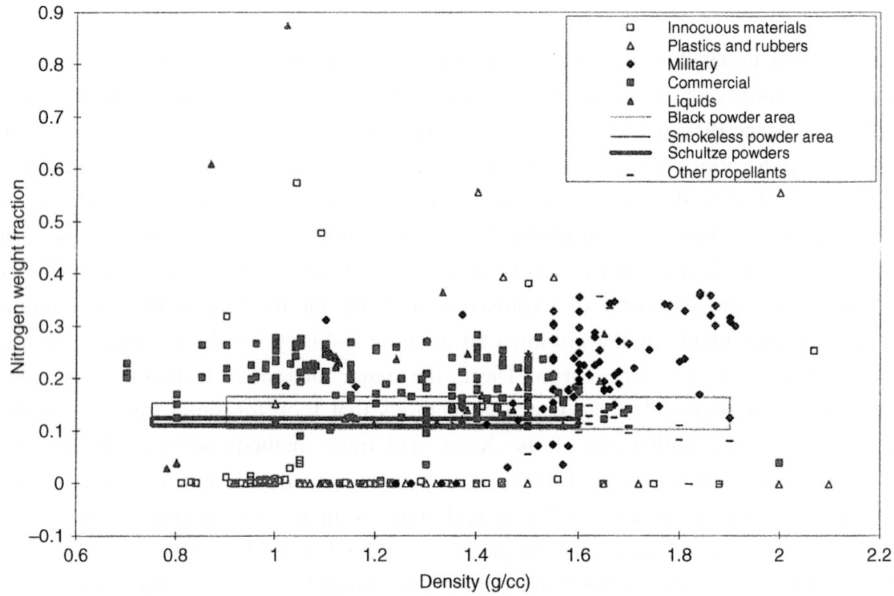

Fig. 1. Density versus nitrogen weight fraction for representative explosives. Reproduced with permission from Makky [2], pp. 18–21.

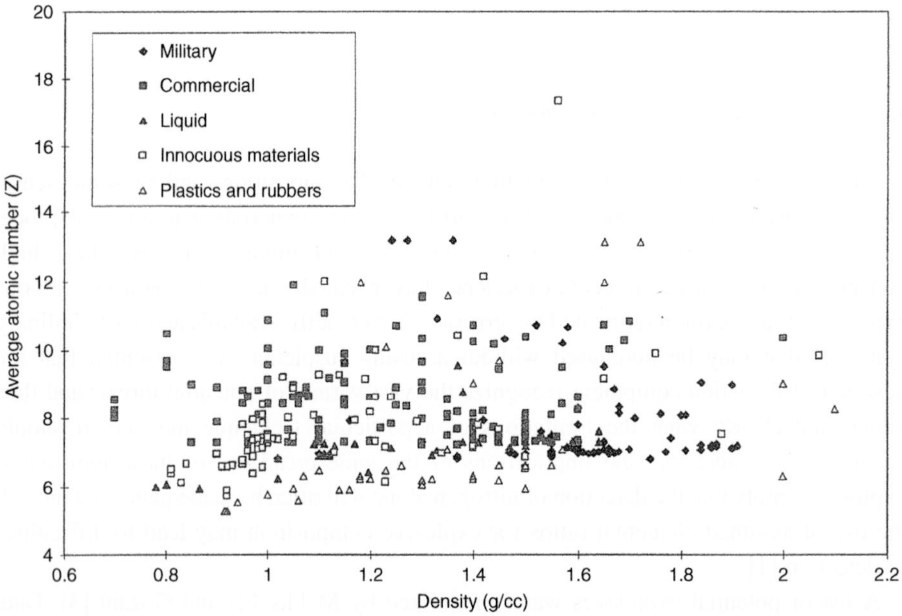

Fig. 2. Density versus average atomic number for representative explosives. Reproduced with permission from Makky [2], pp. 18–21.

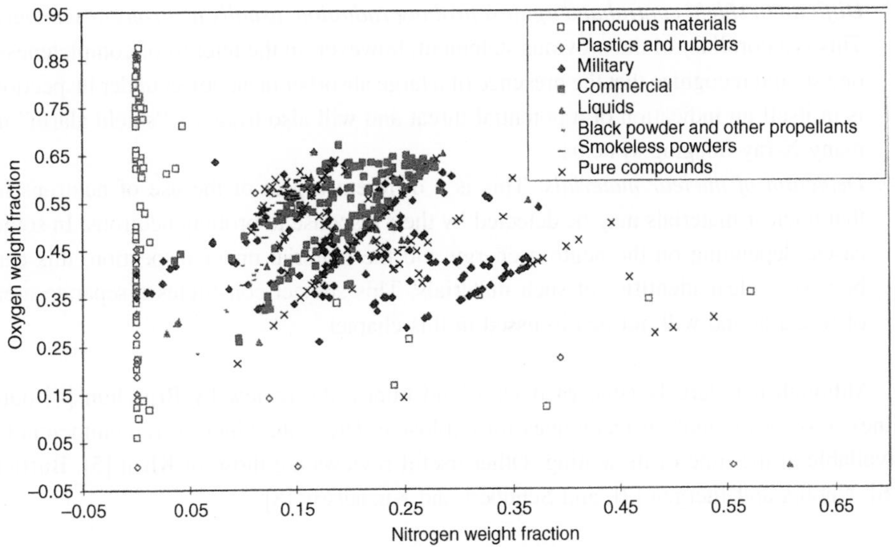

Fig. 3. Nitrogen weight fraction versus oxygen weight fraction for representative explosives. Reproduced with permission from Makky [2], pp. 18–21.

2. Neutron techniques

2.1. Overview

Neutron techniques are based on the detection of the nuclear properties of possible threats rather than their density or chemical properties. One may summarize the advantages of these techniques as follows:

1. *Determine elemental composition not just density*: This is the major advantage of nuclear techniques as compared with conventional X-ray scanners. Conventional X-ray techniques generally measure the apparent density and shape of objects. Using dual-energy X-rays, one also gets an approximation to the average atomic number (Z) of the material, and can it and spatial extent as a clue to the presence of an explosive material. Neutron techniques offer the possibility of determining the elemental composition of a suspicious object.
2. *Great penetration and potential for cargo inspection*: The effective range of neutrons even in metals enables penetration of much greater thicknesses of materials than the usual low-energy (140 kV) X-ray scanner, although it should be noted that high-energy X-ray sources, such as from multi-MeV electron linacs, offer similar penetration. Of course, each technique has its limitations; neutrons are shielded by low-Z materials, particularly organics, and high-energy X-rays by high-Z metals.

3. *Difficult to shield contraband against probing radiation, usually neutrons or gammas*: This is a corollary of the previous statement; however, in the interest of completeness, one should recognize that the presence of a large absorber in an object under inspection is in itself an indication of a potential threat and will also trigger a "shield alarm" in many X-ray imaging systems.
4. *Detection of nuclear materials*: This is a different aspect of the use of neutrons in that nuclear materials may be detected by their response to probing neutrons. In some cases, depending on the neutron energy and the material under inspection, this can be a very clear identifier of such materials. This subject constitutes a separate area of research and will not be discussed in this chapter.

Although it is largely concerned with land mines, the review by Bruschini [4] outlines most of the nuclear techniques for explosive detection, which were commercially available at the time of its writing. Other useful reviews are those of Khan [5], Buffler [6], Nebbia and Juergen [7], and Schubert and Kuznetsov [8].

2.2. *Neutrons in, gammas out*

This is one of the most common approaches using neutrons. One of the earliest approaches to the explosives problem was thermal neutron analysis used for aircraft luggage inspection [9,10]. A piece of luggage is bombarded by neutrons from either an isotopic (^{252}Cf) or an electronic source deuteron-tritium (DT). The neutrons are allowed to thermalize in the luggage, and nitrogen is detected through the reaction $^{14}\text{N} + \text{n(thermal)} \rightarrow {}^{15}\text{N}^* \rightarrow {}^{15}\text{N} + \gamma$ (10.8 MeV). Detection of the 10.8 MeV gamma confirms the presence of nitrogen. Unfortunately, the lack of imaging and the general low sensitivity of the method led to its being abandoned for aircraft security, although there have been some attempts at using it for the detection of land mines.

The more common approach which is now used relies on fast neutrons as probes that excite nuclear processes, generally leading to the production of characteristic gamma rays through inelastic processes of the form (n, n'γ). The technique had its origins in the mining industry, where neutron probes were used to analyze the composition of coal and other materials. The cross-sections for these processes for elements of interest for explosive detection are shown in Fig. 4, along with the energies of the resultant gammas for materials such as nitrogen, carbon, oxygen and elements associated with many common explosives. Examples of spectra from coal analysis are shown in Figs 5 and 6.

One of the most developed of these methods is the technique usually referred to as pulsed fast neutron analysis (PFNA) [3,11–13]. The operation is illustrated in Fig. 7 [11]. Neutrons in the range of ~8 MeV are generated by an accelerator (not shown) by the reaction D(d,n)He. The accelerator is pulsed with a ~1 ns pulse width to produce 1 ns pulses of neutrons with a repetition rate of ~1 MHz. Gamma rays are detected in a series of scintillation detectors. The time difference between the accelerator pulse and the

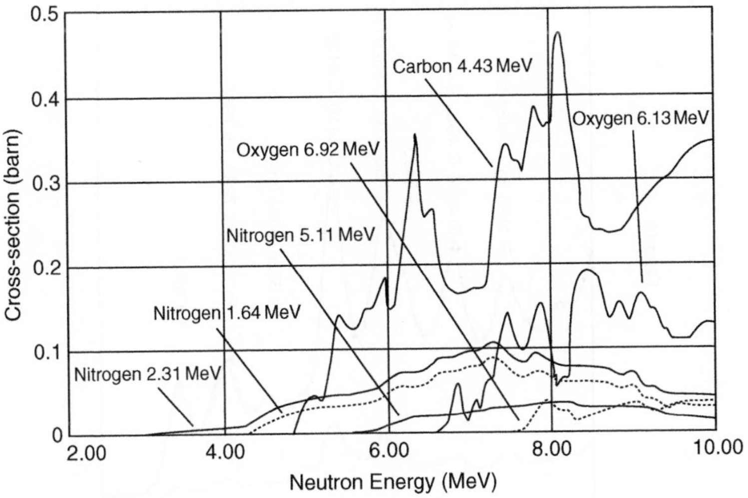

Fig. 4. Cross-sections for inelastic neutron gamma production for various lines in nitrogen, oxygen and carbon.

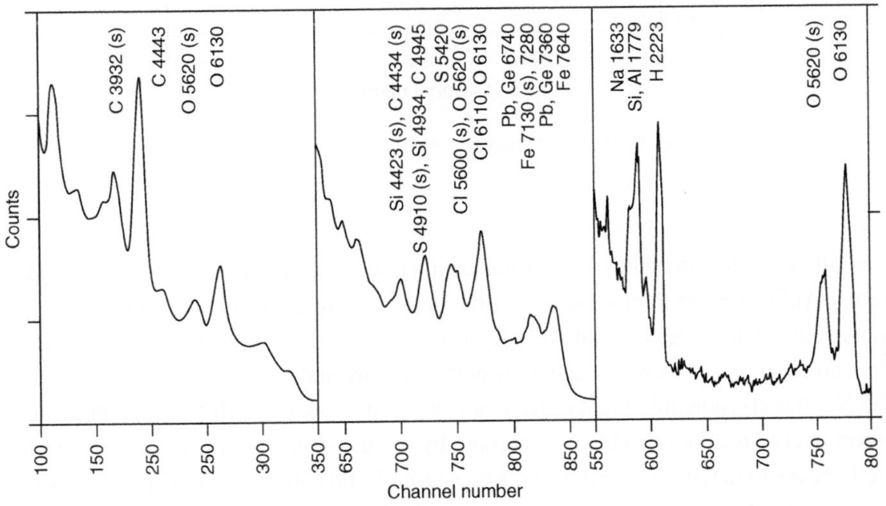

Fig. 5. Spectra from pulsed neutron analysis of coal.

gamma detected is used to determine the position along the beam where the interaction took place. By mechanically scanning the beamline so as to move the beam through the object, the gammas can be correlated with the position of interaction in three dimensions. The resulting gamma ray spectrum is used to determine relative elemental content for "voxels" in the object under inspection. For 8 MeV neutrons, the time-of-flight (ToF) measurement yields an accuracy of $\sim$5 cm in space. The physics for this particular system has been refined over a number of years. One should note that the choice of neutron energy, $\sim$8 MeV, means that scattered neutrons will generally lose enough energy so that

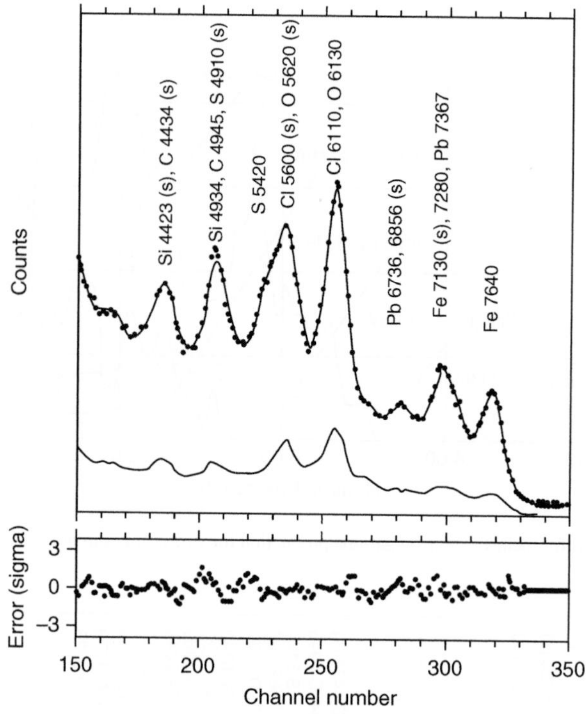

Fig. 6. Details of spectrum of coal showing fitted results.

they fall below the threshold for gamma production, and thus do not contribute to spatial clutter. Another point to observe is that the detectors are conventional scintillators (NaI) rather than high-energy-resolution detectors such as High Purity Germanium (HPGe). The choice of detectors was made after both experimental observation and cost analysis. The NaI has considerably higher efficiency and count rate capability than, HPGe but has poorer energy resolution. This is overcome by the use of the entire spectrum rather than single peaks to determine elemental composition. In this approach, rather than fitting the data to individual peaks, the entire spectrum is fitted to a linear combination of whole spectra from each of the elements. This approach has long been used in areas such as well logging and online coal and mineral analysis [14–16].

In controlled tests in the United Kingdom, the system demonstrated the ability to achieve a high detection probability (98%) for cargo inspection with only one false alarm during the tests. Despite this, the system was not installed, mostly due to concerns over system complexity, unease about downtime, large space requirements, and, finally, cost [17]. Another major study of PFNA came to similar conclusions [18].

An alternative and less complex (and less powerful) approach to the use of probing with neutrons uses a sealed neutron generator based on the DT reaction. This approach is commercially used in fields such as coal analyzers among others, and there have

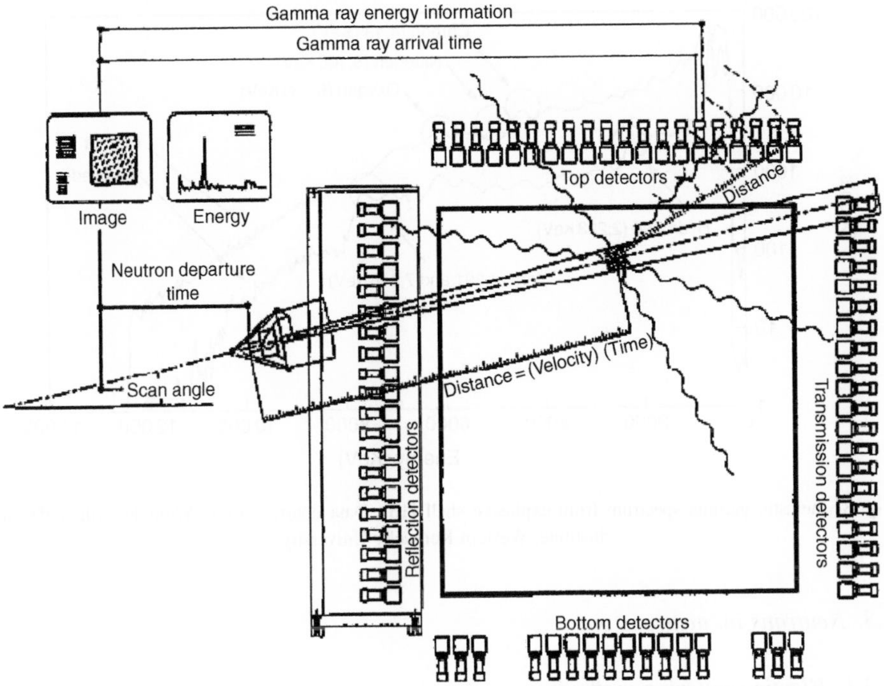

Fig. 7. Pulsed fast neutron analysis system [11].

been a number of attempts at using this method land mine detection and explosive detection [45,46].

The most common implementations of this technique use high-energy neutrons, typically 14 MeV from a sealed DT tube [16]. The resulting gamma rays are then detected by a gamma detector, usually a scintillator, but sometimes a germanium solid-state detector. This approach has been well described in the literature and has led to several commercial implementations as well as many research laboratory studies [7,19–24]. Figure 8 shows the results from probing an explosive shell and indicates the presence of explosive compounds. In a modification of this approach, usually referred to as associated particle interrogation (API) or "tagged" neutrons, the recoil alpha particle associated with a neutron from the target is detected and its direction and timing are used to improve the signal-to-noise ratio (SNR) [21,22].

Despite the apparent simplicity of design obtained by the use of a sealed DT generator, practical considerations often lead to less than satisfactory results, and as a result, a considerable amount of ingenuity and skill must often be employed to build a practical unit. Because the use of 14-MeV sealed DT neutron generators, both tagged and untagged, is so common, the details of operation of such systems will be discussed in more detail later.

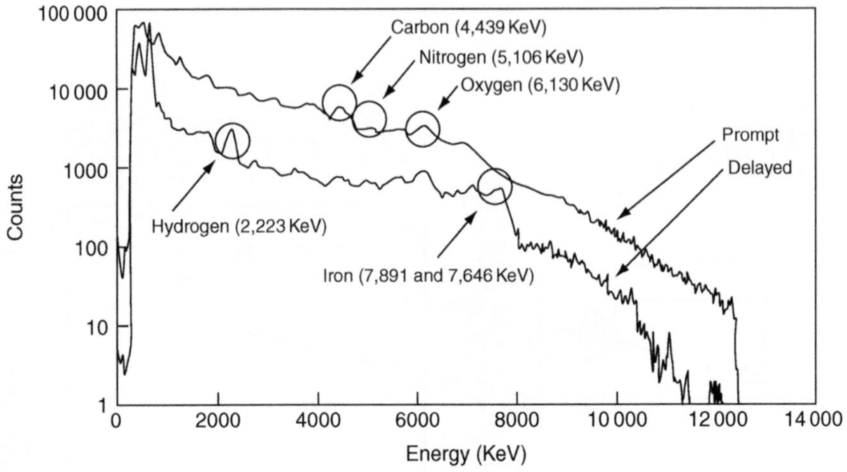

Fig. 8. Inelastic gamma spectrum from explosive shell [16]. Data courtesy of P. Womble, Allied Physics Institute, Western Kentuchy University.

2.3. Neutrons in, neutrons out

2.3.1. Elastic scattering methods

Several variations in this approach have been implemented. In the simplest, the backscattering of polyenergetic neutrons, generally from a small isotope source, is used to identify the presence of a low-Z anomaly, such as may be the case for searching for land mines [25,26]. Considering the simplicity of the approach, this approach has demonstrated potential, but the detection of anomalies in complex environments has proved to be a major limitation. Neutron elastic scattering [23,27], a more complex approach to backscattering, utilizes resonances in elastic neutron backscatter cross-sections to identify elemental composition. As with the simple backscatter approach, complex environments typical of realistic situations are particularly difficult to disentangle, and thus, the reliability of this method remains unproven.

2.3.2. Transmission methods

These methods are essentially multi-energy rediography using fast neutron transmission. By exploiting the energy dependence of neutron cross-sections in the 2–8 MeV range, images may be obtained which measure the elemental composition of the object being inspected [24,28–32]. Another approach is taken by Eberhardt et al. who used a combination of gamma ray and fast neutron transmission imaging to detect contraband. Although it does not explicitly determine elemental composition, it has proven useful in several applications [33].

Several approaches to varying the energy of the probing neutron have been used. Overley et al. [34], Miller et al. [35], and Yule et al. [36] used fast neutron ToF

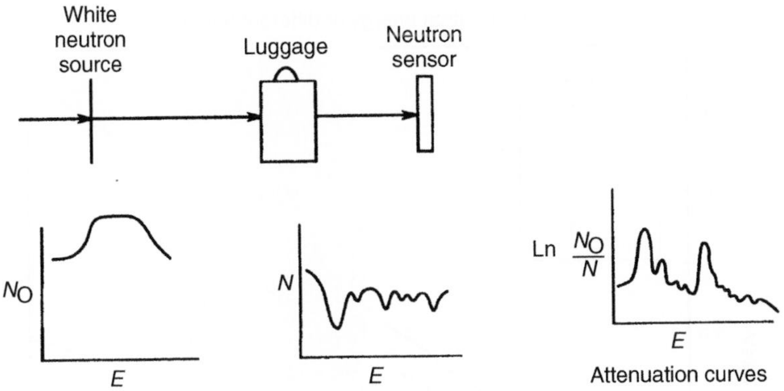

Fig. 9. Principles of pulsed fast neutron transmission analysis. No, open beam intensity with no material in the beam path.

spectroscopy. The accelerator-based source of neutrons is pulsed with a few nanoseconds wide pulse at repetition rates of up to several megahertz. The ToF spectrum can then be translated directly into a transmission energy spectrum, and the elemental composition can be deduced. Figure 9 shows a schematic of this approach. In order to produce images, each "pixel" of the detector must be capable of fast time measurements (ns). The results in laboratory tests were encouraging, but several practical limitations of this method, usually referred to as pulsed fast neutron transmission spectroscopy [37].

Another way to vary the energy of the incoming neutron beam is to utilize the kinematics of the production process, as in Fig. 10. For the deuteron-deuterium (DD) reaction, the energy of the neutron is dependent on the production angle in the center of mass system. For incident deuterons of 2.5 MeV, for example, neutron energies of ~2.2–5.7 MeV can be obtained. This range is adequate to cover most of the resonances in the total cross-section for carbon, nitrogen, and oxygen; an object may be scanned by simple rotation around the production target, as in Figs 11 and 12. Appropriate angles for the neutron are chosen to match resonance peaks and valleys for the major elements. Using spectral decomposition methods, the composite images may be processed (Fig. 13) to yield elementally resolved images for various compounds [47]. The design of imagers for fast neutrons is discussed by Watterson et al. [32], Ambrosi et al. [38,39], and Raas et al. [40].

3. Practical issues

The approaches and methods described here are selected from a much larger set of neutron methods, which have been proposed and, in some cases, experimentally tested. At the time of this writing (2006), neutron methods are still not in common use for contraband, explosive or land mine detection, despite their clear potential for obtaining more detailed information than X-ray methods. The reasons for this are not immediately clear from the physics; however, it is important to examine the practical issues that have arisen with neutron systems. The issues with larger accelerator-based systems are usually

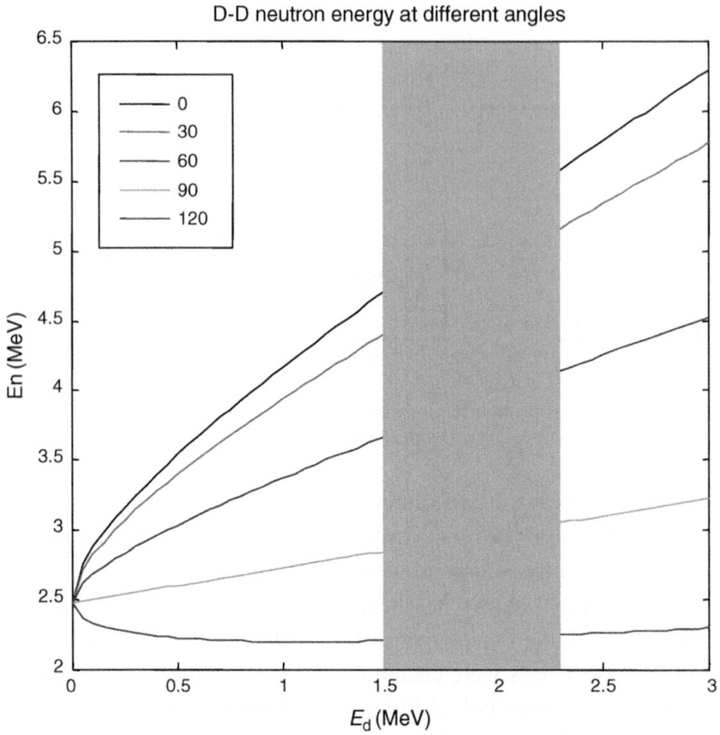

Fig. 10. Operation of neutron resonance radiography showing changing neutron energy with neutron angle. En, emitted neutron energy; Ed, incident deuteron energy.

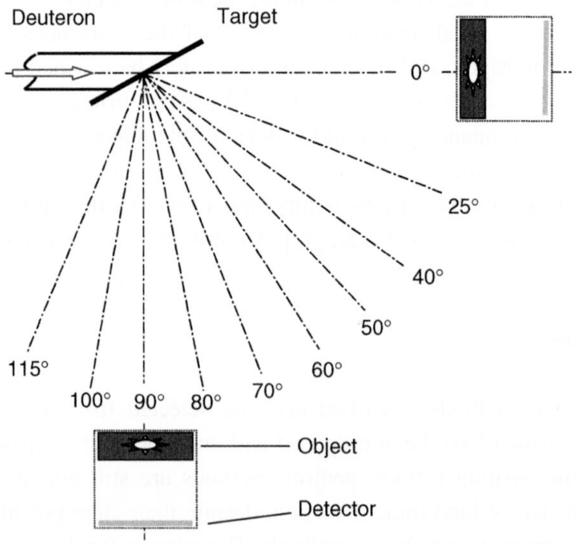

Fig. 11. Schematic of neutron resonance radiography system. The object being scanned moves around the target; the neutron energy changes with angle scanning each pixel of the object at multiple energies.

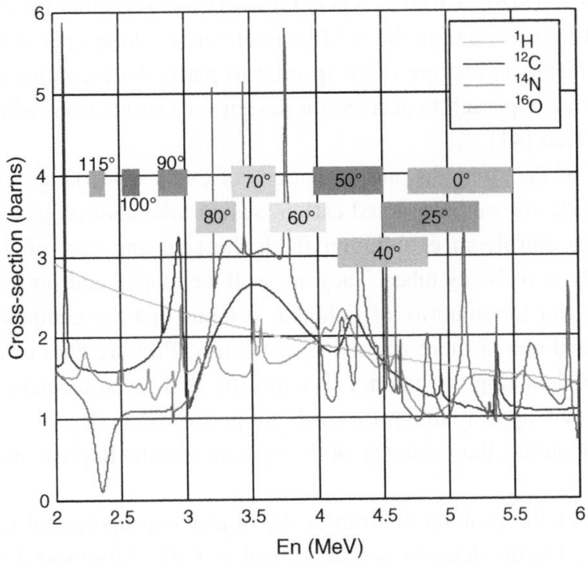

Fig. 12. Neutron resonance radiography absorption spectra for various angles of neutron production. Energy spread due to target thickness is included.

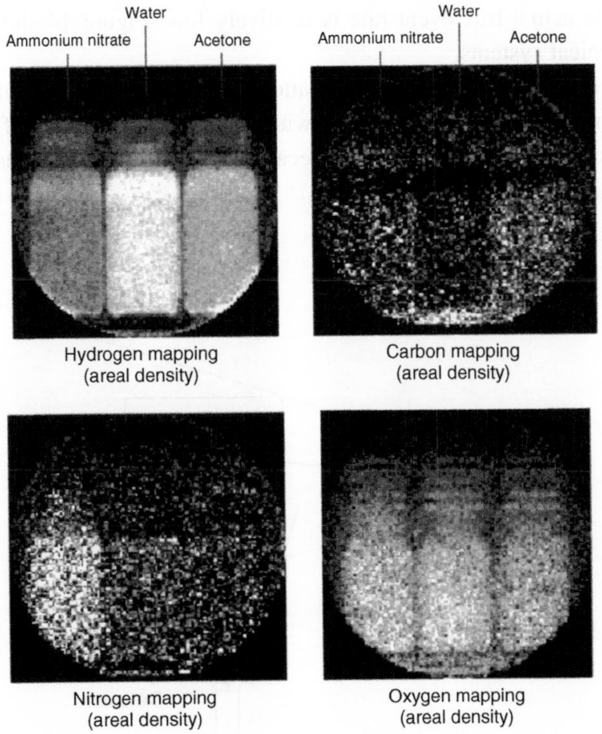

Fig. 13. Neutron resonance radiography images showing elemental decomposition [40].

centered on cost and size, as well as a general worry as to whether these accelerators are indeed practical for operation in the field by relatively unskilled personnel. These issues are not strictly technical and are often specific to particular locations and applications. One example of an approach to accelerator design for fast neutron radiography is given by Rusnak and Hall [41].

A discussion of the strengths and limitations of every neutron system is beyond the scope of this work. As we have noted earlier, sealed tube neutron generators have long been used in the petroleum exploration field; the size and cost of the generator are comparable to those of X-ray tubes. They are well developed and rugged, and thus have formed the basis for the majority of fieldable systems that are composed of a 14-MeV neutron source and one or more gamma ray detectors. Probing with neutrons from such sources and then detecting the gamma rays specific to a given element, such as carbon, nitrogen, or oxygen, appear straightforward; but performance depends on the physics of the neutron interactions, the geometry of the system, and the implementation of the total system.

The intensity of the probing neutron at the object is proportional to $1/S^2$, whereas the signal returned to the detector is proportional to $1/D^2$. More significantly, however, the background from the container and its contents are essentially constant for a given distance from the source to the container and for a constant distance from the detector to the container. Thus, the detectors in this application must deal with high count rates, even though the actual true event rate is relatively low. Figure 14 shows a simplified geometry of typical systems.

It is important to recognize these limitations on the performance of neutron interrogation systems using 14 MeV neutrons with respect to uniformity of sensitivity and potential application of timing methods, because, as we have seen, these sources have essentially isotropic neutron distributions.

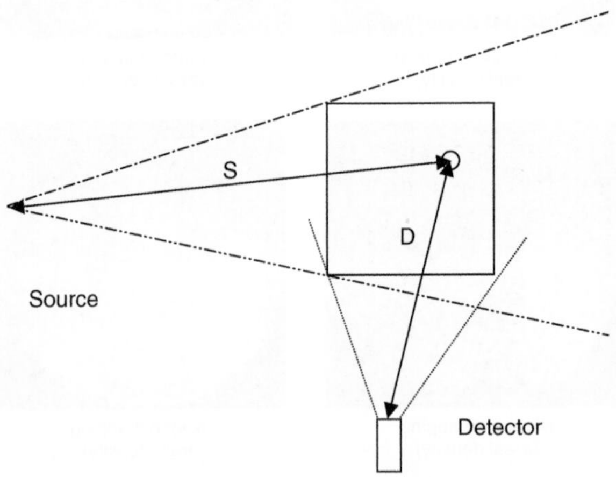

Fig. 14. Geometry for fast neutron analysis showing single source and single detector.

3.1. Physics of gamma production

The processes of interest typically have cross-sections for production of ~100 mb (1 b = 10^{-24} cm^2 per nucleus) (Fig. 4). As a result, a relatively low fraction of the neutrons incident on the target material give rise to gamma rays.

To illustrate this in a simple way, we first compute the mass absorption coefficient, μ (cm^2/g), for these materials.

$$\mu = \sigma \frac{N_{\text{Avogadro}}}{A},$$

where σ is 10^{-25}, $N_{\text{Avogadro}} = 6.025 \times 10^{23}$ and A is the atomic weight, taken as ~14 for these elements. The result is that $\mu \approx 0.0043\,\text{cm}^2/\text{g}$.

We now calculate the number of gammas produced:

$$dN_\gamma = N_\text{n} \mu dm,$$

where N_n is the number of neutrons incident and dm is the areal mass density (g/cm^2).

The number of neutrons actually incident will be given by the product of the flux, Φ (n/cm^2), and the area, A (cm^2), and so, the number of gammas produced will be

$$dN_\gamma = \Phi A \mu dm$$

but since the product of area and areal mass density is just mass

$$dN_\gamma = \Phi \mu dM$$

and assuming that there is very little attenuation through the mass of the target (true for areal thicknesses of <250 g/cm^2), we just get a simple expression for the number of gammas produced by a mass of M:

$$N_\gamma = \phi \mu M.$$

Finally, we can compute the flux from a source of neutrons S_n assuming the source is isotropic. (This is true of sealed tube sources. Note that even if API is used, the source is still isotropic although we may generate a trigger only for the forward-going neutrons.) The flux, at a distance r_s, is just the total source strength divided by the area of a sphere of radius r_s, so we get

$$N_\gamma = \frac{S_\text{n}}{4\pi r_\text{s}^2} \mu M.$$

Of course, we still have to detect the gammas and to include the solid angle of the detectors. If we have a detector of area A_d and efficiency ε at a distance r_d from the material in the target, we obtain the number counted

$$N_\text{c} = \frac{S_\text{n}}{4\pi r_\text{s}^2} \frac{\varepsilon A_\text{d}}{4\pi r_\text{d}^2} \mu M.$$

All of this manipulation is done so that we can get an approximation of the best we can do with this method, and to avoid the details of how we actually implement the method.

Two examples illustrate this with practical source strengths and detectors. Consider an electronic neutron source such as a DT generator with 10^8 neutrons/s, a detector of area 45 cm^2 (typical 3-inch diameter detector) and efficiency 1. For distances of 100 cm and a mass of 1 kg, we get ~1.25 counts/s. Another view might be a truck scanner with 100 kg of material (factor of 10^2 increase in count rate) and distances of 200 cm (a factor of 16 decrease) which would be a factor of about 6 overall, ~8 cps. If we reduce the distance to 50 and 20 cm, we would get a factor of more than 100 or about 123 cps/kg, respectively. Just by way of reference, PELAN operating in those conditions for land mine detection gets about 10 cps for a 100 g mass, so we are about right in this calculation.

This simple-minded approach does not include the other effects that are present with real containers, such as neutron and gamma attenuation, neutron scatter, and a large production of gammas from other reactions within the material, causing high gamma rates in the detectors. The problem here is that although most of these can be rejected by energy resolution, the sheer numbers result in degradation of detector performance due to high count rates. The count rate problem is exacerbated by short-duty-factor neutron sources. For example, there have been proposals to build plasma focus sources with as many as 10^{11} neutrons per pulse and then to inspect large containers with a single pulse of neutrons. Using the numbers from our previous calculation for 1 kg of material with 1 m distances (the first column of data), it would appear that the number of true events would be raised by a factor of 10^3 and that one could achieve detected rates of ~10^3 per pulse. Unfortunately, this is not true in any realistic detector performance scenario. Assuming a neutron pulse of 50 ns, this implies that *all* of the detected gammas will also be in a 50-ns pulse, an impossibly high instantaneous rate (10^7) for any existing energy-resolved detector system. Furthermore, as we have seen, because other gammas are also present, the actual rate in the detector could be a factor of 10^3 higher. The maximum practical rate would be 1 gamma per detector per pulse. But because plasma focus devices are limited in pulse rate, typically no more than a few Hertz, the actual performance of such systems is worse than that of the conventional DT tube.

3.2. Sensitivity variations

The issue of sensitivity variation due to geometry may be significant when single-source–single-detector systems are used. As an example, consider a cargo container 2 m × 2 m × 2 m with a single source located 10 cm from the center of one face, and a detector located 10 cm from a face at 90° to the source face. For simplicity, we will examine a plane that passes through the center, so we are in effect examining a 2 m × 2 m area with a source at the top and a detector on one side. The relative sensitivity for a point (x, y) is given simply by

$$S = \frac{1}{(x-x_s)^2+(y-y_s)^2} \cdot \frac{1}{(x-x_d)^2+(y-y_d)^2}$$

where the source is at (x_s, y_s) and the detector is at (x_d, y_d).

The two terms represent the inverse square of the source distance (because, as we have seen, the neutron source is isotropic) and the inverse square of the detector distance (the solid angle subtended by the detector). In Fig. 15, the range of sensitivity is plotted logarithmically and varies by a factor of 100 over the entire container. The $\log_{10}(S) = 0$ contour is the sensitivity normalized to the center of the container. It is interesting to note that the most sensitive regions are those closest to the source or the detector. As a result, any material close to these regions will produce more than 30 times as much signal as that due to the material in the center. It should also be noted again that the use of an imaging system improves this situation because imaging systems have a sensitivity that goes down less rapidly than $1/D^2$.

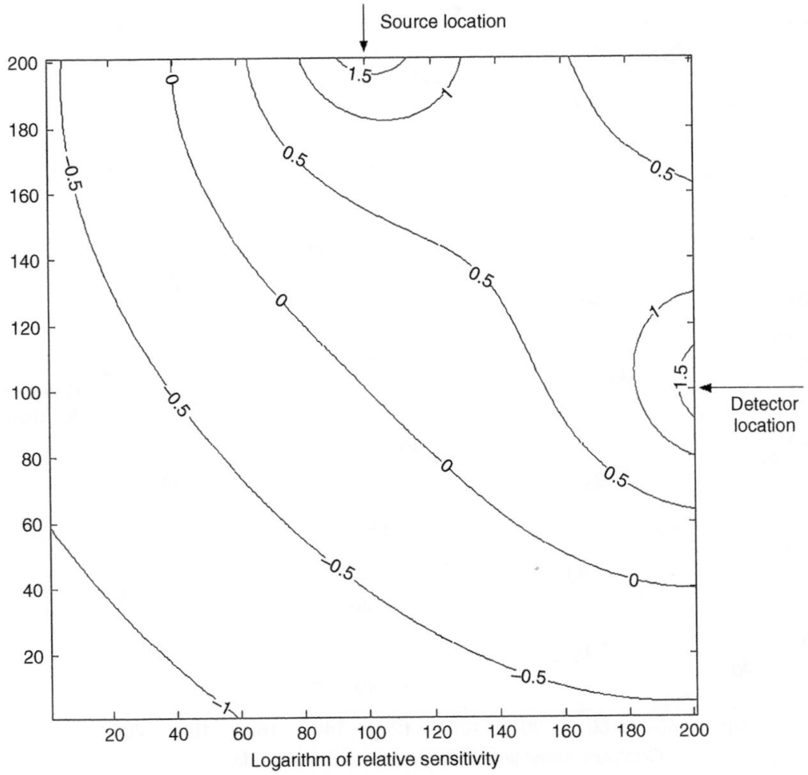

Fig. 15. Variation of sensitivity with position for fast neutron analysis. Distances in centimeters.

The use of a pulsed neutron source can help reduce some of the background by gating the detectors to eliminate material outside of a particular region, or by binning the data in time so as to improve SNR. In this case, we compute the time difference Δt between the neutron and the gamma detection at the detector.

$$\Delta t = \frac{\sqrt{(x - x_s)^2 + (y - y_s)^2}}{\nu_n} + \frac{\sqrt{(x - x_d)^2 + (y - y_d)^2}}{\nu_\gamma},$$

where ν_n and ν_γ are the neutron and gamma velocity, 5 cm/ns for 14 MeV neutrons and 30 cm/ns for gammas, respectively. Figure 16 shows the regions of constant time difference, which shows how even modest time resolution enables the definition of potential regions in the object and hence may reduce background.

Unfortunately, there remains the problem of differing sensitivity for identical time differences. In the case of API, it has been argued that the knowledge of the direction of the neutron will further reduce background. Although this is true in principle (t) with loaded containers a significant fraction of the emitted neutrons will be scattered before producing gammas, and thus, the knowledge of direction is frequently lost.

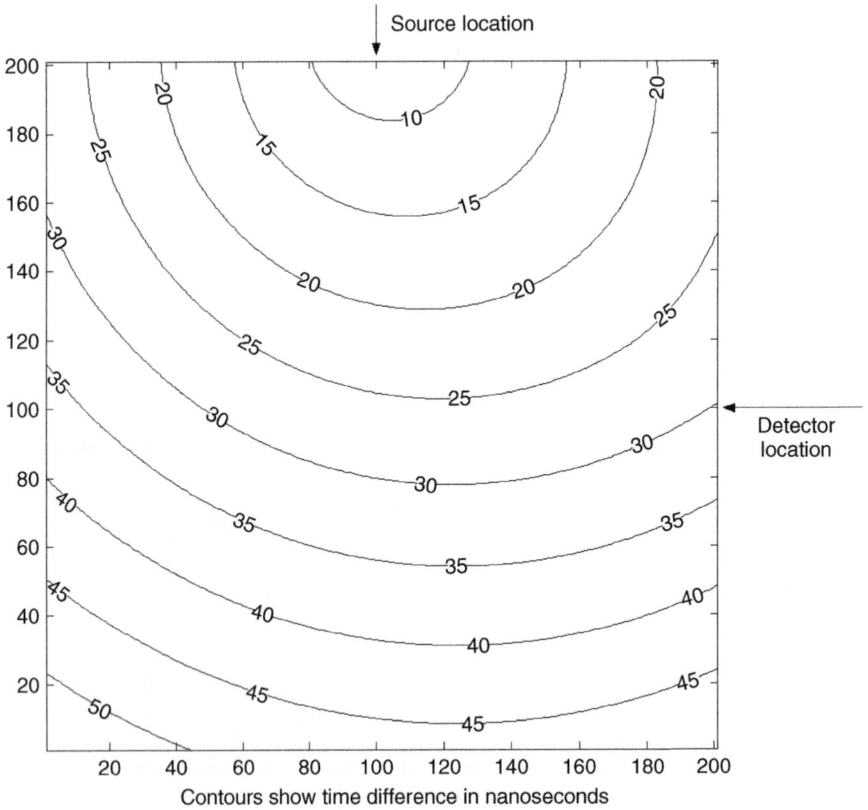

Fig. 16. Time difference using tagged alpha in API system. Distances in centimeters and times in nanoseconds.

API is a variant of 14-MeV DT neutron generators. The reaction that produces neutrons in the usual DT tube is $D(T, n)^4He$; the neutron and ^4He exit essentially back to back. By detecting the time of arrival of ^4He (α particle) and its position on a detector, both the time of emission and exiting angle of the neutron can be determined. If we now put the α signal in time coincidence with the signal from the gamma detector, we reduce the background from non-coincident gammas and therefore improve the SNR of the system; the limitations of the system stem primarily from time accidentals. Recall that we observed earlier that the count rate in the PELAN non-associated particle system with a neutron source strength, S, of 10^8 neutrons/s was a factor of 10^4 greater than the true rate. Typically one sees a count rate in the detector of ~100 kcps as compared with a true rate of ~10 cps. In the usual geometry for such tubes, about 5% of the neutrons are tagged, which results in a count rate for the α detector of 5×10^6. Using the usual relationships for accidental rates:

$$R_{\text{accidental}} = R_{\alpha} R_{\gamma} \tau_{\text{coincidence}}.$$

We find that for a resolving time of 1 ns, this results in an accidental rate of $5 \times 10^6 \times 10^5 \times 10^{-9}$, or 500 cps, compared with a true rate of 10 cps. Because the accidental rate is proportional to S^2, the API systems generally run at source strengths of 10^7 neutrons/s or less, and consequently, the true count rates are low resulting in long inspection times. Whether this is a problem will depend on the inspection scenario and the amount of explosive to be detected.

One might, however, observe that the PFNA system also uses nanosecond timing and wonder whether there is a problem here as well. The simple answer is that there is not a problem with accidentals, as the nanosecond timing pulses in PFNA are not random but periodic and are ~500 ns apart. In general, during a pulse, PFNA systems have only one neutron in the object being inspected, and thus, the issue of accidentals is not a problem despite the high rate in the detectors.

3.3. Detector choice

The choice of detectors depends on many factors. Bismuth germanate (BGO) and NaI are two possibilities that can achieve high efficiency at these energies. Although some claim that the best detector is HPGe, the *significantly* lower efficiency and poorer rate/time characteristics introduce a new set of problems aside from energy resolution. Data from large detectors used in high-energy physics such as EXOGAM (Fig. 17) show that these very large Ge detectors (composed of four 60 mm diameter × 90 mm deep HPGe cylinders) are significantly reduced in timing and photopeak efficiency performance at MeV energies. This makes them inferior in overall performance in this application, despite their better energy resolution.

There are other issues with HPGe detectors, which have made their use for this application problematic. The most obvious operational issue is the requirement for

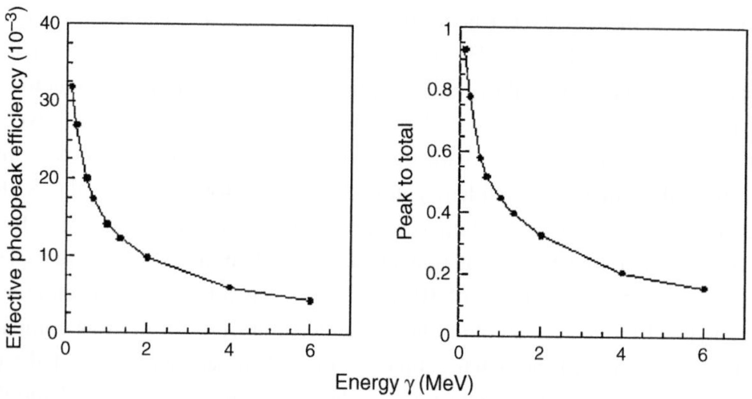

Fig. 17. Efficiency of large germanium detectors at higher energies and photopeak to total ratio.

cryogenic operation, which is a significant issue for units that are field deployable and which must be both portable and rugged. Care must also be taken with the shielding of HPGe detectors when used in intense neutron fields, as HPGe is susceptible to neutron damage; experimental data show that a substantial loss of performance (energy resolution degrading from 2 keV to more than 15 keV) appears at an integrated dose of about 10^7 neutrons/cm^2. Furthermore, it has been experimentally shown that the critical integrated neutron dose to induce damage is not a linear function of the detector volume but is somewhat steeper.

For applications dealing with security issues, we are mainly looking at the decay of oxygen, carbon, and nitrogen, with gamma energy ranging from 1.5 up to 6 MeV, therefore if one wants reasonable efficiency, it is necessary to use large-volume detectors.

The largest detectors readily available on the market are about 70–80% HPGe. Note that the "efficiency" for HPGe is defined *relative* to a 3-inch diameter NaI scintillation detector at ^{60}Co gamma energies (defined as 100%), resulting in a detection efficiency of ~70–80% of a 3 inch × 3 inch NaI for the Co lines (around 1300 keV) but which rapidly decreases with energy, down to 30% for the HPGe at 5–6 MeV. In most applications, the tendency is to move to 4 inch × 4 inch (or even to 5 inch × 5 inch) detectors and to use crystals "denser" than NaI (BGO or BaF or even LaBr). The loss of efficiency using HPGe is therefore unjustified, especially if one considers the efficiency/price ratio. This assumes that the whole-spectrum methods are used for analysis. Extrapolating to 6 MeV for an HPGe detector can be done with the aid of Fig. 18, which shows efficiency as a function of energy for various relative efficiency Ge detectors [42,43].

Even assuming that these limitations are acceptable, the most appropriate and promising techniques for security applications are based on nanosecond neutron analysis (not only in API); therefore, the use of HPGe with its intrinsically slow signal (timing resolutions of tens of nanoseconds at best) and low-rate capability is highly undesirable. The use of HPGe in this application with 14-MeV neutron sources has been described in the literature for more than 15 years with little evidence that such detectors improve real-world system performance and much evidence that they do not.

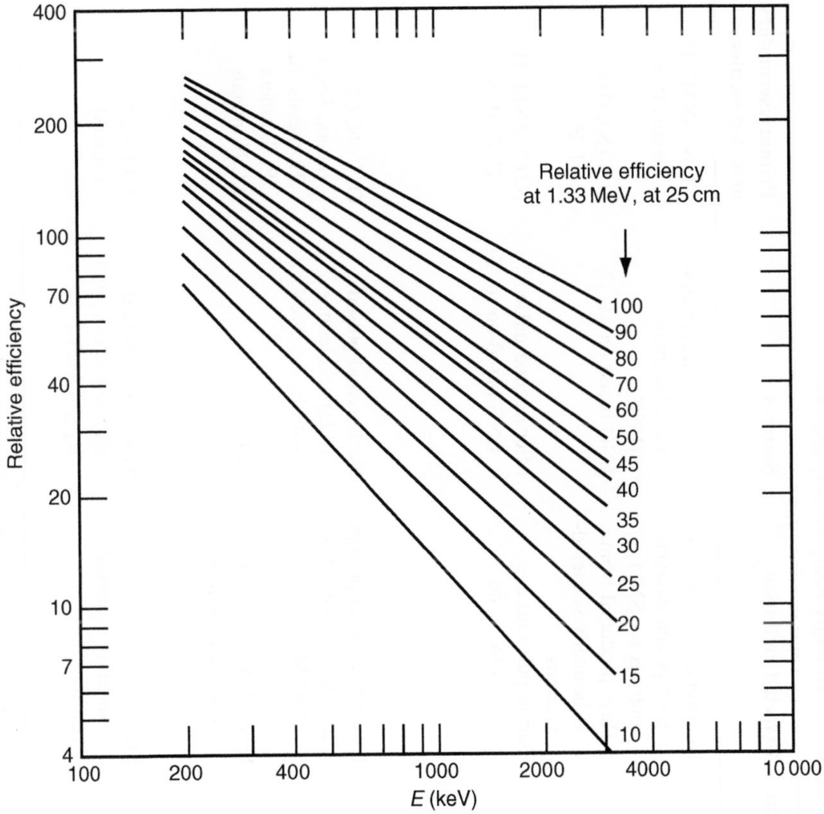

Fig. 18. Efficiency of germanium detectors at high energies for various nominal efficiencies at 1.33 MeV [42,43].

4. Summary

Due to several factors that have been discussed, using neutron techniques for transportable systems for the detection of explosives and other contraband remains only a goal at this time. The attraction of using neutrons lies in both their penetrating ability and the ability to use them to detect elemental composition. These properties are unique to neutron-based systems and have led to continued interest in their use, and multiple systems have been investigated as can be seen in Table 1 [44]. At the present time, the use of neutron techniques as a complement to other inspection methods appears to be the most likely near-term mode of operation.

In source development, there is a need for more compact and reliable accelerator-based neutron systems. The advantage of accelerator-based technology (as distinguished from DD or DT systems) lies in the ability to both vary the energy of the neutrons and, with kinematic focusing, limit the angles of neutron production, i.e. produce neutrons in the forward direction rather than isotropically. However, sealed tube generators, both

Table 1. Summary of neutron techniques currently being investigated [44].

No.	Technique name	Probing radiation	Main nuclear reaction	Detected radiation	Sources	Primary/secondary detected elements
1	TNA	Thermalized neutrons	(n,γ)	Neutron capture γ-rays (prompt and delayed neutrons for SNM)	^{252}Cf also accelerator-based sources (ENG[a])	CI, N, SNM[b], H, metals, P, S
2	FNA	Fast (high energy, usually 14 MeV) neutrons	(n, n′γ)	γ-Rays produced from inelastically scattered neutrons	ENG based on (d, T)	O, C(N), (H) CI, P
3	FNA/TNA	Pulsed neutron source; fast neutrons during the pulse, thermal neutrons between pulses	(n, n′γ) + (n,γ)	During pulse no.2 + after pulse – no.1	Microsecond pulsed ENG based on (d, T)	N, CI, SNM, H, C, O, P, S
4	PFNA	Nanosecond (ns) pulses of fast neutrons	(n, n′γ)	Like FNA (no.2)(prompt and delayed neutrons for SNM)	Nanosecond pulsed (d,D) accelerator with E_d ~6 MeV	O, C, N, CI, others, SNM, H, metals, Si, P, S, others
5	API	14 MeV neutrons in coincidence with the associated α-particles	(n, n′γ)	Like FNA in delayed coincidence with α	(d,T)	O,C,N, metals
6	NRA	ns pulsed fast neutrons (0.5–4 MeV), broad energy spectrum	(n, n)	Elastically and resonantly scattered neutrons	Nanosecond pulsed (d, Be) accelerator, with $E_d \leq 4$ MeV	H, O, C, N (others)

API, associated particle interogation; FNA, fast neutron analysis; PFNA, pulsed fast neutron analysis; TNA, thermal neutron analysis.

conventional and timed (API), will still have the advantage of portability but will suffer from the problems discussed earlier.

It is also interesting to note that the limitations of most neutron systems lie not in the neutron source(s) but rather in the (gamma ray) detectors and subsequent data acquisition system. The obvious need is for development of detectors with higher rate capability and improved energy resolution. All modern digital data processing techniques have allowed significant improvement in the performance of conventional scintillation detectors with respect to rate and pileup rejection.

It is essential that the design of practical systems consider the entire system and the environment in which it will be used; consideration must be given especially to environments which are heterogeneous and contain materials and objects that can lead to erroneous detection, either excessive numbers of false alarms or lowered probability of detection. The high penetration of neutron sources makes investigation of shielding an integral part of the system design. Portable systems whose weight and size are dominated by shielding are less likely to be used.

References

[1] J. Zukas and W. Walters (eds), *Explosive Effects and Applications*, Springer-Verlag, New York (1997).

[2] W. Makky (ed.), *Proceedings of the Second Explosives Detection Technology Symposium and Aviation Security Technology Conference*, 12–15 November, Atlantic City International Airport, FAA Technical Center, NJ. Available from National Technical Information Service (1996).

[3] T. Gozani, Novel applications of fast neutron interrogation methods, *Nucl. Instrum. Methods Phys. Res. Sec. A*, 353(1–3) (1994) 635–640.

[4] C. Bruschini, *Commercial Systems for the Direct Detection of Explosives (for Explosive Ordnance Disposal Tasks)*, ExploStudy, Final Report. EPFL-DI-LAP Internal Note (2001).

[5] S.M. Khan (ed.), *Proceedings of the First International Symposium on Explosive Detection Technology*, 13–15 November, Atlantic City International Airport, FAA Technical Center, NJ. Available from National Technical Information Service (1992).

[6] A. Buffler, Contraband detection with fast neutrons, *Radiat. Phys. Chem.*, 71(3–4) (2004) 853–861.

[7] G. Nebbia and G. Juergen, Detection of buried landmines and hidden explosives using neutron, X-ray and gamma-ray probes, *Europhysics News*, July/August, no. 4, pp. 119–123 (2005).

[8] H. Schubert and A. Kuznetsov, *Detection of Explosives and Landmines: Methods and Field Experience*, Proceedings of the NATO Advanced Research Workshop on Detection of Explosives and Landmines, 9–14 September 2001, St Petersburg, Russia; Kluwer Academic Publishers, Boston (2002).

[9] P. Shea, T. Gozani and H. Bozorgmanesh, A TNA explosives-detection system in airline baggage, *Nucl. Instrum. Methods Phys. Res. Sec. A*, 299(1–3) (1990) 444–448.

[10] W.C. Lee, D.B. Mahood, P. Ryge, P. Shea and T. Gozani, Thermal neutron analysis (TNA) explosive detection based on electronic neutron generators, *Nucl. Instrum. Methods Phys. Res. Sec. B*, 99(1–4) (1995) 739–742.

[11] D.R. Brown and T. Gozani, Cargo inspection system based on pulsed fast neutron analysis, *Nucl. Instrum. Methods Phys. Res. Sec. B*, 99(1–4) (1995) 753–756.

[12] R. Loveman, J. Bendahan, T. Gozani and J. Stevenson, Time of flight fast neutron radiography, *Nucl. Instrum. Methods Phys. Res. Sec. B*, 99(1–4) (1995) 765–768.

[13] D.R. Brown, T. Gozani, R. Loveman, J. Bendahan, P. Ryge, J. Stevenson, F. Liu and M. Sivakumar, Application of pulsed fast neutrons analysis to cargo inspection, *Nucl. Instrum. Methods Phys. Res. Sec. A*, 353(1–3) (1994) 684–688.

[14] L. Dep, M. Belbot, G. Vourvopoulos and S. Sudar, Pulsed neutron-based on-line coal analysis, *J. Radioanal. Nucl. Chem.*, 234(1–2) (1998) 107.

[15] O. Serra, *Fundamentals of Well-Log Interpretation, Volume 1, Acquisition of Logging Data*, Elsevier, New York (1984).

[16] D.L. Chichester and J.D. Simpson, Compact accelerator neutron generators, *Ind. Physicist*, 9(6) (2004) 22–25.

[17] R. Lacey, *IAEA Meeting on Neutron Generators for Security Applications*, Vienna 2005.

[18] Panel on Assessment of the Practicality of Pulsed Fast Neutron Analysis for Aviation Security, National Research Council, *Assessment of the Practicality of Pulsed Fast Neutron Analysis for Aviation Security*, National Academies Press, Washington (2002)

[19] G. Vourvopoulos, P.C. Womble and J. Paschal, PELAN: a pulsed neutron portable probe for UXO and landmine identification, *Proc. SPIE*, 4142 (2000) 142–149.

[20] P.C. Womble, G. Vourvopoulos, J. Paschal, I. Novikov and G. Chen, Optimizing the signal-to-noise ratio for the PELAN system, *Nucl. Instrum. Methods Phys. Res. Sec. A*, 505(1–2) (2003) 470–473.

[21] G. Viesti, S. Pesente, G. Nebbia, M. Lunardon, D. Sudac, K. Nad, S. Blagus and V. Valkovic, Detection of hidden explosives by using tagged neutron beams: status and perspectives, *Nucl. Instrum. Methods Phys. Res. Sec. B*, 241(1–4) (2005) 748–752.

[22] S. Pesente, G. Nebbia, M. Lunardon, G. Viesti, S. Blagus, K. Nad, D. Sudac, V. Valkovic, I. Lefesvre and M.J. Lopez-Jimenez, Tagged neutron inspection system (TNIS) based on portable sealed generators, *Nucl. Instrum. Methods Phys. Res. Sec. B*, 241(1–4) (2005) 743–747.

[23] F.D. Brooks, A. Buffler, M.S. Allie, K. Bharuth-Ram, M.R. Nchodu and B.R.S. Simpson, Determination of HCNO concentrations by fast neutron scattering analysis, *Nucl. Instrum. Methods Phys. Res. Sec. A*, 410(2) (1998) 319–328.

[24] G. Chen and R.C. Lanza, Fast neutron resonance radiography for elemental imaging: theory and applications, *IEEE Trans. Nucl. Sci.*, 49(4) (2002) 1919–1924.

[25] G. Viesti, M. Lunardon, G. Nebbia, M. Barbui, M. Cinausero, G. D'Erasmo, M. Palomba, A. Pantaleo, J. Obhodas and V. Valkovic, The detection of landmines by neutron backscattering: exploring the limits of the technique, *Appl. Radiat. Isot.*, 64(6) (2006) 706–716.

[26] F.D. Brooks, M. Drosg, A. Buffler and M.S. Allie, Detection of anti-personnel landmines by neutron scattering and attenuation, *Appl. Radiat. Isot.*, 61(1) (2004) 27–34.

[27] H.J. Gomberg, G. Charatis, D. Wang and M.R. McEllistrem, Neutron elastic scatter for detection and identification of obscured objects, *Proc. SPIE*, 1942 (1993) 276–288; Underground and Obscured-Object Imaging and Detection (ed. by Nancy K. Del Grande, Ivan Cindrich and Peter B. Johnson).

[28] R.J. Rasmussen, W.S. Fanselow, H.W. Lefevre, M.S. Chmelik, J.C. Overley, A.P. Brown, G.E. Sieger and R.M.S. Schofield, Average atomic number of heterogeneous mixtures from the ratio of gamma to fast-neutron attenuation, *Nucl. Instrum. Methods Phys. Res. Sec. A*, 124(4) (1997) 611–614.

[29] R.C. Lanza, Neutron resonance radiography for security applications, *Proc. SPIE*, 4786 (2002) 40–51.

[30] P.K. Van Staagen, T.G. Miller, B.C. Gibson and R.A. Krauss, High speed data acquisition for contraband identification using neutron transmission, *AIP Conf. Proc.*, 392(pt. 2) (1997) 853–856.

[31] B.C. Gibson, T.G. Miller, P.K. Van Staagen and R.A. Krauss, Background reduction in neutron attenuation studies, *AIP Conf. Proc.*, 392(pt. 2) (1997) 865–868.

[32] J.I.W. Watterson, J. Guzek, U.A.S. Tapper and S.N. Surujblal, The development of a computational model for fast neutron radiography, *Proc. SPIE*, 2867 (1997) 358–361.

[33] J.E. Eberhardt, S. Rainey, R.J. Stevens, B.D. Sowerby and J.R. Tickner, Fast neutron radiography scanner for the detection of contraband in air cargo containers, *Appl. Radiat. Isot.*, 63 (2005) 179–188.

[34] J.C. Overley, M.S. Chmelik, R.J. Rasmussen, R.M.S. Schofield, G.E. Sieger and H.W. Lefevre, Explosives detection via fast neutron transmission spectroscopy, *Nucl. Instrum. Methods Phys. Res. Sec. B*, 251(2) (2006) 470–478.

[35] T.G. Miller, P.K. Van Staagen, B.C. Gibson and R.A. Krauss, Contraband identification in sealed containers using neutron transmission, *Proc. SPIE* 2867 (1997) 215–218; in International Conference Neutrons in Research and Industry (ed. by George Vourvopoulos).

[36] T.J. Yule, B.J. Micklich, C.L. Fink and L. Sagalovsky, Fast-neutron transmission spectroscopy for illicit substance detection, *Proc. SPIE*, 2867 (1997) 239–242; in International Conference Neutrons in Research and Industry (ed. by George Vourvopoulos).

[37] The Practicality of Pulsed Fast Neutron Transmission Spectroscopy for Aviation Safety. *Report by the National Materials Advisory Board of the Commission on Engineering and Technical Systems*, National Research Council, National Academy Press, Washington DC (1999).

[38] R.M. Ambrosi and J.I.W. Watterson, Factors affecting image formation in accelerator-based fast neutron radiography, *Nucl. Instrum. Methods Phys. Res. Sec. B*, 139 (1998) 279–285.

[39] R.M. Ambrosi, J.I.W. Watterson and B.R.K. Kala, A Monte Carlo study of the effect of neutron scattering in a fast neutron radiography facility, *Nucl. Instrum. Methods Phys. Res. Sec. B*, 139(1–4) (1998) 286–292.

[40] W.L. Raas, B.W. Blackburn, E. Boyd, J. Hall, G. Kohse, R.C. Lanza, B. Rusnak and J.I.W. Watterson, Neutron resonance radiography for explosives detection: technical challenges, Nuclear Science Symposium Conference Record, 23–29 October, *IEEE*, 1 (2005) 129–133.

[41] B. Rusnak and J. Hall, An accelerator system for neutron radiography, *AIP Conf. Proc.*, 576 (2001) 1105–1108.

[42] E. Vano, L. Gonzalez, R. Gaeta and J.A. Gonzalez, An empirical function which relates the slope of the Ge(Li) efficiency curves and the active volume, *Nucl. Inst. Meth.*, 123 (1975) 573–574.

[43] E. Somorjai, Note on the relation between efficiency and volume of Ge(Li) Detectors, *Nucl. Inst. Meth.*, 131 (1975) 557–558.

[44] T. Gozani, The role of neutron based inspection techniques in the post 9/11/01 era, *Nucl. Instrum. Methods Phys. Res. Sec. B*, 213 (2004) 460–463.

[45] F.J. Schultz, D.C. Hensley, D.E. Coffey, R.D. Caylor, R.D. Bailey, J.A. Chapman, G. Vourvopoulos and J.T. Caldwell, Pulsed interrogation neutron and gamma inspection system, *Trans. ANS*, 64 (1991) 144–146.

[46] P.C. Womble, G. Vourvopoulos, J. Paschal and P.A. Dokhale, Multi-element analysis utilizing pulsed fast/thermal neutron analysis for contraband detection, *Proc. SPIE*, 3769 (1999) 189–195.

[47] G. Chen, R.C. Lanza and J. Hall, Fast neutron resonance radiography for security applications, *AIP Conf. Proc.*, 576 (2001) 1109–1112.

Chapter 7

Nuclear Quadrupole Resonance Detection of Explosives

Joel B. Miller

Chemistry Division, Code 6120, Naval Research Laboratory, Washington, DC 20375, USA

Counterterrorist Detection Techniques of Explosives
Jehuda Yinon (Editor)

Contents

1. Introduction

The detection of explosive devices can be sorted into two broad schemes: 'seeing' the device and detecting a component common to all devices. X-ray imaging is a classic example of the former scheme. It provides an image of the contents of the item under inspection; it is incumbent upon the operator to recognize the explosive device. Metal detection is a good example of the latter scheme. Nearly all land mines contain at least a small quantity of metal that can be detected by a metal detector. Of course, many innocent objects also contain metal. The problem is separating the innocent from the dangerous.

To unambiguously detect an explosive device, an obvious choice is to detect the explosive itself. Particle and vapor detection strategies offer high specificity but suffer from difficulties in collecting samples of the explosive for analysis. Dual-energy X-ray and thermal neutron analysis give up some specificity, and issues of radiation can be problematic in some scenarios. In this chapter, we will discuss nuclear quadrupole resonance (NQR). NQR excites and detects radio frequency (RF) signals from certain nuclei in molecules of interest, that is, explosives. The frequency of the signal depends on molecular parameters. The inherent advantages of NQR are high specificity, absence of ionizing radiation, geometry-independent bulk detection, and low false alarm rates. Its main disadvantage is low sensitivity.

This chapter is designed to give a basic overview of NQR explosives detection technology, with references that provide detailed information to the interested reader. Section 2 discusses the basic physics behind the NQR phenomenon, with particular emphasis on the points that relate to detection technology. Section 3 discusses standard NQR detection hardware and signal processing for explosives detection. This section is important to understanding the primary issue in NQR detection: signal sensitivity. The latter half of this section discusses non-standard detection hardware. Section 4 discusses signal excitation schemes, both the commonly used schemes and several other techniques that have been proposed. Section 5 starts with a discussion of the NQR parameters of several of the more common explosives targets before moving on to specific NQR explosives detection hardware.

2. Basic NQR Physics

In this section, a brief, qualitative description of the physics behind NQR, with sufficient details to understand its working in the context of explosives detection, is provided. The reader interested in a more detailed discussion is referred to these review articles and the references therein [1–4]. Basic NQR instrumentation and how the physics and hardware relate to sensitivity are also discussed.

2.1. The quadrupole interaction

Nuclear quadrupole resonance is one of the family of magnetic resonance spectroscopies, the most notable members being nuclear magnetic resonance (NMR) [5] and magnetic resonance imaging (MRI) [6]. Magnetic resonance can be observed from nuclei with a magnetic dipole moment, that is, nuclei with a spin quantum number, $\boldsymbol{I} \neq 0$ ($\boldsymbol{I} = n/2$, where n is an integer). In NMR and MRI, a large static magnetic field is applied to the material under investigation. The nuclear magnetic moments tend to align along the static magnetic field (the Zeeman interaction), splitting the nuclear ground state into $2\boldsymbol{I}+1$ energy levels ($m = -\boldsymbol{I}, -\boldsymbol{I}+1, \ldots, \boldsymbol{I}$). The energy difference between adjacent energy levels m and $m+1$ due to the Zeeman interaction is proportional to the applied static magnetic field, B_0, and a constant of the nucleus, the gyromagnetic ratio, γ_{n}:

$$\Delta E_{m \to m+1} = h\gamma_{\mathrm{n}} B_0. \tag{1}$$

From Eq. 1, we see that, in the absence of other interactions, the splitting is independent of m. The difference in population of the energy levels produces a net nuclear magnetic moment aligned along the static magnetic field. This nuclear magnetic moment can be tipped away from the applied static magnetic field using a time-dependent magnetic field, perpendicular to the static field, and oscillating at frequency

$$\nu_0 = \frac{\Delta E}{h} = \gamma_{\mathrm{n}} B_0. \tag{2}$$

If the time-dependent magnetic field is turned off after the nuclear magnetic moment is tipped away from the static magnetic field, the nuclear magnetic moment will precess about the static magnetic field at frequency ν_0. For typical values of B_0, ν_0 falls in the RF range of tens to hundreds of megahertz. The time-dependent (RF) magnetic fields used to excite the magnetic resonance signal are typically of short duration and are commonly referred to as RF pulses.

2.1.1. The NQR spectrum

The value of magnetic resonance as an analytical spectroscopic technique is in the interactions, other than the Zeeman interaction, that affect the spectrum, including the chemical shift, J-coupling, and dipole–dipole interaction [5]. For NQR, the interaction of primary importance is the electric quadrupole interaction. Nuclei with $\boldsymbol{I} > 1/2$ have a non-spherical charge distribution and hence an electric quadrupole moment, Q. An asymmetric charge distribution outside the nucleus will result in an electric field gradient at the nucleus. The nuclear electric quadrupole moment will tend to align in the electric field gradient, splitting the nuclear ground state into several energy levels, just as in the case of the Zeeman interaction. The most important difference between NMR and NQR is that in NQR we get magnetic resonance without the magnet. However, unlike the Zeeman interaction, in NQR, we cannot control the size of the splitting; it is controlled by molecular parameters. With the quantization axis of NQR tied to the molecular frame, the average projection of the quantization axis on to the detector axis is zero for the rapidly tumbling molecules of a liquid; therefore, unlike NMR, NQR can only be observed in rigid solids.

The distribution of bonding, or valence, electrons is the largest contributor to the electric field gradient. In a molecular frame of reference, we can define three orthogonal components of the electric field gradient, V_{xx}, V_{yy}, and V_{zz}, where $V_{xx}+V_{yy}+V_{zz}=0$ and, by convention, $|V_{xx}| \leq |V_{yy}| \leq |V_{zz}|$. From these electric field gradient components, we define two parameters:

$$eq = V_{zz}$$

and

$$\eta = \frac{V_{yy} - V_{xx}}{V_{zz}}, \tag{3}$$

where η is referred to as the asymmetry parameter and $0 \leq \eta \leq 1$. There is no analytical expression for E for the general case of $\boldsymbol{I} > 1/2$, but, for explosive detection, the most important cases are $\boldsymbol{I} = 1$ and 3/2, for which there areanalytical expressions. The rest of this chapter will focus on $\boldsymbol{I} = 1$, discussing $\boldsymbol{I} = 3/2$ only when it differs from our discussion of $\boldsymbol{I} = 1$ in a relevant way. For $\boldsymbol{I} = 1$, the spin states are labeled 0, +, −, as opposed to −1, 0, +1 [7], with the resonance frequencies

$$\begin{aligned} \nu_+ &\equiv \nu_{+\to 0} = \frac{3}{4}\frac{e^2qQ}{h}\left(1+\frac{1}{3\eta}\right), \\ \nu_- &\equiv \nu_{-\to 0} = \frac{3}{4}\frac{e^2qQ}{h}\left(1-\frac{1}{3\eta}\right), \\ \nu_0 &\equiv \nu_{+\to -} = \frac{1}{2}\frac{e^2qQ}{h}\eta, \end{aligned} \tag{4}$$

where e^2qQ/h is referred to as the nuclear quadrupole coupling constant. For the common cases of $\eta \neq 0$ and 1, there are three observable resonance frequencies, with $\nu_0 + \nu_- = \nu_+$ (Fig. 1). In general, the difference in resonance frequencies is sufficiently large that only one can be excited and detected in an experiment. The NQR spectrum for $\boldsymbol{I} = 3/2$ is simpler:

$$\begin{aligned} \nu &\equiv \nu_{\pm\frac{3}{2}\to\pm\frac{1}{2}} = \frac{1}{2}\frac{e^2qQ}{h}\left(1+\frac{1}{3\eta^2}\right)^{\frac{1}{2}}, \\ \nu_{-\frac{1}{2}\to+\frac{1}{2}} &= 0. \end{aligned} \tag{5}$$

Only one resonance frequency is observable, regardless of the value of η.

The discussion so far has centered around the detection of the NQR signals from a quadrupolar nucleus in a molecule. It is often the case that there is more than one

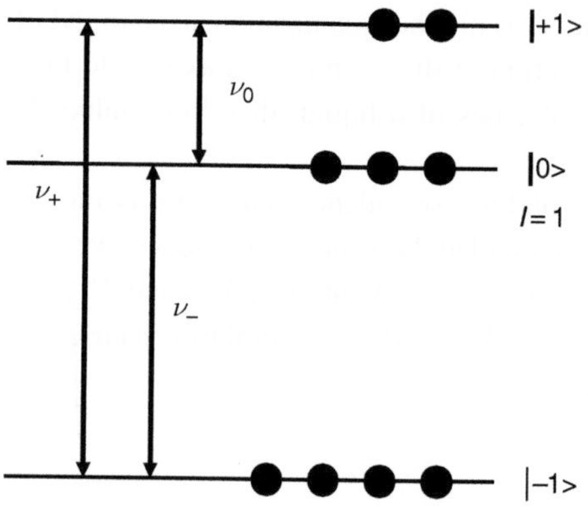

Fig. 1. Energy level diagram for a spin I = 1 nucleus.

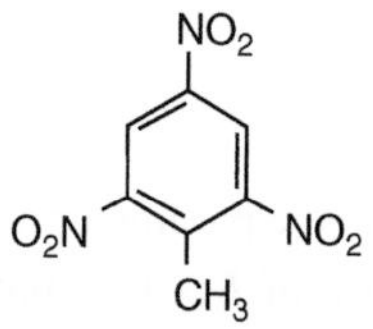

Fig. 2. The chemical structure of 2,4,6-trinitrotoluene (TNT).

quadrupolar nucleus, and hence more than one NQR signal, per molecule. For a good example of how complicated an NQR spectrum can be, consider TNT, whose structure is shown in Fig. 2. The target nucleus for NQR detection of TNT is the ^{14}N isotope of nitrogen, which has $\boldsymbol{I} = 1$. There are three nitrogens per TNT molecule, two of which appear to be chemically equivalent. Although the two nitrogens adjacent to the methyl group appear to be chemically equivalent, they are not magnetically equivalent, having distinguishably different NQR frequencies due to small distortions in molecular symmetry. With three inequivalent nitrogens in the molecule, we expect a total of nine peaks in the NQR spectrum. However, the situation is even more complicated. It turns out that there are two inequivalent molecules per unit cell of the crystal, resulting in 18 peaks for TNT. To further complicate the detection of TNT, there are two crystalline forms, orthorhombic and monoclinic, that can coexist under ambient conditions, each with its own set of 18 resonances. In some samples of TNT, both crystalline forms can be observed together, leading to an NQR spectrum of 36 peaks. Figure 3 shows an NQR spectrum of the ν_+ resonances of a sample of commercial TNT. The sample is a mix of orthorhombic and monoclinic TNT. In most detection scenarios, the NQR excitation/detection bandwidth is not sufficient to observe all 6 (or 12) ν_+ resonances, let alone all 18 (or 36) peaks in the NQR spectrum of TNT.

2.1.2. *Excitation of NQR signals*

In analogy to the Zeeman interaction, NQR signals can be excited and detected by application of a time-dependent magnetic field at the resonance frequency of the nuclei,

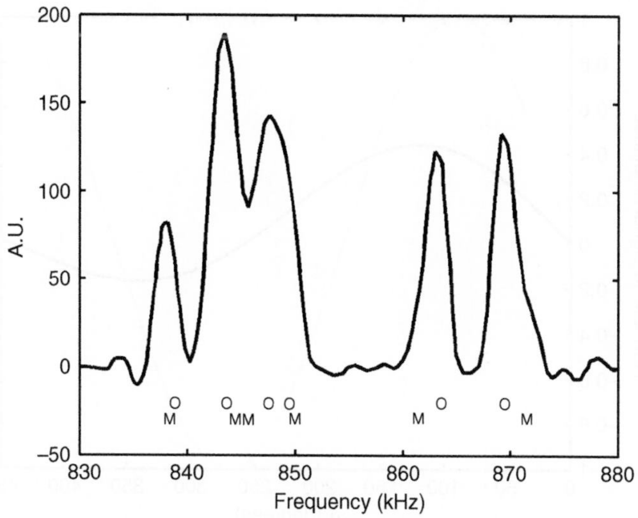

Fig. 3. The ν_+ region of the nuclear quadrupole resonance (NQR) spectrum of mixed orthorhombic and monoclinic 2,4,6-trinitrotoluene (TNT). The positions of the orthorhombic and monoclinic resonances at 13°C are indicated by 'O' and 'M', respectively (data from Fig. 3 of reference [91]).

but without the application of the static magnetic field. Here is another subtle but important difference between NMRand NQR. In NMR, the nuclear magnetic moments are quantized along the applied static magnetic field. By applying an RF pulse orthogonal to the quantization axis, an NMR signal is excited whose amplitude is

$$S_{\text{NMR}} \propto \sin(\alpha), \tag{6}$$

where $\alpha = \gamma_n B_1 t_p$ is the pulse tip angle, B_1 is the magnitude of the time-dependent magnetic field, and t_p is the duration of the pulse. For NMR, the optimal pulse nutation angle is 90°. In NQR, the quantization axis is tied to the molecular frame, and in the case of $\boldsymbol{I} = 1$, there is more than one quantization axis. So what is the optimal direction to apply the RF pulse? For the general case of explosives detection, the sample will consist of a powder of randomly oriented crystallites. The magnitude of the RF pulse 'felt' by the nuclear magnetic moments in a given orientation is equal to the projection of the RF field on the quantization axis of those moments; therefore, nuclei in different crystallites experience different pulse nutation angles. Averaging over all possible orientations in the powder, and assuming that only one of the resonance frequencies can be excited and detected at a time, we arrive at the so-called powder average of the NQR signal:

$$S_{\text{NQR}} \propto \sqrt{\frac{\pi}{2\alpha'}} J_{3/2}(\alpha'), \tag{7}$$

where $J_{3/2}$ is the Bessel function of order 3/2 and $\alpha' = 2\gamma_n B_1 t_p$ is the nutation angle of the RF pulse. (Equation 7 assumes that the NQR signal is detected by the same

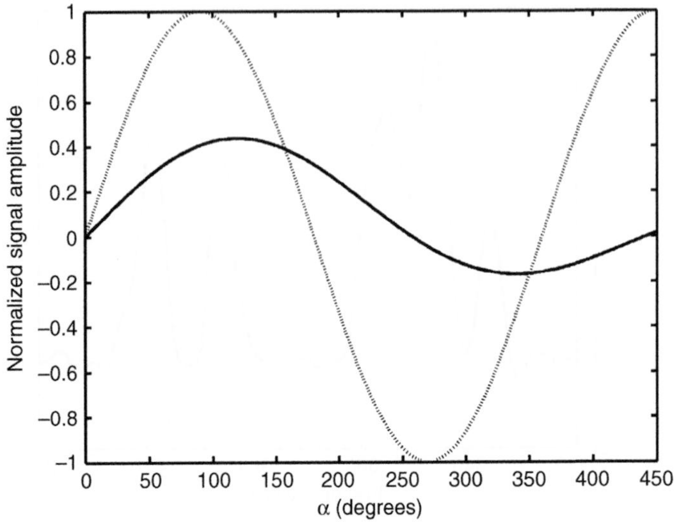

Fig. 4. Signal amplitude as a function of excitation pulse nutation angle for nuclear quadrupole resonance (NQR) of a powder sample (solid line) and single crystal (dotted line).

directional device that emitted the RF pulse. Signal detection will be discussed in more detail in Section 3.) The Besselfunction is similar to a damped sinusoid, with its first peak at 119° and amplitude 0.43 (Fig. 4). Because of the powder average, only a fraction of the quadrupolar nuclei that are present are excited and detected. The powder average also has significant implications for the ability to manipulate NQR signals to enhance sensitivity and reduce the effects of artifacts, as discussed in Section 4.

2.2. Relaxation and spin echoes

Once an NQR signal is excited by an RF pulse, it does not persist indefinitely but decays over a time of tens to hundreds of microseconds. The decay of the signal can usually be reversed, forming a *spin echo*. The amplitude of the spin echoes decays as the time between signal excitation and echo formation increases; in some materials, spin echoes can be formed several seconds after signal excitation. Finally, the spin system relaxes to its equilibrium state after signal excitation with another time constant that typically ranges from tens of milliseconds to tens of seconds. The time constants that govern signal decay and return to equilibrium are collectively referred to as relaxation times. The theory of relaxation in spin systems is well beyond the scope of this chapter; however, these time constants dictate the types of NQR experiments that can be used for sensitive signal detection and determine the ultimate sensitivity of the technique. Only a brief description of the decay constants is provided. The reader interested in a more detailed discussion of spin relaxation may wish to start with these references [7,8]. The effects of relaxation on NQR signal detection are discussed in Section 4.

2.2.1. T_2*

The NQR signal has been discussed in terms of a single frequency determined by the nuclear quadrupole moment and electric field gradients. In real materials, the signal is characterized by a distribution of frequencies. The distribution of frequencies may arise from variations in the electric field gradients throughout the sample. Those variations may be due to strains, temperature gradients, or imperfections in the crystal lattice that distort bond lengths and angles, causing local redistribution of electrons. The distribution of frequencies also may be due to other magnetic interactions such as the magnetic dipole – dipole interaction between neighboring nuclei. In many cases, it will arise from a combination of effects [9]. Observing in the time domain, all the frequencies start with the same initial phase, determined by the RF pulse, but over time, the frequencies lose their phase coherence and destructively interfere with each other, leading to a decay of the signal [called the free induction decay (FID) in magnetic resonance literature]. Although the frequency distribution is not necessarily Lorentzian or Gaussian, it is customary to characterize the decay by a time constant T_2^*. Different explosives exhibit different values of T_2^*, and T_2^* varies somewhat from sample to sample of the same explosive. These variations in T_2^* can be an indication of strain or temperature gradients in the material; NQR has been used as a strain gage in complex materials [10,11] and can be used as a temperature sensor [12]. Sample-to-sample variations in T_2^* also can be an indication of sample purity or crystal imperfection [13,14] and has been proposed as a method of quality control. For explosives, typical values of T_2^* fall in the range of 10–1000 μs.

T_2^* generally does not limit the time over which the signal can be measured: the loss of coherence from a distribution of NQR frequencies can be reversed. When an NQR signal is excited with an RF pulse, a non-equilibrium spin state is created that generates an observable signal. Application of additional RF pulses can change the spin state. With properly chosen pulses, we can create a state that appears to evolve backward in time, retracing its steps such that if the additional, refocusing, RF pulse is applied at a time t after the excitation pulse, a spin echo [5,8,15] forms at a time t after the refocusing pulse. The leading edge of the spin echo is the mirror image of the FID, and the trailing edge matches the FID. There are many schemes for inducing spin echoes [16–23], leading to different types of spin echoes with a variety of names. In theory, for refocusing a distribution of frequencies from a distribution of electric field gradients, the simplest is a single RF pulse whose duration is twice the optimal excitation pulse. This type of echo is often referred to as the 'Hahn' echo in honor of the discoverer of spin echoes in NMR [15]. Keep in mind that only a small fraction of the nuclei see the optimal refocusing pulse because of the orientation dependence of the RF field. In practice, refocusing pulses that are about 1.25 times the optimal excitation pulse length are often found to be the best. Contributions to the distribution of frequencies from other interactions lead to other requirements for the refocusing pulse [9].

2.2.2. T_2

Just as the FID does not last forever, spin echoes cannot be formed at arbitrarily long times after the excitation pulse. The amplitude of the spin echoes decreases as the time between signal excitation and refocusing pulse increases with a time constant T_2. Obviously, $T_2 \geq T_2^*$. Typical values of T_2 range from milliseconds to seconds. If T_2 is much longer than T_2^*, it can be advantageous to form multiple sequential spin echoes. There are a number of RF pulse *sequences* that create a *train* of spin echoes. Trains of hundreds of echoes are possible in some cases. The echoes in the train are observed to decay with a time constant T_{2e}, where $T_{2e} \geq T_2$ [16]. The reasons that $T_{2e} \geq T_2$ lie in the physical causes of T_2, the details of which are beyond the scope of this chapter. For the discussion of explosives detection, it is sufficient to note that T_2 relaxation is caused by low frequency modulation of the electric field gradients or magnetic fields at the nucleus, often caused by molecular motions [24]. The important point is that T_{2e} is sensitive to fields modulated at the inverse of the refocusing pulse spacing. It is generally observed that, using optimal refocusing pulses, T_{2e} increases with decreasing time between the refocusing pulses [20].

2.2.3. T_1

Following saturation by RF pulses, the nuclei return to their equilibrium state at a rate characterized by a time constant T_1, $S(t) = S_0 e^{-t/T_1}$, where $T_1 \geq T_{2e}$. It is not uncommon to observe biexponential signal recovery in materials, that is, signal recovery characterized by two T_1s. In such cases, the shorter T_1 usually accounts for a small fraction of the total signal. Like T_2, T_1 relaxation is caused by modulation of the electric field gradients or magnetic fields at the nucleus, being most sensitive to modulation frequencies at the NQR frequency [25]. Typical values of T_1 range from tens of milliseconds to tens of seconds. T_1 limits how often the measurement of the NQR signal can be repeated, which has a great influence on the sensitivity of explosives detection. The optimal measurement repetition time is $1.25T_1$.

2.3. *Pressure, temperature, and magnetic fields*

As mentioned in Section 2.1.1., the NQR frequency is extremely sensitive to the distribution of electrons about the nucleus, making NQR a good reporter of the chemistry of the material under study. But NQR is also a good reporter of the physics of the material under study. The effects of crystalline morphology have already been mentioned as an example of how the physical structure of the molecules environment affects the NQR frequency. Even small changes in temperature or pressure cause the crystal lattice to expand and contract, resulting in changes in the NQR frequency. The change in resonance frequency with temperature is generally linear over modest temperature ranges, and temperature coefficients of several hundred Hertz per degree Celsius are often observed in ^{14}N NQR. If the temperature change leads to a phase transition, the change in the NQR frequency can be more dramatic. Imprecise knowledge of the temperature translates to imprecise knowledge of the NQR frequency. The effects of directly applied pressure are more subtle than the effects of temperature and do not affect explosives detection, but NQR measurements of 'sensor' molecules embedded in materials have been used to report on stress in the materials [11].

Temperature and pressure can also affect the NQR relaxation times. Thermally activated molecular motions are generally the cause of the modulated fields that give rise to the relaxation discussed in Section 2.2. For crystalline solids, including most explosives, the frequencies of the motions are slow compared with the NQR frequency; therefore, T_1 decreases with increasing temperature. T_2 and T_{2e} are often observed to decrease with increasing temperature as well.

In NQR explosives detection, the exact temperature of the potential target material is never known. In searching for the NQR signal, the detector must be capable of measuring a range of frequencies corresponding to the possible range of temperatures that may be encountered. In addition, a temperature gradient may exist in the target; for example, on a hot, sunny day, a buried land mine may be warmer on the top than on the bottom. This will broaden the NQR signal, effectively shortening T_2^*.

Application of an external static magnetic field can also affect the NQR. For very large magnetic fields, we are in the regime of NMR where the resonance frequency is given by Eq. 2. For nuclei with half-integer spin quantum number such as ^{35}Cl, weak static magnetic fields broaden and shift the NQR. For nuclei with integer spin quantum number such as ^{14}N, the static magnetic field starts to broaden and shift the NQR only when the frequency given by Eq. 2 approaches ν_0 [26]. For many explosives, that translates to a static magnetic field of more than 100 G, therefore the much smaller stray magnetic fields from nearby metallic objects usually have no effect on explosives detection.The potential use of static magnetic fields to enhance the NQR signal will be discussed in Section 4.2.2.

3. NQR detection hardware

Properly designed hardware is important to any successful detection scheme. At first glance, the NQR hardware requirements are deceptively simple: a device to generate a strong RF magnetic field and another to detect a weak RF magnetic field. Clearly, there are parallels between these requirements and those of pulsed radar, and some early NMR systems were built around surplus radar hardware. In this section, NQR hardware and how it affects detection sensitivity is discussed.

3.1. Faraday detection

We cannot discuss the sensitivity of NQR detection without considering the hardware used for detection. As discussed in Section 2.1.2., NQR signals are commonly excited with an RF magnetic fieldpulse. An RF magnetic field is easily generated by flowing alternating current through an inductor. The inductor, or coil, may take the form of a solenoid, a flat loop of wire, or some more sophisticated design, depending on the application [27,28].To increase efficiency, the coil is resonated at the NQR frequency with a capacitor (C), and matched to the impedance of the driving amplifier (and the receiver preamplifier noise-match impedance) with matching impedance X_m, to form the NQR *probe* circuit [29] with resistive impedance, R_0, typically 50Ω, as shown in Fig. 5. The nuclear magnetic moments excited by the RF pulse produce an oscillating RF magnetic field parallel to the direction of the applied RF magnetic field. The most common way to detect this oscillating RF magnetic field is through Faraday detection: the RF magnetic field induces an RF voltage in the nearby coil. Efficient detection of the oscillating nuclear magnetic moments dictates that the detection coil be capable of creating RF magnetic fields parallel to the oscillating moments [27]. The simplest way to achieve this is to use the same coil for excitation and detection. There are some cases where separate excitation and detection coils provide a better solution for NQR detection, but most work to date has employed a single coil. Here, a single coil for Faraday excitation and detection is assumed. In Section 3.2., alternative detection hardware that has been proposedis discussed.

3.1.1. *Sensitivity of Faraday detection*

The RF magnetic field created by the oscillating magnetic moments from an average quantity of explosive is very small and, in turn, induces a very small RF voltage in a typical Faraday detector coil. As shown below, this RF voltage is on the order of the thermal noise voltage from the coilresistance. It is interesting to consider the factors that influence the signal-to-noise ratio (SNR) because much of the research in NQR explosives detection is directed toward improving the SNR. Several authors have addressed the issue of SNR in magnetic resonance, modeling either the sample enclosed in the detection coil [7,27–29] or the sample remote from the detection coil [29,30].

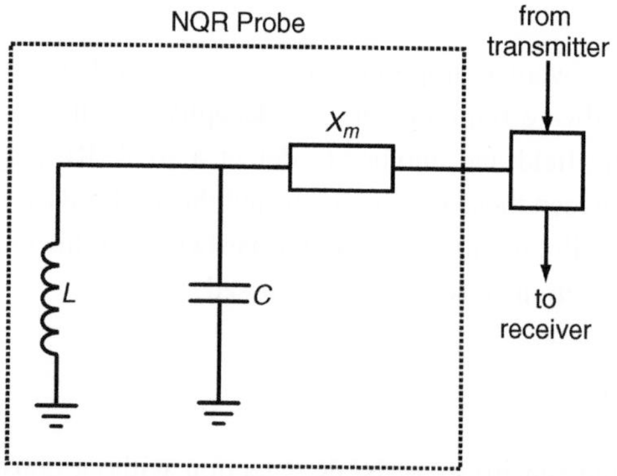

Fig. 5. Basic electronic circuit for a nuclear quadrupole resonance (NQR) probe: L is the detection coil; C is the tuning capacitor; X_m is a component to match the probe impedance to the transmitter/receiver impedance.

With these models, an order-of-magnitude estimate of the signal size is obtained. Both models require calculation of the magnetization of the nuclear spins and the noise due to the detector coil. The rms thermal noise voltage for the probe circuit resistance, R_0, is (Fig. 5)

$$V_N = \sqrt{4kTR_0\Delta\nu}, \tag{8}$$

where $\Delta\nu$, the signal bandwidth, is, for the purposes of NQR, the resonance line width, which is proportional to $1/T_2^*$. For a typical probe circuit resistance of 50Ω and bandwidth of 1000 Hz, at ambient temperature, the noise voltage is approximately 29 nV or −140 dBm.

To calculate the nuclear magnetization produced by the oscillating nuclear magnetic moments, we assume $\boldsymbol{I} = 1$ for ^{14}N and that only one of the three transitions is excited and detected. The nuclear magnetization is then given by [30]:

$$M(\nu) = 0.43\frac{N_s}{V_S}\frac{\gamma_n h}{3}\left(\frac{h\nu}{kT}\right), \tag{9}$$

where V_S is the sample volume. N_s is the number of nuclei in the sample active in the observed transition:

$$N_s = \frac{N_a W_f m}{AW}, \tag{10}$$

where N_a is Avogadro's number, m is the mass of the sample, W_f is the weight fraction of the NQR-active nucleus contributing to the observed transition, and AW is the atomic weight of the nucleus being observed. Note that N_s is not necessarily the total number of NQR-active nuclei in the molecule: from Fig. 2, there are three ^{14}N atoms in the TNT molecule; however, there are six ^{14}N resonances for each crystalline type. If only a single resonance is observed, then that signal is derived from only 1/6 of the ^{14}N atoms in the sample. The population difference between the two energy levels involved in the excited transition is given by the Boltzmann factor in parentheses in Eq. 9. The numerical factor, 0.43, comes from the NQR powder average (Eq. 7) with optimal excitation. For the simple geometry of the sample enclosed in the detector coil (local detection), appropriate to package and baggage screening, the voltage induced in the detector coil by the nuclear magnetization is given by a modified Faraday's law:

$$S_L = \zeta\mu_0 2\pi\nu M(\nu) A\sqrt{\frac{QR_0}{2\pi\nu L}}, \tag{11}$$

where μ_0 is the permeability of free space and A is the turns-area, the cross-sectional area of the coil times the number of turns. Several terms have been included in Eq. 11 that scale the Faraday response to account for the details of NQR detection. We define ζ, the 'filling factor', as the ratio of the volume of the sample to the volume of the coil. For baggage or mail screening, ζ can be as low as a few percent. The term under the square root comes from tuning and matching the coil [30]: Q is the quality factor of the coil and L is the coil inductance. Exciting signal from a single resonance in 100 g of TNT at 840 kHz, using an NQR probe with $\zeta = 1$, $Q = 200$, $L = 1\,\mu\text{H}$, and $A = 1.4 \times 10^{-3}\text{m}^2$, we calculate an rms NQR signal of 47 nV, or −130 dBm, corresponding to an SNR of 1.6.

Recently, the issue of remote detection has garnered a lot of attention. For reasons that will be made obvious from the forthcoming discussion, NQR cannot be a remote detector in the sense of having a large standoff distance. Instead, remote NQR detection refers to detecting materials that are not contained within the detector. Such an NQR apparatus might consist of a simple flat-loop detector coil (Fig. 6), similar in geometry to a metal detector used for land mine detection. A particularly straightforward method to model the signal in remote NQR detection is to treat the coupling of the oscillating nuclear magnetic moments to the detector coil as a mutual inductance. Here, we will follow the presentation in reference [30]; a similar discussion appears in reference [31].

Consider the NQR detection geometry shown in Fig. 6, consisting of a disk-shaped sample of radius r and thickness t at a distance d from the NQR detector coil. The sample containing the NQR-active nuclei has a magnetization, $M(\nu)$, that creates a magnetic flux density, $B(d)$, at the detector coil (M. V. Romalis and K. L. Sauer, personal communication):

$$B(d) = \frac{\mu_0}{2\pi} \frac{V_S}{d^3} M(\nu) . \tag{12}$$

The magnetic flux density created by a single transition from 100 g of TNT at a distance of 10 cm is 1.0 fT. Even at only 10 cm distance, the flux density is approaching the detectability limits of the most sensitive magnetometers. The flux density decreases with d^{-3}, severely limiting the detector standoff distance: increasing the standoff distance to 1 m decreases the flux density by a factor of 1000.

As in the case of the sample surrounded by the detector coil, the oscillating nuclear magnetization induces a voltage in the remote detector coil. In addition to the factors in Eq. 13, the magnitude of the voltage depends on the detector coil geometry and its placement relative to the sample. Assuming that the NQR-active nuclei are uniformly distributed, the nuclear magnetization can be treated as being created by bound surface currents on the sample (Fig. 5). For the sample geometry of Fig. 6, this reduces to a

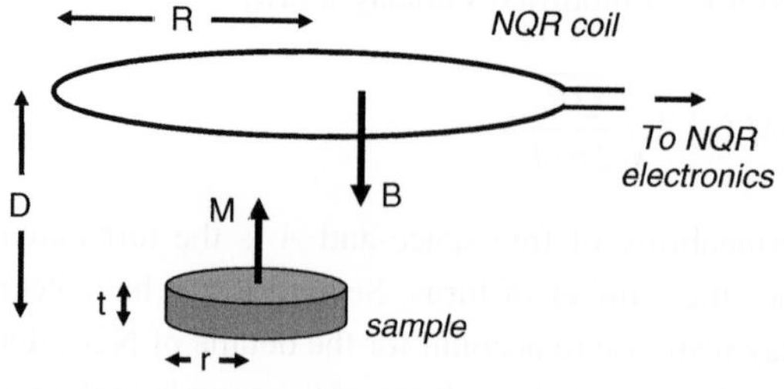

Fig. 6. Detection geometry appropriate to land mine detection.

simple current loop of radius r and thickness t [30]. The current, $I(\nu) = tM(\nu)$, in the fictitious current loop creates a magnetic field at the detector coil, inducing a voltage:

$$S_R = 2\pi\nu MI(\nu)\sqrt{\frac{QR_0}{2\pi\nu L}}, \tag{13}$$

where M, the coupling or mutual inductance, between the detector coil and fictitious current loop, contains all the information about the coil geometry and placement. The mutual inductance can be calculated for any combination of sample and detector coil geometries; for the simple geometry presented here, values of M have been calculated, adjusting the coil radius, R, to optimize M for the values of r and d, and are plotted in Fig. 6 of reference [30]. For the same geometry used to calculate $B(d)$ earlier, and with $t = 7.8 \times 10^{-3}$m, $M = 0.02\,\mu$H, we calculate an rms NQR signal of 2.2 nV or -160 dBm.

From the preceding calculations, we can draw a few general conclusions. First and foremost, NQR signals are weak. TNT was chosen as the example explosive in part because it is something of a worst-case scenario. The multiple resonances in TNT lead to a low value of W_f. Increasing the detector bandwidth can allow detection of several resonances simultaneously, thereby increasing W_f, although the increased bandwidth also increases the noise. Most explosives of interest for NQR detection today have resonance frequencies higher than that of TNT (see Tables in Section 5.1.). From Eq. 13 or 15, and Eq. 8, the NQR signal is proportional to $\nu^{3/2}$, leading to a significantly reduced SNR at lower frequencies.

Some parameters that affect SNR are within the equipment designer's control. Detector coils can sometimes be designed with much higher Q factors than used in the calculations earlier. The SNR improves with the square root of Q, even in the limit where the Q is so large that it results in a detector bandwidth smaller than the NQR line width [32]. Such a 'super Q' is attainable with superconducting detector coils [33]. Unfortunately, the detector coil 'recovery time' also increases with Q [29]. The time taken by the detector coil to dissipate the energy from the excitation pulses, τ_r, so that signal detection can begin is

$$\tau_r = N\frac{Q}{\pi\nu}, \tag{14}$$

where N, a constant of the hardware and the RF pulse amplitude, is typically in the range of 20–30. The recovery time at low frequency can be long compared with $T_2{}^*$, creating difficulties in observing the NQR signal. In practice, this problem is usually dealt with by 'Q-damping', switching the detector coil Q to a low value for a short time after the excitation pulse [34–36]. We should note that even though very high Q coils can be built in the laboratory, the Q cannot always be maintained in the field. When a detector coil is in proximity to a lossy conductor, some of the energy stored in the magnetic field of the coil is dissipated in the lossy conductor, resulting in a decrease in Q [37]. Such effects are well known in NMR of biological samples and MRI of animals, both of which contain significant quantities of saline.

The detector coil area and filling factor also, to some extent, are under the equipment designer's control. In local detection schemes, the volume of the detector coil is often dictated by the item to be scanned for explosives, for example, airline carryon baggage. The coil geometry can be optimized to allow access for the largest carry on bags, but the filling factor is beyond the control of the equipment designer. Likewise, for remote detection, the coil geometry can be optimized for the form of explosive to be detected. When searching for land mines, for example, the detector must be designed for a range of mine sizes and depths. A very rough rule of thumb in remote detection is that the detector coil can detect objects up to one-coil radius away.

From Eq. 11, one might be tempted to increase the number of turns in the coil to increase A; however both A and $\sqrt{L}$ are proportional to the number of turns, resulting in S_L being independent of the number of turns. In practice, L is chosen to give acceptable values of the tuning and matching components and to obtain the desired Q. It is interesting to note that, at constant filling factor, S_L is proportional to the square root of the coil volume ($A \propto V^{2/3}$, $L \propto V^{1/3}$): it is easier to detect large quantities of explosive in a large detector than small quantities in a proportionately smaller detector.

3.1.2. Interference mitigation

Many common materials contain nitrogen or one of the many other nuclei from which NQR can be measured [38]. Fortunately, Resonances are narrow in NQR and are dispersed over a very large bandwidth so that NQR signals from benign materials do not interfere with the detection of explosives. Based on an NQR database of more than 10 000 compounds [39], no NQR frequencies from benign materials are found within the typical detection bandwidth of an NQR system when it is set to the frequency of any of the explosives detected by NQR.

Although there is no interference from other NQR signals, other types of signals may interfere with NQR explosives detection. For unshielded detector coils, the primary culprit is RF interference (RFI). Most NQR frequencies of interest for explosives detection fall in the frequency range of 0.5–5 MHz. The biggest problem in this range is the amplitude modulation (AM) radio band, from 0.5 to 1.5 MHz. Besides radio transmissions, other sources of RFI can include nearby electrical equipment, power transmission lines and lightning strikes.

The most reliable method for reducing or eliminating RFI is shielding the NQR detector coil from distant RF sources. Shielding is straightforward when designing a system for package or baggage scanning but is impractical for applications such as land mine detection. Two approaches have been used when shielding is not a viable option, both relying on the RFI being from a distant source: digital subtraction of the RFI from the detected signal and the use of NQR detector coils that are insensitive to distant RF sources.

The first technique, often referred to as RFI mitigation [40], uses auxiliary detectors that are sensitive to the RFI but not the NQR signal to record the RFI, a weighted version of which is then subtracted from the signal recorded with the NQR detector coil, which is made up of the RFI and, potentially, the NQR signal. The auxiliary detectors must produce a low noise version of the RFI in order not to degrade the SNR of the NQR signal. It is important that the auxiliary detectors be proximal to the NQR detector coil to ensure that both detectors record the same RFI.

The second technique uses an NQR detector coil known as a gradiometer [41–44]. The standard coils used for NQR detection are magnetometers; they possess a dipole moment and are very sensitive to uniform magnetic fields. In contrast, the ideal gradiometer does not possess a dipole moment, is insensitive to uniform magnetic fields, but detects magnetic field *gradients*. RFI, being a distant RF source, presents a uniform magnetic field, whereas the NQR signal emanates from a nearby source and presents a magnetic field gradient. For the purposes of NQR detection, nearby is considered to be within about one-coil diameter. In its simplest form, a gradiometer is a coil with two equivalent sets of counter wound turns (Fig. 7). The offset of the sets of turns determines the magnitude and spatial profile of the RF magnetic field created locally by the coil. At greater distances, the fields created by the two sets of turns cancel each other.

Both of these techniques depend on separating 'near source' signals from 'distant source' signals. From the point of view of the NQR detector coil, distant sources may not appear so distant if their fields are distorted locally, for example, by a nearby metallic object. In such instances, the gradiometer is expected to provide better cancellation than RFI mitigation because it requires spatial uniformity of the RFI only over the area of the coil, as opposed to over the distance between NQR detector coil and auxiliary detectors for RFI mitigation. Gradiometers also have an advantage in canceling the RFI at the NQR detector coil, reducing the overall dynamic range of the NQR system required in high RFI environments. The effectiveness of RFI cancellation with a gradiometer is limited by our ability to balance the two sets of turns in the gradiometer coil. RFI reductions of 30–40 dB have been reported with gradiometers [41,42]; much higher reductions have been reported for RFI mitigation [40]. The overall sensitivity of gradiometers to NQR signals is lower than that for magnetometers, particularly at greater distances from the detector coil. The use of RFI mitigation and gradiometers is not mutually exclusive. In very high RFI environments, the two techniques could be combined.

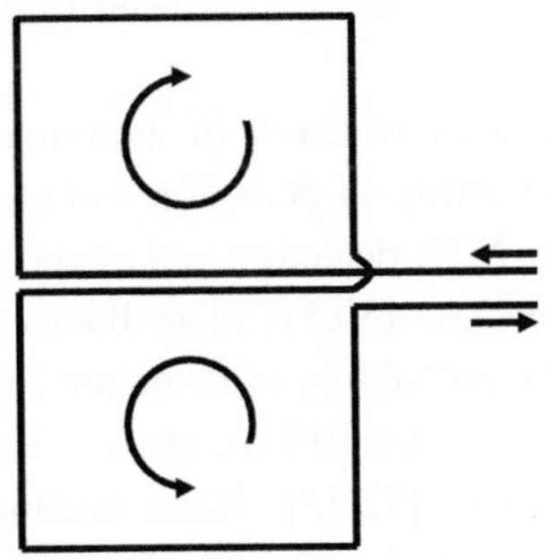

Fig. 7. Schematic of a gradiometer coil. The arrows indicate the direction of current flow.

In addition to RFI, there are two other sources of interference in NQR detection: piezoelectric ringing [45] and magnetoacoustic ringing [46,47]. Both piezoelectric and magnetoacoustic ringing are induced by the RF pulses that excite NQR signals. Piezoelectric crystals such as quartz give rise to the former source of ringing. Small magnetic particles or domains in larger pieces of material give rise to the latter. Piezoelectric ringing differs from magnetoacoustic ringing in that it is excited by the electric field associated with the RF pulse, and the signal is detected as an electric field. Magnetoacoustic ringing, like NQR, is excited by, and detected as, a magnetic field. The frequency and amplitude of ringing from both sources depends on the size and the orientation of the crystals or domains as well as their molecular properties. Both types of ringing often show a strong temperature dependence.

Ringing mitigation is usually accomplished using pulse sequences that take advantage of the different response to RF pulses exhibited by NQR spin echoes and ringing artifacts. Details of the pulse sequences are discussed in Sections 4.2.1. and 4.2.2. Piezoelectric ringing can also be reduced by minimizing the electric fields associated with the NQR detector coil: careful coil design to minimize coil capacitance; careful placement of tuning and matching capacitors; and use of an electric field, or Faraday, shield [41,42,48,49].

3.1.3. Signal processing

The standard method of processing the complex time-domain NQR data recorded with the detector, sometimes referred to as demodulation processing, involves converting the data to the frequency domain with the fast Fourier transform. The echoes in multiple spin-echo data may be summed before Fourier transformation. The echoes may be weighted with the expected T_{2e} before summation. If the appropriate factors are known or can be determined from the data, the phase of the frequency-domain data is adjusted to produce a real absorptive spectrum. If the phase parameters are unknown, the magnitude of the frequency-domain data can be used, at the expense of reduced SNR in the low SNR limit. Within the frequency range of the expected NQR signal, the largest signal is selected and compared with a predetermined threshold value. Signals that exceed the threshold produce an alarm. If the phase parameters are known a priori, the phase of the signal can be used to distinguish NQR signals from interference.

There has been a great deal of research in applying advanced signal processing techniques to increase the probability of detection and reduce false alarms due to RFI. Bayesian signal processing for NQR detection and adaptive filtering for RFI mitigation have been used with some success [50,51]. The Bayesian signal processing showed measurable improvement in the probability of detection for a fixed false alarm rate on a set of TNT data. With hardware-based RFI mitigation, Kalman filtering has been used to further reduce the residual RFI [52,53]. These techniques can operate even when the frequencies of the NQRs are not known precisely. Another technique that operates on NQR signals in the presence of RFI when the NQR frequencies are not known is referred to as frequency-selective approximative maximum likelihood [54]. Results with this technique were also demonstrated with TNT data, using four resonances observed in one experiment. Whereas the Bayesian method operates on frequency-domain data, this method operates directly on the time-domain data.

3.2. *Other detection hardware*

Faraday detection as described in Section 3.1., is the traditional method for recording NQR signals. As discussed earlier, Faraday detection sensitivity is limited by the detector's own thermal noise. The desire for improved detection sensitivity has driven the search for non-traditional methods and lower noise Faraday methods of RF magnetic field detection. In this section, several methods that have been applied to NQR signal detection will be discussed.

3.2.1. Cooled and superconducting detector coils

From Eqs 8, 11, and 15, we see that there are two parameters affecting the signal gain and thermal noise voltage that are under the control of the detector designer, the coil resistance, and temperature, through its Q. Copper is the most commonly used metal for fabricating detector coils. Under ambient conditions, only silver has lower resistance than copper, and only by a few percent [55], therefore reducing the thermal noise by lowering temperature would appear to be the most practical approach to improving sensitivity. Equation 8 predicts that cooling the detector coil to 77K (liquid N_2 boiling point) reduces the detector coil's thermal noise by a factor of 2. There is also some increase in signal gain because cooling the copper to 77K reduces its resistance by a factor of 2 resulting in an equivalent rise in Q. Coil noise can be reduced even more by further reducing the temperature. It is important to keep in mind that as we reduce the coil noise with a cooled coil, other sources of thermal noise will start to dominate: tuning and matching elements in the probe, preamplifier input noise, transmission lines connecting the components, and pickup of noise from nearby 'warm' metal objects [56].

Although the resistance of copper is close to the lowest attainable with a normal metal, it is possible to get very low resistance if a superconductor is used. NMR probes with high-temperature superconducting coils have been commercially available for several years. Recently, reports of their use in NQR detection systems have appeared [33,57–60]. The coil Qs in these NQR probes can exceed 50 000. The $Q^{1/2}$ signal gain coupled with the reduced noise from the lower temperature (the superconducting coils are generally operated at 77K) gives very large sensitivity enhancements. Very high Q probes have extremely narrow detection bandwidths requiring special techniques for dealing with materials whose NQR frequencies are unknown because of uncertainty in the material's temperature. Special methods are also needed to avoid the long probe recovery time associated with high Q.

3.2.2. SQUID magnetometers

A superconducting quantum interference device (SQUID) is a ring of superconducting material interrupted by either one or two Josephson junctions, a thin layer of insulating material [61]. A magnetic flux passing through the SQUID induces a current in the ring proportional to the flux, making it an ideal device for measuring small magnetic fields. SQUID magnetometers used for the detection of magnetic resonance signals can have very high sensitivities [62]; a SQUID magnetometer tuned to 425 kHz has a reported magnetic field sensitivity of 0.08 $fT/\sqrt{Hz}$ [63]. For magnetic resonance detection, it is common to use a 'flux transformer', a pickup coil electrically connected to a second coil that in turn is inductively coupled to the SQUID, to couple the magnetic field from the oscillating magnetic moments to the detector. Several reports of NQR signal detection with SQUID magnetometers have appeared [64–67].

There are several factors that have made SQUID magnetometers less attractive than other methods for NQR explosives detection. The SQUID magnetometers used for NQR detection were cooled to 4.2K for operation, as were the flux transformers. SQUIDs are generally broadband devices and must be shielded from stray magnetic fields. Optimal performance is obtained at low frequency, with sensitivity decreasing with increasing frequency.

3.2.3. Atomic magnetometers

A relatively new entry in the field of magnetometry is the atomic magnetometer [68]. Atomic magnetometers operate on the principles of magnetic resonance: they measure optically the spin precession frequency of highly spin-polarized gas-phase alkali-metal atoms in the presence of a magnetic field. From Eq. 2, the precession frequency is proportional to the magnetic field. The high spin polarization is obtained by saturating an appropriate optical transition. Like SQUID magnetometers, these atomic magnetometers are most sensitive at low frequency, with their sensitivity being proportional to the inverse of the frequency at higher frequencies. They do have the advantage of operating at ambient temperature, except for the cell containing the alkali metal, which must be heated to vaporize the metal. Flux transformers are not needed with the atomic magnetometer.

Recently, a variant of the atomic magnetometer, designed for the detection of RF magnetic fields, has been demonstrated [69]. This RF atomic magnetometer is tuned to the frequency of interest with a small static magnetic field. The RF magnetic field to be detected acts like an excitation pulse to tip the alkali-metal spins away from the static field. The tip angle, on the order of nanoradians, is detected optically. The theoretical sensitivity of the RF atomic magnetometer is on the order of 0.01 $fT/\sqrt{Hz}$. Detection of both NMR [70] and NQR [71] signals has been reported.

3.2.4. Force detection

Nanotechnology has brought new levels of sensitivity to detection technology. A prime example is the atomic force microscope (AFM), which detects small variations in the forces between interacting atoms. The technology behind the AFM is being developed to detect the magnetic forces between nuclear spins and a sensor, the magnetic resonance force microscope (MRFM) [72]. The MRFM is designed to detect magnetic resonance from microscopic samples at close range. Detection of NMR and EPR signals has been reported. Recently, it was proposed that MRFM-like technology be extended to the detection of NQR from large samples [T. Rayner, personal communication]. The proposal is unique in that the detection of the RF electric field accompanying the RF magnetic field from the oscillating magnetic moments is proposed.

4. Signal excitation methods

In this section, we will discuss signal excitation methods. We start with the commonly used pulse sequences before moving on to new and hybrid excitation schemes.

4.1. Pulse sequences

Even with the hardware optimizations discussed in Section 3.1.1., the typical SNR in an NQR detection measurement is very low. Fortunately, we have seen that the SNR is limited by the thermal noise in the detector. For white noise, the noise amplitude increases with the square root of the number of added measurements, whereas the signal increases linearly with the number of added measurements. This means that we can repeat the measurement as many times as the interrogation time and T_1 allow, adding the results to improve the SNR. In addition, for each T_1 period, we may be able to generate and combine multiple spin echoes to further improve the SNR. The NQR FID signal tracks the phase of the excitation pulse, but spin-echo signals, being formed by two or more RF pulses, can have a more complicated phase relationship with the pulse sequence. This complex phase behavior can be used to separate NQR signals from other interfering signals discussed in Section 3.1.2. Excitation schemes that generate many intense spin echoes with specified phases are the goal of much of the NQR explosives detection research.

The excitation schemes, or pulse sequences, that optimize the SNR depend on the relaxation times and frequencies inherent to the quadrupolar nuclei under observation and on the characteristics of the excitation/detection hardware. Focusing first on relaxation times, we define two broad categories: (i) $T_2 << T_1$ and (ii) $T_2 \approx T_1$. Most materials fall in the former category, and the pulse sequences discussed in Sector 4.1.1. all involve the creation and detection of multiple spin echoes, which generally requires that T_{2e} be at least several tens of milliseconds. The most notable exception is RDX, which is in the second category under ambient conditions. For materials with very short T_1s such as RDX at elevated temperatures, a single echo pulse sequence, or even a single-pulse FID acquisition, may be required.

For optimal performance of all pulse sequences, parameters such as irradiation frequency, pulse phase, duration, and timing, and RF magnetic field amplitude must be accurately set. Conditions in the field can make this difficult. Although small deviations from the optimal settings do not lead to large changes in the SNR, in most applications of NQR detection, even a small change in SNR can be important. The use of feedback NQR has been suggested to address this problem [73,74]. Feedback NQR is not a signal processing method. Rather, it creates a software feedback loop with the detected signal to optimize pulse program parameters in real time.

4.1.1. Pulse sequences for $T_2 << T_1$

Many different pulse sequences have been proposed for generating spin echoes in NQR, and most are applicable to the regime where $T_2 << T_1$. The most commonly encountered, and simplest, pulse sequence in NQR is historically referred to in the NQR literature as the spin-lock-spin-echo (SLSE) sequence [16]. The name refers to the fact that the decay time constant of the spin-echo train, T_{2e}, can approach the spin-locking time constant, $T_{1\rho}$ [5]. It is functionally the same as the Carr–Purcell–Meiboom–Gill (CPMG) pulse sequence [75,76] described in the NMR literature. The SLSE sequence comprises an initiating or excitation pulse followed by a delay $\frac{1}{2}\tau$, and then a series of refocusing pulses, each separated by a delay τ (Fig. 8(a)). The refocusing pulses all have the same phase and are phase-shifted by $\pi/2$ relative to the excitation pulse. As in single-pulse excitation discussed in Section 3.1.2., the optimal excitation pulse nutation angle in the SLSE sequence is 119°. By analogy to the CPMG sequence in NMR, one might predict a refocusing pulse nutation angle of 240°. In NQR, the situation is complicated by the powder average effect and the contributions of magnetic interactions other than the quadrupole interaction. A recent detailed study has found that the optimal refocusing pulse nutation angle can vary from one material to another over a range of more than 100° depending on the types, strengths, and relative orientations of the magnetic interactions in the material [9]. In practice, the refocusing pulse nutation angle is most often optimized experimentally; it is frequently found to be near 150°. The echo amplitudes and T_{2e} generally have only a weak dependence on the refocusing pulse nutation angle in the vicinity of 150°.

The selection of the interpulse delay, τ, is very important for optimizing the SNR with SLSE. In many materials, the spin-echo train decay constant, T_{2e}, is observed to increase with decreasing τ. Increasing T_{2e} allows the sampling of more spin echoes, enhancing the SNR; however, there are fundamental limitations on the length of τ. Foremost is the probe recovery time, τ_r (Eq. 14). Following each refocusing pulse, the probe must dissipate the energy stored in the RF magnetic field before the weak NQR signal can be detected. Because the echo peak occurs at $1/2\tau$ after the pulse, τ is often set to approximately $2\tau_r$, to allow the detection of the echo peak. The second limitation is of a more practical nature. During the echo, the signal rises and falls with the time constant T_2^*. Except in cases where T_{2e} is a strong function of τ, there is no advantage to $1/2\tau << T_2^*$: for times much shorter than T_2^*, there is no appreciable decay of the

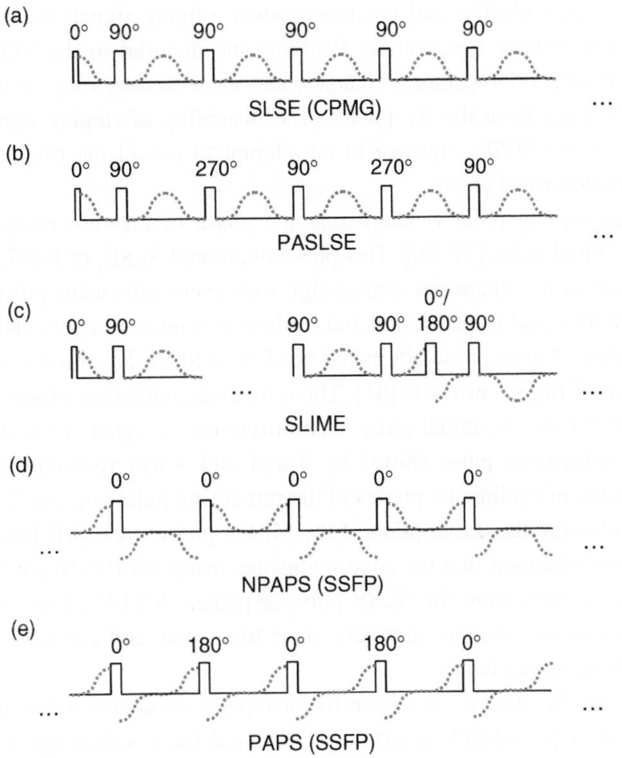

Fig. 8. Pulse sequences for nuclear quadrupole resonance (NQR) excitation: (a) spin-lock-spin-echo (SLSE); (b) phase-alternated spin-lock-spin-echo (PASLSE); (c) spin-lock-inversion-midecho (SLIME); (d) non-phase-alternated pulse sequence (NPAPS); (e) phase-alternated pulse sequence (PAPS). Numbers above the pulses indicate the relative phase of the RF; dashed lines represent expected NQR signals.

NQR signal and hence nothing to refocus. In fact, taking into account probe recovery, the SNR can drop as τ becomes too short because the fraction of time during the pulse sequence devoted to signal measurement decreases. The optimal SNR is obtained from an exponentially decaying signal with Gaussian white noise when the signal is integrated over 1.25 times the decay constant.

The SLSE pulse sequence can be used to reduce the effects of ringing artifacts in addition to enhancing the SNR. Exciting with the SLSE sequence, the phase of the NQR signal follows the initial pulse and is independent of the refocusing pulses. In contrast, the phase of piezoelectric and magnetoacoustic ringing signals follows the phase of the most recent pulse, or in the case of long-lived ringing the phases of a few previous pulses, so that after a very few refocusing pulses the phase of the ringing artifacts is determined by the refocusing pulses. Ringing can be reduced or canceled by phase-shifting the initial pulse *or* the refocusing pulses by 180° in a second experiment and properly co-adding the data from the two experiments.

Complete cancellation of ringing is limited by the phase and amplitude stability of the RF pulses and by the reproducibility of the ringing signals. The amplitude, frequency, and duration of piezoelectric and magnetoacoustic ringing signals are affected by the environment, particularly temperature. Bringing the material to the NQR detector, or vice versa, can induce temperature changes, and local heating may be induced by the absorption of energy from the RF pulses. The instability of ringing signals has led to modifications of the SLSE sequence in an attempt to cancel the ringing on a shorter time scale than described above.

The obvious starting point is alternating the phase of the refocusing pulses $\pm\pi/2$ relative to the initial pulse [77–80]. This phase-alternated SLSE, or PASLSE (Fig. 8(b)) causes the phase of the ringing to change sign with every refocusing pulse, whereas the phase of the NQR signal is unchanged, but leads to less intense echoes and unacceptable reduction in SNR. A four phase-alternated version of SLSE has also been proposed for the cancellation of ringing artifacts [81]. The refocusing pulses are phase-shifted by odd multiples of $\pi/4$ from the initial pulse. The difference of signals from two sequences, with the first refocusing pulse shifted by $+\pi/4$ and $-\pi/4$ respectively, cancels the ringing. A scheme of cycling the phases of the refocusing pulses by $\pm\pi/2$ relative to the initial pulse following the prescription of the MLEV phase cycle [81] has recently been proposed. It was observed that the echo intensities using MLEV-16 are approximately one-half that of echoes from the SLSE pulse sequence. MLEV phase cycling has the advantage of canceling ringing on a very short time scale and can cancel ringing that persists for a time longer than τ.

Another successful strategy has been to intersperse inversion pulses in the train of refocusing pulses to periodically invert the NQR signal, the so-called spin-lock-inversion-mid-echo (SLIME) pulse sequence [83]. As the name suggests, the inversion pulses are inserted into the SLSE pulse sequence to coincide with the echo maxima (Fig. 8(c)). The inversion pulses do cause some loss of signal intensity; therefore, the number of refocusing pulses between inversion pulses, typically tens to hundreds, is chosen as a compromise between maintaining adequate SNR and reducing the ringing artifacts.

4.1.2. Pulse sequences for $T_2 \approx T_1$

The pulse sequences that work in the regime where $T_2 << T_1$ also work in the regime where $T_2 \approx T_1$; however, there is a class of experiments based on steady-state free procession (SSFP) that can be particularly advantageous in the latter regime [84,85]. In SSFP, a dynamic equilibrium, or steady state, is established between the NQR relaxation processes and pulse sequence excitation. Once the steady state is established, in a time on the order of T_1, the echo amplitudes no longer decay, and the echo train can be infinitely long, allowing for very efficient signal acquisition. The echo amplitudes depend on T_2/T_1 and the pulse spacing [85]. SSFP pulse sequences differ from SLSE sequences in that there is no initial, or excitation, pulse (Fig. 8(d)). With no initial pulse, the echoes form during the refocusing pulses. The part of the echo that immediately follows a refocusing pulse is called the beginning echo, or FID, and is usually not observable because of the probe recovery time. The observable part of the echo comes immediately before the refocusing pulse and is referred to as the end echo.

There are two versions of the basic SSFP pulse sequence: the first is a train of identically phased pulses, commonly referred to as the strong off-resonance comb (SORC) [17,18] or the non-phase-alternated pulse sequence (NPAPS), and the second is a train of pulses that alternate phase by π, hence the phase-alternated sequence (PAPS) [87]. The optimal nutation angle in both cases is 119°. NPAPS has the property that the phases of the beginning echoes have the same phase as the RF pulse, but the end echoes are phase-shifted by π (Fig. 8(d)). Following any refocusing pulse, both the beginning and the end echoes in PAPS have the same phase as the RF pulse (Fig. 8(e)). In contrast to the SLSE and phase-alternated SLSE pulse sequences, the PAPS sequence tends to give a large signal and the NPAPS a smaller signal.

Ringing artifacts can be reduced or canceled by combining echoes from four SSFP sequences: NPAPS, PAPS, and PAPS and NPAPS with RF pulses phase-shifted by π relative to the first sequences. As with the SLSE sequence, phase-cycling SSFP sequences to cancel ringing can be inefficient in cases where the T_1 is long. The long T_1s result in a long time to reestablish equilibrium when switching between pulse sequences, making it difficult to cancel time-varying ringing signals. In cases of long T_1, $T_2 \approx T_1$ implies long T_{2e}; therefore, the ringing cancellation variants of the SLSE pulse sequence may provide better results than SSFP.

4.1.3. Resonance offset effects on spin-echo pulse sequences

The discussion of pulses and pulse sequences to this point assumes that the RF pulses are applied at the NQR frequency of the explosive to be detected. However, as discussed in Sections 2.3. and 4.1., the NQR frequencies are temperature dependent and, therefore, are generally not known precisely. We can get a qualitative understanding of the effects caused by the RF pulse frequency being offset from the NQR frequency by recognizing that the nuclear spin *phase evolution* has a similar effect as *phase shifts* of the RF pulses in the opposite direction. Resonance offset can turn the SLSE pulse sequence into phase-alternated PASLSE, the PAPS pulse sequence into NPAPS. All of the pulse sequences discussed in this chapter behave in a similar manner. A pulse sequence such as SLSE will produce an NQR response that periodically varies between strong and weak signals, the period increasing as the refocusing pulse spacing decreases [87].

The depth of the 'valleys' in the NQR response becomes shallower as the NQR line width increases. The NQR line width produces a distribution of signal phases not all of which can be in the valley simultaneously, resulting in a shallower valley. By the same argument, not all the signal phases of a broadened NQR line can be at the peak simultaneously, leading to lower peak signals. These effects point to the need for the smallest refocusing pulse spacing possible to move the valleys away from the RF pulse frequency and to broaden the peak. When time permits the use of more than one pulse sequence, combinations of SLSE and phase-alternated SLSE or PAPS and NPAPS can be used to level out the peaks and valleys at the expense of somewhat reduced SNR overall [87].

4.2. Alternate detection methods

The techniques and pulse sequences described in Section 4.1. comprise the NQR detection methods in use in explosives detection today. There are a number of other methods for exciting and detecting NQR signals that have been proposed and demonstrated in the laboratory. In this section, the potential advantages to explosives detection of these methods are discussed.

4.2.1. Multiple-frequency excitation and detection

In the NQR of spin 1 nuclei, the three transitions are interconnected by the three energy levels. Irradiating one transition will affect the signal of a second excited and detected transition [88–90]. For explosives detection, inverting or saturating one transition can increase the population difference, and hence the SNR, of a second excited and detected transition, if the second transition is not v_+ [91]. For example, inverting v_0 will result in the v_- population difference becoming that of v_+ (Fig. 1). Of course, from Fig. 4, we can see that in a powder sample, RF pulses can never completely invert a transition so the full signal enhancement is not realized. Nonetheless, this is a useful technique when the properties such as relaxation times or temperature coefficients make observing one of the lower frequency transitions preferable to v_+.

Another intriguing possibility is to irradiate two transitions while observing the third transition [92,93], often referred to as three-frequency NQR. The ability to detect an NQR signal without applying RF pulses at its resonance frequency offers the possibility of performing the detection without exciting piezoelectric or magnetoacoustic ringing and with no probe recovery issues. FIDs and spin echoes can be excited in three-frequency NQR [93], and like the two-frequency technique, SNR enhancement is possible when one of the lower frequency transitions is observed.

Three-frequency NQR requires a more complicated probe design than the standard single-frequency NQR techniques discussed in this chapter. For any given nitrogen, the quantization axes for the three transitions are orthogonal to each other. Optimal excitation and detection requires three orthogonal RF fields; therefore, the three-frequency technique cannot be applied with a single RF coil. The RF pulses on two transitions can be applied simultaneously or sequentially. Laboratory experiments using three orthogonal RF coils demonstrated the ability to detect NQR FID and spin-echo signals without direct excitation at the observation frequency [92,93]. It was observed that when the pulses are applied simultaneously, there is enough interaction between the two RF fields to induce substantial probe and magnetoacoustic ringing at the third frequency.

4.2.2. Double resonance

In NMR spectroscopy, double resonance refers to experiments or techniques involving irradiation of two different isotopes, for example, ^{13}C and ^{1}H. Such experiments are used to manipulate spin couplings or to enhance the SNR of a lower gyromagnetic ratio isotope. It is this latter application of double resonance that is of interest to NQR explosives detection. Most double-resonance techniques in NQR use the high-abundance, high-gyromagnetic ratio ^{1}H nucleus as the second isotope, although the use of another quadrupolar isotope is also possible.

There are two basic schemes for improving detection sensitivity through double resonance between nitrogen and protons [94]. One is to create a large polarization on the ^{1}H at high static magnetic field, transfer it to the ^{14}N, and detect the NQR signal at zero field. The second is by transferring the ^{14}N polarization to ^{1}H and detecting the ^{1}H nuclei at a frequency higher than the ^{14}N NQR frequency to take advantage of the increased Faraday detection sensitivity at higher frequency (Section 3.1.). The nitrogen polarization may be through T_1 relaxation or by polarization transfer from the protons as in the first scheme. In the latter case, the polarization on the nitrogen is saturated with RF pulses at the NQR frequency before transfer back to the proton. Then, the explosive is detected as a reduction in the proton NMR signal. The sensitivity gain of high-field ^{1}H detection depends on the number of nitrogen NQRs and the proton line width. The relative advantage of obtaining polarization from the protons depends on the T_1s of the two isotopes.

Both of these schemes require a high static magnetic field for proton polarization or detection. The mechanism of polarization transfer is somewhat complicated. Experimentally, the polarization transfer is accomplished by matching the ^{1}H NMR frequency to the ^{14}N NQR frequency. Under the matched frequency condition, magnetic dipole coupling between the isotopes allows for efficient, energy-conserving polarization transfer between the isotopes [7]. To obtain the best frequency match, the static magnetic field is usually swept to cover a range of proton NMR frequencies. The field sweep technique is often referred to as 'field cycling' in the magnetic resonance literature.

The potential signal enhancement available from these double-resonance techniques depends on the available polarization (^{1}H or ^{14}N), the detection frequency, and the polarization transfer efficiency. The transfer efficiency depends not only on the coupling between the isotopes and the time spent with the frequencies matched but also on the number of ^{14}N transitions matched and the order in which the matching conditions occur [91]. Irradiating multiple transitions, as described in Section 4.2.1., will also affect the choice of ^{14}N transitions to polarize [94,95].

Nearly all the efforts toward the application of double-resonance NQR to explosives detection have been driven by the problem of TNT detection [91,96,97], although reports on its application to RDX [98] and PETN [99] detection have appeared recently. Both of the double-resonance detection schemes described earlier have been applied to TNT detection, as it does not fit neatly into either category: around the ^{1}H frequency of 1 MHz, the proton and nitrogen T_1s are similar, with the proton T_1 becoming much longer at higher frequencies; there are multiple ^{14}N NQRs; the proton line width is greater than 10 000 Hz.

Potentially, extremely large SNR enhancements are obtainable with double resonance: NMR spectrometers with ^{1}H frequencies 1000 times greater than the ^{14}N NQR frequencies in TNT are available. Of course, such equipment is completely unsuitable for explosives detection in the field because of its size and complexity, not to mention the difficulties in dealing with very strong magnetic fields in uncontrolled environments. But very strong magnetic fields and complex equipment are not necessary for significant SNR enhancement through double resonance. Polarization transfer can be performed with a small permanent inhomogeneous magnet moved by hand to sweep the field [98]. The SNR of TNT can be enhanced by a factor of 2.2 using only a 26-mT static field for proton polarization (proton NMR frequency 30% higher than the nitrogen NQR frequency), and detection at zero field, under optimal polarization transfer conditions [91].

It is also possible to use double resonance with a second isotope other than ^{1}H. If the second isotope is quadrupolar with NQR frequency greater than the nitrogen NQR frequency, the SNR can be enhanced, with the polarization transfer occurring at zero field through the 'solid effect' [94]. The SNR enhancement depends on the difference in NQR frequencies of the two isotopes and on the difference in T_1s, with the requirement that the higher frequency isotope have the shorter T_1. Of course, it is possible to detect the higher frequency isotope directly; however, if its spin quantum number is odd half integer, it is likely to have a very broad line width. In cases where the dipolar coupling between the two quadrupolar isotopes is weak, the polarization transfer efficiency will be low. Then, a triple resonance technique using protons to mediate polarization transfer has been suggested [94,100].

4.2.3. Circular polarization

In NMR, the excited nuclear magnetization precesses about the static magnetic field, creating a circularly polarized signal in the plane perpendicular to the field. In most NMR spectrometers, only one of the two linearly polarized components comprising the circularly polarized signal is detected, but in systems designed to detect the full circularly polarized signal, a $\sqrt{2}$ SNR advantage is realized. A detector for circularly polarized signals may consist of two orthogonal detector coils or a specially designed device such as the aptly named 'birdcage' coil [101].

In contrast, nuclear magnetization oscillating along the direction of the exciting RF field results in a linearly polarized NQR signal. In a powder sample, only some of the crystallites have the correct orientation of their quantization axis to contribute to the signal; the resulting signal is 43% of that possible from an oriented sample (Section 2.1.2.). If, however, the NQR signal is excited with a circularly polarized RF pulse, a pulse whose direction rotates in a plane at the NQR frequency, many more crystallites have the correct quantization axis part of the time, and 74% of the total possible signal is recovered when the optimal pulse flip angle, now 102°, is used [102–104]. All the crystallites oscillate at the same frequency, but the phase of the oscillation depends on their orientation in the plane of the RF excitation. The net result is a circularly polarized NQR signal with the SNR enhancement of 21%.

Like NQR signals, piezoelectric and magnetoacoustic ringing signals are linearly polarized. Unlike NQR, there are rarely more than one or a few crystals or domains contributing to the ringing signal; therefore, under circularly polarized excitation, one expects ringing signals to be substantially linearly polarized. Circularly polarized NQR can be used to differentiate NQR from ringing signals, at the expense of reducing the SNR by $1.21\%/\sqrt{2} = 0.86\%$ [103,104].

4.2.4. Shaped pulses

In the preceding discussions, the RF pulses were assumed to be rectangular, that is either on or off with a constant amplitude while on. The RF pulses need not be rectangular, however. 'Shaped' pulses are well known in NMR spectroscopy for enhancing or restricting the excitation bandwidth or compensating for inhomogeneity in the RF field. In form, shaped pulses fall into two categories: composite pulses [105] composed of several back-to-back rectangular pulses of varying phase, amplitude, and/or duration and pulses whose phase and/or amplitude vary continuously [106].

In NQR, composite pulses have been proposed for the compensation of resonance offset [107–111] and RF field inhomogeneity [107,108,112–114]. Several reports on pulses of continuously variable shape for use in NQR have also appeared [115–117]. Shaped pulses that compensate for RF field inhomogeneity also partially compensate for the powder average effect (Section 2.1.2.). Theoretically, shaped excitation pulses can enhance SNR by up to 15% in powder samples [117]. Combinations of shaped excitation and refocusing pulses may also provide better response from multi-echo pulse sequences [9].

5. NQR explosives detection

Before turning our attention to some examples of NQR explosives detection systems, we discuss the NQR parameters of several common explosives.

5.1. Nuclei and explosives

Of the first 96 nuclei in the periodic Table, 74 have one or more isotopes with spin quantum number greater than 1/2 [38]. Because of an isotope's natural abundance and quadrupole moment, some are more easily studied by NQR than others. Fortunately for explosives detection, quadrupolar nuclei are found in many explosives. For example, nitrogen, chlorine, and bromine, all have isotopes amenable to NQR detection. By far, the nitrogen-containing explosives have received the most attention in NQR explosives detection work, primarily due to their military importance and relative availability. In the following section, the NQR properties of some of the nitrogen-containing explosives are discussed. This section is not intended as an all-inclusive catalog of explosives and their NQR properties; rather, the intent is to give an indication of the range of properties that are encountered and how they affect NQR detectability. The properties for each explosive, ν_+ and ν_- frequencies, T_1, T_2, T_2^*, T_{2e}, and temperature coefficient are summarized in Tables[1] for each explosive in its respective section.

5.1.1. RDX

RDX (hexahydro-1,3,5-trinitro-1,3,5-triazine) contains six nitrogens, all in magnetically inequivalent sites in the crystal. The ν_+ lines are above 5 MHz for the three amine nitrogens and at 500 kHz or below for the three nitro nitrogens. Compared with the other explosives discussed here, T_1 is short, and with $T_2 \approx T_1$, RDX is an ideal candidate for the SSFP-type pulse sequences (Section 4.1.2.). The temperature coefficients are moderately large for most of the lines. Much emphasis is placed on detecting RDX because it is one of the main ingredients in plastic explosives such as Semtex and C-4. The combination of high frequencies and short T_1s makes RDX one of the easiest explosives to detect. Frequencies and relaxation time values for the amino sites are listed in Table 1 [118,119].

5.1.2. TNT

TNT (2,4,6-trinitrotoluene) contains three nitrogens per molecule and two molecules per unit cell, resulting in six magnetically inequivalent nitrogens. The ν_+ lines are all grouped between 800 and 900 kHz. There are two crystal forms at ambient temperature, monoclinic and orthorhombic. The difference in resonance frequencies of the two crystal forms is small but distinguishable. Monoclinic appears to be the most common form in commercial materials. The T_1s are moderately long, and line widths vary from sample to sample but tend to be broad. It has been suggested that in some cases the broadening may be due to the coexistence of the two crystal forms. The combination of long T_1s, broad lines, and low frequencies makes TNT one of the more difficult explosives to detect by NQR. Frequencies and relaxation times for the monoclinic form of TNT are listed in Table 2 [120].

Table 1. NQR properties for selected lines of RDX

Number	**ν (kHz)**	**T_2^* (ms)**	**T_1 (ms)**	**T_2 (ms)**	**$d\nu/dT$ (kHz/°C)**
2	5240	0.74	12.3	7.1	–0.47
1	5192	1.59	12.6	8.2	–0.43
3	5047	0.71	13.3	6.8	–0.43
2	3458	0.59	12.1	5.7	–0.33
1	3410	0.80	11.1	6.2	–0.06
3	3359	0.74	14.6	6.3	–0.27

[1] The data in these Tables were compiled by Dr. Michael Buess.

Table 2. NQR properties for selected lines of TNT

Number	ν (kHz)	T_2^* (ms)	T_1 (s)	$T_{2e}{}^a$ (ms)	$d\nu/dT$ (kHz/°C)
6	870	0.64	4.0	34	−0.109
2	859	0.24	3.0	24	−0.223
4	848	0.80	9.6	29	−0.151
5	844	0.40	4.7	28	−0.121
1	842	0.40	3.5	29	−0.181
3	837	0.35	2.1	24	−0.122
2	768	0.45	9.8	33	−0.190
1	751	0.45	2.2	31	−0.241
3	743	0.80	3.0	40	−0.148
4	740	0.32	5.5	45	−0.169
5,6	714	0.45	4.3	42	−0.094

[a] Values for refocusing pulse spacing of 1.1 ms.

5.1.3. PETN

PETN (1,2-bis[(nitroxy)methyl]-1,3-propanediol-dinitrate, pentaerythritol-tetranitrate) contains four nitrogens, all magnetically equivalent. Like TNT, the PETN frequencies are low, and its T_1 is very long, however its T_{2e} is also long (Table 3), improving detection sensitivity significantly [119,121].

5.1.4. HMX

HMX (octahydro-1,3,5,7-tetranitro-1,3,5,7-tetrazocine) chemically is very similar to RDX. It has eight nitrogens, four amino and four nitro. Of the four amino nitrogens, there are two magnetically inequivalent sites in the molecule. For both sites, ν_+ is above 5 MHz, but unlike RDX, T_1 is quite long and the lines are broad. Even so, with the high frequencies and long T_{2e}, HMX is not difficult to detect. NQR parameters for the nitro groups have not been determined. Values for the ν_- NQR parameters of the amino sites are given in Table 4 [119,122].

5.1.5. Ammonium nitrate

Ammonium nitrate has two magnetically inequivalent nitrogens. The ammonium nitrogen frequencies have not been observed by NQR; from NMR measurements, the quadrupole

Table 3. NQR properties for selected lines of PETN

Number	ν (kHz)	T_2^* (ms)	T_1 (s)	T_2 (s)	$T_{2e}{}^a$ (s)	$d\nu/dT$ (kHz/°C)
1	890	0.80	32	0.055	4.59	−0.07
1	495	0.80	33	—	—	−0.04

[a] Values for refocusing spacing of 1.0 ms

Table 4. NQR parameters for selected lines of HMX

Number	ν (kHz)	T_2^* (ms)	T_1 (s)	T_2 (s)	$T_{2e}{}^a$ (s)	$d\nu/dT$ (kHz/°C)
1	3737	0.23	12	0.125	—	−0.02
2	3623	0.14	10	0.160	3.95	−0.04

[a] Values for refocusing pulse spacing of 1.2 ms.

coupling constant for the ammonium nitrogen under ambient conditions is 245 kHz and the asymmetry parameter is 0.85 [123], leading to a predicted ν_+ of 236 kHz. For the nitrate ion under ambient conditions, ν_+ is 497 kHz. The quadrupole coupling constant and asymmetry parameter of both nitrogens change near 35°C because of the transition between phases III and IV [123]. Ammonium nitrate detection suffers from low frequencies and long T_1s, but the lines are narrow and T_{2e} is very long. Table 5 gives the NQR parameters for the nitrate ion in phase IV (under ambient conditions) [124].

5.1.6. Potassium nitrate

Potassium nitrate contains a single nitrogen. Its frequency is higher than that of the nitrate ion in ammonium nitrate. The T_1s and T_{2e}s of the two nitrate ion transitions are similar; therefore, their detection sensitivities are similar. The potassium ion also can be detected by NQR. Its resonance is higher in frequency than the nitrogen and its T_1 is shorter, but its line is broader and T_{2e} is shorter. Table 6 lists NQR parameter for nitrogen (1) and potassium (2) [119,125].

5.2. Explosives detection

Below, a brief overview of NQR explosives detection effort prior to 1990, followed by several forms of NQR explosives detectors currently being developed, is given.

Table 5. NQR parameters for selected lines of ammonium nitrate

Number	ν (kHz)	T_2^* (ms)	T_1 (s)	$T_{2e}{}^a$ (s)	$d\nu/dT$ (kHz/°C)
1	497	6.37	14.0	3.5	−0.46
1	423	5.31	16.6	14.6	+0.12

[a] Values for refocusing spacing of 1.1 ms.

Table 6. NQR parameters for selected lines of potassium nitrate

Number	ν (kHz)	T_2^* (ms)	T_1 (s)	$T_{2e}{}^a$ (s)	$d\nu/dT$ (kHz/°C)
1	567	2.89	20.1	15.5	−0.23
1	559	2.89	24.5	14.6	−0.19
2	665	0.40	1.9	1.35	−0.19

5.2.1. Explosives detection prior to 1990

Although often viewed as a new technology in the explosives detection field, the idea of using NQR for detecting nitrogenous explosives was conceived more than 50 years ago. In fact, the idea of explosives detection, particularly with regard to land mines, drove the earliest development of NQR by Garroway and coworkers [30]. Their success was ultimately limited by hardware issues. Over the intervening years, numerous groups have pursued NQR for explosives detection [30,126]. In the 1960s, J. D. King of Southwest Research Inc. examined NQR [127] and NMR explosives detection [128,129]. During the 1970s, Marino and coworkers [120,121,130,131], and Hirschfeld and Klainer [132] pursued NQR for explosives detection. They, with Block Engineering, investigated NQR for land mine detection. Much of this early work focused on determining the NQR parameters for the explosives of interest, particularly TNT and RDX [118,120–122]. V. S. Grechishkin began work in NQR in the 1970s. During the 1980s and into the early 1990s, he and coworkers at several Russian laboratories worked on land mine detection and baggage screening [133–137]. The land mine detection efforts were largely driven by wars in Vietnam and Afghanistan and were severely curtailed at the end of those wars.

5.2.2. Baggage screening

In the late 1980s, following the bombing of Pan Am flight 103 over Lockerbie, Scotland, there was renewed interest in NQR explosives detection, now for airport security. The airport security scenario involved interrogating a large volume, a piece of baggage, for a relatively small amount of explosive. Initially, the explosives of interest were RDX and PETN. The main difficulty to overcome, aside from the ever-present issue of sensitivity, was generating a sufficient RF magnetic field in a large volume. Most laboratory work on small-scale NQR systems used RF magnetic field strengths of many tens of Gauss requiring hundreds of watts to irradiate a sample volume of a few tens of cubic centimeters. Scaling the coil volume to hundreds of liters would require a transmitter capable of delivering hundreds of kilowatts. The problem of scalability was addressed, in part, by the introduction of large volume coils with $Q >> 1000$ [119]. These high-Q coils, typically constructed from a single sheet of copper to form a one-turn solenoid, also provided very uniform detection sensitivity over their entire volume. The realization of good detection efficiency with RF magnetic field strengths on the order of one Gauss [132,138], combined with the large-volume high-Q coils, has allowed detection volumes of 1000 l and larger [119,139,140]. Initial tests on airline baggage with a 300-l RDX-only detector demonstrated a 97% probability of detection[2] and 3.8% false alarm rate [141]. Similar results were predicted for PETN in the same 300-l coil, based on laboratory measurements [142].

[2] The probability of detection was determined for one threat quantity of RDX, as defined by the U.S. Federal Aviation Administration, in six seconds of data accumulation.

A number of groups have pursued NQR for baggage screening in recent years [140,142–145], working to improve sensitivity and develop robust hardware. Improvements in Q-damping [146,147] and pulse sequences for ringing suppression have increased sensitivity [146]. Although these baggage scanners are shielded, it is found that in some environments RFI mitigation techniques can help further reduce noise [146]. Work on increasing the information content from the NQR measurement has lead to threat localization [55,148], an NQR imaging [149] method related to MRI. Threat localization can help reduce false alarms and correlate NQR data with other imaging techniques. Threat localization can also be implemented through data fusion, combining NQR data with data from another sensor such as X-ray imaging [56]. Increasing the number of explosives baggage can be screened for is an ongoing issue. Although the primary application of these devices has been for airport security, many other applications have been considered, including mail scanning [150].

5.2.3. Personnel screening

The threat to airport security from explosives is not limited to bombs smuggled in baggage. Explosives can easily be hidden on a person's body, as is evident from the frequent news reports of suicide bombings. NQR has adequate sensitivity for screening people for carried explosives [142]. A walk-through NQR portal prototype has been demonstrated for RDX detection [151]. It utilizes a whole-body multi-turn solenoid for signal excitation and detection, and auxiliary flat-loop coils for crude signal localization and false alarm reduction. The portal, at 400 l in volume, is similar in size to the baggage screeners.

A concern unique to the application of NQR to personnel screening is the RF power deposition in the body. The electrically conductive fluids of the body absorb RF energy, producing local heating. One solution to the power deposition problem is to use a coil designed to produce an RF magnetic field that is rapidly attenuated from the coil. A transmitter such as the meanderline coil [86] produces a strong RF magnetic field near the surface of the coil where the explosive might be, but very little field farther away, in the interior of the body where it is not needed. These *surface* coils can be used to scan the body for explosives in a manner similar to handheld metal detectors.

Gradiometer coils have RF magnetic field spatial profiles similar to the meanderline coil. A prototype device designed around a gradiometer showed very good explosives detection sensitivity [152]. The presence of metal near the coil causes it to detune; therefore, the automatic tuning circuitry could be used as a metal detector. It was found that in noisy environments, the gradiometer coil was not sufficient to cancel RFI pickup; additional RFI mitigation coils were used.

5.2.4. Land mine detection

It is evident from the discussion in Section 5.2.1. that the detection of land mines has been a long-time goal for NQR explosives detection. In the mid-1990s, the interest in land mine detection was renewed, and several groups began new work in the area [30,142,153,154]. The main impetus for new land mine technologies is plastic-cased mines. These mines have very little metal and are thus problematic for the commonly used metal detector. When the sensitivity of the metal detector is adjusted to detect these low metal mines, the false alarm rate increases tremendously because of the large numbers of metal fragments found in the ground in combat areas. NQR was expected to be insensitive to the metal fragments and therefore a good alternative to metal detection.

As was the case with baggage screening, the first prototypes were designed to detect RDX [155,156]. Initially, gradiometer coils were used [157], but some later designs switched to magnetometers to take advantage of the increased sensitivity to deeply buried mines [156]. These handheld NQR detectors demonstrated high probability of detection and low sensitivity to metallic clutter compared with electromagnetic detection schemes such as metal detection and ground-penetrating radar. The great majority of land mines in the world contain TNT. TNT detection capabilities were added to the RDX detection systems [157,158] at the expense of scan time. These prototypes demonstrated the capabilities of NQR but consisted of a handheld detector coil connected to a large rack of electronics [156]. As the sensitivity of the detectors improved, the issues of RFI mitigation became important [39,42]. Work has continued on land mine detection with NQR, focusing on hardware development for a truly handheld device (Fig. 9), data fusion with other sensors including metal detection and ground-penetrating radar, and addition of other explosives to the detection capabilities.

The foregoing discussion focused on handheld NQR land mine detectors, but a vehicle-mounted format has also been investigated [159]. This device consisted of a large magnetometer coil, approximately 1 m in diameter, mounted on the front of a vehicle. It was designed to detect TNT and RDX in large antitank mines. All NQR hardware,

Fig. 9. Fully man-portable NQR land mine detector (from Dr P. Prado, GE Infrastructure, Inc.).

including RFI mitigation antennae, was integrated into the vehicle. In field tests, a 95% probability of detection and 4–7% false alarm rate were measured.

5.2.5. Vehicle screening

Recently, the topic of improvised explosive devices (IEDs) has garnered a lot of attention. IEDs may take the form of roadside bombs, suicide bombers, or vehicle-borne bombs. Roadside bombs share some detection issues with land mines, and in principle, the NQR land mine detectors discussed in the previous section can be used to detect roadside IEDs. However, many IEDs are remotely controlled, and the short standoff distance available with NQR makes IED detection dangerous for the NQR operator. The personnel screening devices already discussed in Section 5.2.3. are applicable to the suicide bomber.

Because of the detection volume involved, vehicle screening is a somewhat different problem than baggage screening. Generating sufficient RF magnetic field is an important consideration. The problem is exacerbated by the difficulty in getting RF magnetic fields in and out of the vehicle. On the positive side, vehicle-borne IEDs usually contain large quantities of explosive, with large NQR signals, if they can be detected outside the vehicle.

Barras and coworkers [160,161] built and demonstrated a prototype NQR vehicle screener to detect ammonium nitrate. Ammonium nitrate-based bombs were the vehicle-borne IED of choice in terrorist attacks in the United Kingdom at the time the research started. In tests on two small vehicles, they determined that cross-axial RF magnetic fields (parallel to the wheel axles) gave the best penetration into the vehicle. Maximum attenuations of 14 and 20 dB were measured for the two vehicles.

For the detector coil, a single-turn toroid with a gap for the vehicle was chosen (Fig. 10). This design provides good RF magnetic field homogeneity, high Q (1150 unloaded in this case), and reduced sensitivity to RFI. Additional shielding around the coil further reduced RFI pickup. Introducing a vehicle into the coil gap reduces the Q by 15–20%. Peak RF magnetic field strength in the coil gap, in the absence of a vehicle, was 1.3 G.

The inherently low RF magnetic field strength of the toroid, and the large field inhomogeneity within the vehicles, required the use of some specialized pulse sequences. Typical phase-cycling schemes were not expected to adequately preserve signal intensity and simultaneously cancel ringing artifacts. SLSE and PASLSE (called pPAPS in reference [160,161]) pulse sequences were used with the modification that after a fixed number of pulses, the NQR signal was saturated and then the pulse sequence continued. The data acquired after the saturation were subtracted from the data acquired before in order to cancel ringing. With these pulse sequences, very good SNR was obtained from several samples of ammonium nitrate, with weights between 250 and 475 kg, inside the vehicles, using 1-min scans. Better response is expected with higher transmitter power.

Fig. 10. Toroid coil for vehicle screening under construction (Fig. 6 of reference [161], by permission).

6. Outlook

NQR was first proposed for explosives detection more than 50 years ago. In the intervening time, significant progress has been made in the parallel areas of our understanding of how to maximize the SNR per unit time and in the design of fieldable hardware. Progress in these areas continues today. Even so, sensitivity continues to be the primary concern for NQR. For detection scenarios that require high throughput rates, the low sensitivity can be a problem. However, for detection scenarios where other techniques suffer from high false alarm rates, NQR's main advantage, its high specificity for the target materials of interest, still makes it an attractive candidate for explosives detection.

References

[1] J. A. S. Smith, *J. Chem. Educ.*, 48 (1971) 39.
[2] Y. K. Lee, *Concepts Magn. Reson.*, 14 (2002) 155.
[3] D. R. Vij (Ed.), *Handbook of Applied Solid State Spectroscopy*, Springer, New York (2006).
[4] T. P. Das and E. L. Hahn, *Nuclear Quadrupole Resonance Spectroscopy*, Academic Press, New York (1958).
[5] E. Fukushima and S. B. W. Roeder, *Experimental Pulse NMR, a Nuts and Bolts Approach*, Addison-Wesley, Reading (1981).
[6] P. Mansfield and P. G. Morris, *NMR in Biomedicine*, Academic Press, New York (1982).
[7] A. Abragam, *The Principles of Nuclear Magnetism*, Oxford University, Oxford (1961).
[8] C. P. Slichter, *Principles of Magnetic Resonance*, 2nd ed., Springer Verlag, Berlin (1980).
[9] K. L. Sauer and C. A. Klug, *Phys. Rev. B*, 74 (2006) 174410.
[10] P. Nickel, H. Robert, R. Kimmich, and D. Pusiol, *J. Magn. Reson.*, 111 (1994) 191.
[11] S. A. Vierkötter, C. R. Ward, D. M. Gregory, S. M. Menon, and D. P. Roach, *Proc. SPIE*, 5046 (2003) 176.
[12] R. Labrie, M. Infantes, and J. Vanier, *Rev. Sci. Instrum.*, 42 (1971) 26.

[13] M. L. Buess and S. M. Caulder, *Appl. Magn. Reson.*, 25 (2004) 383.
[14] S. M. Caulder, M. L. Buess, A. N. Garroway, and P. J. Miller, Shock Compression in Condensed Matter – 2003, *AIP Conf. Proc.* In M. D. Furnish, Y. M. Gupta, and J. W. Forbes, eds., *American Institute of Physics*, Melville, 706 (2004).
[15] E. L. Hahn, *Phys. Rev.*, 80 (1950) 580.
[16] R. A. Marino and S. M. Klainer, *J. Chem. Phys.*, 67 (1977) 3388.
[17] S. M. Klainer, T. B. Hirschfeld, and R. A. Marino, *Fourier, Hadamard, and Hilbert Transforms in Chemistry*, Plenum Press, New York (1982).
[18] S. S. Kim, J. R. P. Jayakody, and R. A. Marino, *Z. Naturforsch.*, 47a (1992) 415.
[19] D. Ya. Osokin, *J. Exp. Theor. Phys.*, 88 (1999) 868.
[20] T. N. Rudakov and V. T. Mikhaltsevich, *Phys. Lett. A*, 309 (2003) 465.
[21] T. N. Rudakov, *Chem. Phys. Lett.*, 398 (2004) 471.
[22] T. N. Rudakov, P. A. Hayes, and V. T. Mikhaltsevitch, *Phys. Lett. A*, 330 (2004) 280.
[23] J. A. S. Smith and J. D. Shaw, *Improvements in NQR Testing*, International Patent Application No. PCT/GB92/02254 (1993).
[24] N. A. Sergeev, A. M. Panich, and M. Oiszewski, *Appl. Magn. Reson.*, 27 (2004) 41.
[25] C. A. Meriles, S. C. Pérez, and A. H. Brunetti, *J. Chem. Phys.*, 107 (1997) 1753.
[26] G. W. Leppelmeier and E. L. Hahn, *Phys. Rev.*, 141 (1966) 724.
[27] C. N. Chen and D. I. Hoult, *Biomedical Magnetic Resonance Technology*, Hilger, New York (1989).
[28] D. I. Hoult and R. E. Richards, *J. Magn. Reson.*, 24 (1978) 71.
[29] J. B. Miller, B. H. Suits, A. N. Garroway, and M. A. Hepp, *Concepts Magn. Reson.*, 12 (2000) 125.
[30] A. N. Garroway, M. L. Buess, J. B. Miller, B. H. Suits, A. D. Hibbs, G. A. Barrall, R. Matthews, and L. J. Burnett, *IEEE Trans. Geosci. Remote Sens.*, 39 (2001) 1108.
[31] V. S. Grechishkin and N. Ya. Sinyavskii, *Physics-Uspekhi*, 36 (1993) 960.
[32] B. H. Suits, A. N. Garroway, and J. B. Miller, *J. Magn. Reson.*, 132 (1998) 54.
[33] C. Wilker, J. D. McCambridge, D. B. Laubacher, R. L. Alvarez, J. S. Guo, C. F. Carter III, M. A. Pusateri, and J. L. Schiano, *2004 IEEE MTT-S International Microwave Symposium Digest*, 1 (2004) 143.
[34] D. I. Hoult, *Rev. Sci. Instrum.*, 50 (1979) 193.
[35] E. R. Andrew and K. Jurga, *J. Magn. Reson.*, 73 (1987) 268.
[36] C. Anklin, M. Rindlisbacher, G. Otting, and F. H. Laukien, *J. Magn. Reson. B*, 106 (1995) 199.
[37] B. H. Suits, A. N. Garroway, and J. B. Miller, *J. Magn. Reson.*, 135 (1998) 373.
[38] *NMR Periodic Table*, http://arrhenius.rider.edu/nmr/NMR_tutor/periodic_table/nmr_pt_frameset.html.
[39] *Nuclear Quadrupole Resonance Spectra Database*, Japan Association for International Chemical Information, Tokyo (2003).
[40] G. Liu, Y. Jiang, J. Li, and G. A. Barrall, *Proc. SPIE*, 5415 (2004) 834.
[41] B. H. Suits, A. N. Garroway, and J. B. Miller, *J. Magn. Reson.*, 131 (1998) 154.
[42] B. H. Suits, *Appl. Magn. Reson.*, 25 (2004) 371.
[43] K. Long, R. Deas, D. Riley, and M. Gaskell, *Proc. SPIE*, 5089 (2003) 107.
[44] G. E. Poletto, T. M. Osán, and D. J. Pusiol, *Hyperfine Interact.*, 159 (2004) 127.
[45] K. Choi and I. Yu, *Rev. Sci. Instrum.*, 60 (1989) 3249.
[46] M. L. Buess and G. L. Petersen, *Rev. Sci. Instrum.*, 49 (1978) 1151.
[47] E. Fukushima and S. B. W. Roeder, *J. Magn. Reson.*, 33 (1979) 199.
[48] A. Stensgaard, *J. Magn. Reson. A*, 122 (1996) 120.
[49] A. Stensgaard, *J. Magn. Reson.*, 128 (1997) 84.
[50] S. Tantum, L. Collins, L. Carin, I. Gorodnitsky, A. Hibbs, D. Walsh, G. Barrall, D. Gregory, R. Matthews, and S. Vierkötter, *Proc. SPIE*, 3710 (1999) 474.

[51] F. Liu, S. Tantum, L. Collins, and L. Carin, *Proc. SPIE*, 4038 (2000) 572.
[52] Y. Y. Tan, S. L. Tantum, and L. M. Collins, *Proc. SPIE*, 5415 (2004) 822.
[53] Y. Y. Tan, S. L. Tantum, and L. M. Collins, *IEEE Signal Process. Lett.*, 11 (2004) 490.
[54] A. Jacobsson, M. Mossberg, and J. A. S. Smith, *IEEE Trans. Geosci. Remote Sens.*, 43 (2005) 2659.
[55] *Electric Resistance*, http://hypertextbook.com/physics/electricity/resistance/.
[56] A. Hudson, Y. Lee, P. Prado, and T. Rayner, *Proceedings of the Third International Aviation Security Technology Symposium* (2001) 277.
[57] D. B. Laubacher, *Portable Nuclear Quadrupole Resonance Detection System for Detecting Presence of Explosives by Scanning Mail and Luggage, Uses High Temperature Superconductor Self Resonant Planar Transmit and Pick-Up Coil*, US Patent Application No. US2004245988-A1 (2004).
[58] D. B. Laubacher and C. Wilker, *Resonance Frequency Adjusting Circuit in Frequency Detection System, has Switch to Connect and Disconnect Reactance to Single Loop Coil Which Couples Circuit to High Temperature Superconductor Self-Resonant Transmit-and-Receive Coil*, US Patent Application No. US2006012371-A1 (2005).
[59] J. D. McCambridge, *Nuclear Quadrupole Resonance Detection System for Luggage, Has Q-Damping Circuit Comprising High-Temperature Superconductor Single Loop, for High-Temperature Superconductor Self-Resonant Transmit and Receive Coil*, US Patent Application No. US2006082368-A1 (2006).
[60] R. L. Alvarez and C. Wilker, *Nuclear Quadrupole Resonance Detection System for e.g. Security System, Has High Temperature Superconductor Sensors Detecting NQR Signals, Where Each Sensor Has High Temperature Superconductor Self-Resonant Planar Coil*, US Patent Application No. US2005270028-A1 (2005).
[61] J. Clarke, *The New Superconducting Electronics*, H. Weinstock and R. W. Ralston, eds., Kluwer Academic, Dordrecht (1993).
[62] Ya. S. Greenberg, *Rev. Mod. Phys.*, 70 (1998) 175.
[63] C. Seton, J. M. S. Hutchinson, and D. M. Bussell, *IEEE Trans. Appl. Supercond.*, 7 (1997) 3213.
[64] C. Hilbert and J. Clarke, *SQUID '85, Proceedings of the International Conference on SQUID Devices and Their Applications*, H. D. Hahlbohm and H. Lubbit, eds., Walter de Gruyter, Berlin 1985.
[65] R. E. Sager, A. D. Hibbs, D. N. Shykind, and B. D. Thorson, *The NQR Newsletter*, 1 (1993) 15.
[66] J. Clarke, *Z. Naturforsch. A*, 49A (1994) 5.
[67] J. P. Yesinowski, M. L. Buess, A. N. Garroway, M. Ziegeweid, and A. Pines, *Anal. Chem.*, 34 (1995) 2256.
[68] I. K. Kominis, T. W. Kornack, J. C. Allred, and M. V. Romalis, *Nature*, 422 (2003) 596.
[69] I. M. Savukov, S. J. Seltzer, M. V. Romalis, and K. L. Sauer, *Phys. Rev. Lett.*, 95 (2005) 063004.
[70] I. M. Savukov and M. V. Romalis, *Phys. Rev. Lett.*, 94 (2005) 123001.
[71] S.-K. Lee, K. L. Sauer, S. J. Seltzer, O. Alem, and M. V. Romalis, *Appl. Phys. Lett.*, 89 (2006) 214106.
[72] A. Suter, *Prog. NMR Spectrosc.*, 45 (2004) 239.
[73] A. J. Blauch, J. L. Schiano, and M. D. Ginsberg, *Proc. SPIE*, 3710 (1999) 464.
[74] A. J. Blauch, J. L. Schiano, and M. D. Ginsberg, *J. Magn. Reson.*, 144 (2000) 305.
[75] H. Y. Carr and E. M. Purcell, *Phys. Rev.*, 94 (1954) 630.
[76] S. Meiboom and D. Gill, *Rev. Sci. Instrum.*, 29 (1958) 688.
[77] D. Ya. Osokin, *J. Mol. Struct.*, 83 (1982) 243.
[78] D. Ya. Osokin, *Mol. Phys.*, 48 (1983) 283.

[79] D. Ya. Osokin, V. L. Ermakov, R. H. Kurbanov, and V. A. Shagalov, *Z. Naturforsch. A*, 47 (1992) 439.
[80] A. K. Dubey and P. T. Narasimhan, *J. Mol. Struct.*, 192 (1989) 321.
[81] V. T. Mikhaltsevitch, T. N. Rudakov, J. H. Flexman, P. A. Hayes, and W. P. Chisholm, *Appl. Magn. Reson.*, 25 (2004) 449.
[82] M. H. Levitt, R. Freeman, and T. A. Frenkiel, *J. Magn. Reson.*, 47 (1982) 328.
[83] G. A. Barrall, L. J. Burnett, and A. G. Shelton, *Method and System for Cancellation of Extraneous Signals in Nuclear Quadrupole Resonance Spectroscopy*, US Patent No. 6 392 408 (2002).
[84] H. Y. Carr, *Phys. Rev.*, 112 (1958) 1701.
[85] M. L. Gyngell, *J. Magn. Reson.*, 81 (1989) 474.
[86] M. L. Buess, A. N. Garroway, and J. B. Miller, *J. Magn. Reson.*, 92 (1991) 348.
[87] M. L. Buess, A. N. Garroway, and J. P. Yesinowski, *Removing the Effects of Acoustic Ringing and Reducing Temperature Effects in the Detection of Explosives by NQR*, US Patent No. 5 365 171 (1994).
[88] V. S. Grechishkin and N. Ja. Sinjavsky, *Z. Natureforsch.*, 45A (1992) 559.
[89] G. V. Mozjoukhine, *Appl. Magn. Reson.*, 22 (2002) 31.
[90] D. Ya. Osokin, R. R. Khusnutdinov, and V. A. Shagalov, *Appl. Magn. Reson.*, 25 (2004) 513.
[91] K. R. Thurber, K. L. Sauer, M. L. Buess, C. A. Klug, and J. B. Miller, *J. Magn. Reson.*, 177 (2005) 118.
[92] K. L. Sauer, B. H. Suits, A. N. Garroway, and J. B. Miller, *Chem. Phys. Lett.*, 342 (2001) 362.
[93] K. L. Sauer, B. H. Suits, A. N. Garroway, and J. B. Miller, *J. Chem. Phys.*, 118 (2003) 5071.
[94] R. Blinc, T. Apih, and J. Seliger, *Appl. Magn. Reson.*, 25 (2004) 523.
[95] J. Seliger, V. Zagar, and R. Blinc, *J. Magn. Reson. A*, 106 (1994) 214.
[96] R. Blinc, J. Seliger, D. Arcon, and V. Zagar, *Phys. State Solid A*, 180 (2000) 541.
[97] M. Nolte, A. Privalov, J. Altmann, V. Anferov, and F. Fujara, *J. Phys. D: Appl. Phys.*, 38 (2002) 939.
[98] J. Luznik, J. Pirnat, and Z. Trontelj, *Solid State Commun.*, 121 (2002) 653.
[99] V. T. Mikhaltsevitch and A. V. Beliakov, *Solid State Commun.*, 138 (2006) 409.
[100] J. Seliger, R. Blinc, T. Apih, and G. Lahajnar, *Triple Resonance Enhanced Nuclear Quadrupole Resonance (NQR) Detection of TNT and Other Explosives*, Slovenian Patent No. 21 715 (2005).
[101] M. C. Leifer, *J. Magn. Reson.*, 124 (1997) 51.
[102] Y. K. Lee, H. Robert, and D. K. Lathrop, *J. Magn. Reson.*, 148 (2001) 355.
[103] J. B. Miller, B. H. Suits, and A. N. Garroway, *J. Magn. Reson.*, 151 (2001) 228.
[104] J. B. Miller, A. N. Garroway, and B. H. Suits, *Nuclear Quadrupole Resonance (NQR) Method and Probe for Generating RF Magnetic Fields in Different Directions to Distinguish NQR from Acoustic Ringing Induced in the Sample*, US Patent No. 6 522 135 (2003).
[105] M. H. Levitt, *Prog. NMR Spectrosc.*, 18 (1986) 61.
[106] M. Garwood and L. DelaBarre, *J. Magn. Reson.*, 153 (2001) 155.
[107] A. Ramamoorthy and P. T. Narasimhan, *J. Mol. Struct.*, 192 (1989) 333.
[108] G.-Y. Li and X.-W. Wu, *Chem. Phys. Lett.*, 204 (1993) 529.
[109] A. Ramamoorthy, N. Chandrakumar, A. K. Dubey, and P. T. Narasimhan, *J. Magn. Reson.*, 102 (1993) 274.
[110] A. Ramamoorthy, *Mol. Phys.*, 93 (1997) 757.
[111] V. T. Mikhaltsevitch, T. N. Rudakov, J. H. Flexman, P. A. Hayes, and W. P. Chisholm, *Solid State Nucl. Magn. Reson.*, 25 (2004) 61.
[112] G.-Y. Li, Y. Jiang, and X.-W. Wu, *Chem. Phys. Lett.*, 202 (1993) 82.
[113] V. P. Anferov and G. V. Mozzhukhin, *Russ. Phys. J.*, 42 (1999) 826.
[114] K. L. Sauer, C. A. Klug, J. B. Miller, and A. N. Garroway, *Appl. Magn. Reson.*, 25 (2004) 485.

[115] S. Z. Ageev and B. C. Sanctuary, *Chem. Phys. Lett.*, 225 (1994) 499.
[116] C. Schurrer and S. C. Perez, *Appl. Magn. Reson.*, 16 (1999) 135.
[117] J. B. Miller and A. N. Garroway, *Appl. Magn. Reson.*, 25 (2004) 475.
[118] R. J. Karpowicz and T. B. Brill, *J. Phys. Chem.*, 87 (1983) 2109.
[119] A. N. Garroway, M. L. Buess, J. P. Yesinowski, and J. B. Miller, *Proc. SPIE*, 2092 (1993) 318.
[120] R. A. Marino and R. F. Connors, *J. Mol. Struct.*, 111 (1983) 323.
[121] R. A. Marino, *Final Technical Report for Task 7–09*, Department of the Army, Battelle, NC (1987).
[122] R. A. Landers, T. B. Brill, and R. A. Marino, *J. Phys. Chem.*, 85 (1981) 2618.
[123] T. Giavani, H. Bildsøe, J. Skibsted, and H. J. Jakobsen, *J. Phys. Chem. B*, 106 (2002) 3026.
[124] J. Seliger, V. Zagar, and R. Blinc, *Z. Phys. B Con. Mat. Quant.*, 25 (1976) 189.
[125] T. J. Barstow, *J. Chem. Soc. Faraday Trans.*, 87 (1991) 2453.
[126] J. Shaw, *NQI Newsletter*, 1 (1994) 26.
[127] J. D. King, *Nuclear Quadrupole Resonance of Chlorine-35 in Ammonium Chlorate and Ammonium Perchlorate*, M.S. Thesis, St. Mary's University, San Antonio, TX (1963).
[128] W. L. Rollwitz, J. D. King, and S. D. Shaw, FAA Rep. FA-RD-76-29 (1975).
[129] W. L. Rollwitz, J. D. King, and G. A. Matzkanin, *Proceedings of the New Concepts Symposium and Workshop on Detection and Identification of Explosives* (1978).
[130] R. A. Marino, R. F. Conners, and L. Leonard, Final Report on US Army Research Office Contract DAAG29 79 C 0025 (1982).
[131] J. C. Harding Jr., D. A. Wade, R. A. Marino, E. G. Sauer, and S. M. Klainer, *J. Magn. Reson.*, 36 (1979) 21.
[132] T. Hirschfeld and S. M. Klainer, *J. Mol. Struct.*, 58 (1980) 63.
[133] V. S. Grechishkin, N. Ya. Sinyavskii, and G. V. Mozzhukhin, *Izv. Vyssh. Uchebn. Zaved. Fiz.*, 35 (1992) 58.
[134] V. S. Grechishkin, *Appl. Phys.*, A55 (1992) 505.
[135] V. S. Grechishkin, *Russ. Phys. J.*, 35 (1993) 637.
[136] V. S. Grechishkin, *Appl. Phys. A*, 58 (1994) 63.
[137] V. S. Grechishkin and N. Ya. Sinyavskii, *Usp. Fiziol. Nauk*, 163 (1993) 95.
[138] M. L. Buess, A. N. Garroway, and J. B. Miller, *Detection of Explosives and Narcotics by Low Power Large Sample Volume Nuclear Quadrupole Resonance (NQR)*, US Patent No. 5 233 300 (1993).
[139] L. J. Burnett, A. D. Hibbs, and B. D. Thorson, *Proceedings of the 2nd Explosives Detection Technology Symposium and Aviation Security Conference* (1996) 270.
[140] T. J. Rayner, L. J. Burnett, and S. Beevor, *Proceedings of the 2nd Explosives Detection Technology Symposium and Aviation Security Conference* (1996) 275.
[141] A. N. Garroway, M. L. Buess, J. P. Yesinowski, J. B. Miller, and R. A. Krauss, *Proc. SPIE*, 2276 (1994) 139.
[142] A. N. Garroway, M. L. Buess, J. B. Miller, K. J. McGrath, J. P. Yesinowski, B. H. Suits, and G. R. Miller, *Proceedings of the 6th International Symposium on Analysis and Detection of Explosives*, P. Mogfdk, ed. (1999).
[143] M. Ostafin and B. Nogaj, *Appl. Magn. Reson.*, 19 (2000) 571.
[144] E. L. Shanks, *Proceedings of the Third International Aviation Security Technology Symposium* (2001) 195.
[145] J. H. Flexman, T. N. Rudakov, P. A. Hayes, N. Shanks, V. T. Mikhaltsevich, and W. P. Chisholm, *NATO Science Series, Series II: Mathematics, Physics and Chemistry*, 138 (2004) 113.
[146] P. J. Prado, Y. Lee, K. Degenhardt, and A. Hudson, *Proceedings of the Third International Aviation Security Technology Symposium* (2001) 190.
[147] S. Peshkovsky, J. Forguez, L. Cerioni, and D. J. Pusiol, *J. Magn. Reson.*, 177 (2005) 67.

[148] H. Robert and P. J. Prado, *Appl. Magn. Reson.*, 25 (2004) 395.

[149] R. Kimmich, E. Rommel, P. Nickel, and D. Pusiol, *Z. Naturforsch. A*, 47 (1992) 361.

[150] T. N. Rudakov, J. H. Flexman, V. T. Mikhaltsevitch, P. A. Hayes, W. P. Chisholm, C. N. Aitken, and J. H. Feldman, *Nuclear Quadrupole Resonance Scanner for Detecting Explosives in Mail, Has Low Equivalent Series Resistance Switch to Switch Capacitance of Tunable circuit Enclosed by Electromagnetic Shield*, US Patent Application No. US2006012366-A1 (2006).

[151] A. Singsaas and C. Crowley, *Proceedings of the Third International Aviation Security Technology Symposium* (2001) 283.

[152] G. A. Barrall, D. A. Taussig, and T. G. Owens, *Proceedings of the Third International Aviation Security Technology Symposium* (2001) 183.

[153] M. D. Rowe and J. A. S. Smith, *Proceedings of the EUREL International Conference* (1996).

[154] J. A. S. Smith, *Proceedings of the Conference on Demining Technologies* (1998), 314.

[155] A. D. Hibbs, G. A. Barrall, P. V. Czipott, D. K. Lathrop, Y. K. Lee, E. E. Magnuson, R. Matthews, and S. A. Vierkötter, *Proc. SPIE*, 3392 (1998) 522.

[156] A. D. Hibbs, G. A. Barrall, P. V. Czipott, A. J. Drew, D. Gregory, D. K. Lathrop, Y. K. Lee, E. E. Magnuson, R. Matthews, D. C. Skvoretz, S. A. Vierkötter, and D. O. Walsh, *Proc. SPIE*, 3710 (1999) 454.

[157] R. M. Deas, I. A. Burch, and D. M. Port, *Proc. SPIE*, 4742 (2002) 482.

[158] A. D. Hibbs, G. A. Barrall, S. Beevor, L. J. Burnett, K. Derby, A. J. Drew, D. Gregory, C. S. Hawkins, S. Huo, A. Karunaratne, D. K. Lathrop, Y. K. Lee, R. Matthews, S. Milberger, B. Oehmen, T. Petrov, D. C. Skvoretz, S. A. Vierkötter, D. O. Walsh, and C. Wu, *Proc. SPIE*, 4038 (2000) 564.

[159] J. B. Miller and G. A. Barrall, *Am. Sci.*, 93 (2005) 50.

[160] J. Barras, M. J. Gaskell, N. Hunt, R. I. Jenkinson, K. R. Mann, D. A. G. Pedder, G. N. Shilstone, and J. A. S. Smith, *NATO Science Series, Series II: Mathematics, Physics and Chemistry*, 138 (2004) 125.

[161] J. Barras, M. J. Gaskell, N. Hunt, R. I. Jenkinson, K. R. Mann, D. A. G. Pedder, G. N. Shilstone, and J. A. S. Smith, *Appl. Magn. Reson.*, 25 (2004) 411.

Chapter 8

X-ray Diffraction Imaging for Explosives Detection

Geoffrey Harding and Adrian Harding*

GE Security Germany, D-22453 Hamburg, Germany

**In memoriam,* 18 October 2006

Counterterrorist Detection Techniques of Explosives
Jehuda Yinon (Editor)

Contents

1. Introduction and history of X-ray diffraction imaging

This chapter describes a novel radiographic modality, X-ray diffraction imaging (XDI), which is currently finding increasing application in various fields of public security, notably in the detection and identification of explosives, whether plastic, amorphous or liquid, and drugs. XDI is the imaging counterpart to the analytical technique of X-ray diffraction (XRD) in the same way as magnetic resonance imaging (MRI) is the imaging counterpart to magnetic resonance (MR).

This chapter is organized along the themes of physical principles, technological realization and security applications of XDI, whose historical development is traced in the remainder of this section. Section 2 reviews the principles underlying the two fields of physics that are involved in XDI: XRD on the one hand and X-ray tomography on the other.

In order to experience continued attention, any physical technique must satisfy a pressing need. The development of XDI has been stimulated by a so-called 'killer application', that of detecting and identifying explosives (and drugs) whose illegal use represents a grave danger to society. Physical parameters provided by XDI diffraction profiles that are useful in explosives detection are detailed in Section 3.

Section 4 describes two alternative XDI techniques, namely direct XRD tomography and computed XRD tomography. Representative results from both are presented, and the two techniques are compared and contrasted.

Advances in XDI over the years have been enabled by technological innovations, particularly in the fields of multi-foci, high-radiance X-ray sources and pixellated energy-resolving semiconductor detectors. The technological realization of a proposed next-generation XDI device will be discussed in Section 5.

The chapter concludes with a brief outlook in Section 6 on future developments in security applications for XDI.

1.1. XDI principles

X-ray diffraction imaging is the synthesis of two branches of X-ray physics, both originating in the early days of the twentieth century. The first branch refers to radiography, as exemplified in the celebrated X-ray image of the hand of Roentgen's wife (Fig. 1). The second branch is XRD (Fig. 2) and is illustrated by a diffraction pattern of copper sulphate, presented in the Nobel Prize lecture of Max von Laue (Nobel Physics Prize in 1914), that displays the characteristic interference peaks (spots) resulting from the interaction of a beam of X-rays with atoms arranged in a regular lattice structure.

It seems to have taken a long time to realize that these two branches of X-ray physics can profitably be integrated in XDI [1]. What factors were of importance in prompting the synthesis of XDI from XRD and X-ray imaging? There has traditionally been a cultural divide separating physicists active in the fields of X-ray imaging on the one

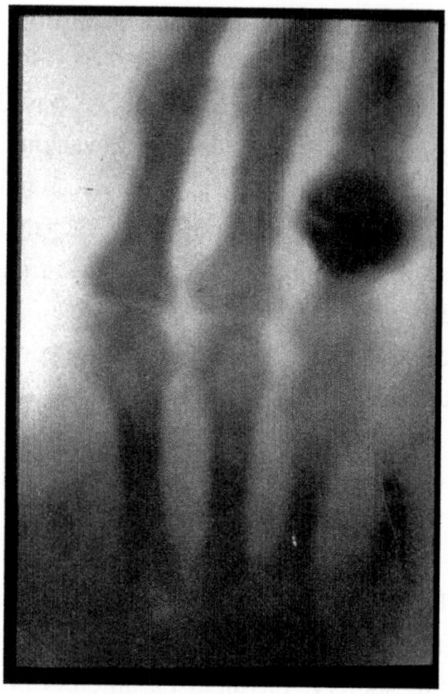

Fig. 1. Radiograph of the hand of Roentgen's wife.

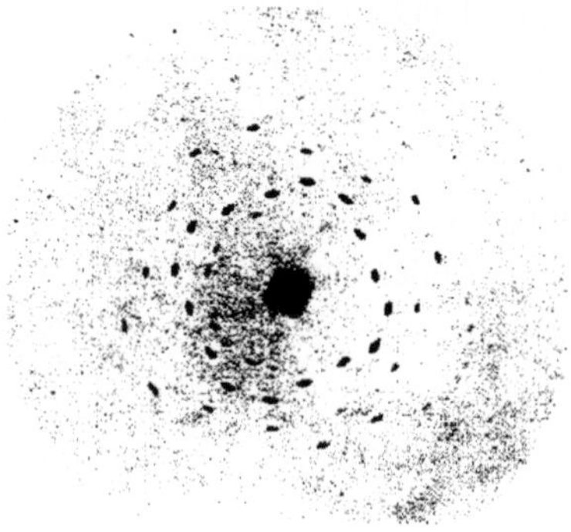

Fig. 2. X-ray diffraction pattern of copper sulphate, from von Laue's Nobel lecture.

hand and XRD on the other. The advent of computed tomography (CT) certainly played a major role in the development of XDI. It was in CT that the dominant contribution of coherent X-ray scattering, which is the basis of XRD, to the small-angle signal of X-rays in the diagnostic radiology energy range was first observed and correctly interpreted [2].

This conclusion appears to be in conflict with the relative magnitudes of the coherent and Compton interaction coefficients, illustrated for an organic explosive such as trinitrotoluene (TNT) in Fig. 3. It is apparent that the cross-section for Compton scatter dominates over the energy range above 10 keV.

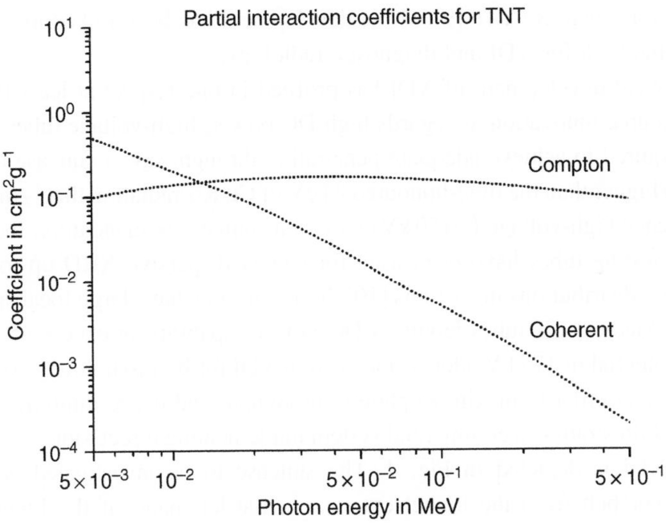

Fig. 3. Compton and coherent scatter interaction coefficients from XCOM program [?].

It fell to Neitzel et al. [3] to show that, even when multiple scatter is considered, coherent scatter forms the major component of low-angle scattering when a collimated beam of photons propagates through an extended object. In the low-angle regime of interest, the coherent scatter differential (angular) cross-section is maximum just where single Compton scatter is inhibited by electron-binding effects.

The last twenty years have seen the introduction of novel XDI geometries based on pencil beams [4], cone beams [5], fan beams [6], parallel sheet beams [7] and inverse fan beams [6]. These are all, indeed, special cases of a general 3-D arrangement synthesized from a generic 2-D section [6].

1.2. XDI technology

Technological convergence of XRD and diagnostic radiology at the component level has concerned, perhaps surprisingly, X-ray detectors rather than radiation sources. Development in the radiological field has tended to concentrate on X-ray sources having a high pulse power, permitting short exposure times. XRD sources on the other hand tend to be stationary anode devices with only moderate power-handling capability, though capable of prolonged exposure times.

X-ray film has been traditionally the mainstay both of diagnostic radiology and XRD. Efforts to replace it with digital detectors, such as the photo-stimulated luminescence plate [8], have benefited both fields. There has been enormous investment in silicon-based X-ray detectors enabled by the semiconductor industry, and these are found in increasing measure in conventional X-ray diffractometers. Room temperature energy-resolving semiconductors such as CdTe [9] whose development has been linked to that of silicon offer potential both for XDI and diagnostic radiology.

The technical development of XDI has profited in one respect at least from medical CT X-ray source innovation as regards high DC power, high-voltage tubes. The photon energies required to achieve adequate penetration through, say, a suitcase are at least a factor of 10 higher than the time-honoured 8 keV of Cu Kα radiation that is the workhorse of XRD. Hence high-voltage ($\geq$150 kV) radiation sources are mandatory. Industrial non-destructive testing tubes have been used for energy-dispersive XRD up to 300 keV to analyse stress distributions in castings [10]; however, they have large focal spots and low DC power. Medical CT tubes having a DC power capability in excess of 10 kW at an operating potential of 175 kV address the need in XDI for bremsstrahlung sources having high radiance in order to maximize photon throughput and hence minimize scan time.

By way of illustration, a commercial system implementing direct tomographic, energy-dispersive XDI is depicted in Fig. 4. The suitcase to be investigated is transported on a conveyor belt from the tunnel entrance at the left hand of the Figure to a pre-scanner. This supplies projection transmission images that are used to derive structural landmarks of the suitcase contents for registration purposes. The XDI system is to the centre-right of the Figure. Here the suitcase is irradiated with a primary cone beam that executes a 2-D meander scan movement relative to the suitcase. Small-angle scatter is recorded at constant deflection angle in a multi-segmented germanium detector. The scatter collimator images many separate regions of varying depths in the suitcase to their corresponding detector segments. A diffraction profile of TNT obtained with good counting statistics using this device is shown in Fig. 5.

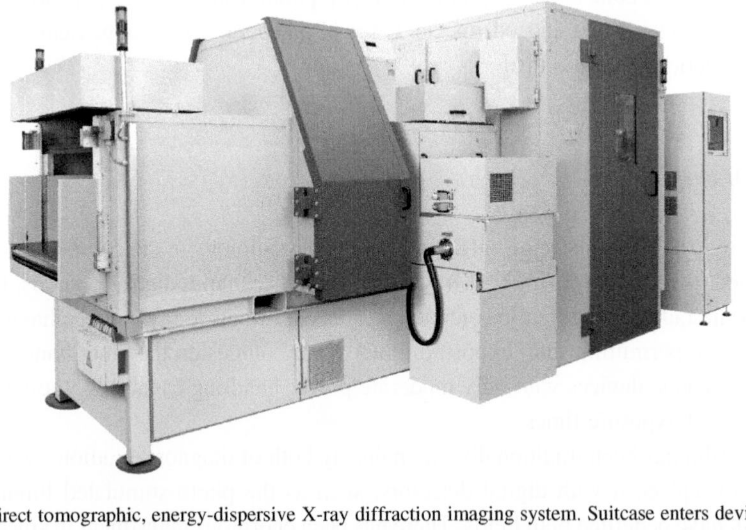

Fig. 4. Direct tomographic, energy-dispersive X-ray diffraction imaging system. Suitcase enters device at left of Figure where spatial landmarks for registration purposes are measured by pre-scanner. In main housing centre-right of Figure, a primary cone-beam executes a meander scan, either of a region-of-interest or suitcase in its entirety. Illustration courtesy of GE Security, Germany.

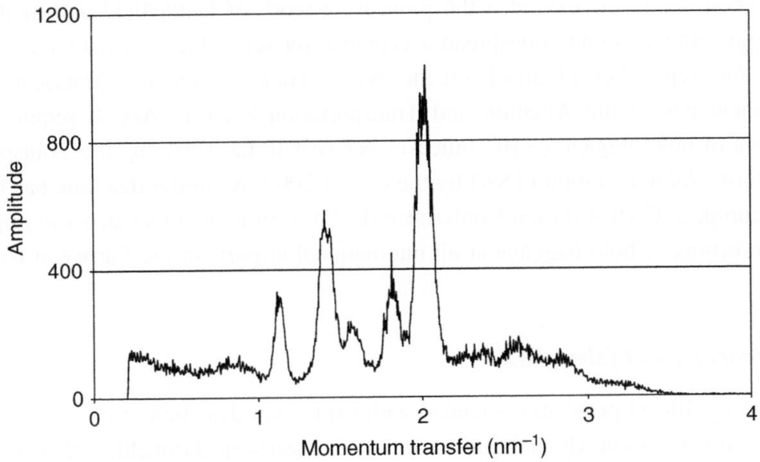

Fig. 5. Diffraction profile of TNT obtained from X-ray diffraction imaging device depicted in Fig. 4.

1.3. XDI applications

As far as applications of XDI, the comparison with MRI is illuminating. MRI has from its beginnings found a ready application for soft-tissue imaging in diagnostic radiology, driven by the powerful market forces at work in the health-care sector.

X-ray diffraction imaging in its early days followed a similar path to target medical diagnostic applications. Several investigators found remarkable differences between the diffraction profiles of healthy and diseased tissue in the skeleton [11] and breast [12], to name just two examples. Dose considerations however, particularly compared with MRI, have prevented up to now the widespread application of XDI in diagnostic radiology.

It is an empirical finding that the diffraction patterns of many organic explosives display prominent diffraction peaks that lend themselves to material identification [13]. XDI is sensitive to a wide range of explosives, and its low false-alarm rate (FAR) when confronted with the harmless materials that comprise the vast majority of suitcase contents is unsurpassed by alternative bulk detection techniques.

This section would not be complete without giving some idea of the path that XDI has gone to become a modality of major importance in contemporary security screening. The first hard X-ray (~100 keV) measurements of diffraction profiles of explosives in an XDI system that was in principle scalable in size to be able to analyse checked baggage were performed in the mid-1980s [14].

The Lockerbie disaster in 1988, in which 259 passengers died when Pan Am flight 103 disintegrated shortly after taking off from Heathrow Airport, brought the threat of bombs concealed in aeroplanes firmly into the public consciousness. Explosives detection systems employing a variety of different physical principles were suggested, developed and in some cases applied to the task. The interested reader is referred to the article of Singh and Singh [15] containing a useful set of references. In particular, the CT 'bomb-scanner' extended the pioneering work of Hounsfield and medical CT technology and has found widespread acceptance for screening checked baggage.

After the September 11 attack on the World Trade Center, the American Federal Government passed the Aviation and Transportation Security Act. It requires 100% inspection of hold luggage ($\sim 10^9$ suitcases per year in the USA) by the Transportation and Security Administration (TSA) by the end of 2002. A similar deadline has been set by the European Civil Aviation Conference (ECAC) for its member states to implement 100% screening of hold baggage at all international airports in the European Union.

1.3.1. Economics of false alarms

The most significant problem associated with explosives detection in passenger luggage is that of false alarms in which the detection system issues an alarm although no explosive is present. Exact Figures for the FAR in security screening are not publicly available. A recent article [16] suggests a value of around 30% for the CT systems certified by the TSA. When the large number of bags per year (10^9) in the domestic flights in the USA is recalled, this FAR translates into an enormous number of false alarms raised. The economic cost of resolving these false alarms is considerable, currently requiring the intervention of a large number of human operators.

It is in this context that the current and future development of XDI-based devices for bulk detection of explosives in checked baggage is justified [17] as it represents an accurate and general molecular-specific analysis modality for bulk detection of illicit substances. There are now indeed several manufacturers of equipment for XRD detection of explosives. A relatively new development in the public safety field is commercial development of equipment for XRD-based drug detection. Although this topic is beyond the scope of this chapter, it is noteworthy that drug abuse places an enormous financial burden on society. It is estimated for the year 2000 that in the USA alone the cost of drug abuse to society was over $150 billion. It is likely that XDI will have an increasing role to play in detecting illicit narcotic substances.

It is no exaggeration to say that the high cost of resolving false alarms in checkpoint and hold baggage screening is a significant driving force for continued development of XDI.

2. Physical principles of XDI

2.1. Energy- and angular-dispersive XRD

Consider the schematic illustration of a coherent scatter experiment illustrated in Fig. 6. A well-collimated beam of X-rays strikes a scatter sample initiating coherent scatter events that are recorded at a full angle of scatter, θ, in a detector.

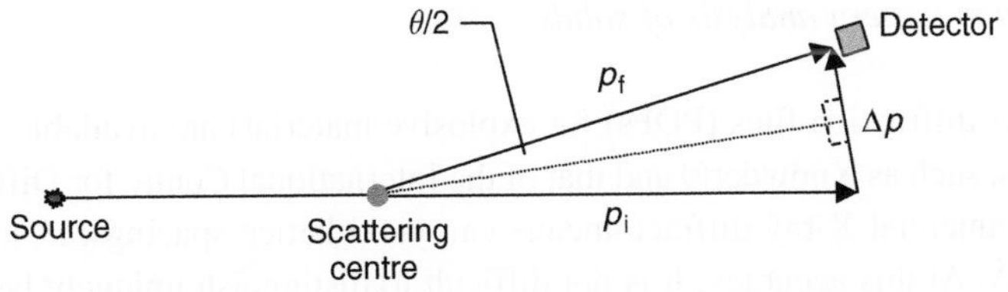

Fig. 6. Schematic illustration of a coherent scatter experiment.

As the scattered X-rays have the same energy as the primary rays, the momentum transferred from the crystal lattice to the photon beam is:

$$\left|\frac{\Delta p}{2}\right| = |p_f| \cdot \sin\left(\frac{\theta}{2}\right) \tag{1}$$

Thus:

$$\Delta p = 2\hbar k \sin\left(\frac{\theta}{2}\right) = \frac{hk}{\pi} \cdot \sin\left(\frac{\theta}{2}\right) \tag{2}$$

It is customary in X-ray scatter physics to express the reduced coherent momentum transfer in the following form:

$$x = \frac{\sin\left(\frac{\theta}{2}\right)}{\lambda} \tag{3}$$

There are hence two fundamentally analogous ways of probing a crystal lattice by varying the momentum transfer. In the first (classical) case the wave vector, k, is held constant and the angle of scatter, θ, is varied (angular-dispersive). Alternatively, it is possible to hold θ constant and to allow k to vary (energy-dispersive).

It transpires that, when measurement speed is a critical requirement, energy-dispersive XRD is the technique of choice. This conclusion is drawn from both the faintness of monochromatic relative to polychromatic (bremsstrahlung) X-ray sources and also the high bandwidth of large area, pixellated energy-sensitive detectors. The bandwidth is a measure of the counting efficiency of the system and is defined as the product of detector area and the number of independent energy channels through which data can be simultaneously acquired. Although angular-dispersive XDI explosives detection systems may find application in certain restricted circumstances, they are much slower than their energy-dispersive counterparts [18] and hence will not be considered further in this chapter.

The next section relates structural characteristics of typical explosives to their counterparts in features displayed by XRD profiles.

2.2. *Diffraction pattern analysis of solids*

X-ray powder diffraction files (PDFs) for explosive materials are available as subsets of PDF libraries, such as Xpowder® and that of the International Centre for Diffraction Data (ICDD). Commercial X-ray diffractometers can yield lattice spacings, d, to a precision of ~0.0001 Å. At this accuracy, it is not difficult to distinguish uniquely between many thousands of chemical compounds. Hence, XRD is a standard analytical technique for precisely identifying crystalline materials.

2.2.1. *Peak positions*

The well-known Bragg diffraction formula (Eq. 4) relates the angle of scatter at which a diffraction peak may occur to the spacing, d, of the lattice planes.

$$n\lambda = 2d \sin\left(\frac{\theta}{2}\right) \tag{4}$$

The wavelength, λ, is often the Cu Kα line at 0.154 nm. Primary information for material identification is provided by relative peak heights and peak positions. The latter is illustrated in Eq. 5, in which x is the momentum transfer, inversely proportional to the d-spacing at constant order, n.

$$x = \frac{1}{2nd} \tag{5}$$

It shall be assumed in this chapter that molecular arrangement in the bulk of solid explosives, and all amorphous and liquid explosives, has no preferred orientation direction. The diffraction patterns in this case are isotropic around the primary X-ray beam, and the vector quantity, $\mathbf{x}$, can be replaced by its scalar magnitude. It is customary to speak of diffraction profiles, rather than patterns, when isotropy obtains and the diffraction profiles are derived by integration of the (circularly-symmetric) diffraction pattern over the azimuthal component of the scattering angle.

Generally speaking, the diffraction profiles available in baggage screening applications are 'photon starved' owing to the limited measuring time; and this may present problems in peak position determination. An example is given in Fig. 7 of the XRD profile of TNT containing only 500 photons. This corresponds to the profile shown with good photon statistics in Fig. 5.

There are several ways of detecting peaks in such noisy signals. The Wiener–Hopf filter minimizes the expectation value of the noise power spectrum and may be used to optimally smooth the original noisy profile [19]. An alternative approach described by Hindeleh and Johnson employs knowledge of the peak shape. It synthesizes a simulated diffraction profile from peaks of known width and shape, for all possible peak amplitudes and positions, and selects that combination of peaks that minimizes the mean square error between the synthesized and measured profiles [20]. This procedure is illustrated

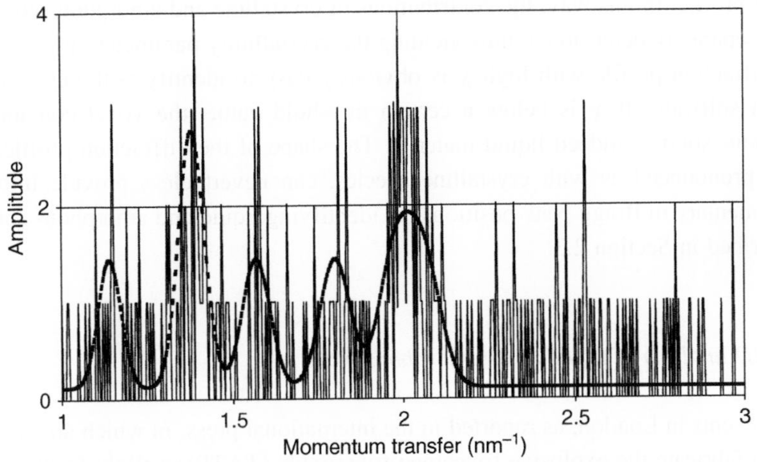

Fig. 7. Energy-dispersive diffraction profile of TNT with 500 counts total.

in Fig. 7, for which the fitting procedure assumed an elementary peak form of Gaussian shape. The diffraction profile of TNT in this Figure is to be compared with the high signal-to-noise version presented in Fig. 5.

The amorphous component of the diffraction profile can often be fitted with acceptable accuracy using a third order polynomial.

Blaffert [21] has addressed the question of optimum peak widths. Broad peaks imply a high photon throughput as the detector acceptance angle may be increased when the severe restrictions on scatter angle are relaxed. This advantage is however bought at the price of sacrificing the ability to distinguish materials with similar diffraction patterns from one another. A compromise has to be sought and is usually found when the peaks of the diffraction profile have a full width at half maximum (FWHM) of several per cent of their position.

2.2.2. *Crystallinity*

A further parameter of interest is the crystallinity, χ, representing the ratio of the number of photons scattered into the diffraction profile peaks to the total number of scatter photons:

$$\chi = \frac{A_{\text{crystalline}}}{A_{\text{crystalline}} + A_{\text{amorphous}}} \tag{6}$$

The Hindeleh–Johnson procedure derives the amorphous component of the diffraction profile in a natural way as described in Section 2.2.1. An alternative technique is to determine the positions, amplitudes and widths of diffraction lines with a peak-finding algorithm (e.g. based on the second differential) and to subtract lines so found from the original profile. In this way, the contributions of crystalline and amorphous components can be separately determined, thus yielding the crystallinity parameter, χ.

A diffraction profile with high χ is obviously easy to identify with customary peak position software. If χ is below a certain threshold value, the voxel contains quasi-amorphous solid or indeed liquid material. The shape of the diffraction profile, though not so pronounced as with crystalline species, can nevertheless provide information complementary to Bragg peak positions for identifying liquid and amorphous substances as described in Section 2.3.

2.3. *XRD analysis of liquids and amorphous materials*

Recent events in London, as reported in the international press, in which an attempt was made to fabricate the explosive triacetone triperoxide (TATP) in-flight from the liquids acetone and a concentrated hydrogen peroxide solution, have forced the topic of the identification of liquids firmly into the public consciousness. XRD identification of liquids is thus the topic of this section. Several parameters by which XRD profiles of liquids may be characterized will be described here.

2.3.1. *Effective atomic number determination*

The total differential scatter cross-section is often written in the form factor approximation as a product of three factors describing three levels, electron, atom and molecule, of complexity of the scattering system:

$$\frac{d\sigma_{\text{total}}}{d\Omega} = \left[\frac{r^2 e}{2}\left(1+\cos^2\frac{\theta}{2}\right)\right]\cdot\left[F^2(x)+S(x)\right]\cdot\left[s(x)\right] \tag{7}$$

The first factor in square brackets represents the Thomson cross-section for scattering from a free electron. The second square bracket describes the atomic arrangement of electrons through the atomic form factor, F, and incoherent scatter function, S. Finally, the last square bracket contains the factor $s(x)$, the molecular interference function that describes the modification to the atomic scattering cross-section induced by the spatial arrangement of atoms in their molecules.

Figure 8 shows some total free atom scatter cross-sections ($F^2 + S$) derived from the tabulation of Hubbel et al. [22]. At zero momentum transfer, the ordinate is simply Z^2. For all species apart from hydrogen, the cross-section falls (note the logarithmic scale) rapidly with increasing x.

Consider Fig. 9 showing the diffraction profile for the polymer Lucite ($C_5H_8O_2$) taken from Tartari et al. [23]. The dashed line represents the function $F^2 + S$ calculated from the mixture rule on the basis of the Independent Atom Model (IAM) using the tabulated atom form factors for H, C and O data presented in Fig. 8. The IAM curve is seen to approach the measured profile at high x values.

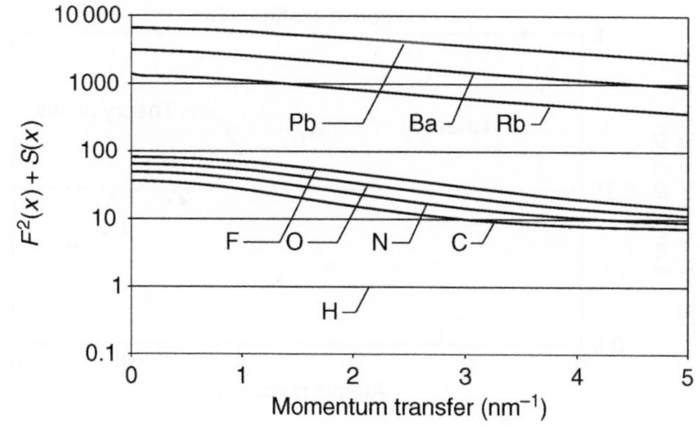

Fig. 8. Free atom scatter functions for a selection of elements.

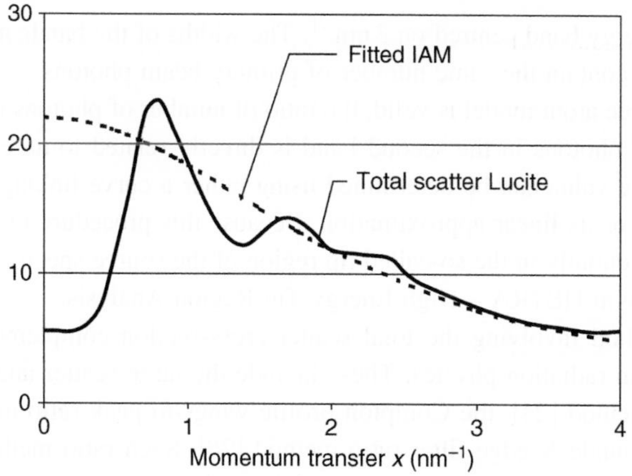

Fig. 9. Diffraction profiles for Lucite ($C_5H_8O_2$). Solid curve is measurement whereas dashed plot is fitted from Independent Atom Model (IAM).

The next Figure (Fig. 10) shows the ratio of the atomic factors ($F^2 + S$) taken at x values of 3 and 4 nm^{-1} as a function of atomic number Z. Both the theoretical points and a linear fit are shown. Apparently, the ratio is a monotonically falling function of Z for elements relevant in plastic explosives. Conversely, measurement of this ratio from experimental profiles allows the effective atomic number of the scattering sample to be determined.

Assume for example that energy-dispersive diffraction profiles have been acquired covering the momentum transfer region $3 \leq x \leq 4\,\text{nm}^{-1}$. This might correspond through the $\sin(\theta)$ scaling relationship of Eq. 3 to photons in energy bands centred on 150 and 200 keV at an angle of scatter of ~ 0.05 radians. It is assumed further that these photons can be assigned to one of two energy bands: an energy band corresponding to 4 nm^{-1} in the tip region of the X-ray source spectrum (directly below the tube potential) and a

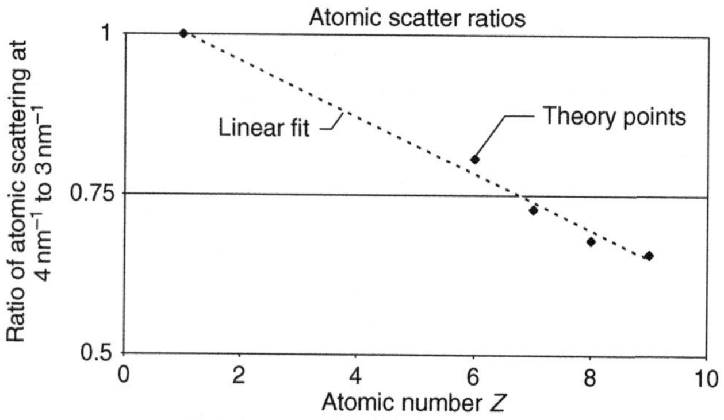

Fig. 10. Ratio of total atom scatter at 4 to $3\,\mathrm{nm}^{-1}$. Trapezoids are calculated from theory [22] whereas dashed line indicates linear fit for $Z = 1$ and $6 \leq Z \leq 9$.

slightly lower energy band centred on $3\,\mathrm{nm}^{-1}$. The widths of the bands may be adjusted so that they both contain the same number of primary beam photons.

Provided the free atom model is valid, the ratio of number of photons in the first band to the number of photons in the second band is directly related to an effective atomic number, $\dot{Z}$, whose value can be determined using either a curve linking the theoretical points of Fig. 10 or its linear approximation. Because this procedure measures photons originating preferentially in the so-called tip region of the source spectrum, it is denoted here by the acronym HETRA – High Energy Tip Region Analysis.

This ratio method involving the total scatter cross-section complements other ratio methods known in radiation physics. These include the large scatter angle coherent-to-Compton ratio method [24], the Compton profile wings-to-peak ratio method [25] and the small scatter angle K-edge filter ratio method [26]. Such ratio methods are known to be capable of high systematic accuracy, as the paths in the object taken by the two signals whose ratio is formed are identical.

It has to be verified that the IAM model is valid for a certain substance in the x regime envisaged above. It should be remembered that the main application area for this determination of effective Z is the identification of low crystallinity (liquid or amorphous) substances, for which peaks arising from crystal structure should either have been removed previously or are in any case absent.

2.3.2. Molecular interference function

Once the effective atomic number of the scattering species is known, for example using the HETRA method described in Section 2.3.1., it is possible to account for the IAM component of scattering in the diffraction profile, allowing the molecular interference function to be extracted. For this purpose, a universal free atom scattering plot, which can be extrapolated to non-integral values of the effective atomic number, is needed. The IAM total scattering curves for the elements with $6 \leq Z \leq 9$ normalized to unit $1/e$

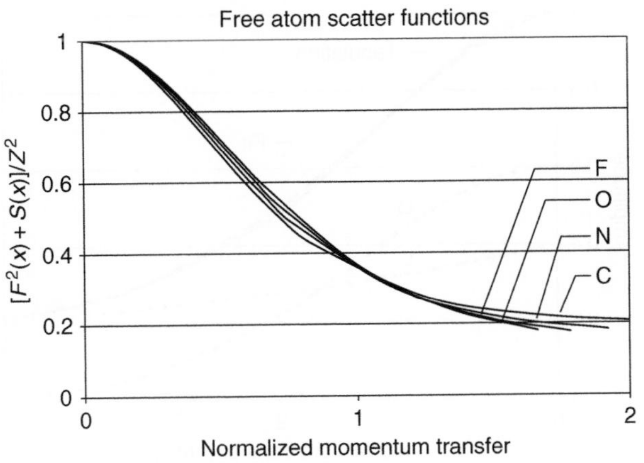

Fig. 11. Total free atom scatter functions for the elements C, N, O and F. The amplitudes of these plots are normalized by the factor Z^2. Moreover, they are scaled to have unit width at their $1/e$ height.

width (that width at which the curve has fallen to $1/e$ of its zero abscissa amplitude) are reproduced in Fig. 11.

The $1/e$ widths are found to follow the linear relationship $X_{1/e} = 0.02\ Z + 0.12$. It is apparent that the normalized plots are fairly similar to one another. The IAM curves may be fitted to an rms deviation of below 5% by a 'universal free atom' curve comprising a Gaussian (accounting for the central region) and a Lorenzian (describing the wings).

$$\mathrm{IAM}(x_{\mathrm{scale}}) = \frac{0.4}{\left(1 + \frac{x_{\mathrm{scale}}}{10}\right)^2} + 0.6 \cdot \exp\left(\frac{-x_{\mathrm{scale}}^2}{4}\right) \tag{8}$$

In this equation, the scaled width, x_{scale}, is related to the momentum transfer, x, by the following equation:

$$x_{\mathrm{scale}} = \frac{x}{x_{1/e}} \tag{9}$$

The 'universal' IAM fit for nitrogen is compared with the tabulated values from [22] in Fig. 12.

Knowledge of the effective atomic number allows the true width and height of the IAM curve to be determined and hence permits the molecular interference function, $s(x)$, to be uniquely extracted on dividing the measured diffraction profile by the IAM function (cf. Eq. 7).

2.3.3. Radial distribution function

The radial distribution function, $g(r)$, generally forms the starting point for analysis of the liquid structure once the molecular interference function, $s(x)$, is known. As discussed

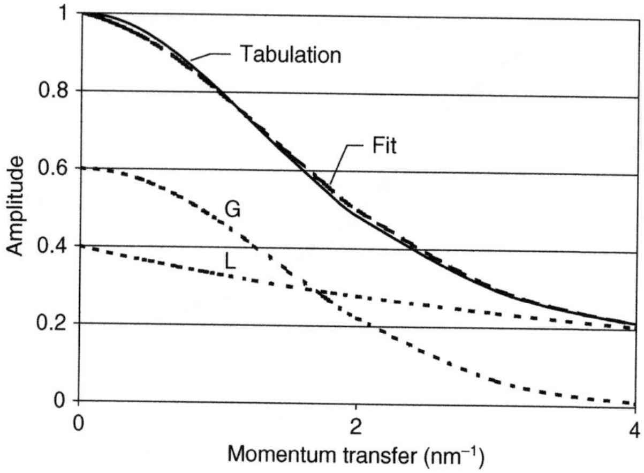

Fig. 12. 'Universal' fit (long dashed line) to tabulated total Independent Atom Model (IAM) cross-section (unbroken line) for N. The Gaussian (G) and Lorenzian (L) components of the fit are also shown.

by Hukins [27], there is a Fourier transform relationship between $s(x)$ and $g(r)$, the radial distribution function, after they have been properly normalized. The latter describes the probability per unit volume of finding a particle at distance r given a first particle at the origin. The function $g(r)$ is zero at small separations where Coulomb repulsion prevents atoms closely approaching one another. There is a distance from the origin molecule at which the probability of finding a neighbouring molecule is greatest, leading to a first peak in $g(r)$. This represents the most likely nearest neighbour distance. At large radii, the oscillations in $g(r)$ dampen down, and it approaches a value of unity when normalized to the mean density. The Fourier transform relationship between $s(x)$ and $g(r)$ can be exploited to extract the latter from which, hopefully, mean molecular separations, coordination numbers, etc. may be derived. The molecular interference function is useful for the purposes of liquid identification in its own right, even without the Fourier transform operation. In the next section, some representative results from XRD investigations of liquids are presented.

2.3.4. Examples of liquid profiles

The plots of alcohol and water mixtures presented here serve to illustrate the usefulness of XRD for liquid identification from the molecular interference function $s(x)$. The curves shown here differ from the $s(x)$ discussed in Section 2.3.1. in that they portray the square of the ratio of $s(x)$ of the sample to $s(x)$ of a 'white scatterer', a calibration object of proprietary composition whose scattering characteristics are fairly constant over the x range of interest. Notwithstanding these manipulations, the plots have absolute ordinate scale.

The alcohol plot (Fig. 13) shown top right reveals a peak centred on $\sim 1.2\,\text{nm}^{-1}$ and having a FWHM of $0.25\,\text{nm}^{-1}$. Relative to its comparatively low effective atomic number, estimated using HETRA to be 5.3, it scatters profusely as shown by the peak

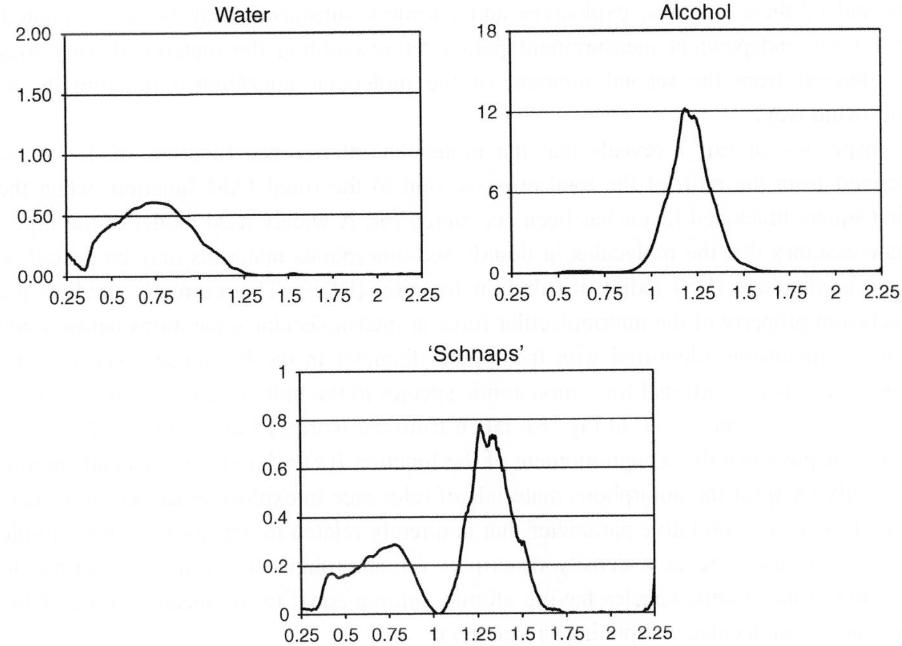

Fig. 13. Diffraction plots of water, alcohol and a German liquor sample of 'Schnaps'.

height of 12, indicating a high degree of long-range spatial order. Parameters such as peak position, peak area, peak width and effective atomic number are related to such characteristics of the molecular structure as nearest neighbour separation, coordination number, degree of long-range order and mean atomic number and hence prove useful for liquid identification.

The water plot (Fig. 13, top left) shows a much broader peak of lower peak amplitude at the lower x value of $0.75\,\text{nm}^{-1}$. The effective atomic number derived from HETRA is 6.3. Of particular interest is that these two peak structures are maintained in the plot of German liquor (centre bottom) indicating that the molecular structure of water and alcohol are preserved even when they are dissolved in one another. This possibly surprising conclusion is well known in the literature [28].

The work presented in this section is of a preliminary nature, and extensive trials are required to determine the limits of applicability of XRD for identification of liquids and to evaluate the significance of FARs for checkpoint and hold baggage screening.

2.3.5. Density descriptor

In Section 2.3.1., the diffraction profile was fitted to IAM atomic scatter cross-sections in the region where the molecular interference function is practically unity. This procedure yields the effective atomic number and, particularly for liquids, several further parameters derived from peaks in the molecular interference function. With the aid of these features, explosives and harmless substances may be differentiated. A further, independent measurement parameter resembling the material density may be derived from the second moment of the molecular interference function in the following way.

Inspection of Eq. 7 reveals that the molecular interference function, $s(x)$, can be derived from the ratio of the total cross-section to the fitted IAM function, when the first square bracketed factor has been accounted for. A widely used model of the liquid state assumes that the molecules in liquids and amorphous materials may be described by a hard-sphere (HS) radial distribution function (RDF). This correctly predicts the exclusion property of the intermolecular force at intermolecular separations below some critical dimension, identified with the sphere diameter in the HS model. The packing fraction, η, is proportional for a monatomic species to the bulk density, ρ. The variation of $s(x)$ on η is reproduced in Fig. 14, taken from the work of Pavlyukhin [29].

It transpires that the second moment of the function $I(x) = [s(x) - 1]^2$ depends monotonically on η for the amorphous materials of relevance in explosives detection. Hence, it represents a quantitative parameter that is directly related to density for a monatomic species. Its use here as a density descriptor for amorphous materials corresponds to defining a monatomic species having atomic volume equal to the mean volume of the polyatomic molecules comprising the material.

The wide array of independent features available for characterizing both solid and liquid explosives, as described in the Sections 2.3.1.–2.3.5., is largely responsible for the high detection rate and low FAR of explosives detection by XDI.

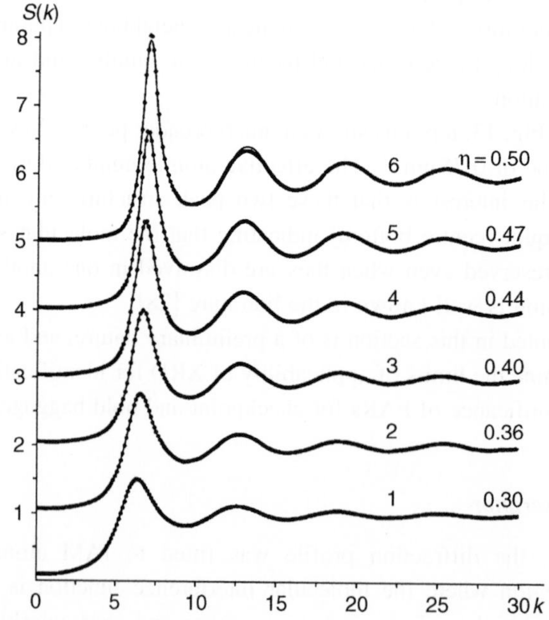

Fig. 14. Variation of $s(x)$ on packing fraction, η, for a hard-sphere fluid [29].

2.4. Spatially resolved XRD imaging

The physical principles of XRD have to be complemented with those underlying radiological imaging in order to complete the description of XDI for extended objects. Attenuation effects are much more significant as a source of signal degradation in XDI than in XRD, which only deals with small samples, and multiple scatter effects have to be explicitly accounted for as described in this section.

Consider Fig. 15 showing a well-collimated beam of photons having wave number k_i incident on a small sample of side length dl and surface area normal to the incident beam of A. The differential cross-section $\mathrm{d}\sigma(\theta)/\mathrm{d}\Omega$ is defined as the probability of scattering into a small solid angle dΩ around the scatter angle θ. The momentum transfer for this case was derived in Eq. 2.

The infinitesimal element of scatter signal recorded in a detector subtending a solid angle of $\Delta\Omega_{\mathrm{det}}$ at a small sample having volume $\Delta V = A\mathrm{d}l$ on irradiation by N_0 photons is:

$$\mathrm{d}N(\theta) = N_0 A n_0 \Delta\Omega_{\mathrm{det}} \left[\frac{\mathrm{d}\sigma_{\mathrm{inc}}}{\mathrm{d}\Omega} + \frac{\mathrm{d}\sigma_{\mathrm{coh}}}{\mathrm{d}\Omega} \right] \mathrm{d}l \tag{10}$$

The symbol n_0 refers to the number density of scatter centres.

The extension of XRD from small samples to extended objects requires two main modifications of Eq. 10: to account for the self-attenuation of X-rays within the object and also to describe the component of scatter signal that suffers more than one interaction within the object.

Consider one arbitrary voxel of the extended object located at the position $\boldsymbol{r}$ relative to some coordinate scheme. If the voxel is small relative to the dimensions of the measuring arrangement, solid angle factors can be assumed to be constant and the total scatter signal can be represented in the following form:

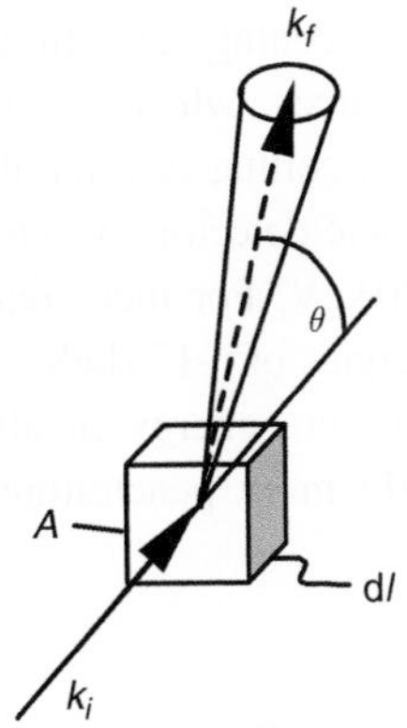

Fig. 15. Geometric relations of importance in signal calculation.

$$\frac{N(\theta)}{N_0} = F_{\text{prim}}(\boldsymbol{r})S(\boldsymbol{r})F_{\text{scatter}}(\boldsymbol{r})\Delta\Omega_{\text{det}}(\boldsymbol{r}) + M(\boldsymbol{r}) \tag{11}$$

$S(\boldsymbol{r})$ relates to the scattering strength of the voxel derived from integrating Eq. 10 over the voxel length. F_{prim} relates to attenuation of the primary beam between the radiation source and the voxel under investigation, whereas F_{scatter} similarly describes attenuation between the measurement voxel and the detector. The term, $M(\boldsymbol{r})$, represents multiple scattering into the detector. It is necessary to account for both these effects in XDI as described in Sections 2.4.1. and 2.4.2.

2.4.1. Attenuation correction

The X-ray beam attenuation depends naturally on the range of photon energies, E, with which the XDI investigation is to be conducted. In practice, the choice of optimum photon energy is derived from a judicious compromise of several factors as detailed in Sections 2.4.1.1.–2.4.1.4.

2.4.1.1. Dark alarms

There will be a certain proportion of suitcases whose contents are so optically thick that it is impossible to achieve sufficient penetration to acquire a useful number of photons in the diffraction profile. This renders evaluation impossible. Such an impenetrable suitcase raises a 'dark' (sometimes called 'shield') alarm. Clearly, the most effective way of reducing the dark alarm rate is to raise the mean photon energy. It is in this respect that the HETRA method of diffraction profile analysis (Section 2.3.1.) may prove helpful as it utilizes photons at the high-energy limit, where the suitcase is most transparent.

2.4.1.2. Source and detector efficiency

As discussed in Section 5 below, XDI benefits from technological advance in the sectors of radiation sources as well as in energy-resolving pixellated detectors. It is nevertheless a fact of life that 200 keV represents a major upper limit to the photon energy that can be used in XDI for baggage screening. This limit is set on the one hand by high-radiance rotating anode X-ray sources, which are invariably developed for medical (CT) applications and experience a ceiling beyond about 180 keV. On the other hand, commercially available spectroscopic detectors, whether based on Ge or CdTe, become increasingly inefficient above 100 keV. For these reasons, XDI has traditionally been confined to maximum photon energies of ~175 keV.

This is not necessarily an optimum energy in all situations, and cargo screening applications requiring significantly more penetrating radiation will be considered in Section 6.

2.4.1.3. Angle of scatter

The photon energy range chosen for XDI influences the angle of scatter required to probe a certain momentum transfer value, x, implicitly through Eq. 3. As indicated in

Fig. 5, prominent diffraction peaks generally occur in the momentum transfer range of 1–3 nm^{-1}. An angle of scatter, θ, of ~ 0.04 radians is required to scale this momentum transfer range into an energy range of several tens of keV. Equation 12, relating the uncertainty in momentum transfer to instrumental (angular) broadening, indicates that an angular acceptance below 1 mrad is required to attain peak broadening of $\leq 2\%$ FWHM. The photon throughput at constant diffraction peak resolution falls off with $\sim \theta^{-4}$ as θ varies. This arises from scale reductions in focus size, the acceptance of the primary and scatter collimators and the detector pitch, needed to keep the angular uncertainty constant as the angle of scatter decreases.

2.4.1.4 Small scatter angle attenuation approximation

As the angle of scatter is quite small, the combined attenuation of the primary and secondary beams is given to a good approximation by the attenuation of the undeflected radiation. For the case indicated in Section 2.2.1.3. which $\theta = 0.04$, the undeflected beam exits a suitcase of 500 mm height only 10 mm away from a beam scattered from its centre. Errors in determining peak positions arising from differences in the paths through the suitcase taken by the scattered and transmission beams are usually very small and can be further reduced by implementing a symmetrical measurement arrangement, in which signals are recorded from detectors on both sides of the transmitted beam.

Moreover, the coherent scatter interaction is also elastic, and hence photons have identical energy before and after the interaction. Hence, a first-order attenuation correction simply normalizes the coherent scatter spectrum against the spectrum of transmitted rays. Further refinement of the first-order correction is possible, but its discussion is beyond the scope of this chapter.

2.4.2. Multiple scatter correction

Several authors have investigated multiple scatter in spatially resolved photon-scattering applications [3,30]. As the coherent cross-section is very small, the mean free path for coherent interactions is larger than typical object dimensions, and coherent multiple scatter can be ignored. Multiple Compton events rapidly decrease the photon energy; hence, the high-energy region of the coherent scatter spectrum is practically free from multiple scatter degradation [30].

Multiple scatter by its very nature arises in several uncorrelated scatter interactions, and its amplitude is often insensitive to the precise composition of the object under investigation. Its spatial variation by the same token is often small [3]. A position-independent multiple scatter component can be judiciously estimated from calibration measurements or from Monte Carlo photon transport simulation programs and subtracted as necessary from experimental profiles.

3. XRD characterization of explosives

This section is mainly anecdotal in nature and a reference work such as *Chemistry of Explosives* [31] should be consulted for fuller information. The story of modern explosives can be said to have started with the accidental discovery of nitrocellulose by Christian Schoenbein in 1846, when his wife's apron, with which he wiped up a spilled mixture of acids, exploded and vanished in a puff of smoke. The significance of nitration is evident in the structures of 'classical' explosives shown in Fig. 16.

A primary structural characteristic of explosives is generally that they feature a copious supply of oxygen in the molecular vicinity of combustible species such as hydrogen and carbon. They thus tend to be oxygen-rich. Explosive reactions are usually solid-state, in which the crystal structures and molecular neighbourhoods play an important role. This is incidentally true also of so-called liquid explosives in which a solid crystalline explosive is prepared from liquid precursors. A further advantage of solid-state explosives (i.e., used at a temperature below the melting point) is their enhanced stability against accidental detonation compared with liquid explosives. Crystalline solids, with very few exceptions, are chemically more stable than amorphous solids of the same molecular composition. Moreover, other things being equal, the detonation velocity of an explosive is proportional to its density. Hence, important explosives tend to have a higher physical density than harmless organic material of similar composition. Finally, explosives are often composed of randomly oriented grains with small sizes and hence large surface areas. Whether owing to these factors or others, it seems to be an empirical fact that most useful explosives are polycrystalline, have a density that is high for organic material and are rich in oxygen. This statement applies incidentally also to explosive slurries and emulsions, which are generally saturated aqueous solutions of nitrates, with additives.

The physical parameters whose measurement from diffraction profiles was considered above in Section 2.2. are directly applicable to the challenge of identifying and characterizing organic and other explosives in checked baggage. Table 1 summarizes the relationship between the diffraction profile analysis procedures described in Section 2.2. and physical parameters of significance for explosives detection discussed in this section.

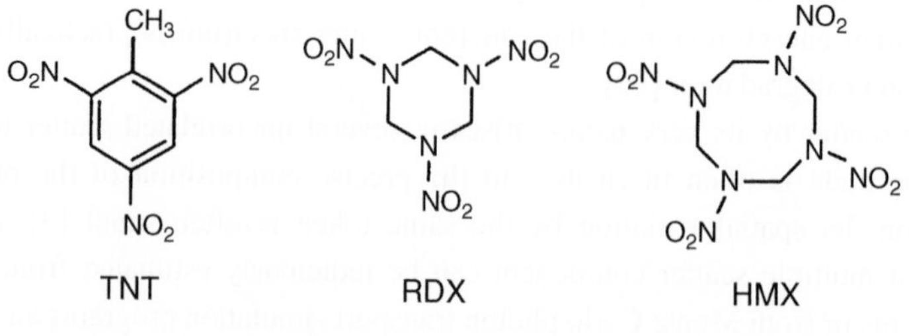

Fig. 16. Illustration of molecular structure of 'classical' organic explosives.

Table 1. Characterization parameters for plastic explosives and their relation to diffraction profiles

Parameter	Diagnostic potential	XDI analysis
Density	Organic explosives tend to have higher density than equivalent harmless plastic materials	Diffraction profile yields density descriptor based on analysis of molecular interference function
Volume	Critical volume exists under which explosive is non-threatening	Tomographic imaging capability supplies 3-D volume data set with volume estimation of alarm
Mean $\dot{Z}$	Organic explosives tend to be oxygen-rich leading to enhanced mean atomic numbers	HETRA analysis of diffraction profile yields mean $\dot{Z}$. Section 2.4.1.1. details significance for 'dark' alarms
Crystallinity	Descriptor of sample age, particularly for freshly-prepared slurries and emulsions	Profile decomposition procedures yield contributions of peak and continuum components
Lattice constants	Many important commercial, military and home-made explosives have diffraction profiles with prominent Bragg peaks	Material-specific identification procedure as diffraction profile represents explosives 'signature'

4. Tomographic imaging techniques

Spatial localization of a certain region in the object under investigation is simpler in scatter X-ray than in transmission X-ray imaging owing to the change in photon propagation direction that accompanies scattering. To determine the origin point of a scatter photon, it is merely necessary to find the intersection point of the primary and scatter beams. Both direct and CT solutions to the imaging problem have been proposed and tested as described in this section.

4.1. Spatial resolution

X-ray scatter imaging can be performed in principle with such components as lenses and mirrors, though these suffer from poor efficiency beyond the soft X-ray region. In practice, X-rays having the penetrability ($E_0 \geq 40\,\mathrm{keV}$) needed for explosives XDI can be collimated only by arrangements of pinholes, slits, etc. in otherwise absorbing structures. The collimation scheme to be used in a certain situation for XDI depends sensitively on the choice of imaging parameter to be optimized, and this optimization process involves detailed consideration of several factors, including:

- the angular width of the tight, forward-scatter cone;
- the angular resolution required to determine diffraction profiles with a certain peak width;
- the spatial resolution (pitch) of the detector;
- the size of the voxels into which the object is to be segmented;
- the maximum dose that may be applied;
- the physical characteristics of the radiation source and detector (e.g. a reverse imaging geometry, comprising an extended source and a point-like detector, permits imaging of extended objects with the use of only a small detector area).

As Eq. 12 indicates, uncertainty in the momentum transfer corresponding to a certain diffraction line can arise both in the limited angular resolution of an XDI system and in its limited energy resolution. Assuming these two sources of line broadening are uncorrelated and applying the customary law for the addition of relative errors yields for angles of scatter in the small-angle approximation:

$$\frac{\partial x_{\text{total}}}{x} = \sqrt{\left(\frac{\partial \theta}{\theta}\right)^2 + \left(\frac{\partial E}{E}\right)^2} \tag{12}$$

The second term under the square root is limited by the detector energy resolution. Fortunately, for a good Ge detector, the energy resolution can have a value of 1 keV at 60 keV corresponding to 1.3%. The relative angular uncertainty is often more critical. The angle of scatter in energy-dispersive XDI is typically ~0.04 radians. An angular broadening of 2% or better implies an angular precision of 0.8 mrad. This should be achieved in a collimator that is mechanically stable, inexpensive and easy to align.

As the large number of patents in the X-ray collimator field reveals, it is by no means a trivial exercise to design a collimator optimally satisfying the constraints imposed by the above factors. It is unfortunately beyond the scope of this chapter to discuss the science of collimator design ('collimatology') except that it is customary to group X-ray collimation techniques into two main categories, i.e. direct and reconstructive. These possibilities are summarized in Sections 4.2 and 4.3.

4.2. Direct X-ray scatter imaging

In its simplest form, direct X-ray scatter imaging relies on the use of simple mechanical collimation elements such as pinholes, Soller slits and the like to determine the origin coordinates of a scattered photon. They all achieve spatial resolution of the scatter field at the detector by restricting the angular range over which radiation can reach the detector. Examples of direct tomography in the explosives detection field include the

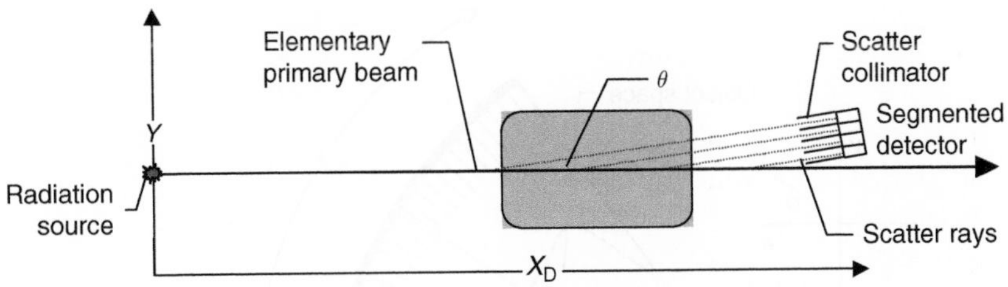

Fig. 17. Two-dimensional generic cross-section for energy-dispersive direct tomographic X-ray diffraction imaging.

Energy-Dispersive X-ray Diffraction Tomography (EDXDT) geometry [32], the Sheet Explosive X-ray Imager (SEXI) system [33], the Low Angle X-ray Scatter (LAXS) device of Luggar et al. [34] and a combined angular and energy-sensitive device [18].

In its simplest 2-D form, the energy-dispersive technique for direct tomography resembles the section shown in Fig. 17. A beam from a polychromatic radiation source (e.g. electron impact bremsstrahlung) is well collimated before being incident on the object under investigation. A segmented semiconductor detector performs energy analysis on those photons reaching it after scattering through a constant angle θ. By virtue of the geometry of the scatter collimator, the origin coordinate of single X-ray scatter impinging on the detector is simply related to the location of the corresponding detector element.

The direct tomographic technique can be further subdivided according to the detector area into which a certain elemental volume in the object can radiate. For the case of a primary beam confined to a 1-D pencil beam, these include the point-to-point, point-to-line, line-to-line and point-to-plane variants [35].

For a 2-D primary beam emanating from a point source ('fan beam'), point-to-point imaging can be achieved with simple pinholes or one of the several multi-hole collimator variants (parallel, converging and diverging) known from radionuclide imaging. A mechanical collimator has typically a solid angle efficiency of only 10^{-4} sr and hence achieves imaging capability only at the expense of a significant loss in efficiency. This has prompted investigation of other, possibly more efficient collimation schemes, as discussed in the next section.

4.3. *XRD computed tomography*

The introduction of transmission CT by Godfrey Hounsfield stimulated interest in similar reconstruction schemes for X-ray scatter imaging. Hence, these techniques are grouped under the general heading of reconstructive X-ray scatter imaging.

As far as the application to explosives detection is concerned, the reconstructive technique that has been most widely applied is that of XRDCT, also termed coherent scatter CT (CSCT) [36–38].

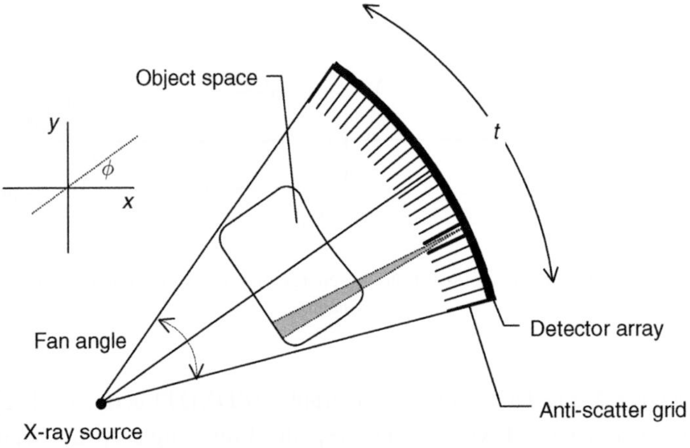

Fig. 18. Geometry of fan-beam coherent scatter computed tomography. A fan beam is used to illuminate one object slice. The central detector row (in the plane of the drawing) receives the transmitted radiation while the out-of-plane rows receive only scattered radiation. The coordinate ϕ represents the projection angle whereas t signifies a detector column.

It is beyond the scope of this chapter to discuss filtered back-projection (FBP) reconstruction procedures or their 3-D extensions to cone-beam CT for which reference should be made to other works [39].

This section is based on the description published in reference [40]. The geometry of the data acquisition for XRDCT is shown in Fig. 18. A fan beam illuminates the object, and the transmitted fan (in the plane of the drawing) reaches the 2-D detector at its central row. These data can be used for attenuation correction and also for a conventional transmission-CT image. The detector columns are indexed by the variable t, whereas the relative angle between measurement system (source, detector, etc.) and suitcase is described by the variable ϕ. In conventional CT, a 2-D data set, $I(t, \phi)$, is measured for many angular positions, ϕ, of the object with respect to the device.

Out-of-plane detector rows receive solely scattered radiation through an anti-scatter grid (ASG) indicated in Fig. 18. The segment of object which a certain detector column 'sees' through the ASG is shown in the Figure, and the signal a certain detector element records approximates a line integral of the coherent scatter function at the mean photon energy of the fan and the mean angle of scatter. A 3-D data set, $I(t, \phi, a)$, can now be measured where, in addition to the transmission projection, the scatter data in the vertical dimension, a, (perpendicular to the plane of the drawing in Fig. 18) are acquired. From this 3-D data set, the molecular coherent scatter function $\sigma(x, y, q)$ can be reconstructed, where the coordinates x, y refer to the scan plane and q is the momentum transfer.

4.3.1. Description of CSCT system

Experiments to verify the principle of CSCT have been performed using a demonstrator shown in Fig. 19. This set-up has a geometry equivalent to a medical CT scanner. Instead

Fig. 19. Coherent scatter computed tomography demonstrator set-up [37]. The Philips X-ray tube is located on the left side, opposite to the detector. The primary beam slit unit and the rotational stage can be seen between tube and detector.

of a rotating gantry it only has a rotating sample Table, which fits objects up to 24 cm diameter. It is equipped with a 4.5-kW DC power X-ray tube emitting a bremsstrahlung spectrum incorporating tungsten characteristic lines. A double-slit collimator was used to form the primary beam into a horizontal fan. A 0.5-mm Cu filter reduced the spectral width of the tube output to $\Delta E/E = 26\%$, with an average energy of 77 keV at 150 kV high voltage.

Because a 2-D photon-counting detector was not available at the time of construction of the demonstrator, the detector was realized as a single 1-D vertical detector column based on scintillator crystals and photomultipliers, which can be rotated around the focus point, thereby acquiring one projection. Spatial resolution was ensured by a collimator made of thin tungsten lamellae placed in front of the detector.

The acquisition mode (detector pixel width, number of projections and scan time) was optimized for CSCT acquisition. Assuming a point source, in CT the spatial resolution is mainly given by the size of the detector pixels, whereas in CSCT it is given by the detector pixel size, the length of the collimator lamellae in front of the detector and the distance between the object and the detector. In the experiment presented here, the combined effects of detector and collimation resulted in an average spatial resolution of ~8 mm. Therefore, only a low-resolution image can be expected from the reconstruction, and consequently a low number of projections is sufficient.

4.3.2. Data processing

Before the data can be reconstructed, an attenuation correction has to be performed by normalizing the scatter signal for a certain column of detectors against the primary beam signal from the central detector of that column. This correction assumes that for small scatter angles, the attenuation of the transmitted beam is the same as the attenuation of the scattered radiation (see Section 2.4.1.4.). This assumption is valid for good collimation (long lamellae) and objects with no structure variation in the out-of-fan-plane direction.

The Radon transform permits reconstruction of a 2-D slice of an object from a complete set of its line integrals. Reconstruction is performed on a 3-D object array consisting of the two spatial coordinates $(x,\ y)$ in the illuminated slice and one momentum transfer (or angular) coordinate, q. This has to be calculated from the distance of an object voxel (x, y) from the detector, d, and the vertical distance, a, of the corresponding detector pixel from the central detector row. From Eq. 13, the calculation of q for all object positions along a ray results in curved trajectories described by:

$$q \approx \frac{1}{\overline{\lambda}} \cdot \frac{a}{2d} \tag{13}$$

Here $\overline{\lambda}$ is the average wavelength of the used radiation. The reconstruction algorithm is based on FBP. Filtering is carried out in the same manner as in 3-D (cone-beam) transmission CT, and back-projection is performed along curved trajectories. Details of the FBP algorithm can be found elsewhere [40].

An obvious extension to the CSCT technique as described so far is to replace the scintillator detectors with room temperature semiconductors, thus markedly improving the energy and thus the momentum transfer resolution [41].

An illustration of the image quality of the energy-dispersive XRDCT is afforded by the work of Delfs and Schlomka [41], who used a single energy-resolving detector that was mechanically scanned in the t direction (Fig. 18) each time the projection angle, ϕ, was altered. The authors used an aluminium (Al) test object of 250 mm outer side length comprising five compartments each having a volume of 2500 mm^2. Plastic materials of (1) Lucite, (2) polyethylene, (3) polytetrafluoroethylene (PTFE), (4) PVC, and (5) polyamide were placed in this object. The container walls were 0.5 mm thick Al (material 6).

The reconstructed diffraction profiles are given in Fig. 20. There is broad agreement between the reconstructed profiles and those measured from small samples. All the plastics appear to have artefactual peaks at ~2.3 and 2.7 nm^{-1}. As explained by the authors, this results from the intense Al peak at 2.5 nm^{-1} that negatively affects the scatter signal from neighbouring materials and can be expected to diminish in importance as more projections are measured during the scan.

Further work in the energy-dispersive CSCT area is needed to determine the limits to performance set by photon noise and reconstruction artefacts. Comparison of the plastic profiles of Fig. 20 with the diffraction profile of TNT (Fig. 5) derived from a direct tomographic energy-dispersive XDI device shown in Fig. 4 suggests that the latter has currently an advantage owing to its avoidance of reconstruction artefacts.

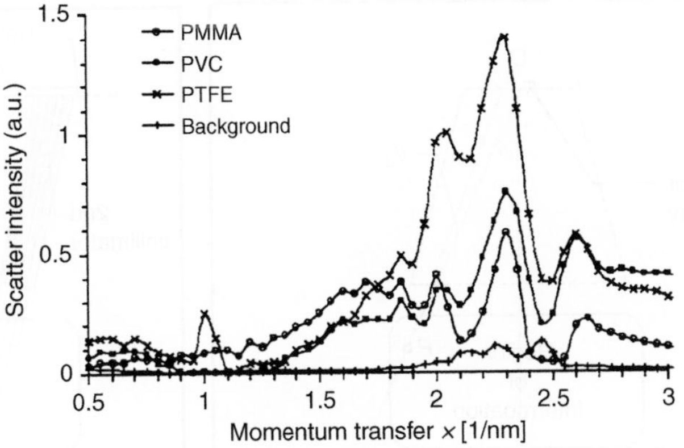

Fig. 20. Energy-dispersive coherent scatter computed tomography reconstructions of diffraction profiles of miscellaneous plastics [41].

5. Next-generation XDI

Experience gained with previous generations of XDI devices has led to a concept for technological realization of direct tomographic, energy-dispersive next-generation XDI that shall be described in some detail in this section.

It is based on the measurement of predominantly coherent scatter at constant deflection angle excited in an object on irradiation with polychromatic X-rays from a bremsstrahlung electron impact X-ray source. The scatter is recorded in a linearly segmented semiconductor array. This geometry is derived from a general design procedure for energy-dispersive, direct tomographic, XDI systems [6]. In the notation of that article, the cone angle, α, is 90°, the source radius is equal to the source–detector separation, X_D (cf. Fig. 17), and the range of the azimuthal angle is $-20° \leq \phi \leq 20°$. This corresponds to an inverse fan beam with point-like detector and extended source of arcuate form. The main feature of the next-generation device relative to its predecessors is the avoidance of mechanical scanning movements with the exception of the conveyor belt on which the suitcase is transported. This enables a compact design with very fast scan speed and response time limited only by the electronic control of the scanning electron beam X-ray source.

The geometrical configuration is shown in Fig. 21. The left-hand side illustrates the geometry of the scan plane. This is defined by the transmission plane of the primary collimator. A Multi-Focus X-ray Source (MFXS), at the bottom of the Figure, features an electron beam focus that can be deflected along the whole circumference of the anode. Such radiation sources are known from cardiac imaging applications, where they are known under the acronyms of SBDX (Scanning Beam Digital X-ray) [42] and EBCT (Electron Beam CT). They are conventionally realized as in cathode-ray televisions by magnetic deflection of a collimated electron beam from a thermionic emitter (dispenser) cathode.

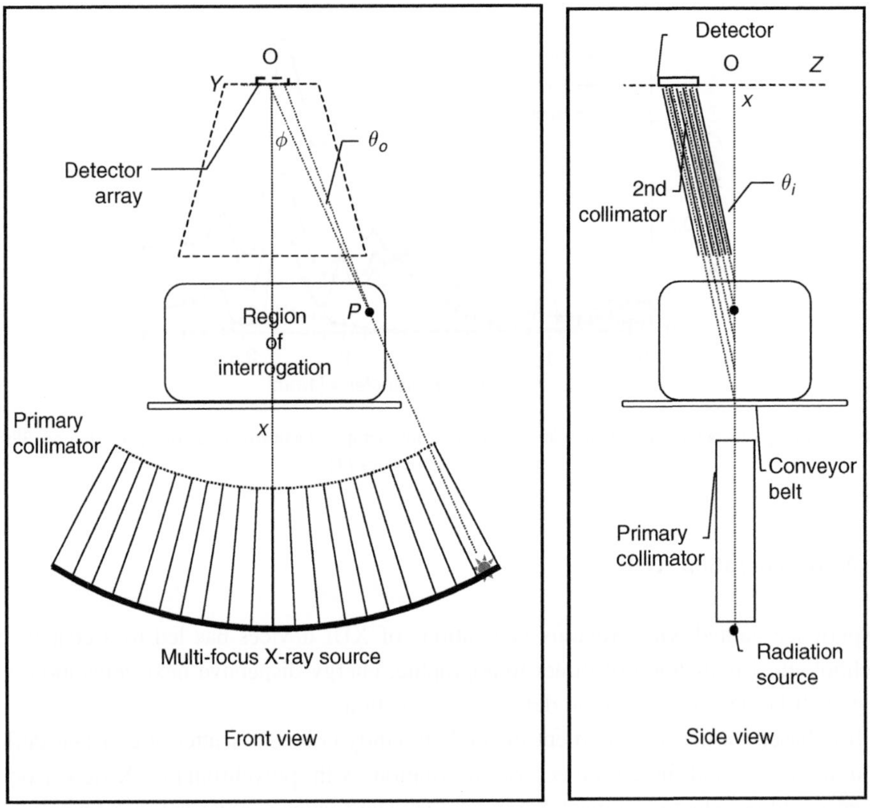

Fig. 21. Geometrical configuration for next-generation X-ray diffraction imaging.

Radiation from the MFXS is confined by the primary collimator to a 'pencil' beam focussed on O and having a deflection angle, ϕ, that depends on the location of the X-ray focus that is activated in the MFXS. The angle, ϕ, represents the deflection of the primary beam relative to the vertical X-axis. As the pencil beam propagates through the interrogation region, it induces scatter events, for example at the point, P, having a radius R_P from the origin. By virtue of the secondary collimator, X-ray scatter at the angle θ_Z (Eq. 14) arrives at the detector at a distance $R_P \tan(\theta)$ from the scan plane. The secondary collimator thus codes the point of origin of the scatter on to the Z coordinate of the detector.

The energy spectrum of constant angle, polychromatic coherent scatter may be transformed with the help of Eq. 3 into a diffraction profile depending on the momentum transfer, x, variable. The position of the active X-ray focus yields one spatial coordinate (Y position) of the voxel irradiated for the 2-D scan data. This is complemented by the X voxel coordinate ($= R_P \cos(\theta)$) derived from the Z detector coordinate. The third spatial dimension of the scattering volume in the object under investigation in the Z direction is known from the position of the conveyor belt on which the object is translated and from the Z coordinate of the scan plane as defined by the primary collimator. Hence, next-generation XDI implicitly yields without the need for reconstruction procedures' 4-D data (three space coordinates and a momentum transfer dimension).

5.1. Radiation source

The need for an angular collimation of better than 1 mrad in energy-dispersive XDI was described in Section 4.1. An angular resolution of 2% for $\theta = 0.04$ radian can only be achieved when sub-millimeter focus dimensions are used. Employing small focus dimensions reduces the first term of Eq. 12, as a large focal spot leads naturally to a poor angular resolution. The source focus dimension in the Z direction has a first-order effect on the angle of scatter θ, whereas θ is independent of the Y extent of the focus. A high electron beam power, needed to minimize measurement time, concentrated on a small focus, needed to achieve high angular resolution, implies a high electron bream power density.

The relationship between allowable electron beam power density and anode material characteristics, such as melting point, thermal conductivity and irradiation time, was first derived by Mueller [43] for stationary anode tubes and by Oosterkamp [44] for rotating targets. The only way to drive high-power performance operation from an anode having given thermophysical constants is to reduce the local dwell time of the focus by arranging for the focus to sweep rapidly across the anode. The power loading varies approximately with the inverse square root of the local dwell time. An additional benefit of the swept-beam realization of MFXS is that the electron power is distributed over a large area, which is consequently easy to cool, thus permitting high DC power. The anode is preferentially at ground potential and is of stationary design, permitting the focus location to be precisely located in the Z dimension.

5.2. Primary and secondary collimators

The primary collimator has the function of transmitting only those X-rays from the MFXS that converge on the YZ detector plane in the neighbourhood of O. It thus corresponds to the converging collimators known in radionuclide imaging [45] except that it acts only in 1-D transforming a line of radiation into a point. An angular broadening of only 10^{-4} radians in the XZ plane can be tolerated compared with the more generous value of 10^{-2} radians in the XY plane.

The secondary collimator ensures that only constant angle scatter reaches the detector. As indicated in Fig. 21, it consists of a series of thin lamella that are conical surfaces of resolution around the Z axis. Each of these has a surface area of $\sim 0.25\,\mathrm{m}^2$ and requires stabilization using a low-density foam filling and/or radial ribs positioned at regular intervals in the ϕ sense. The scatter detector and collimator shown in Fig. 21 can be replicated on both sides of the XY plane for increased signal and reduced attenuation artefacts.

5.3. Scatter detector

The scatter detector is either linearly segmented in the Z direction with sub-millimeter pitch or has a pixellated structure in the YZ dimensions. The latter devices are becoming increasingly common in radionuclide imaging applications where they combine the advantages of high spatial resolution with energy resolution below 5% at 60 keV [46]. They are generally bump-bonded to custom application-specific integrated circuits (ASICs) permitting photon spectra from each pixel to be acquired at a maximum counting rate for the whole device of in excess of 1 MHz.

Consider a central primary beam ($\phi = 0$). The angle of scatter, θ, will have components θ_Y in the Y direction and θ_Z in Z. Assuming θ and its components are small and measured in radians, then:

$$\theta = \sqrt{\theta_Y^2 + \theta_z^2} \tag{14}$$

The detector acceptance may be increased from Eq. 14 without affecting significantly the value of θ using linear segments of length in the Y direction, small compared with the Z coordinates of the segments.

It is necessary to ensure that all components of angular broadening are matched to one another. Moreover, the total broadening is constrained by the need to achieve diffraction peaks of several per cent relative FWHM. A proprietary ray tracing simulation program has been developed to analyse and optimize the contributions to peak broadening as described in the next section.

5.4. Simulation results

A ray-trace program has been developed to permit analysis and optimization of the measurement arrangement in next-generation XDI. It traces a set of X-ray beams defined by their point of origin in the MFXS and their exit coordinates at the primary collimator.

The program follows these ray paths through the object, of which each point acts as a source of scatter. For an arbitrary object point P, the program forms the group of scatter rays connecting P with a grid of points covering the detector area. If the scatter ray suffers no interaction with the secondary collimator, its angle of scatter, coordinates of the point P and arrival coordinate at the detector are stored.

This procedure is repeated for the next object point and the next focus point and so on, until a complete set of primary rays from all foci has been analysed.

The final data set is a collection of angles of scatter, $\theta(Sy, Sz, Cy, Cz, Px, Py, Pz, Dy, Dz)$, where the independent variables refer to the source coordinates (S), the primary collimator exit coordinates (C), the object point coordinates (P) and the detector coordinates (D), respectively.

Various forms of data rearrangement and presentation are possible as illustrated in Figs 22 and 23. The former shows the detector array response to a point scatterer at

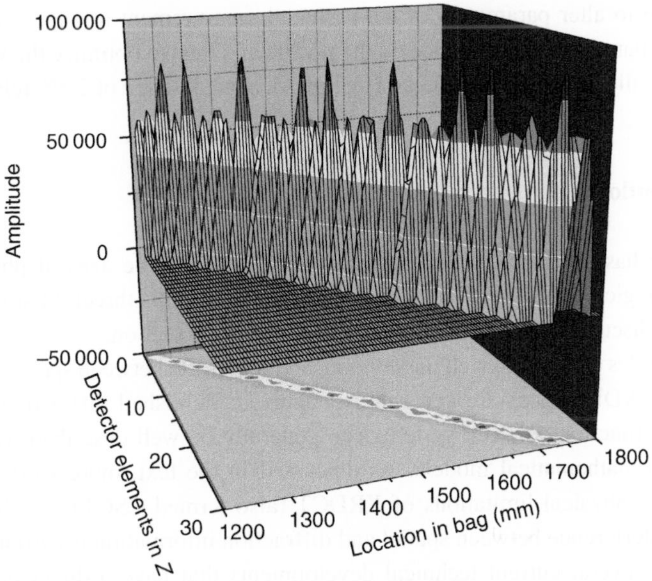

Fig. 22. Simulated detector response to point scatterer positioned at all locations in object space of next-generation X-ray diffraction imaging.

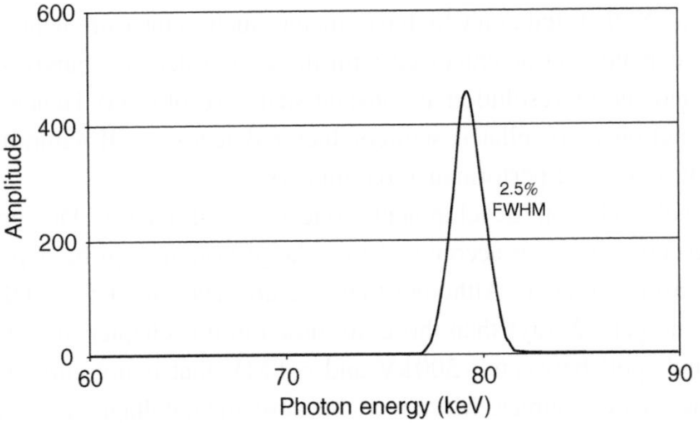

Fig. 23. Simulated energy-dispersive X-ray diffraction imaging profile for diffraction line at $1.2\,\text{nm}^{-1}$.

all possible locations in the object space. As expected, each detector segment is only sensitive to scatter from a unique region of the object. The gradient, dR/dN_D, of the distribution shown in Fig. 22 is $w_Z/\tan(\theta)$, where w_Z is the detector pitch in Z and N_D is the detector element number.

The program also outputs energy spectra for each detector segment. It is helpful to simulate energy spectra from a fictitious material having only a single, well-defined diffraction line [47]. A typical example is presented in Fig. 23 showing the peak from a material having a single line at $x = 1.2\,\text{nm}^{-1}$. With the aid of the ray-trace program, it is possible to alter parameters of the assumed measurement arrangement in order to determine separately their influence on the profile and thus to optimize the whole system. The example illustrated in Fig. 23 has a symmetrical peak shape of 2.5% relative FWHM.

6. Future outlook

This chapter has summarized developments in XDI along the axes of physical principles, technological realization and security applications. This threefold structure will be adopted for discussing the future outlook of XDI in this section.

The principles of XDI are well understood, and trade-offs between speed and resolution in designing XDI devices are generally recognized. Published diffraction profiles and point-spread functions of XDI systems can generally be well described within existing physical and mathematical models. As discussed in the text, more work is needed to determine the physical limitations of XRDCT (also termed CSCT), in which a certain amount of interference between spatial and diffraction information is currently observed.

There are several current technical developments that have a direct bearing on the performance of XDI systems. These include nano-technological research into cold (e.g. carbon nanotube) electron emitters as a replacement for thermionic (mainly tungsten filament) cathodes [48]. The widespread availability of reliable cold electron emitters would spur the development of compact, fast, affordable MFXS, Moreover, the photon throughput of an XDI system, on which parameters such as measurement speed, detection rate and FAR depend, can be enhanced with increasing detector bandwidth (product of detector area and energy resolution at constant spatial resolution). Hence, improvements in energy resolution of pixellated semiconductor detectors will naturally lead to XDI systems having improved performance parameters.

An additional technical development concerns high-energy DC operation X-ray sources. Cargo containers on account of their large volume can be expected to pose a severe false-alarm challenge. Although they are an ideal target for XDI analysis, they require more energetic X-rays than those discussed in this chapter. It is well known that there is a 'blind spot' between $\sim$500 kV and $\sim$1 MV that is not currently accessed by electron impact X-ray sources. Moreover, because megavoltage sources are generally pulsed, they are not useful for XDI requiring energy-sensitive photon-counting detection. Owing to increasing interest in cargo container screening [49], it is expected that X-ray sources operated in continuous current mode in the 'blind spot' energy range will become a topic for future development.

As far as security applications of XDI are concerned, it is likely that the future will see an extension to other targets. These will include illicitly used materials including narcotics in addition to organic explosives that can be detected at the photon energy ranges of interest in this chapter (several tens of keV). A separate issue concerns substances designated as special nuclear materials (SNMs) under the 1954 US Atomic Energy Act. The availability of high-power, high-energy DC X-ray sources is mandatory when these are to be considered as target materials.

Acknowledgments

As time goes on, it is becoming increasingly difficult to acknowledge the work of all those known to me who have contributed to the success of XDI for explosives detection.

Without the unflagging support of Dr Hermann Weiss and other members of the Philips Research Labs in Hamburg this endeavour would not have survived its infancy. Dr Helmut Strecker and colleagues at GE Security in Hamburg have been instrumental in the development of XDI for explosives detection, and it is a pleasure to warmly acknowledge the time and advice they generously contributed in the preparation of this chapter. Mark Vermilyea and members of GE Global Research in Niskayuna kindly hosted several visits, and conversations at their facility were unfailingly stimulating.

We would like to acknowledge the continued help afforded by our families.

References

[1] G Harding, J Kosanetzky and U Neitzel (1987) X-ray diffraction computed tomography. *Med. Phys.* 14, 515–525.

[2] P C Johns and M J Yaffe (1982) Scattered radiation in fan-beam imaging systems. *Med. Phys.* 9, 231–239.

[3] U Neitzel, J Kosanetzky and G Harding (1985) Coherent scatter in radiographic imaging: a Monte Carlo simulation study. *Phys. Med. Biol.* 30, 1289–1296.

[4] G Martens, H Bomsdorf, G Harding, J Kanzenbach and R Linde (1993) Coherent scatter x-ray imaging for foodstuff contaminant detection. *SPIE* 2092, 387–398.

[5] H Strecker (1999) Automatische Gepäckkontrolle mit Röntgenstreustrahlung. *Physik* 30, 31–34.

[6] G Harding (2005) The design of direct tomographic, energy-dispersive x-ray diffraction imaging (XDI) systems. *SPIE* 59230R.

[7] M C Green and L D Partain (2003) High throughput baggage scanning employing x-ray diffraction for accurate explosives detection. *SPIE* 5048, 63–71.

[8] M Sonoda, M Takano, J Miyahara and H Kato (1983) Report on the development of a CR system using an imaging plate. *Radiology* 148, 833–839.

[9] G F Knoll (2000) *Radiation Detection and Measurement* (3rd edition, Wiley: New York).

[10] R D Black, C J Bechtoldt, R C Placious and M Kuriyama (1985) Three dimensional strain measurements with x-ray energy dispersive spectroscopy. *J. Nondestr. Eval.*, 5, 21–25.

[11] G Harding, K M Kosanetzky, D W L Hukins and A J Freemont (1987) Potential of x-ray diffraction computed tomography for discriminating between normal and osteoporotic bone. *The Ageing Spine* (Manchester University Press: Manchester), 157–162.

[12] K D Rogers, R A Lewis, C Hall, E Towns-Andrews, S Slawson, A Evans, A Pinder, I Ellis, C Boggis and A Hufton. (1999) Preliminary observations of breast tumor collagen. *Synchr. Radiat. News* 12, 15–20.

[13] H Strecker (1995) Simulation-based training and testing of classification schemes for explosives detection. *SPIE* 2511, 88–98.

[14] G Harding and J Kosanetzky (1989) Scattered beam non-destructive testing. *Nucl. Instrum. Methods* A280, 517–528.

[15] S Singh and M Singh (2003) Explosives detection systems (EDS) for aviation security. *Signal Processing* 83, 31–55.

[16] B Pinsker (2003) Confessions of a baggage screener. *Wired Magazine* 11, 32–35.

[17] H Strecker, G Harding, H Bomsdorf, J Kanzenbach, R Linde and G Martens (1993) Detection of explosives in airport baggage using coherent x-ray scatter. *SPIE* 2092 (Eds: Harding, Lanza, Myers and Young), 399–410.

[18] I D Jupp, P T Durrant, D Ramsden, T Carter, G Dermody, I B Pleasants and D Burrows (2000) The non-invasive inspection of baggage using coherent x-ray scattering. *IEEE Trans. Nucl. Sci.* 47, 1987–1994.

[19] G Harding and B Schreiber (1999) Coherent x-ray scatter imaging and its applications in biomedical science and industry. *Radiat. Phys. Chem.* 56, 229–245.

[20] A M Hindeleh and D J Johnson (1971) The resolution of multipeak data in fibre science. *J. Phys. D: Appl. Phys.* 4, 259–263.

[21] T Blaffert (1995) Theory of stochastic signal processing for the optimization of explosives detection. *SPIE* 2511, 108–119.

[22] J H Hubbell, W J Veigele, E A Briggs, R T Brown, D T Cromer and R J Howerton (1975) Atomic form factors, incoherent scattering functions and photon scattering cross-sections. *J. Phys. Chem. Ref. Data* 4, 471; Errata in 1977, 6, 615.

[23] A Tartari, E Casnati, C Bonifazzi and C Baraldi (1997) Molecular differential cross sections for x-ray coherent scattering in fat and polymethyl methacrylate. *Phys. Med. Biol.* 42, 2551–2560.

[24] P Puumalainen, P Sikanen and H Olkkonen (1979) Measurement of stable iodine content of tissue by coherently and Compton scattered photons. *Nucl. Instrum. Methods* 163, 261–263.

[25] A Tartari, E Casnati, J Felsteiner, C Baraldi and B Singh (1992) Feasibility of in vivo tissue characterisation by Compton scattering profile measurements. *Nucl. Instrum. Methods* B71, 209–213.

[26] G Harding, R Armstrong, S McDaid and M J Cooper (1995) A K edge filter technique for optimization of the coherent-to-Compton scatter ratio method. *Med. Phys.* 22, 2007–2014.

[27] D L Hukins (1981) *X-ray Diffraction from Ordered and Disordered Systems* (Pergamon: London).

[28] J H Gyuo, Y Luo, A Augustsson, S Kashtanov, J E Rubensson and J Nordgren (2003) The molecular structure of alcohol-water mixtures. *Phys. Rev. Lett.* 91, 157401.

[29] Y T Pavlyukhin (2000) Radial distribution function of a hard-sphere fluid. *J. Struct. Chem.* 41, 809–824.

[30] A Tartari, C Bonifazzi, J Felsteiner and E Casnati (1996) Detailed multiple-scattering profile evaluations in collimated photon scattering techniques. *Nucl. Instrum. Methods* B117, 325–332.

[31] J Akhavan (2004) *The Chemistry of Explosives* (2nd edition, RSC: London).

[32] G Harding, M Newton and J Kosanetzky (1990) Energy-dispersive x-ray diffraction tomography. *Phys. Med. Biol.* 35, 33–41.

[33] R D Luggar, J A Horrocks, R D Speller, G J Royle and R Lacey (1995) Optimization of a low angle x-ray scatter system for explosives detection. *SPIE* 2511, 46–55.

[34] R D Luggar, J A Horrocks, R D Speller and R J Lacey (1997) Low angle x-ray scatter for explosives detection: a geometry optimization. *Appl. Radiat. Isot.* 48, 215–224.

[35] G Harding (1997) Inelastic photon scattering: effects and applications in biomedical science and industry. *Radiat. Phys. Chem.* 50, 91–111.

[36] U Kleuker, P Suortti, W Weyrich and P Spanne (1998) Feasibility study of x-ray diffraction computed tomography for medical imaging. *Phys. Med. Biol.* 43, 2911–2923.

[37] J-P Schlomka, S M Schneider and G Harding (2000) Novel concept for coherent scatter X-ray computed tomography in medical applications. *Proc. SPIE* 4142, 218–224.

[38] S M Schneider, J-P Schlomka and G Harding (2001) Coherent Scatter Computed Tomography applying a fan-beam geometry. *Proc. SPIE* 4320, 754–763.

[39] L A Feldkamp, L C Davis and J W Kress (1984) Practical cone beam algorithms. *J. Opt. Soc. Am. A* 6, 612–619.

[40] U van Stevendaal, J-P Schlomka, A Harding and M Grass (2003) A reconstruction algorithm for coherent scatter CT based on filtered back-projection. *Med. Phys.* 30, 2465–2474.
[41] J Delfs and J P Schlomka (2006) Energy-dispersive coherent scatter computed tomography. *Appl. Phys. Lett.* 88, 243506.
[42] E G Solomon, B P Wilfley, M S Van Lysel, A W Joseph and J A Heanue (1999) Scanning-beam digital x-ray (SBDX) system for cardiac angiography. *Proc. SPIE* 3659, 246–257.
[43] A Mueller (1929) On the input limit of an x-ray tube with circular focus. *Proc. Roy. Soc. A* 117, 30–42.
[44] W J Oosterkamp (1948) Heat dissipation in the anode of an x-ray tube. *Philips Res. Rep.* 3, 49–59.
[45] H H Barrett and W Swindell (1981) *Radiological Imaging* (Academic: New York).
[46] K Oonuki, H Inouea, K Nakazawaa, T Mitania, T Tanakaa, T Takahashia, C M H Chen, W Cook and F A Harrison (2004) Development of uniform CdTe pixel detectors based on Caltech ASIC. *SPIE* 5501, 218–228.
[47] H Bomsdorf, T Mueller and H Strecker (2004) Quantitative simulation of coherent x-ray scatter measurements on bulk objects. *J. X-ray Sci. Technol.* 12, 83–96.
[48] G Z Yue, Q Qiu, B Gao, Y Cheng, J Zhang, H Shimoda, S Chang, J P Lu and O Zhou (2002) Generation of continuous and pulsed diagnostic imaging x-ray radiation using a carbon-nanotube-based field-emission cathode. *Appl. Phys. Lett.* 81, 355–357.
[49] Air Cargo Security (2003) Congressional Service Report RL32022.

Chapter 9

Detection of Explosives by Millimeter-wave Imaging

David M. Sheen, Douglas L. McMakin and Thomas E. Hall

Pacific Northwest National Laboratory, Richland, WA 99354, USA

Counterterrorist Detection Techniques of Explosives
Jehuda Yinon (Editor)

Contents

1. Introduction

Millimeter-wave imaging has emerged as an effective method for screening people for nonmetallic weapons, including explosives. Millimeter waves are effective for personnel screening, because the waves pass through common clothing materials and are reflected by the human body and any concealed objects. Completely passive imaging systems have also been developed that rely on the natural thermal emission of millimeter waves from the body and concealed objects. Millimeter waves are nonionizing and are harmless to people at low or moderate power levels. Active and passive imaging systems have been developed by several research groups, with several commercial imaging sensors becoming available recently. These systems provide images revealing concealed items and, as such, do not specifically identify detected materials. Rather, they provide indications of unusual concealed items. The design of practical, effective, high-speed (real-time or near real-time) imaging systems presents a number of scientific and engineering challenges, and this chapter will describe the current state of the art in active and passive millimeter-wave imaging for personnel screening. Numerous imaging results are shown to demonstrate the effectiveness of the techniques described. The authors have been involved in the development of active wideband millimeter-wave imaging systems at Pacific Northwest National Laboratory (PNNL) since 1991.

Traditional screening techniques for personnel typically consist of metal detectors for personnel coupled with X-ray systems for hand-carried baggage. Ion-mobility spectrometer (IMS)-based trace explosive detection systems are sometimes added to the screening procedures. Manual pat-downs may also be used in some cases. There are limitations in all of these techniques. Metal detectors can only detect metallic items and will not detect nonmetallic objects, such as bulk hand-carried explosives. IMS systems may not detect all types of explosives or be effective in all scenarios. Manual pat-downs are offensive to most people and are time consuming. X-ray systems are effective for baggage, but can have difficulty separating explosives from other materials, and may not be acceptable for personnel screening. These systems are all effective, but do not completely solve the personnel screening requirements.

Additional complementary systems are needed which can detect nonmetallic concealed items carried by personnel. Threats in the current environment include explosives, metallic and nonmetallic handguns, and edged weapons, as well as other tools and devices that may be incorporated into weapons. Potential solutions are provided by active and passive millimeter-wave and backscatter X-ray imaging systems. All of these techniques provide the ability to form an image of a person that reveals concealed items with moderate to high resolution. Imaging systems are a good match with metal and trace detectors, because they can be used to detect all other concealed items and provide discrimination of the objects based on their size and location on the body. Backscatter X-ray imaging systems can provide very high-resolution images for effective detection, but their use of ionizing radiation (X-rays) may hinder their acceptance for personnel screening. Additionally, these systems are large and relatively slow – scanning just one side of a person at a time. Millimeter-wave imaging systems utilize nonionizing radiation and may be more readily accepted by the public. These systems can scan quickly – video frame rates in some cases – and can effectively detect most concealed items.

The current state of the art of these active and passive millimeter-wave imaging systems is the topic of this chapter. Section 2 will describe properties of millimeter waves and give the historical development for some of this technology. Section 3 will provide a description of millimeter-wave imaging system architectures and components, and Section 4 will detail some imaging system design considerations. Section 5 will describe several millimeter-wave imaging technologies and systems and provide numerous imaging results that highlight the current and potential performance of these systems. Section 6 will provide concluding remarks and discuss some possibilities for future development of millimeter-wave imaging systems.

2. Background

Millimeter waves have two important properties that enable their use in personnel screening applications, such as weapon and explosive detection. The waves can pass through common clothing materials with little distortion or attenuation, and the wavelengths are short enough to allow moderate- to high-resolution imaging. Sections 2.1 and 2.2 will provide more details on the properties of millimeter waves and the historical development of millimeter-wave imaging systems.

2.1. Properties of millimeter waves

Millimeter waves are electromagnetic (radio) waves typically defined to lie within the frequency range of 30–300 GHz. The microwave band is just below the millimeter-wave band and is typically defined to cover the 3–30-GHz range. The terahertz band is just above the millimeter-wave band and is typically defined to cover the 300 GHz to 3+ THz range. The wavelength of electromagnetic radiation is given by $\lambda = c/f$, where $c = 3 \times 10^8$ m/s is the speed of light and f is the frequency (in Hz). The millimeter-wave band thus corresponds to a wavelength range of 10 mm at 30 GHz decreasing to 1 mm at 300 GHz.

Millimeter waves are effective for explosive detection on personnel because the waves readily pass through common clothing materials and reflect from the body and any concealed items. These reflected wavefronts can be focused by an imaging system that will reveal the size, shape, and orientation of the concealed object. Diffraction generally limits resolution to spot sizes of $\lambda/2$ or larger, so resolution spot sizes of <10 mm are readily achievable at millimeter wavelengths.

There is little spectral (frequency) variation in the reflection or emission of millimeter waves from most bulk materials, including the human body and most concealed objects. This means that millimeter-wave imaging systems cannot uniquely identify specific materials, such as explosives. They can, however, form high-resolution images that will reveal discrepancies from the expected image of a person and reveal the shape and position of the concealed items, which enables the development of high-performance and versatile concealed weapon detection imaging systems.

Millimeter waves can be used for both active and passive imaging systems. Active imaging systems primarily image the reflectivity of the person/scene including the effect of the object's shape and orientation. Passive systems measure the thermal (black-body) emission from the scene, which will include thermal emission from the environment that is reflected by objects in the scene (including the person).

For both active and passive personnel screening systems to be effective, it is necessary that most clothing be relatively transparent at the frequency of operation of the system, so that concealed items will be detected. Fabrics can be considered to be a thin layer of dielectric material. The thickness of most materials will be much less than the wavelength throughout most of the millimeter-wave band. Additionally, most materials have relatively low attenuation losses over the millimeter-wave band. The combination of thinness (thickness $\ll \lambda$) and low loss means that fabrics will cause only slight absorption and reflection losses to the millimeter-wave signals. Bjarnason et al. [1] have published a number of fabric attenuation measurements covering the millimeter-wave, terahertz, and infrared (IR) frequency bands. These measurements confirm the relative transparency of most materials in the frequency range below 300 GHz.

In contrast to most fabrics, the human body can be considered a good conductor and strongly reflects and absorbs waves in the millimeter-wave range. Concealed objects can generally be classified as dielectrics with unknown shape and dielectric properties. Metals can be considered to be a limiting case of a highly conductive dielectric. Dielectric objects including metals, the human body, and concealed items will all produce reflections based on the Fresnel reflection at each air–dielectric or dielectric–dielectric interface [2]. Additionally, these reflections will be altered by the shape, texture, and orientation of the surfaces. This complexity renders it difficult to directly measure dielectric properties of concealed items. However, it does create significant variation in the reflectivity which provides significant contrast in active imaging systems.

Passive systems exploit the natural thermal emission of radiation that emanates from all warm objects (above absolute zero). For objects or bodies near room temperature, these emission spectra peak near wavelengths of 10 μm, which is in the long-wave IR region of the spectrum [3]. IR imaging cameras typically operate near this wavelength or at shorter wavelengths, closer to visible light. For longer wavelengths, such as in the millimeter-wave band, this radiation is at much lower intensity – but is still present and can be used to form passive millimeter-wave imaging systems. These systems are analogous to IR imaging camera systems but are tuned to take advantage of the unique properties of millimeter waves, which includes effective clothing penetration to detect concealed objects. Owing to the significantly reduced signal levels available in the millimeter wave, it is considerably more difficult to develop sensitive imaging systems. Sensitive receivers employing advanced integrated low-noise amplifiers have allowed the development of effective systems [4–9].

Passive systems form an image of the emitted millimeter-wave radiation that is the sum of energy directly emitted from the target or scene and energy that originates elsewhere and is reflected by the target or scene. This emission increases directly with temperature; therefore, imaging systems frequently display their imaging results calibrated to an effective temperature scale, with contrast represented as differential temperature. Noise in the image is also characterized as a noise-effective differential temperature.

The emission of millimeter waves from concealed objects is complicated somewhat by the environment in which they are employed. Targets within the image, including the human body and any concealed items, emit millimeter waves based on both their temperature and their emissivity. Objects with high emissivity radiate at close to the black-body limit, whereas objects with low emissivity radiate proportionally less. Metals and other good reflectors have low emissivity, whereas good absorbers have relatively high emissivity. The human body has both moderate emissivity and reflectivity – so it is easily visible in both active and passive systems. These differences in target/scene emissivity provide contrast in images even if the temperatures of different components of the image are all close to the same value.

Objects in passive images that have moderate to high reflectivity will typically contain signals due to both thermal emission and reflected radiation from the background. Outside, the sky represents a relatively cold background, whereas indoors the background is relatively warm. These factors can significantly reduce the thermal contrast available in passive imagery, particularly for systems operated inside. Passive systems rely on effective temperature contrast in the images, which is altered by the environment in which the systems are used. Active systems essentially measure reflectivity and are not significantly affected by the environment.

Atmospheric attenuation properties of millimeter waves can be important, especially in specific bands. Electromagnetic waves effectively pass through the atmosphere without significant losses over much of the spectrum, including many portions of the microwave, millimeter-wave, IR, and optical bands. However, significant absorption because of water vapor or other atmospheric constituents does occur over several narrow frequency bands in the millimeter-wave band and is extremely significant over much of the terahertz band. Atmospheric attenuation is covered in detail in Section 4.6.

2.2. *Historical development*

Active and passive millimeter-wave imaging systems have been under development for many years for a wide variety of applications including passive radiometric remote sensing [10], all-weather enhanced synthetic vision [6–8], synthetic aperture radar (SAR) imaging [11–13], and concealed weapon detection [5,9,14–17]. However, only relatively recently have practical systems been realized for personnel screening. The following subsections highlight some of the developments in this field that are relevant to concealed weapon detection and personnel security screening.

2.2.1. Active millimeter-wave imaging for concealed weapon detection

Active millimeter-wave imaging was originally proposed for concealed weapon detection by Farhat et al. [18–20]. They developed an active imaging technique that used a fixed millimeter-wave illumination source from outside the imaging aperture and a mechanically scanned receiver/detector. The amplitude and phase of the scattered wavefront was recorded optically onto film as an interferogram or hologram. This recorded hologram was then focused or reconstructed using laser optical holographic reconstruction techniques. This technique is similar to acoustical holographic imaging techniques that were being developed during the 1960's and early 1970's. Descriptions of optical and acoustical holographic imaging are contained in the references [21–26]. The millimeter-wave holographic imaging technique showed significant potential in this early development but was very slow because of the mechanical scanning and film-based optical reconstruction necessary to form the images.

The holographic technique was improved significantly by Collins et al at the PNNL by incorporating a scanned transmitter–receiver (transceiver) system and digital computer-based image reconstruction techniques [27]. This scanning configuration, referred to as simultaneously scanned source receiver, uses a source that coincides with, or is adjacent to, the receiver as it is scanned over a two-dimensional planar aperture. Data were then recorded in a computer system, and a digital holographic image reconstruction algorithm was used to mathematically focus the imagery. Scanning the source results in a significant improvement in the quality of the image because of the wide diversity of angles over which the target is illuminated, which reduces or eliminates shadowing that is caused by specular reflection. The digital image reconstruction technique eliminated the need for film, thereby reducing potential distortions and increasing the speed of the imaging process.

Active, holographic millimeter-wave imaging was further improved significantly by Sheen et al. [17,28–31] at PNNL by incorporating a wideband swept-frequency transceiver and improved three-dimensional image reconstruction technique. In this technique, a wideband transceiver was scanned over a two-dimensional planar aperture with sequential illumination and coherent detection at each point in the aperture over a wide frequency bandwidth. These wide bandwidth three-dimensional data can then be used to synthesize sharp range/depth resolution and allow a true fully focused, three-dimensional image reconstruction [17]. This overcomes the narrow depth of focus that is observed in single frequency results and improves the sensitivity of the transceivers. This dramatically improved the fidelity of the images produced using the holographic millimeter-wave imaging technique.

The technique was made practical for personnel screening applications by incorporating sequentially switched linear antenna arrays to improve the data collection speed to near real-time rates [17,27–31]. The linear arrays allowed one axis of the two-dimensional planar aperture to be scanned electronically at very high speed (on the order of several microseconds per element). This reduced total data collection for practical systems to <1 s. Because the human body is not transparent to millimeter-waves, personnel will need to be inspected from multiple aspects (front and back at a minimum).

Further increases in imaging speed and elimination of the need for multiple scans were realized at PNNL by the development of a cylindrical imaging system [32–36]. This technique utilizes a vertical linear sequentially switched antenna array that is swept about the person in a cylindrical fashion. This allows for acquisition of data from all aspect angles and can be used to form a set of reconstructed images that allow inspection of the subject from any desired angle. Using this technique, we can present the results from a single scan to an operator for inspection from all angles.

A number of other enhancements to the active holographic imaging technique have also been developed. These include the development of preliminary techniques for automated detection, the use of polarimetric properties of the imaging target, and other techniques [37–39].

Recently, researchers at Agilent have developed a novel active real-time millimeter-wave imaging technique based on a digitally controlled reflecting antenna array [14]. This system is described in detail in Section 5.3.

2.2.2. *Passive millimeter-wave imaging for concealed weapon detection*

Passive millimeter-wave imaging systems detect the natural thermal (black-body) emission that emanates directly from the imaging target and the thermal emission from the environment that is reflected from the imaging target [3]. Thermal emission of electromagnetic radiation is most intense at wavelengths in the long-wave IR around $10\,\mu m$. At longer wavelengths, such as in the millimeter wave, the emission is greatly reduced but is readily detectable using sensitive receivers (radiometers).

Passive millimeter-wave imaging systems are analogous to common IR cameras, with several notable challenges. The longer wavelength requires larger receive array antenna elements and larger lens/reflector apertures for acceptable resolution, which makes the systems much less amenable to the integrated circuit arrays common in IR/optical camera systems. The signal strengths are also considerably weaker compared with the IR; therefore, each receiver element will often require its own low-noise amplifier for acceptable signal-to-noise performance. This requirement makes fully populated focal plane array (FPA) imaging systems (analogous to most IR cameras) very difficult and expensive to develop and fabricate.

Huguenin et al. [40–43] proposed passive millimeter-wave imaging for concealed weapon detection. These researchers used mechanically scanned radiometer-based imaging to demonstrate the potential of concealed weapon detection using millimeter-wave imaging. They also proposed a two-dimensional FPA imaging system design for concealed weapon detection. This system has been partially realized in subsequent years but has not been fully developed into a high-performance imaging system. One such passive FPA system that was fully realized, although primarily for all-weather synthetic vision applications, was developed by TRW (now Northrop Grumman Space Technologies) [6–8].

Owing to the difficulty and expense of building FPAs, other researchers have developed passive millimeter-wave imaging systems for concealed weapon detection which do not attempt to build an entire two-dimensional FPA. These systems utilize various architectures, but they typically incorporate a high-speed mechanical/conical scanning system with a one-dimensional or sparse linear array of receive antennas to form a real-time or near real-time imaging system [4,5,9]. Several of these systems have been commercially developed and are described in detail in Section 5.

3. Millimeter-wave imaging system architectures and components

A wide variety of millimeter-wave imaging system architectures have been developed for many applications, including personnel screening/explosive detection. Millimeter-wave imaging systems differ significantly from conventional IR/optical systems for several reasons. The longer wavelength typically demands a much larger aperture (0.20–2 m is common) than is typical in optical systems. Active illumination and coherent (amplitude and phase) detection is often employed. Two-dimensional FPAs are difficult to realize in the millimeter-wave band. These challenges have led to a wide variety of different imaging system architectures used by researchers and in commercial sensors. A number of these are discussed in this section along with a description of sources and detectors that are used in these systems or are current areas of research.

3.1. Imaging modalities

Millimeter-wave and terahertz imaging systems are frequently considered for applications where there is a need to measure or look through an optical obscurant, such as clothing for the personnel screening/explosive detection application. Penetration through most obscurants typically improves as the frequency is lowered, with microwave and millimeter-wave systems often having better penetration than terahertz systems. Imaging system resolution typically improves with increasing frequency. Thus, there is a fundamental tradeoff in the imaging system design between resolution and penetration through lossy materials. Many other variables apply to the design of the imaging system; however, resolution and penetration are among the most important. These and other design considerations are discussed further in Section 4.

A wide variety of imaging system architectures have been designed or proposed for millimeter-wave and terahertz imaging systems. These include passive FPAs, passive mechanically-scanned systems, active mechanically-scanned focused lens systems (both through transmission and reflection), active holographic scanned imaging, and SAR imaging.

Passive FPAs are perhaps the most familiar imaging system architecture because this architecture is used for most optical and IR cameras. In this configuration, an array of small detectors is placed at the focal plane of a lens or reflector system. The received energy in this case is derived from thermal (black-body) emission or reflected radiation from the scene. Thermal emission near room temperature peaks in the long-wave IR (near 10 μm wavelength), and reflected light is significant in the visible light frequency range. However, at terahertz and millimeter-wave frequencies the thermal emission is much reduced. At IR/optical frequencies, advanced integrated circuit technology has allowed for sensitive, high-resolution arrays to be incorporated into these systems, and real-time (video rate) imaging is possible in many systems. Millimeter-wave imaging systems have been fabricated using this architecture; however, the FPA is typically much less integrated and considerably more expensive than in IR systems. One such system was developed by TRW for all-weather enhanced synthetic vision for military and civilian aircraft [6–8]. Results from this system appear somewhat similar to those of an IR camera, but with improved penetration of dust and fog and with reduced resolution. Similar systems have also been proposed for concealed weapon detection [43]. There are a number of difficulties in designing and fabricating arrays at millimeter-wave and terahertz frequencies. For sensitive detection, a heterodyne detector is highly desirable, which requires a significant amount of local oscillator (LO) power (milliwatts) for each detector element. A heterodyne architecture also greatly increases the electrical complexity of each detector element. Furthermore, individual antennas and receiver low-noise amplifiers are often required to achieve high sensitivity. These, and other factors, have prevented widespread development of this type of imaging system. Currently, Defense Advanced Research Projects Agency (DARPA) Microsystems Technology Office (MTO) is working on research to improve the performance and practicality of sources, detectors, and FPAs in the sub-millimeter and terahertz bands. National Institute of Standards and Technology (NIST) has also conducted research on FPA development in the millimeter-wave and terahertz bands [44–46].

Passive mechanically scanned systems operate similarly to the FPA system, except that a single detector (or small array) is mechanically scanned to build up an image of the scene. The lens (or reflector) can be kept stationary while the detector is scanned, or the lens-detector system can be scanned together. Use of a linear array (or other small array) can be used to improve the speed of image acquisition by acquiring multiple pixels in parallel. The use of mechanical scanning can greatly reduce the complexity of the detector/array at the expense of increased mechanical size and complexity. Another drawback to using mechanical scanning for passive systems is that the signal integration time available for each image pixel is reduced. A full FPA can integrate the signal over the entire image frame time, whereas the mechanically scanned system can integrate over a much shorter time. Compensating for this requires increasing the sensitivity of each detector element. This type of system has been developed at millimeter wavelengths for airborne imaging applications and concealed weapon detection [4,5,10,47]. The systems developed by Qinetic utilize a sophisticated folded path, rotating reflector, and detector array to achieve real-time video scanning (multiple frames/second) [4,5].

It should be noted that passive imaging systems can, of course, be enhanced by using active sources of illumination. This is primarily useful for shorter range applications and can improve the performance of the system in situations where the signal is weak or the contrast is low.

Synthetic aperture radar imaging is an airborne imaging configuration that is widely used at microwave frequencies and has also been applied at millimeter-wave frequencies. This imaging modality uses a side-looking transmitter–receiver (transceiver) mechanically scanned over a linear aperture by the motion of the aircraft. Data are recorded over a relatively wide bandwidth at discretely sampled positions along the aperture. As the recorded data are coherent (amplitude and phase are both recorded), the data can be mathematically focused using an image reconstruction algorithm [12]. This image is two-dimensional, but in a cross-range by range format rather than two lateral dimensions commonly found in FPA imaging systems. The SAR imaging technique is nearly indispensable for many medium and long-range military imaging applications. Two important advantages of SAR imaging techniques are that the frequency can be made relatively low to allow for penetration of clouds, rain, fog, dust, smoke, etc. and that the resolution can be made fairly high by using a very long linear aperture.

Active wideband holographic imaging is a technique, developed at PNNL, to form high-resolution, high-fidelity images at short to moderate ranges [17,33,34]. This technique, depicted in Fig. 1, uses a wideband transceiver scanned mechanically (or electronically) over a two-dimensional planar aperture. The coherent reflected signal from the target is recorded over the full bandwidth of the system and at discretely sampled positions over the full two-dimensional planar aperture. As with the SAR imaging modality, the recorded data are coherent and can be mathematically focused, or reconstructed, to form an image of the target [17]. This image is fully three-dimensional with resolution in both lateral (cross-range) dimensions and range. Sequentially switched arrays have been developed which allow for electronic scanning across one or both dimensions of the aperture. This has allowed the development of several real-time and near-real-time imaging systems. This technique has been applied to numerous applications including concealed weapon detection, inner-wall and through-wall imaging, radar cross-section (RCS) imaging, and others. This technique is described in greater detail in Section 5.1.

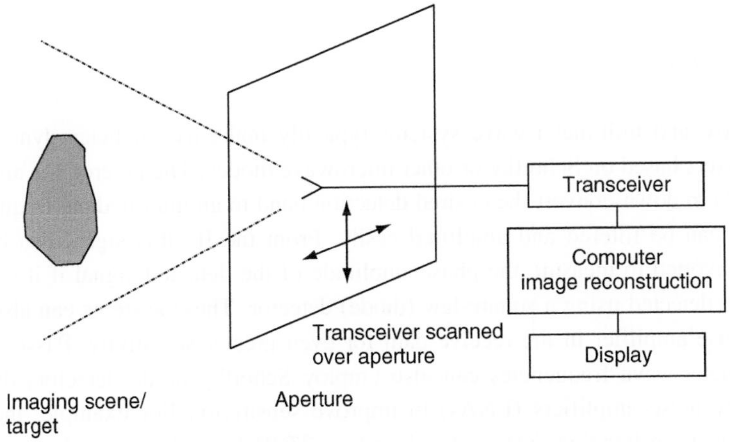

Fig. 1. Active wideband holographic imaging configuration.

3.2. Sources

Sources for microwave and millimeter-wave systems have been well developed for many years; however, solid-state millimeter-wave sources have been steadily increasing in operating frequency and performance. Electronic (solid-state) sources are widely available with power levels ranging from tens of milliwatts to tens of watts. These include Gunn and IMPATT diode-based oscillators that are frequently used in the millimeter-wave ranges as well as monolithic microwave integrated circuit (MMIC) amplifiers that are available at frequencies up to, or in excess of, 140 GHz [48,49]. Above 140 GHz, MMIC power amplifiers become increasingly difficult to fabricate and have lower performance because of fundamental semiconductor limitations; however, MMICs are currently being developed at frequencies up to or exceeding 220 GHz. The frequency/phase control of many of these millimeter-wave sources can be made very fine by phase locking the oscillators to stable reference sources or by deriving the source frequency from a lower-frequency (often synthesized) microwave source.

Electronic sources, such as MMIC power amplifiers, are pushing up in frequency to near 220 GHz – not far from the lower end of the terahertz range [49]. Excellent sources in the 300 GHz to 1+ THz range can be obtained using frequency multiplication techniques. These sources use high-power microwave or millimeter-wave power amplifiers to drive planar GaAs Schottky diodes, which are non-linear devices, producing harmonics of the drive frequency [50–52]. Usually these are configured as doublers or triplers. These frequency multipliers can be cascaded to provide the desired output frequency. Power levels of 10–25 mW at ~300 GHz and 0.4–1 mW at ~600 GHz are currently available (http://www.vadiodes.com). Advantages of these sources include reasonable power output and bandwidth (frequency tuneability), compactness, low cost (potential), solid-state construction, room-temperature operation, and modest line power requirements. Disadvantages include limited power output and power output that falls quickly as the operating frequency is increased.

3.3. Detectors

Microwave and millimeter-wave systems typically make use of heterodyne detection using mixers based on Schottky or other microwave diodes. The heterodyne architecture uses a LO to down-convert the desired detection band to an intermediate frequency (IF) where it can be filtered and amplified easily. From the IF, this signal can be further down-converted to measure the phase/amplitude of the detected signal if it is coherent or can be detected using a square-law (diode) detector. These systems can also employ a low-noise amplifier in the receive path for even greater sensitivity. Passive systems at millimeter-wave frequencies can also employ Schottky diode detectors directly or using low noise amplifiers (LNAs) to improve sensitivity. For example, the passive millimeter-wave imaging system developed by TRW for airborne applications used a MMIC LNA coupled to each antenna in the FPA [6–8].

While heterodyne detection is typically the most sensitive, it is problematic to extend its use to FPA systems. Each detector element in the FPA requires LO power, on the order of 1 mW for Schottky diode-based mixers. For large FPAs operating in the millimeter-wave or terahertz bands, this level of power is currently impractical. Various components including LNAs and mixers are available as MMICs at up to about 140 GHz and may eventually extend to 220 GHz or possibly higher [49]. Schottky diode-based mixers are available at frequencies extending well into the terahertz range, to 2.5 THz and possibly higher, and can be used wherever a suitable LO source can be obtained [53].

4. Imaging system design considerations

The design of an imaging system must consider a wide variety of factors that will determine the quality of images and ultimately the success of the instrument. These include active or passive modes of operation, lateral image resolution, range resolution, aperture sampling, image acquisition speed, attenuation or penetration through the atmosphere and obscurants, and signal-to-noise ratio (SNR) or sensitivity. Depending on the specific application for the imaging system, some of these factors may be more important than others. For example, for some applications, it may not matter if the image-acquisition time is somewhat long (perhaps minutes), whereas for others, real-time video rate imaging may be required. Adequate resolution, contrast, and sensitivity are generally required for all applications.

4.1. Active versus passive

The choice of active or passive operation depends on several factors. These factors can include sensitivity (SNR), resolution, and image contrast.

Passive systems operating in the millimeter-wave and terahertz bands operate analogously to IR cameras. The received signal increases with increasing temperature and is proportional to the product of the emissivity and the Planck black-body function. The total received power level for passive systems is quite low in the millimeter-wave and terahertz frequency ranges, and this can make it difficult to achieve practical imaging system noise performance with low noise equivalent temperature differentials (NEΔT). In addition, the scene itself may have relatively low contrast. That is, the emissivity and temperatures in the scene may cover a fairly narrow span, decreasing the ability of the imaging system to distinguish important features. For example, passive millimeter-wave imaging systems for concealed weapon detection have relatively low-contrast indoors. There are significant emissivity differences between the human body and concealed (often metal) items, but temperature differences are relatively slight and thermal radiation from the room surroundings reflects from concealed objects, making them appear similar to the human body. Performance is improved for these systems by operating outdoors, which reduces the reflected radiation and improves the thermal contrast. Alternatively, the system can be augmented by using external illuminators, which essentially makes the system active, although the imaging portion of the instrument is still essentially passive [47].

Active systems offer a number of advantages. Sensitivity can be increased by using higher power sources and coherent heterodyne detection. Coherent heterodyne detection uses a LO that is phase and frequency locked to the transmit oscillator. This allows for substantial reduction in noise and increases in sensitivity relative to passive systems that often use direct detection. Image contrast can be very good, because shape differences of the objects will influence the signal scattered back to the active imaging system. This can create difficulties for active systems for very smooth specular reflection targets. For smooth targets, the incident wave may reflect away from the target without returning to the imaging system. Thus, portions of these smooth targets that have their surface normal oriented away from the imaging system may not be seen. Small features and rough surfaces tend to scatter in all directions and can be successfully observed by the imaging system.

4.2. *Lateral image resolution*

The lateral image resolution is defined by the ability of the imaging system to differentiate between two point objects separated by a small distance. The minimum distance at which the objects can be successfully differentiated is defined to be the resolution. The resolution for millimeter-wave imaging systems is limited by diffraction of the electromagnetic waves. Two objects can be resolved by an imaging system if the phase variation across the aperture differs by at least 2π radians or one cycle [54]. This definition leads to an angular resolution of

$$\Delta\theta = \frac{\lambda}{D} \tag{1}$$

where D is the diameter or width of the aperture. The spatial resolution is therefore

$$\delta_{lateral} = \lambda\frac{R}{D} \tag{2}$$

where R is the range to the target. The ratio of R/D is often referred to as the F-number of the imaging system.

An active system that scans the transmitter along with the receiver, as is done in synthetic aperture (SAR and holographic) imaging systems has twice the phase variation across the aperture compared with a passive system (or active systems with a fixed source or sources). This doubling of the phase shift results in a factor of two improvement in the image resolution (one half the spot size) for

$$\Delta\theta = \frac{\lambda}{2D} \tag{3}$$

and

$$\delta_{lateral} = \frac{\lambda}{2}\frac{R}{D} \tag{4}$$

This increase in image resolution can be very significant in the millimeter-wave and terahertz frequency ranges because of the relatively long wavelengths involved compared with optical systems.

The resolution required varies by application; however, most imaging applications require that targets be identifiable to the operator of the imaging system. This requirement has been studied in detail for IR imaging applications where it has been determined that approximately 12 resolved pixels across the characteristic dimension of the target are required for 50% identification [55]. Therefore, identifying a handgun of size 18 cm would require a lateral resolution of ~1.5 cm.

4.3. Range resolution

Active systems add an additional dimension to the imaging system because of their ability to resolve the depth of the imaging target. Similar to lateral resolution, two objects at two different ranges can be distinguished if their received phase variations differ by at least 2π radians or one cycle as the frequency is swept over a bandwidth, B. This leads to a depth resolution of

$$\delta_{depth} = \frac{c}{2B} \tag{5}$$

where $c = 3 \times 10^8$ m/s is the speed of light. For example, a millimeter-wave system with a bandwidth of 20 GHz has a depth resolution of 7.5 mm.

This depth resolution is utilized to an extreme degree in SAR systems, in which it becomes one axis of the image, i.e., the image has axes of cross-range and range rather than the more conventional two lateral axes. In wideband holographic imaging, depth resolution is used to make the imaging system completely three-dimensional [17]. The holographic imaging technique in particular can be used to generate an image that is fully three-dimensional and optimally focused at each depth in the three-dimensional reconstructed image [17]. Lens-based, fixed-focus imaging systems can have high lateral resolution, but the depth of focus is very limited, particularly for a high-resolution (low F-number) imaging configuration. All systems with depth resolution benefit from the ability to time or range gate the received signal. This can allow for removal of strong returns from the surfaces of obscuring materials.

4.4. Focal plane or aperture sampling

Focal plane array imaging systems should ideally have at least one detector per resolution cell within the field of view of the system. For example, if a system has a 30° azimuth ×30° elevation field of view and the theoretical angular resolution for the system is 0.5°, then the focal plane would consist of at least a 60 × 60 array of detectors. Synthetic aperture systems such as the SAR and holographic imaging techniques need to sample the scattered wavefront with at least two samples per 2π radians of phase variation, i.e., they require at least two samples per optical fringe. If the system's antennas are essentially omnidirectional (180° beamwidth), then this would require sampling the aperture at a spatial sampling interval of $\lambda/4$ or less for a system that scans the transmitter along with the source. If the beamwidth is reduced, then this sampling density can also be reduced and can be approximated by the formula

$$\Delta < \frac{\lambda}{4\sin(\theta/2)} \tag{6}$$

where θ is either the lesser of the full antenna beamwidth or the angle subtended by the aperture with respect a single point on the target. For an $F-1$ ($R = D$) system, the required sampling is $\Delta < 0.35\lambda$ and for an $F-2$ ($R = 2D$) system, the required sampling is $\Delta < 0.56\lambda$.

4.5. Image acquisition speed

The speed of image acquisition depends primarily on the time required for mechanical scanning (if used) and the detector integration time required to achieve the desired SNR or sensitivity requirement. A highly desirable goal for image acquisition is real-time video rate imaging, typically defined to be 30 frames/second. Achieving this will likely require a fully populated FPA that can integrate each detector signal in parallel. Additionally, the detectors will need to have extremely low noise to allow for sensitive imaging within the relatively short integration time available (e.g., 33 ms). Owing to the expense and difficulty of building two-dimensional FPAs in the millimeter-wave and terahertz bands, mechanical scanning is often used to replace the FPA or to augment a smaller array, as described in Section 5. This mechanical scanning can dramatically reduce the number of detectors required; however, the available integration time for each pixel is reduced. For example, if a 100×100 element sampling is realized by scanning a 100-element linear array, then each detector will only have $1/100^{\text{th}}$ of the total available integration time. The reduced number of detectors may allow for improving the performance of each detector to at least partially compensate for this loss of sensitivity. An example of full FPA millimeter-wave imaging system is described in [6–8], and an array coupled with mechanical scanning is described in [4,5,47].

The wideband holographic imaging technique is very sensitive because of the active, coherent, heterodyne detection that is employed in typical systems [15–17]. Therefore, the primary limitation on image acquisition is the time required to mechanically scan the transceiver. Two-dimensional raster scans are often used for laboratory experiments, and these typically require scan times on the order of 1–30 min. Practical systems employing linear switched antenna arrays and mechanical scanning reduce this time to 1–10 seconds [15–17].

4.6. Attenuation

Atmospheric attenuation, due primarily to water vapor absorption lines, is very significant in many spectral regions in the millimeter-wave and terahertz bands as shown in [56]. Most of the millimeter-wave band has relatively low losses over moderate path lengths, whereas frequencies above 1 THz suffer fairly extreme attenuation. There are a number of atmospheric windows in the spectral regions less than about 350 GHz, near 400 GHz, near 650 GHz, and near 850 GHz. Imaging systems operating away from these window regions will have very short ranges of operation because of the severe attenuation.

For some applications, various materials such as paper, cardboard, or clothing will need to be relatively transparent. Clothing samples have been measured and reported by Bjarnason et al. [1] whose results show that most clothing is reasonably transparent (i.e., one-way attenuation is less than about 6 dB) at frequencies less than about 600 GHz. The attenuation of various paper, cardboard, and fabric samples has been measured at PNNL. These results indicate that the paper samples are reasonably transparent (i.e., one-way attenuation is less than about 6 dB) at frequencies up to $\sim$2 THz. The results from the fabric samples generally agree with those from Bjarnason et al., with losses generally low (typically $<$3 dB) below 500 GHz and increasing markedly above 500 GHz.

In general, lower frequencies will offer improved penetration for most optical obscurants, creating a tradeoff between penetration and resolution that must be considered when designing an imaging system for a particular application.

4.7. Signal-to-noise/sensitivity

The sensitivity of the imaging system will ultimately depend on a large number of factors. These include the imaging system operating modality, received signal strength, and detector noise. Passive imaging systems can improve sensitivity by decreasing detector noise through improved detectors and by increasing integration time. Active imaging systems can improve sensitivity by increasing transmitter power and by reducing detector noise. This ability to use transmitter power to increase sensitivity and the low noise performance of coherent heterodyne receiver architectures give active systems a sensitivity advantage in many situations. Detailed mathematical modeling of various system signal-to-noise performance expectations is described by E. R. Brown [57].

4.8. Reflection from human skin

The reflectivity of human skin is expected to decrease with increasing frequency. This is due to the reduction in the real part of the refractive index as well as an increase in absorption with frequency. In vivo studies have been conducted and used to derive parameters fitting a double Debye relaxation model for the impedance of human skin [58].

Using the double Debye parameters given in Pickwell et al. [58], the reflection coefficient expected from human skin has been calculated by the authors and monotonically decreases from ~0.75 at low frequencies to 0.35 at 1 THz. These results show a general trend of decreasing reflectivity as the frequency is increased due to increased absorption and decreased refractive index.

5. Millimeter-wave imaging systems and results

In this section, a cross-section of systems and results that are of interest for explosive detection and personnel screening are shown and/or discussed. These include numerous results from the authors' team at PNNL, other researchers, and a number of commercial companies. This is not meant to be a comprehensive review but rather a sampling of exciting systems and results obtained using millimeter-wave imaging. Results are shown for several technologies; however, many of the results are drawn from the authors' team's laboratory results. The reader is encouraged to read the reference papers for additional results from other researchers.

5.1. Holographic millimeter-wave imaging technique

Researchers at PNNL have developed three-dimensional wideband millimeter-wave imaging techniques and systems for a wide variety of applications including personnel screening/explosive detection (concealed weapon detection), inner-wall imaging, through-wall imaging, RCS imaging, ground penetrating radar (GPR) imaging, through-concrete imaging, and other applications. These systems make use of coherent wideband microwave, millimeter-wave, and/or terahertz transceivers to capture scattered wavefront data from the imaging targets. The scattered wavefront data can then be mathematically reconstructed to form fully focused, diffraction-limited, high-resolution images of the imaging target(s) [15–17]. Results obtained from laboratory and fieldable prototype imaging systems are detailed below.

PNNL researchers have demonstrated full-body, high-throughput personnel surveillance systems that can detect and identify body-worn, concealed threats that are presently undetected by screening technology in use at airports. These novel personnel screeners use wideband holographic reconstruction algorithms and millimeter-wave technology that can scan and reconstruct an image of a person in as little as 3–5 s.

The personnel scanner uses harmless coherent radar waves from a millimeter-wave array to illuminate the person under surveillance. The radar waves penetrate through the low dielectric clothing barriers and reflect off the weapons and other objects hidden on the body. These reflected signals are received by the arrays, digitized, and sent to high-speed computers for processing to form high-resolution three-dimensional radar images. Millimeter-wave imaging is well suited for the detection of concealed weapons or other contraband carried on personnel, because low-powered millimeter waves are nonionizing (safe for the operator and person under surveillance), readily penetrate common clothing material, and are reflected from the human body and any concealed items.

The wideband holographic imaging technique is a wavefront image reconstruction, or focusing, technique that eliminates the need for physical focusing elements such as lenses or curved reflectors. This technique is essentially a way to mathematically focus the imaging data that are completely unfocused in the actual hardware implementation.

The imaging process begins by scanning a diverging (spherical) beam antenna over a two-dimensional planar aperture, as depicted in Fig. 1. This diverging beam interacts with the imaging target that scatters the wavefront back to a receive antenna. The receive antenna usually coincides with the transmit antenna (monostatic) or is placed adjacent to the transmit antenna (quasi-monostatic). The receiver portion of the transceiver then amplifies the signal and measures its amplitude and phase (or in-phase and quadrature components). The amplitude and phase are sampled as a relatively wide frequency bandwidth is swept by the transceiver. This process is repeated at each discrete point sampled across the planar aperture, resulting in a three-dimensional data set (frequency and two spatial dimensions). Owing to the diverging beam, these data are unfocused in the raw format and must be mathematically focused using a computer-based image reconstruction algorithm. Descriptions of the transceiver and image reconstruction are summarized below.

The imaging system transceiver is shown in a simplified schematic form in Fig. 2. This transceiver consists of two oscillators. The transmit, or RF, oscillator is swept across the millimeter-wave frequency band of interest, and the LO is offset from this frequency by a fixed IF. The received signal is mixed with the LO to frequency down-convert the data to the IF. Once down-converted, this signal can be easily amplified and filtered, resulting in a high SNR. The received IF signal is then mixed with in-phase (nonphase shifted) and quadrature (90° phase shifted) reference IF signals. The outputs of this final

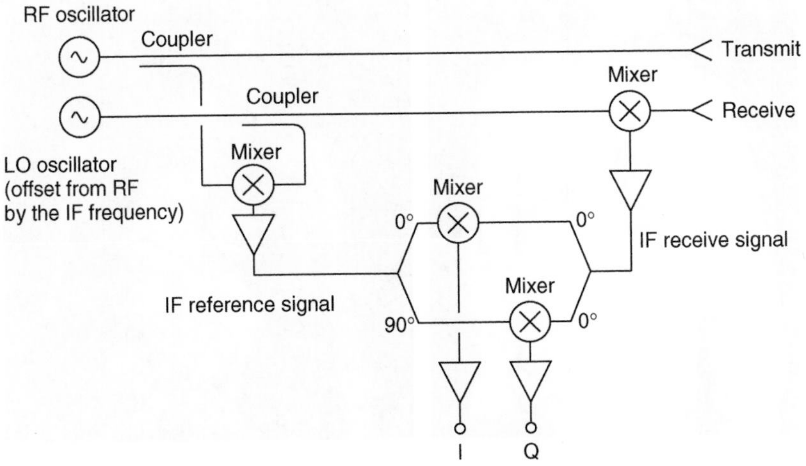

Fig. 2. Heterodyne millimeter-wave transceiver – simplified schematic (drawing from [17] ©IEEE).

down-conversion are the in-phase (I) and quadrature (Q) signals. Mathematically, for a single point target at a distance R from the transceiver, these signals will be given by

$$I = A\cos\phi \tag{7}$$

$$Q = A\sin\phi \tag{8}$$

where A is the amplitude, $\phi = 2kR$, $k = 2\pi/\lambda$, and λ is the wavelength. The I and Q signals essentially capture the magnitude and phase of the scattered wavefront.

Holographic, or Fourier optics based, image reconstruction techniques can be applied to these data directly because the phase of the scattered wavefronts across the aperture have been measured. The image reconstruction process is given in detail in Sheen et al. [17] and is summarized as follows. Two-dimensional spatial Fourier transforms are performed at each frequency to decompose the data into a superposition of plane wave components at known propagation angles. These plane waves are then back-propagated to the object plane through a phase shift and spatial frequency domain interpolation process. Finally, an inverse three-dimensional Fourier transform is used to transform the data back to spatial coordinates, which after computing the image magnitude results in the focused or reconstructed three-dimensional image.

The effectiveness of the wideband holographic millimeter-wave imaging technique is shown in Fig. 3 that shows a 100–112 GHz image of a mannequin carrying a concealed Glock-17 handgun. This image reveals the high resolution that is achievable using this technique, which is on the order of 1.5–3 mm. The image shows some details of the

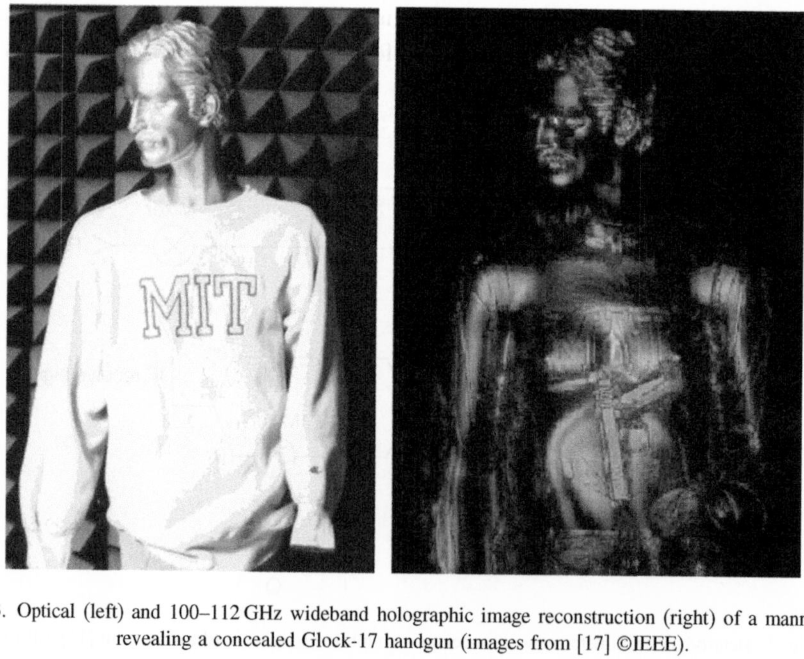

Fig. 3. Optical (left) and 100–112 GHz wideband holographic image reconstruction (right) of a mannequin revealing a concealed Glock-17 handgun (images from [17] ©IEEE).

clothing – note the embroidery on the sweatshirt is visible in the image – and the duct tape used to hold the weapon on the mannequin is visible near the end of the barrel.

The image shown in Fig. 3 was acquired using a laboratory transceiver that was raster scanned over planar apertures using a mechanical x-y scanner. Scan times are typically on the order of 10 min or more. This process is suitable for laboratory evaluation of static targets, but much higher speed is required to freeze motion on personnel for concealed weapon/explosive detection applications. The holographic imaging process can be hastened significantly by replacing one axis of the mechanical scan with a sequentially switched linear antenna array.

One of the concealed weapon detection systems developed at PNNL is shown in Fig. 4. In this system, a 128-element linear millimeter-wave array shown on the right is scanned vertically in about 1 s using the high-speed linear scanner system shown on the left. Subsequent computer reconstruction of the data allows formation of the image in near real time. The antenna array operates from 27 to 33 GHz and consists of 128 antenna elements configured as two interleaved 64-element sub-arrays. Each sub-array consists of eight single-pole eight throw (SP8T) finline pin diode waveguide switches driven by another SP8T switch. The array is sequentially switched beginning at one end of the array and allows for 127 independent spatial samples to be recorded. At each position, the transceiver sweeps the 27–33 GHz band in ~4–6 μs and analog-to-digital converters sample the output and deliver it to a high-speed computer system.

Fig. 4. Millimeter-wave holographic concealed weapon detection system (left) and close-up of the 128-element, 27–33 GHz linear array and transceiver system (right) (images from [17] © IEEE).

The mechanical scanner simply scans the array downward in ~1 s to capture a full aperture (~0.75 × 2.0 m) three-dimensional data set. This data set is then reconstructed and displayed using the computer system.

Imaging results with humans have shown the effectiveness of this system to detect body-worn, concealed threats fabricated with plastic materials such as C-4 and RDX bulk explosives and plastic flare guns. Wideband images from this system of a man carrying two concealed handguns – one in left pant pocket and one on abdomen tucked into belt – as well as several innocuous items including a wallet and checkbook are shown in Fig. 5. The resolution in these images is on the order of 1 cm.

The explosive detection performance of this system is shown in Fig. 6. In this Figure, wideband images of a man with and without a concealed block of RDX explosive are shown. The RDX block is clearly shown along the spine of the man. Additional images taken at slightly different angles help to differentiate the block from the surface of the man's back. A simulated sheet explosive was fabricated using duct putty as a convenient explosive simulant and concealed on the back of a man in the images shown in Fig. 7. In this case, the simulated explosive is clearly distinguished by the highly speckled return. This speckle pattern is likely due to a combination of slight irregularities in the surface of the duct putty and air gap variations between the duct putty and the man's back.

A lower frequency imaging system prototype was also developed that operates from 12.5 to 18 GHz. This system has lower resolution than the 27–33 GHz prototype but still offers excellent imaging capability and would be less expensive to produce. Example images from this system are shown in Fig. 8. These images reveal concealed blocks of explosive simulant (obtained from the Federal Aviation Administration) and C-4. Note that the edges of the explosive are clearly distinguished in the image despite relatively similar brightness of the main portion of the explosive image and the body of the man.

Experience using the planar imaging technique, for example, the imaging results shown in Fig. 5, indicated that it is highly desirable to view the person from multiple angles. An absolute minimum of two views (front and back) is required because the body is opaque to millimeter waves, but many more views allow for more nearly optimal imaging of concealed objects. To achieve multiple viewing angles while still maintaining the near real-time speed of the imaging system, a cylindrical imaging technique was developed at PNNL [33–36].

The cylindrical imaging technique is conceptually similar to the planar imaging technique, except that a full cylindrical aperture enclosing the person (or imaging target) is used. In this technique, wideband coherent data from an inward-oriented transceiver are sampled across the full 360° aperture with height h. A novel image reconstruction algorithm was developed which enables an arc segment portion of these data to be reconstructed into a fully focused three-dimensional image. Experience has indicated that 90–120° arc segments result in the highest quality images. Shifting the center of the arc segment used allows an arbitrary number of views of the subject to be reconstructed from the overall 360° data set. Typically, 32–64 reconstructions are performed, and the

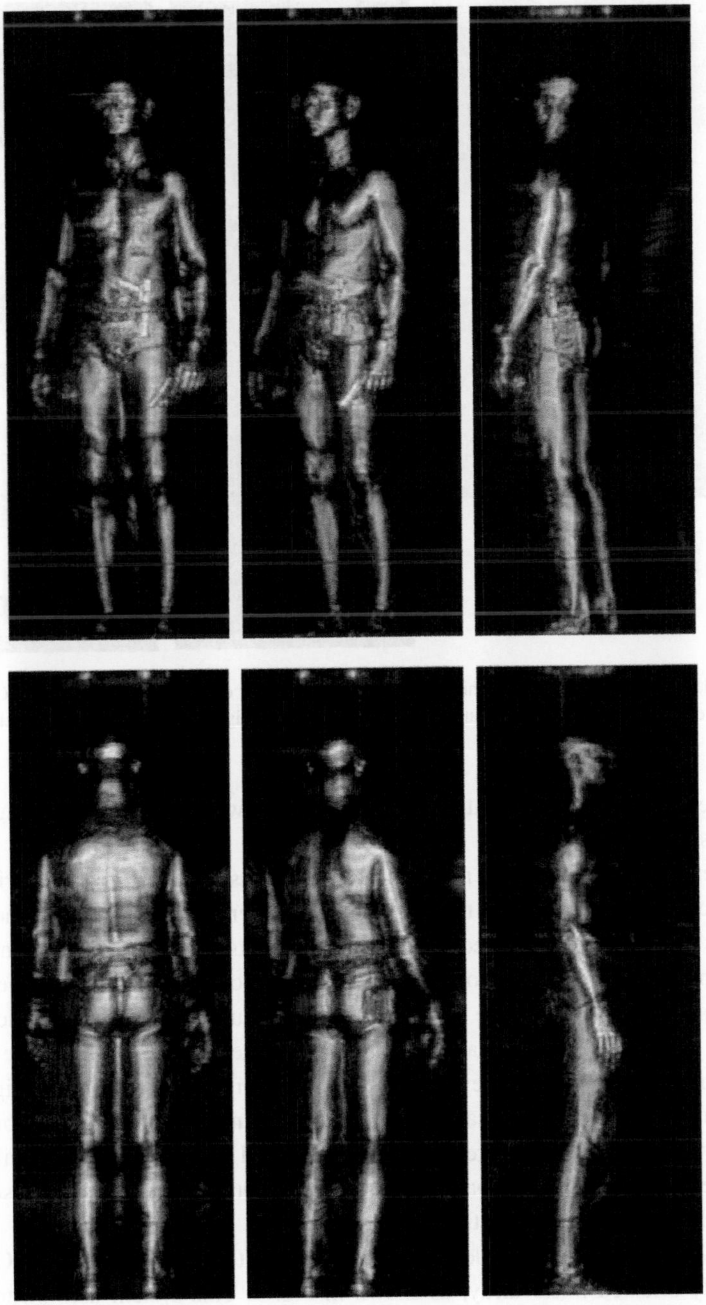

Fig. 5. Wideband images of man carrying two concealed handguns – one in left pants pocket and one on abdomen tucked into belt – as well as several innocuous items including a wallet and checkbook (images from [17] © IEEE).

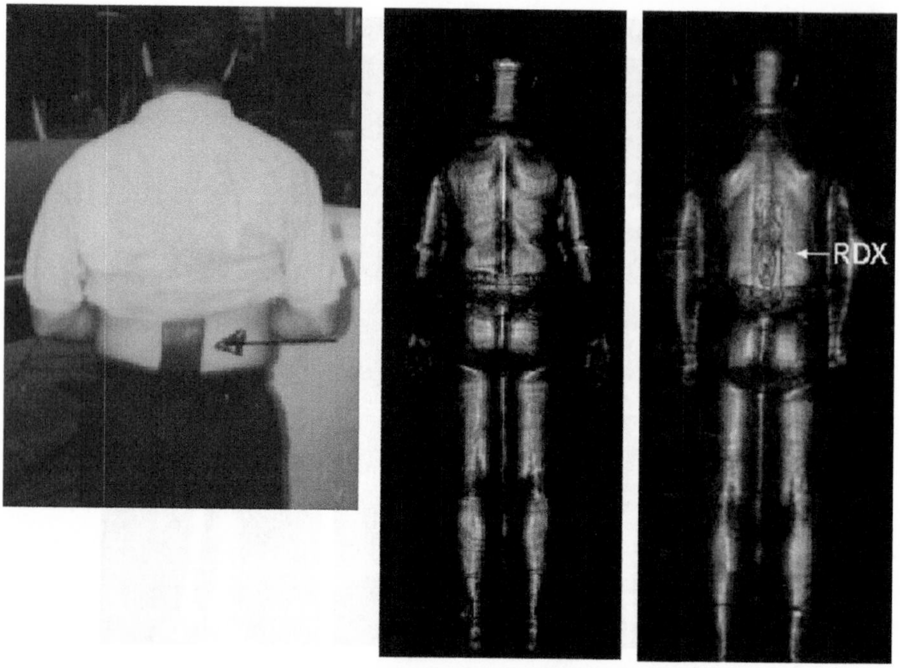

Fig. 6. Wideband 27–33 GHz holographic images of a man carrying concealed RDX explosive. Optical (left), no explosive image (middle), and concealed RDX image (right) (images from [17] ©IEEE).

results are displayed to an operator as an animation in which the person's image appears to rotate slowly.

An alternative reconstruction method is referred to as the combined cylindrical image reconstruction technique. With this method, eight overlapping 90° arc segment reconstructions are combined as three-dimensional data sets. This results in a single three-dimensional data set that incorporates all the illumination information from the full 360° scan. This data set can then be viewed at any angle using computer graphics volume rendering techniques. The primary advantage of this technique is that all the illumination information is present in every frame of the animation, resulting in high-quality images in many cases. The disadvantage of this technique is that the combination of multiple views results in range resolution being viewed transversely, which will degrade the image quality unless the bandwidth is large enough that the range resolution is comparable to the lateral resolution.

A cylindrical prototype was developed at PNNL that operates from 27–33 GHz using a vertically oriented linear array, as shown in Fig. 9. This system uses switching and transceiver technology similar to the prototype shown in Fig. 4. However, this system is composed of 384 antenna elements versus 128 for the previous system. This system collects a 360° scan in ~6–10 s. Image reconstruction is then performed using a conventional PC coupled to an array of high-speed co-processors. Example cylindrical images collected with a laboratory transceiver and scanner are shown in the lower half

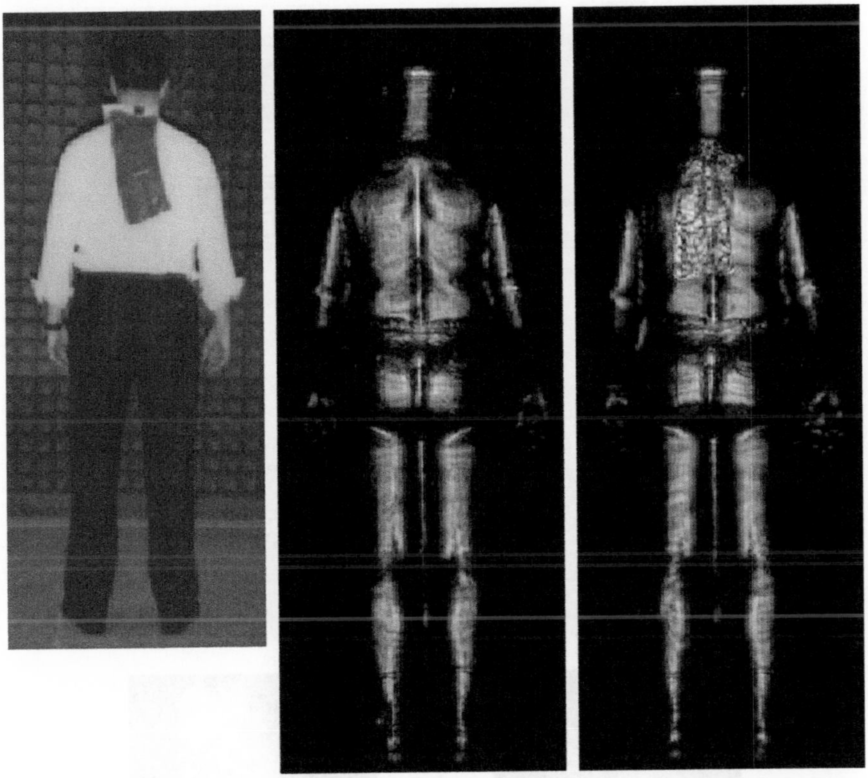

Fig. 7. Wideband 27–33 GHz holographic images of man carrying concealed simulated sheet explosive (duct putty). Optical (left), no explosive image (middle), and concealed simulant explosive (right).

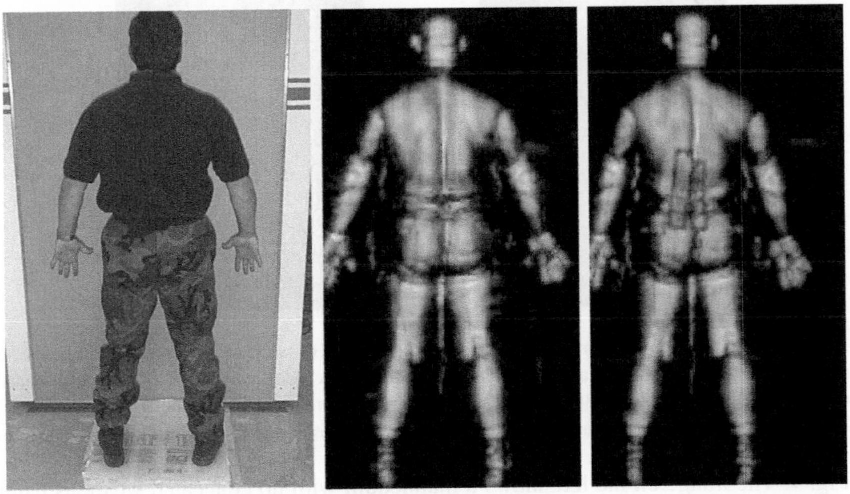

Fig. 8. Wideband 12.5–18 GHz holographic images of man carrying concealed FAA explosive simulant (left side of back) and C-4 explosive (right side of back). Optical image (left), no explosive image (middle), and concealed explosives image (right).

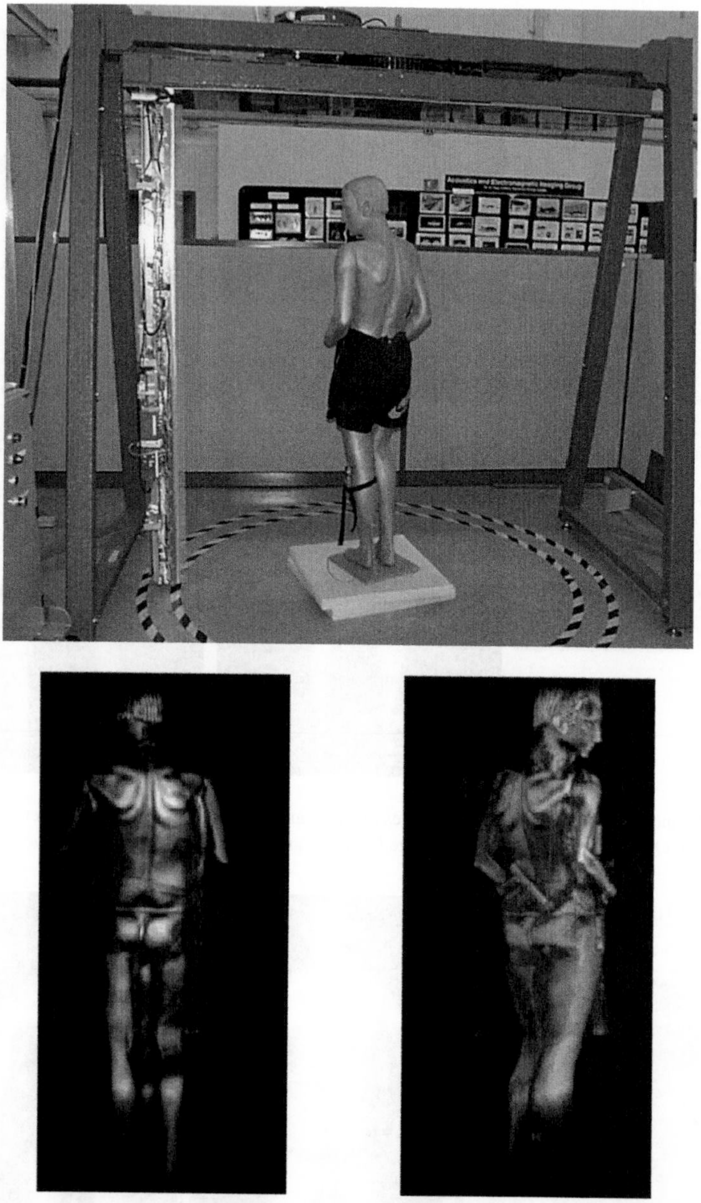

Fig. 9. Cylindrical imaging system prototype (top) and 24–40 GHz laboratory cylindrical imaging results. Lower left image is a single 90° arc segment reconstruction of a mannequin with no concealed weapons. Lower right image is a combined 360° cylindrical reconstruction projected into an individual image showing a concealed handgun on the lower back.

of Fig. 9. The image on the left shows a mannequin with no concealed weapons using a single 90° arc segment image reconstruction. The image on the right shows a mannequin with a concealed handgun at the lower back using the combined image reconstruction algorithm (all angles used in the reconstruction).

This technology can also be applied to longer-range standoff imaging scenarios, especially at higher frequencies [59]. Figure 10 shows holographic images of a metal pellet gun at ranges of 10, 15, 20, and 25 ft. These images were acquired using a laboratory scanner coupled with a 98–100 GHz transceiver with a 1×1 m aperture. The resolution is seen to degrade slightly at the longer ranges because of the increase in the F-number (i.e., an increase in the theoretical, diffraction-limited, resolution size). Figure 10 also shows similar images of a mannequin carrying a concealed metal pellet gun.

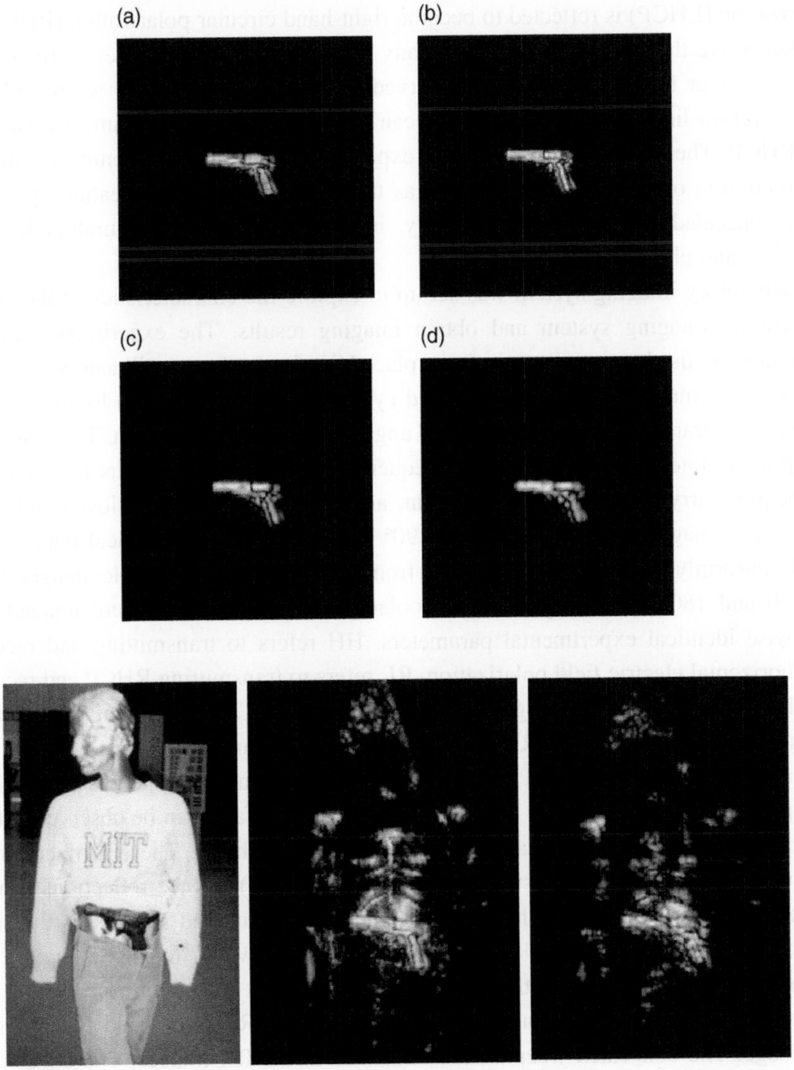

Fig. 10. Standoff 98–100 GHz millimeter-wave images of metal pellet gun at (a) 10 ft, (b) 15 ft, (c) 20 ft, (d) 25 ft, and standoff 98–100 GHz millimeter-wave images of mannequin with concealed metal handgun – optical (lower left), 10 ft (lower center), and 20 ft (lower right).

These results are interesting because they show the effects of specular reflection. Portions of the mannequin's surface which are nearly normal to the scanned imaging aperture remain visible in the images, and portions that are curved away tend to reflect the illumination away from the aperture and are dark in the imagery. This effect may decrease at even higher frequencies, because smaller features will reflect more significantly as the frequency is increased (wavelength is decreased).

Circular polarimetric imaging can be employed to obtain additional information from the target. Circularly polarized waves incident on relatively smooth reflecting targets are typically reversed in their rotational handedness, for example, left-hand circular polarization (LHCP) is reflected to become right-hand circular polarization (RHCP). An incident wave that is reflected twice (or any even number of times) before returning to the transceiver has its handedness preserved. Sharp features such as wires and edges tend to return linear polarization, which can be considered to be a sum of both LHCP and RHCP. These characteristics can be exploited for personnel screening by allowing differentiation of smooth features, such as the body, from sharper features present in many concealed items [39]. Additionally, imaging artifacts due to multipath can be identified and eliminated.

A laboratory imaging system was set up to explore the characteristics of the circular polarization imaging system and obtain imaging results. The experimental imaging configuration used a rotating platform placed in front of a rectilinear x-y scanner. This system emulates a linear array-based cylindrical imaging system by mechanically scanning the transceiver at each rotational angle of the rotating platform. The system was set up to operate over the 10–20 GHz frequency range. Imaging results from a clothed mannequin carrying a concealed handgun, and simulated plastic explosive are shown in Fig. 11. Images were obtained using 90° arc segments of cylindrical data centered at 64 uniformly spaced angles ranging from 0 to 360°, with sample images shown at $\sim$30 and 180° in the Figure. Three polarization combinations were imaged using otherwise identical experimental parameters. HH refers to transmitting and receiving with horizontal electric field polarization. RL refers to transmitting RHCP and receiving LHCP. RR refers to transmitting RHCP and receiving RHCP. The HH images are shown on the left side of Fig. 11, RL images in the center, and RR images on the right. An interesting feature of these images is that the polarization properties can highlight multipath signal returns. In the HH images, multipath artifacts can be observed between the thighs and between the mannequin's right arm and body. These artifacts are not present in the RL image that suppresses double (or even)-bounce reflections. The RR image highlights, or isolates, the double-bounce reflections and suppresses the single-bounce return. Similar results are observed in the back-view images in Fig. 11. The primary multipath artifacts in the HH image are between the upper arm and the body of the mannequin. These artifacts are eliminated in the RL image and isolated in the RR image. The concealed weapons are enhanced in the RR images. The edges of the concealed handgun are highlighted in the RR images because of the dihedral (double-bounce) reflection formed around the perimeter of the handgun as placed on the body of the mannequin. Similarly, the edges of the simulated plastic explosive are highlighted in

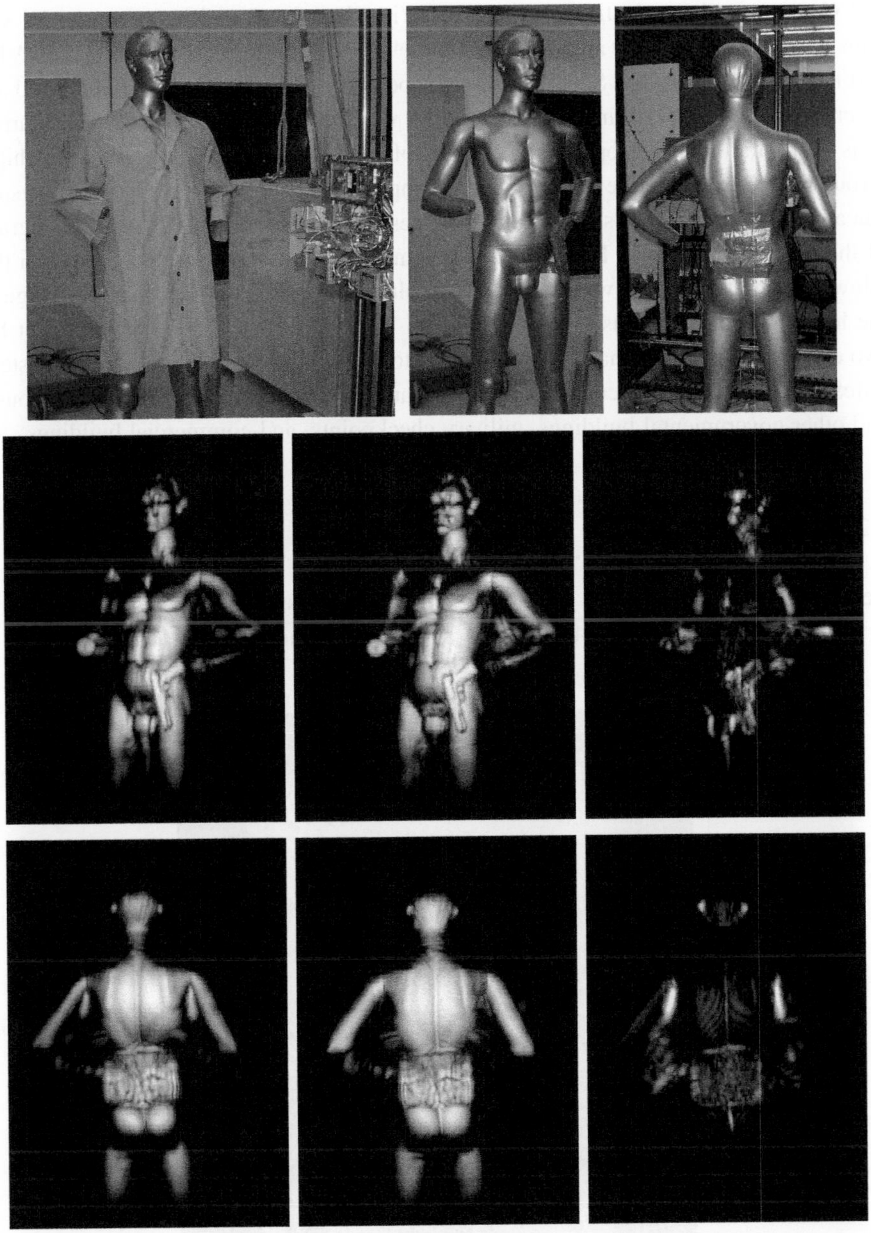

Fig. 11. Laboratory scanner and mannequin setup (upper images) and 10–20 GHz imaging results (center and lower images). Left-side images are HH polarization, center images are RL polarization, and right side images are RR polarization.

the RR image of the back of the mannequin. These polarization properties in the images may be exploited to enhance detection of concealed objects and reduce privacy concerns.

PNNL's wideband holographic millimeter-wave imaging technology has been licensed to SafeView, (http://www.safeviewinc.com). SafeView has been recently acquired by

L-3 Communications. At this time, L3 SafeView has developed two commercial systems, the SafeScout™ 100 and the SafeScout™ 360 (also referred to as Provision™ 100 and Provision™ 360). The SafeScout™ 100 operates within the 24.25–30 GHz band and incorporates two vertical linear antenna arrays with 384 antenna elements in each array. This configuration has an open entry and exit of the scanner and allows convenient high-throughput operation while still providing complete imaging of the subject. This system can acquire a scan in ~1.5 s and presents images to the operator within 2 s. A photograph of the system is shown in Fig. 12. This system is also available in a configuration that allows for multiple lanes with or without multiple operators to take full advantage of the high throughput of this screening system. The SafeScout™ 360 is similar but has two entry and exit doors that are used mainly to control entry and egress for the system. SafeScout™ systems are currently deployed around the world at airports, courthouses and other governmental buildings, military checkpoints, and commercial buildings.

5.2. *Passive millimeter-wave*

Researchers at Qinetic have developed innovative passive millimeter-wave imaging systems at 35 and 94 GHz [4,5]. These systems use linear arrays of direct detection

Fig. 12. SafeView SafeScout™ 100 (Provision™ 100) security screening system (Image courtesy of L3/SafeView).

receivers with MMIC low-noise amplifiers used to enhance sensitivity. The number of detector elements required varies by design with 28–58 elements used in the 35 GHz systems and on the order of 100 used for the 94 GHz system. The linear arrays in these systems are coupled to folded optics that use a scanning rotating mirror to achieve real-time (up to 25 Hz) frame rates. The spot-size (resolution) is nominally 20 mm for the 35 GHz system. An additional active illumination source chamber has also been tested that improves the contrast and sensitivity of the system for indoor concealed weapon detection [47].

Trex Enterprises (http://www.trexenterprises.com) has developed a novel passive millimeter-wave imaging technology for concealed weapon detection applications. This technology employs a frequency scanned two-dimensional pupil plane waveguide aperture array to allow rapid imaging [9]. This technology allows for real-time (multiple frames/second) scanning in some cases. Trex has developed this technology into a commercial instrument, the Sago ST150. This system, shown in Fig. 13, is $\sim 60 \times 60 \times 90$ cm in size and weighs ~ 30 kg. The system can operate at standoff ranges of 5 m or more and collects an image in 2 s. Example images are shown in Fig. 14, with optical images included on the left side of the Figure for comparison with the millimeter-wave images on the right. The upper images reveal a concealed handgun in the rear belt, and the lower images reveal a pipe bomb vest.

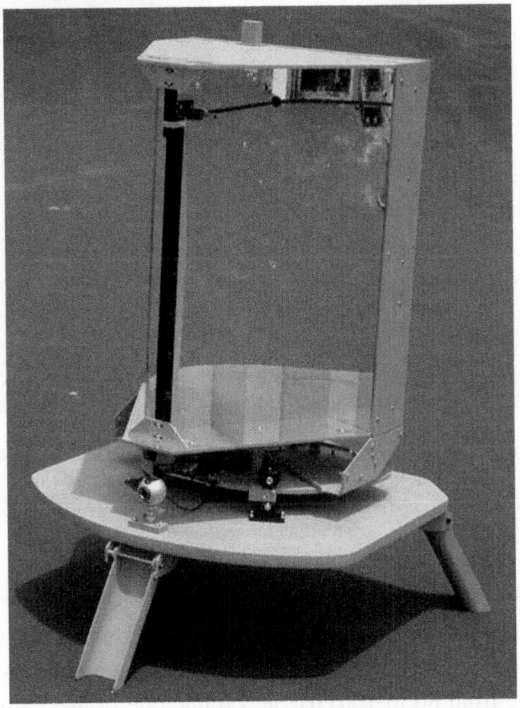

Fig. 13. Sago ST150 passive millimeter-wave imaging concealed weapon detection system (images courtesy of Trex Enterprises/Sago).

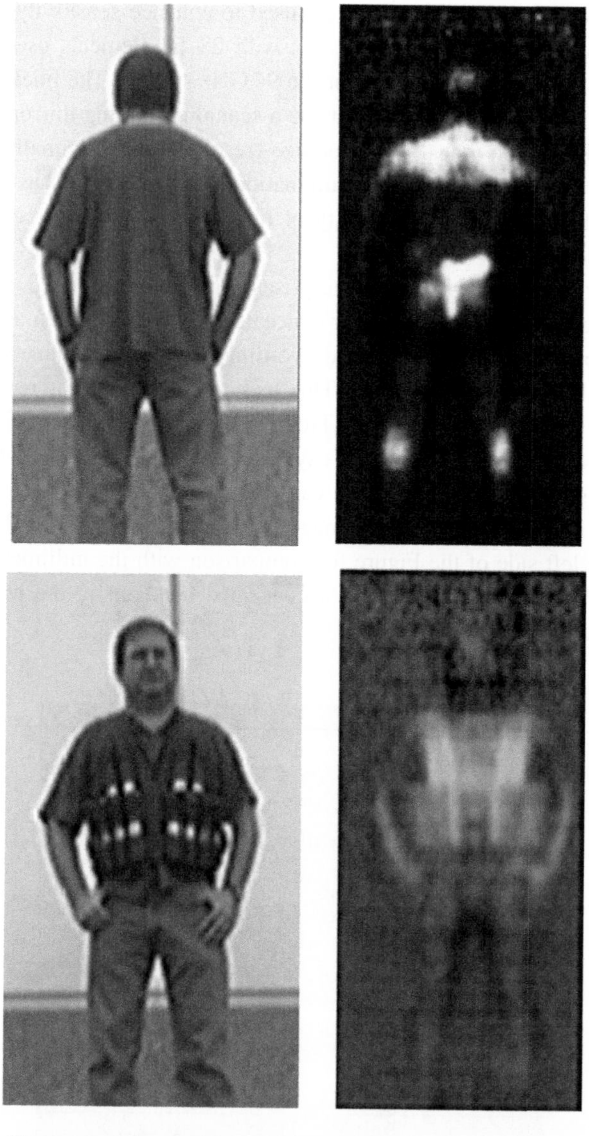

Fig. 14. Example millimeter-wave images from the Sago ST150 which show a subject at 5-m range carrying a concealed handgun at the rear belt line (upper images) and a pipe bomb vest (lower images). Images on the left are optical images used for comparison with the millimeter-wave images on the right (images courtesy of Trex Enterprises/Sago).

Brijot Imaging Systems, Inc. (http://www.brijot.com) has developed the BIS-WDS™ Prime passive millimeter-wave imaging camera for personnel screening applications. With imaging capabilities in indoor and outdoor environments, this system operates over the frequency range from 80 to 100 GHz at a frame rate of 4–12 frames/second with a differential temperature sensitivity of 1 K. This system has a 16-pixel imaging capability that allows for a 5 × 5 cm minimum object size threshold. Automated image processing

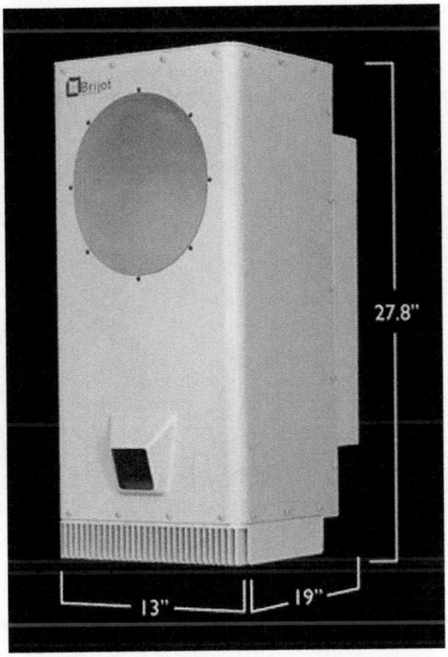

Fig. 15. Brijot BIS-WDS™ Prime passive millimeter-wave imaging concealed weapon detection system (image courtesy of Brijot Imaging Systems).

software detects concealed objects in real-time and indicates their location on a full-motion video. This system, shown in Fig. 15, is relatively compact with dimensions of 70 cm high × 32 cm wide × 48 cm deep. An example image from this system is shown in Fig. 16, which reveals a concealed handgun at the front belt line.

Passive millimeter-wave imaging systems have been developed for aircraft low-visibility applications and concealed weapon detection. Early systems used raster-scanned radiometers at microwave and millimeter-wave frequencies and have been used to form non-real-time imaging systems, as described by Hollinger [10]. A full two-dimensional FPA real-time imaging system was developed by TRW (now Northrop Grumman Space Technology) for airborne synthetic vision in low-visibility conditions [6–8]. This system consists of a FPA of 1040 MMIC direct detection receivers operating at 89 GHz with a 10 GHz bandwidth. The system uses an 18-inch primary lens and has a 15° × 10° field-of-view and a frame rate of 17 Hz. Low-noise MMIC amplifiers precede each detector element to improve the sensitivity of the system.

5.3. *Other active millimeter-wave imaging systems*

Agilent Technologies (http://www.agilent.com) has developed a novel active real-time millimeter-wave imaging system for concealed weapon detection [14]. This system

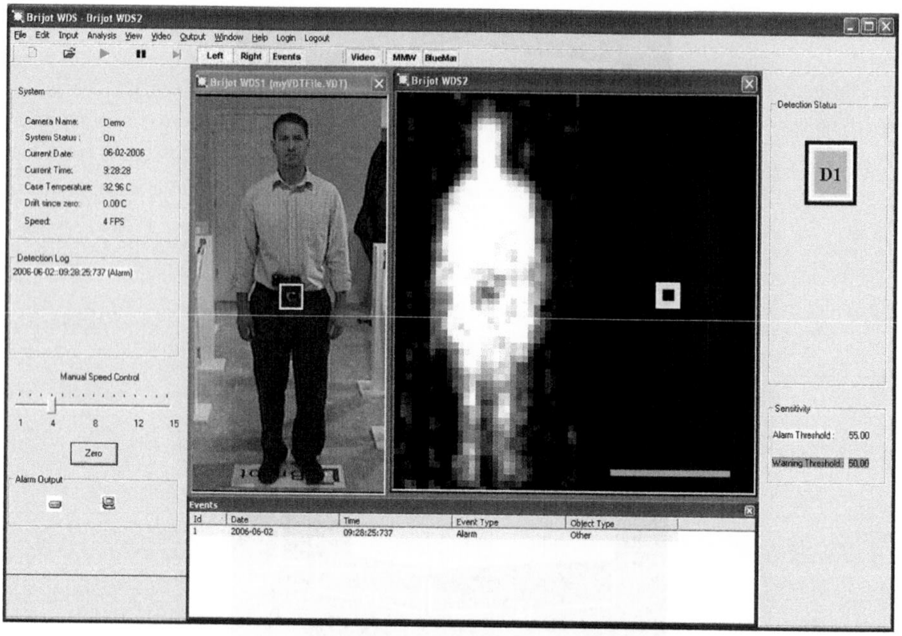

Fig. 16. Example image from the Brijot BIS-WDS™ Prime that shows a concealed handgun at the front belt line (images courtesy of Brijot Imaging Systems).

utilizes a two-dimensional array of patch antennas configured as a programmable reflector array to focus the wavefront from a millimeter-wave transceiver antenna onto a specific target voxel. The operational concept for this system is shown in Fig. 17. The signal reflected from this voxel is subsequently refocused onto the millimeter-wave antenna for detection. Focusing is accomplished by altering the phase of the signal that is reflected from each patch antenna array element under programmed digital control.

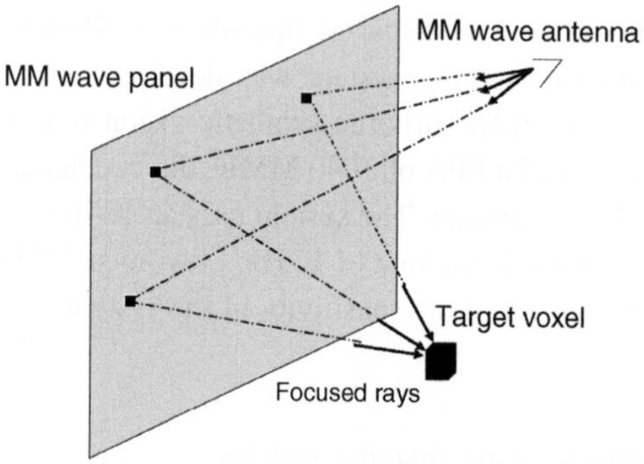

Fig. 17. Agilent reflector array operation. Millimeter-wave reflector array panel alters the phase of the transmitted wavefront to allow high-speed digitally controlled focusing over a range of target voxel locations (images courtesy of Agilent).

As there are no moving parts, the system's scan rate is limited only by the capability of the electronics, which currently allow 10 million voxels/second to be scanned. The three-dimensional data captured by the imaging system are projected to form a focused two-dimensional image. The system is currently configured as two $1\,\text{m} \times 1\,\text{m} \times 7.5\,\text{cm}$ panels each composed of 32 $15 \times 15\,\text{cm}$ sub-arrays. The system operates at 24 GHz with a frame rate of 15 frames/second. Resolution is 0.5 cm lateral and 1.0 cm depth. Example images from this system are shown in Fig. 18. In this Figure, a simulated suicide vest is shown from two different viewing angles.

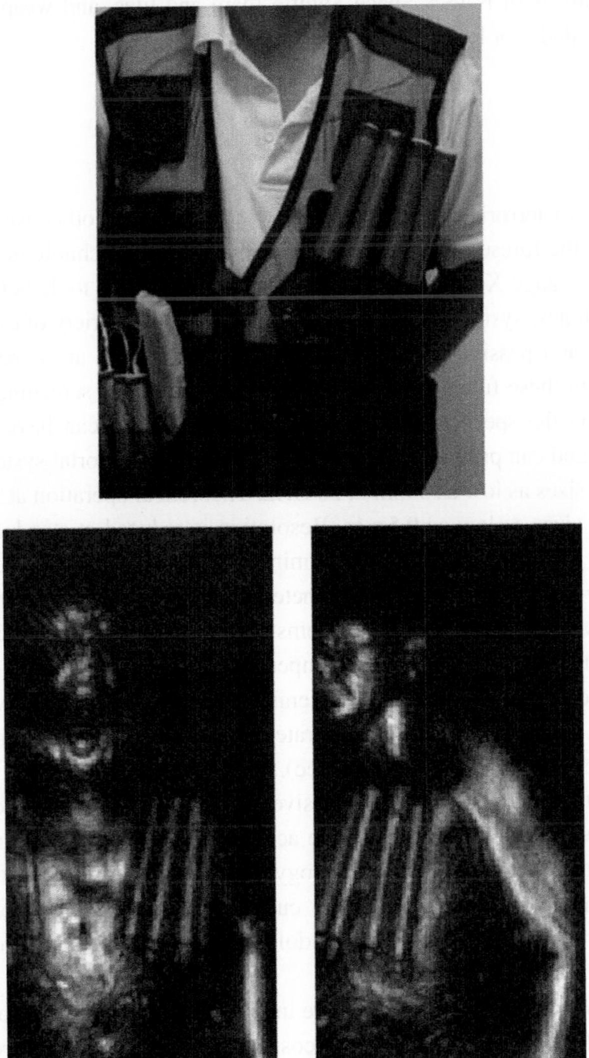

Fig. 18. Example images from the Agilent active millimeter-wave imaging system. A person is shown carrying simulated explosives (images courtesy of Agilent).

A simple focused millimeter-wave imaging system has been demonstrated at 200 GHz by Karpowicz et al. [60]. This system uses a frequency doubled Gunn diode millimeter-wave source that operates at 200 GHz with an output power of ~12 mW and a direct Schottky diode detector. This system has been configured using both an off-axis paraboloid mirror and a polyethylene Fresnel lens to focus the millimeter-wave radiation in both a through-transmission and short-range reflection configuration. The reflection configuration uses a metal plate behind the sample to reflect the illumination that passes through the sample. Both configurations use a direct Schottky diode detector, and raster scan the target to form the images. Imaging results from this system were shown demonstrating the analysis of NASA Space Shuttle foam and tiles, and weapon detection for weapons concealed in a leather briefcase.

6. Conclusion

Attempted acts of terrorism involving explosive devices and other weapons are likely to continue for the foreseeable future. Widely implemented technologies, such as metal detectors and baggage X-ray systems, are important screening tools but will need to be augmented with new system technologies to detect the wide variety of current and future threats. Active and passive millimeter-wave imaging systems are expected to play an important role in these future explosive detection and security screening scenarios.

Depending on the specific implementation, these systems can have high resolution and sensitivity and can produce remarkable imagery. Current portal systems can achieve resolution spot sizes as low as 5 mm at 30 GHz. Theoretical operation at 300 GHz allows resolution spot sizes as low as 0.5 mm. Resolution is reduced at standoff ranges, but is still acceptable for concealed object discrimination. Active systems generally have very high sensitivity because of the low-noise heterodyne receiver architecture and moderate transmitter powers available. Passive systems, however, are more fundamentally limited in their signal-to-noise, or differential temperature sensitivity, because the signal level cannot be artificially increased. These systems have a direct tradeoff between sensitivity and frame rate, i.e., increasing the frame rate will decrease the sensitivity (increase the minimum detectable temperature difference).

In this chapter, several active and passive imaging technologies and systems have been highlighted. These have included the active holographic millimeter-wave imaging technology, active reflector-array technology, and several implementations of passive imaging technology. Several systems are currently available commercially, including both portal and standoff systems. The standoff systems have ranges currently exceeding 5 m in some of the passive systems.

Future improvements to millimeter-wave imaging systems are likely to be achieved in several areas. If widely implemented, the cost of these systems will be reduced through engineering improvements and economies of scale. Improved electronic integration will reduce the cost of all of the antenna arrays and may eventually allow for fully populated two-dimensional FPA imaging systems to become cost effective. FPAs will allow for improved operation of passive millimeter-wave imaging systems, because it will allow increased integration time at each pixel and therefore substantially improve the differential temperature sensitivity. Higher frequency operation and automation of the detection process are additional areas of future research.

Future millimeter-wave imaging systems may also push toward higher frequency operation. Most current active systems operate near 30 GHz or below, and most passive systems operate near 100 GHz or below. Increasing the frequency of operation will improve the image resolution and/or increase the standoff range at which the system can operate effectively. Increasing the frequency range into the sub-millimeter wave, or terahertz, bands may also provide some spectroscopic information that will aid in material discrimination or identification. Operation in the terahertz will be difficult, however, due to atmospheric attenuation, clothing attenuation and reflection, and source/detector limitations.

Automation of these imaging systems will also be an important area of future research. Throughput of current systems is limited by the time required for an operator to inspect the imagery. Additionally, many people object to the invasiveness of this type of imaging and feel that it is an invasion of their privacy. Developing image processing software that can automatically analyze the imagery may alleviate this concern and will improve the throughput and convenience of the screening systems. Operational costs would also be reduced substantially by reducing the number of human screeners required to operate the systems.

Millimeter-wave imaging systems will be an important complement to existing and future trace explosive detection systems, because millimeter-wave imaging can, in principle, detect many types of suspicious concealed objects regardless of their specific chemical composition. This is both an important advantage and disadvantage of the technology. The primary advantage is that virtually any concealed objects can be detected. The disadvantage is that the systems will not be able to explicitly identify the material as an explosive. The performance of various millimeter-wave imaging technologies and systems has been demonstrated with numerous imaging examples throughout this chapter.

Acknowledgments

The authors gratefully acknowledge the ingenuity, hard work, and commitment from the PNNL Radar Imaging Laboratory staff, especially the late Dr H. Dale Collins and R. Parks Gribble, the founders of this technology and laboratory at PNNL, and Ron Severtsen, Jim Prince, Larry Reid, and Jeff Griffin. The Federal Aviation Administration (now Transportation Security Laboratory) funded much of the PNNL research on concealed weapon detection. Also, the authors thank Rick Rowe and Tom Grudkowski (L-3/SafeView), Chris Martin (Trex Enterprises), Marty Neil (Agilent), and Jo Roberts (Brijot) for providing images and information about their imaging systems.

The PNNL work described in this chapter was conducted under Contract DE-AC06-76RL01830 with the U.S. Department of Energy. The PNNL is operated by Battelle for the U.S. Department of Energy.

References

[1] J. E. Bjarnason, T. L. J. Chan, A. W. M. Lee, M. A. Celis, and E. R. Brown, "Millimeter-wave, terahertz, and mid-infrared transmission through common clothing," *Applied Physics Letters*, vol. 85, pp. 519, 2004.

[2] J. A. Kong, *Electromagnetic Wave Theory*. New York: John Wiley and Sons, 1986.

[3] R. W. Boyd, *Radiometry and the Detection of Optical Radiation*. New York: John Wiley and Sons, 1983.

[4] R. Appleby, R. N. Anderton, S. Price, N. A. Salmon, G. N. Sinclair, P. R. Coward, A. R. Barnes, P. D. Munday, M. Moore, A. H. Lettington, and D. A. Robertson, "Mechanically scanned real-time passive millimeter wave imaging at 94 GHz," *Proceedings of the SPIE – The International Society for Optical Engineering*, vol. 5077, pp. 1, 2003.

[5] R. Appleby, R. N. Anderton, S. Price, G. N. Sinclair, and P. R. Coward, "Whole-body 35-GHz security scanner," *Proceedings of the SPIE – The International Society for Optical Engineering*, vol. 5410, pp. 244, 2004.

[6] P. Moffa, L. Yujiri, K. Jordan, R. Chu, H. H. Agravante, and S. Fornaca, "Passive millimeter-wave camera flight tests," *Proceedings of the SPIE – The International Society for Optical Engineering*, vol. 4032, pp. 14, 2000.

[7] L. Yujiri, H. Agravante, M. Biedenbender, G. S. Dow, M. Flannery, S. Fornaca, B. Hauss, R. Johnson, R. Kuroda, K. Jordan, P. Lee, D. Lo, B. Quon, A. Rowe, T. Samec, M. Shoucri, K. Yokoyama, and J. Yun, "Passive millimeter-wave camera," *Proceedings of the SPIE – The International Society for Optical Engineering*, vol. 3064, pp. 15, 1997.

[8] L. Yujiri, M. Shoucri, and P. Moffa, "Passive millimeter wave imaging," *IEEE Microwave Magazine*, vol. 4, pp. 39, 2003.

[9] C. A. Martin and V. G. Kolinko, "Concealed weapons detection with an improved passive millimeter-wave imager," *Proceedings of the SPIE – The International Society for Optical Engineering*, vol. 5410, pp. 252, 2004.

[10] J. P. Hollinger, J. E. Kenney, and B. E. Troy, Jr., "A versatile millimeter-wave imaging system," *IEEE Transactions on Microwave Theory and Techniques*, vol. MTT-24, pp. 786, 1976.

[11] D. C. Munson, J. D. O'Brien, and W. K. Jenkins, "A tomographic formulation of spotlight-mode synthetic aperture radar," *Proceedings of the IEEE*, vol. 71, pp. 917–925, 1983.

[12] M. Soumekh, "A system model and inversion for synthetic aperture radar imaging," *IEEE Transactions on Image Processing*, vol. 1, pp. 64–76, 1992.

[13] M. Soumekh, *Fourier Array Imaging*. Englewood Cliffs, NJ: Prentice Hall, 1994.

[14] P. Corredoura, Z. Baharav, B. Taber, and G. Lee, "Millimeter-wave imaging system for personnel screening: scanning 10^7 points a second and using no moving parts," *Proceedings of SPIE*, vol. 6211, 2006.

[15] D. M. Sheen, D. L. McMakin, and T. E. Hall, "Cylindrical millimeter-wave imaging technique for concealed weapon detection," *Proceedings of the SPIE – 26th AIPR Workshop: Exploiting New Image Sources and Sensors*, vol. 3240, pp. 242–250, 1997.

[16] D. M. Sheen, D. L. McMakin, and T. E. Hall, "Combined illumination cylindrical millimeter-wave imaging technique for concealed weapon detection," *Proceedings of the SPIE – Aerosense 2000: Passive Millimeter-Wave Imaging Technology IV*, vol. 4032, pp. 52–60, 2000.

[17] D. M. Sheen, D. L. McMakin, and T. E. Hall, "Three-dimensional millimeter-wave imaging for concealed weapon detection," *IEEE Transactions on Microwave Theory and Techniques*, vol. 49, pp. 1581–1592, 2001.

[18] N. H. Farhat, "Microwave holography and its applications in modern aviation," *Engineering Applications of Holography Symposium Proceedings, SPIE*, pp. 295–314, 1972.

[19] N. H. Farhat and W. R. Guard, "Millimeter wave holographic imaging of concealed weapons," *Proceedings of the IEEE*, vol. 59, pp. 1383–1384, 1971.

[20] G. Tricoles and N. H. Farhat, "Microwave holography: applications and techniques," *Proceedings of the IEEE*, vol. 65, pp. 108–121, 1977.

[21] Boyer, "Reconstruction of ultrasonic images by backward propagation," *Acoustical Holography*, vol. 3, 1970.

[22] D. Gabor, "A new microscope principle," *Nature*, vol. 161, 1948.

[23] J. W. Goodman, *Introduction to Fourier Optics*. McGraw-Hill, San Francisco, 1968.

[24] B. P. Hildebrand and B. B. Brenden, *An Introduction to Acoustical Holography*, Plenum Press, New York, 1972.

[25] B. P. Hildebrand and K. A. Haines, "Holography by scanning," *Journal of the Optical Society of America*, vol. 59, pp. 1–6, 1969.

[26] E. N. Leith and J. Upatnieks, "Reconstructed wavefronts and communication theory," *Journal of Optical Society of America*, vol. 52, pp. 1123–1130, 1962.

[27] H. D. Collins, D. L. McMakin, T. E. Hall, and R. P. Gribble, "Real-time holographic surveillance system." United States Patent No. 5,455,590, 1995.

[28] D. L. McMakin, D. M. Sheen, H. D. Collins, T. E. Hall, and R. H. Severtsen, "Wideband, millimeter-wave, holographic weapons surveillance system," *Proceedings of the SPIE – EUROPTO European Symposium on Optics for Environmental and Public Safety*, vol. 2511, pp. 131–141, 1995.

[29] D. L. McMakin, D. M. Sheen, H. D. Collins, T. E. Hall, and R. H. Severtsen, "Detection of concealed weapons and explosives on personnel using a wide-band holographic millimeter-wave imaging system," presented at American Defense Preparedness Association Security Technology Division Joint Security Technology Symposium, Williamsburg, VA, 1996.

[30] D. M. Sheen, H. D. Collins, T. E. Hall, D. L. McMakin, R. P. Gribble, R. H. Severtsen, J. M. Prince, and L. D. Reid, "Real-time wideband holographic surveillance system." United States Patent No. 5,557,283, 1996.

[31] D. M. Sheen, D. L. McMakin, H. D. Collins, T. E. Hall, and R. H. Severtsen, "Concealed explosive detection on personnel using a wideband holographic millimeter-wave imaging system," *Proceedings of the SPIE – AEROSENSE Aerospace/Defense Sensing and Controls*, vol. 2755, pp. 503–513, 1996.

[32] D. L. McMakin, D. M. Sheen, T. E. Hall, and R. H. Severtsen, "Cylindrical holographic radar camera," *Proceedings of the SPIE – The International Symposium on Enabling Technologies for Law Enforcement and Security*, vol. 3575, pp. 79–88, 1998.

[33] D. M. Sheen, D. L. McMakin, and T. E. Hall, "Cylindrical millimeter-wave imaging technique for concealed weapon detection," *Proceedings of the SPIE – The International Society for Optical Engineering*, vol. 3240, pp. 242–250, 1997.

[34] D. M. Sheen, D. L. McMakin, and T. E. Hall, "Combined illumination cylindrical millimeter-wave imaging technique for concealed weapon detection," *Proceedings of the SPIE – The International Society for Optical Engineering*, vol. 4032, pp. 52–60, 2000.

[35] D. M. Sheen, D. L. McMakin, and T. E. Hall, "Cylindrical millimeter-wave imaging technique and applications," *Proceedings of SPIE*, vol. 6211, 2006.

[36] D. M. Sheen, D. L. McMakin, T. E. Hall, and R. H. Severtsen, "Real-time wideband cylindrical holographic surveillance system." United States Patent No. 5,859,609, 1999.

[37] P. E. Keller, D. L. McMakin, D. M. Sheen, A. D. McKinnon, and J. W. Summet, "Privacy algorithm for cylindrical holographic weapons surveillance system," *IEEE Aerospace and Electronic Systems Magazine*, vol. 15, pp. 17–24, 2000.

[38] D. L. McMakin, D. M. Sheen, J. W. Griffin, and W. M. Lechelt, "Extremely high-frequency holographic radar imaging of personnel and mail," *Proceedings of SPIE*, vol. 6201, 2006.

[39] D. M. Sheen, D. L. McMakin, W. M. Lechelt, and J. W. Griffin, "Circularly polarized millimeter-wave imaging for personnel screening," *Proceedings of the SPIE – International Society for Optical Engineering*, vol. 5789, pp. 117, 2005.

[40] P. F. Goldsmith, C. T. Hsieh, G. R. Huguenin, J. Kapitzky, and E. L. Moore, "Focal plane imaging systems for millimeter wavelengths," *IEEE Transactions on Microwave Theory and Techniques*, vol. 41, pp. 1664–1675, 1993.

[41] G. R. Huguenin, P. F. Goldsmith, N. C. Deo, and D. K. Walker, "Contraband detection system." United States Patent No. 5,073,782, 1991.

[42] G. R. Huguenin, "A millimeter wave focal plane array imager," *Proceedings of the SPIE – The International Society for Optical Engineering*, vol. 2211, pp. 300–301, 1994.

[43] G. R. Huguenin, "Millimeter-wave video rate imagers," *Proceedings of the SPIE – The International Society for Optical Engineering*, vol. 3064, pp. 34–45, 1997.

[44] E. N. Grossman, A. K. Bhupathiraju, A. J. Miller, and C. D. Reintsema, "Concealed weapons detection using an uncooled millimeter-wave microbolometer system," *Proceedings of the SPIE – The International Society for Optical Engineering*, vol. 4719, pp. 364, 2002.

[45] E. N. Grossman, A. Luukanen, and A. J. Miller, "Terahertz active direct detection imagers," *Proceedings of the SPIE – The International Society for Optical Engineering*, vol. 5411, pp. 68, 2004.

[46] E. N. Grossman and A. J. Miller, "Active millimeter-wave imaging for concealed weapons detection," *Proceedings of the SPIE – The International Society for Optical Engineering*, vol. 5077, pp. 62, 2003.

[47] P. R. Coward and R. Appleby, "Development of an illumination chamber for indoor millimeter-wave imaging," *Proceedings of the SPIE – The International Society for Optical Engineering*, vol. 5077, pp. 54, 2003.

[48] H. Eisele and R. Kamoua, "Submillimeter-wave InP Gunn devices," *IEEE Transactions on Microwave Theory and Techniques*, vol. 52, pp. 2371, 2004.

[49] V. K. Paidi, Z. Griffith, W. Yun, M. Dahlstrom, M. Urteaga, N. Parthasarathy, S. Munkyo, L. Samoska, A. Fung, and M. J. W. Rodwell, "G-band (140–220 GHz) and W-band (75–110 GHz) InP DHBT medium power amplifiers," *IEEE Transactions on Microwave Theory and Techniques*, vol. 53, pp. 598, 2005.

[50] A. Raeisaenen and M. Sironen, "Capability of Schottky-diode multipliers as local oscillators at 1 THz," *Microwave and Optical Technology Letters*, vol. 4, pp. 29, 1991.

[51] T. W. Crowe, D. W. Porterfield, J. L. Hesler, W. L. Bishop, and D. S. Kurtz, "Integrated terahertz transmitters and receivers," *Proceedings of the SPIE – The International Society for Optical Engineering*, vol. 5268, pp. 1, 2003.

[52] G. Chattopadhyay, E. Schlecht, J. S. Ward, J. J. Gill, H. H. S. Javadi, F. Maiwald, and I. Mehdi, "An all-solid-state broad-band frequency multiplier chain at 1500 GHz," *IEEE Transactions on Microwave Theory and Techniques*, vol. 52, pp. 1538, 2004.

[53] P. H. Siegel, R. P. Smith, M. C. Graidis, and S. C. Martin, "2.5-THz GaAs monolithic membrane-diode mixer," vol. 47, pp. 596, 1999.

[54] D. L. Mensa, *High Resolution Radar Cross-Section Imaging*. Norwood, MA: Artech House, 1991.

[55] E. Jacobs, R. G. Driggers, K. Krapels, F. C. De Lucia, and D. Petkie, "Terahertz imaging performance model for concealed weapon identification," *Proceedings of SPIE – The International Society for Optical Engineering*, vol. 5619, pp. 98, 2004.

[56] P. H. Siegel, "Terahertz technology," *IEEE Transactions on Microwave Theory and Tech.*, vol. 50, pp. 910–928, 2002.

[57] E. R. Brown, "Fundamentals of terrestrial millimeter-wave and terahertz remote sensing," in *Terahertz Sensing Technology*, vol. 2, D. L. Woolard, W. R. Leorop, and M. S. Shur, Eds. New Jersey: World Scientific, 2003.

[58] E. Pickwell, B. E. Cole, A. J. Fitzgerald, M. Pepper, and V. P. Wallace, "In vivo study of human skin using pulsed terahertz radiation," *Physics in Medicine and Biology*, vol. 49, pp. 1595–1607, 2004.

[59] D. L. McMakin, D. M. Sheen, and H. D. Collins, "Remote concealed weapons and explosive detection on people using millimeter-wave holography," *IEEE International Carnahan Conference on Security Technology*, pp. 19–25, 1996.

[60] N. Karpowicz, Z. Hua, Z. Cunlin, I. L. Kuang, H. Jenn-Shyong, X. Jingzhou, and X. C. Zhang, "Compact continuous-wave subterahertz system for inspection applications," *Applied Physics Letters*, vol. 86, pp. 54105, 2005.

Chapter 10

Laser-based Detection Methods of Explosives

Chase A. Munson, Jennifer L. Gottfried, Frank C. De Lucia, Jr., Kevin L. McNesby and Andrzej W. Miziolek

US Army Research Laboratory, AMSRD-ARL-WM-BD, Aberdeen Proving Ground, MD 21005-5069, USA

Contents

1. Introduction

Lasers offer multiple approaches for explosive detection that are not possible with other techniques. In general, these can be separated into two types, (i) those based on the unique properties of lasers for long-distance propagation of intense energy and (ii) those that are based on the actual molecular and atomic spectroscopy, and as such utilize the high wavelength specificity that most lasers offer. Of course, the field of laser explosive detection is somewhat young, given the fact that lasers were invented fairly recently in 1958. As such, it is fair to say that laser explosive detection is still a work in progress, with much having been discovered in recent years and still more to be discovered in the near future, particularly as more exotic laser sources (e.g., femtosecond lasers) become more common, less expensive, more rugged, and generally more readily available.

These are very exciting times for the use of lasers for explosive detection. Great progress has been made in particular in solid-state lasers and in the extension of laser radiation throughout the infrared (IR), near-IR, visible, and near-ultraviolet (UV) regions with regard to decrease in size and cost for various systems. The Nd:YAG laser, for example, has become very mature with improvements over many generations, so that in a fairly compact and not-too-expensive package, one can get reliable laser radiation in the near-IR fundamental wavelength of 1.06 μm, as well as visible and UV radiation at the second to fifth harmonic.

One particular area where the laser appears to be uniquely capable is in the standoff detection of explosives where the laser properties of long-distance radiation propagation are providing capabilities not possible with other techniques. Still, although very promising, standoff explosive detection using lasers is an emerging application area requiring time to mature. One other area where lasers offer intriguing potential is in the fusion of orthogonal laser-based techniques, such as laser-induced breakdown spectroscopy (LIBS) and Raman. A number of researchers have started to pursue this avenue because of the recognition that an integrated LIBS/Raman system can use the same laser and spectrometer components. The expected dramatic improvements in probability of detection and reduction of false alarm rates suggest that laser-based explosive detection methods may evolve into a major new technology area in the next 1–3 years.

2. Detection of explosives using laser-based vibrational spectroscopy

Applications of laser-based vibrational spectroscopy to explosive detection have been widely studied. The literature on this topic was summarized by Steinfeld and Wormhoudt [1] and by Henderson [2] in 1998. Instrumentation for explosive detection was summarized by Moore [3] in 2004.

Detection of explosives depends on several factors, including the physical state of the sample to be detected (solid, liquid, and gas), the vapor pressure of the solid/liquid (if vapor is being detected), a knowledge of spectral characteristics, limited sample size,

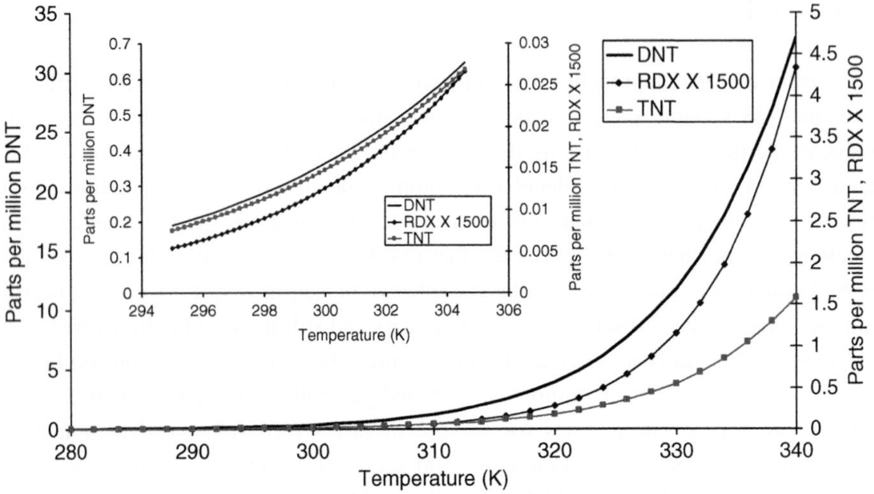

Fig. 1. The vapor pressures of the neat explosives 2,4-dinitrotoluene (DNT), 2,4,6-trinitrotoluene (TNT), and cyclotrimethylenetrinitramine (RDX) over the temperature range from 280 to 340 K.

effects of concealment, spectral interferences, thermal stability, vapor adsorption characteristics, and sampling methods. Figure 1 shows the calculated vapor pressures of the explosives 2,4,-dinitrotoluene (DNT), 2,4,6-trinitrotoluene (TNT), and cyclotrimethylenetrinitramine (RDX) over the temperature range 280 to 340 K [4,5]. There has been some discussion in the literature as to whether detection of concealed explosives is best accomplished by vapor detection or by detection of particles/explosive-laden fingerprints left behind during placement. For example, a single 5-μm diameter particle of the solid explosive RDX contains as many molecules as a liter of air saturated with RDX vapor at STP. Additionally, concealment in plastic containers and some formulation ingredients designed to make explosive materials 'plastic' may reduce partial pressures of explosive vapors by up to three orders of magnitude compared with laboratory values [1].

Solid- or vapor-phase vibrational spectra of (unreacted) explosives are usually measured with the bulk sample in the solid state, mainly because most explosives are solids at room temperature (nitroglycerin being the most well-known exception). For many measurements of vibrational spectra of vapors from solid explosives, the solid sample is heated to increase the vapor pressure [3]. For formulations of high explosives (HE), in which the main ingredient(s) are often crystalline when pure (e.g., C-4, its main ingredient nitramine RDX), samples may be powders or semi-malleable (plasticized) solids. For low explosive formulations and propellants, often containing ingredients which are polymeric when pure (e.g., nitrocellulose), samples are often in the form of grains (compressed or formed in the shape of a right circular cylinder), coarse or fine powders, slurries, or solid solutions.

Development of laser-based explosive detection methods employing vibrational spectroscopy begins with characterization of the spectral signature of the explosive to be detected. Spectral signatures of interest are usually those of the neat condensed-phase explosive (e.g., residue from a fingerprint) or the vapor emanating from the explosive material (e.g., concealed explosives). Because the vapor pressures of many pure explosive materials are exceptionally low (Fig. 1), the vapor above a solid explosive formulation may consist mainly of the most volatile components. Figure 2 shows the IR absorption spectrum of vapor above solid mil-spec TNT at 340 K (measured by the authors). The measured spectrum is actually of vapor-phase DNT, which is an impurity present in several percent in most samples of TNT, but with a much higher vapor pressure than TNT (Fig. 1) [6].

Vibrational spectroscopic studies of explosives may be grouped roughly into studies of unreacted materials and into studies of products of reaction. Most pre-event detection methods use unreacted material, although for some explosive detection methods, decomposition products may provide for increased sensitivity (see discussion that follows and Section 4.3.). Fundamental molecular vibrations ($\Delta\nu = 1$) exhibit characteristic frequencies in the mid-IR spectral region, lying in the wavelength region from 2 to 30 μm (~300–5000 cm^{-1}). Table 1 summarizes the IR spectral regions where many solid explosives and their vapors exhibit features. In general, the most recognizable feature of the vibrational spectrum of an explosive is associated with the symmetric and anti-symmetric vibrations of the almost ubiquitous $-NO_2$ group, between about 1260 and 1375 cm^{-1}.

A significant challenge in using vibrational spectroscopy for explosive detection (especially in the vapor phase) arises because of the combination of low vapor pressures and relatively low cross-section for absorption in the IR and the low scattering cross-section for Raman spectroscopy. For example, typical peak absorption cross-sections, α, for the NO_2 stretching modes are near 1×10^6 cm^2/mole in the IR (for comparison, peak UV absorption cross-sections for TNT approach 50×10^6 cm^2/mole). For Raman spectroscopy, scattering cross-sections in the UV may approach 1×10^{-2} cm^2/mole [3,7].

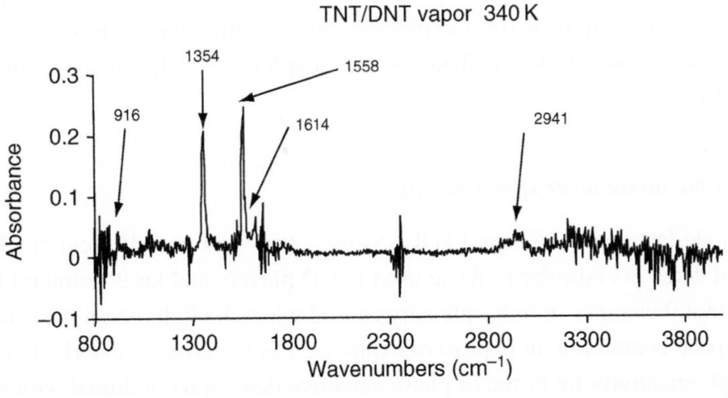

Fig. 2. The infrared absorption spectrum of vapor above solid mil-spec 2,4,6-trinitrotoluene (TNT) at 340 K. The spectrum is identical to that measured above solid DNT at the same temperature (measured by the authors).

Table 1. Wave number ranges and vibrational mode assignments for spectral features commonly observed (650–3100 cm^{-1}) in the infrared absorption spectra of explosives

Vibrational mode assignment	Explosive (type)	Wavenumbers (cm^{-1})
NO_2 deformation and ring stretch	Nitramine (RDX), TNT	650–850
Ring torsion	Nitramine (RDX),TNT	1000–1080
N–N stretch	Nitramine (RDX)	1200–1230
NO_2 symmetric stretch	Nitramine (RDX)	**1260–1320**
CH_2 bend	Nitramine (RDX), TNT	1300–1450
NO_2 asymmetric stretch	Nitramine (RDX),TNT	1450–1600
C–H stretch	Nitramine (RDX), TNT, nitrocellulose	2900–3100
N–O stretch	Nitrate ester (PETN)	850–950
C–C stretch	TNT	1620–1700
NO_2 symmetric stretch	TNT	**1325–1375**
NO_2 bend	Nitrocellulose	800–900
NO_2 symmetric stretch	Nitrocellulose	1200–1300
NO_2 asymmetric stretch	Nitrocellulose	1600–1700
C–O stretch	Nitrate ester (PETN)	1000–1040

Note: NO_2 symmetric stretches at 1260–1320 and 1325–1375, indicated in bold, are the strongest IR absorption features of explosive materials in the mid-infrared region of the spectrum.

2.1. Laser infrared absorption spectroscopy of explosives

Most solid explosives are composed of fairly large molecules with large inertial moments, causing their rotational energy levels to be closely spaced [8]. The IR absorption spectra of many neat solid explosives appear as fairly broad features resulting from the blending together of rovibrational lines corresponding to a given vibrational transition. The broad spectral features of many solid explosives (including their vapors) in the IR lend themselves to measurement by broadband techniques such as Fourier transform (FT) IR spectroscopy [9]. Laser-based methods of detection by IR absorption techniques are often limited by the bandwidth of the light source (this is not necessarily the case with Raman spectra of many neat solid explosives, and for this reason, Raman spectroscopy has been used extensively for analysis of solid explosives [10] – also see discussion in Section 2.2.).

2.1.1. Tunable diode laser spectroscopy

Tunable diode laser spectroscopy (TDLAS) uses mid- and near-IR semiconductor light sources and detectors (similar to those used in CD players and laser pointers) to measure (usually minute) changes in light intensity caused when the light beam passing through a region of space containing an explosive/explosive gas is partially absorbed. TDLAS can achieve high sensitivity by virtue of phase-sensitive detection combined with modulation techniques that discriminate against $1/f$ noise of the laser source [11]. Light sources are commercially available throughout the mid- and near-IR spectral region. Recent

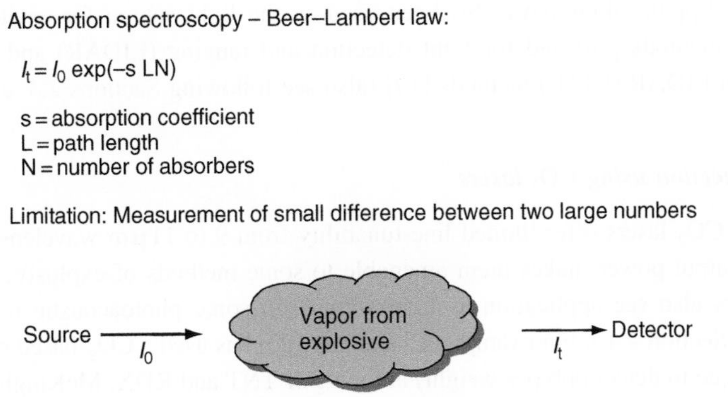

Fig. 3. A schematic representation of the application of the Beer–Lambert law to explosive vapor sensing.

developments in the last decade of quantum cascade (QC) and interband cascade (IC) lasers offer the promise of room temperature, continuous wave operation throughout the IR fingerprint region (3–16 μm) [12]. A schematic representation of the measurement process using the Beer–Lambert law is shown in Fig. 3.

Applications to explosive sensing using semiconductor light sources have been reviewed by Allen [13]. Traditional detection methods (Beer–Lambert law-type experiments) are somewhat limited because the broad spectral features of many neat explosive vapors make phase-sensitive detection methods difficult [9,11]. Because of this, TDLAS is often used to detect and measure light gases (e.g., NO and NO_2) produced by decomposition of the parent explosive. Kolb and Wormhoudt [14] used a near-IR diode laser to measure NO produced during pyrolysis of TNT in soils. Riris et al. [15] used a lead-salt TDL to detect NO_2 produced following thermal decomposition of 5–10 pg of RDX. TDLAS is often used in tandem techniques to detect explosive fragments produced by photofragmentation (PF) of parent explosive. For example, Bauer et al. [16] have used a QC mid-IR laser to detect NO produced by 1.55 μm laser PF of TNT and RDX.

2.1.2. Optical parametric oscillators

Optical parametric oscillators (OPOs) provide an alternative method of generating coherent radiation in the IR spectral region and may exhibit a broad tuning range. An OPO converts an input laser wave (ω_p, pump frequency) into two output waves of lower frequency (ω_s, signal frequency; ω_i, idler frequency) by means of nonlinear (usually crystal borne) optical interaction. The sum of the output wave frequencies is equal to the input wave frequency: $\omega_s + \omega_I = \omega_p$. Employing a nonlinear optical crystal for frequency conversion, quasi-phase-matching (QPM) may be accomplished by periodically changing the nonlinear optical properties of the crystal (periodical poling). For example, output wavelengths from 700 to 5000 nm can be produced in periodically poled lithium niobate (PPLN). Common pump sources are neodymium lasers at 1.064 or 0.532 μm. Effenberger [17] used an OPO-based system that used both idler (mid-IR output) and signal (near-IR output) in a differential absorption experiment and was able to detect explosive vapors to 1 ppm. OPOs have also been used as the light source for cavity-enhanced detection methods [18] and for light detection and ranging (LIDAR) and differential absorption LIDAR (DIAL) methods [19] (also see following Sections 2.3. and 4.5.).

2.1.3. Detection using CO_2 lasers

Although CO_2 lasers offer limited line-tunability from 9 to 11 μm wavelength, the significant output power makes them amenable to some methods of explosives detection. CO_2 lasers also see application as light sources in some photoacoustic measurement schemes (Section 4.4.). For example, Chaudhary [20] has used a CO_2-based photoacoustic technique to detect ppb (by weight) amounts of TNT and RDX. McKnight et al. [21] has used the acoustic pulse from a focused CO_2 laser employing different spot sizes to identify buried objects.

2.1.4. Difference frequency generation spectroscopy

Kim et al. [22] have used vibrational sum-frequency generation spectroscopy (SFG) to characterize the surfaces of β-HMX single crystals, as well as the interface between HMX and the copolymer Estane. SFG is a nonlinear vibrational spectroscopic technique, related to optical parametric amplification that selectively probes vibrational transitions at surfaces and interfaces. Compared with bulk HMX, the surface vibrational features are blueshifted and observed splittings are larger. The technique may have application to detection of explosive residues on surfaces.

2.1.5. Cavity ringdown spectroscopy

Cavity ringdown spectroscopy (CRDS) is a technique used to enhance measured absorption of light by a chemical species by greatly increasing the light path through the sample (Figs 3 and 4). This is achieved by placing the sample within an optical cavity that uses two highly reflective mirrors to create a stable optical resonator, such that the alignment of the reflective mirrors on each end of the cavity serves to 'trap' light within the cavity. When a pulse of light enters the cavity, it can make thousands of round trips before its intensity dies off, resulting in effective path lengths of kilometers. The decrease in intensity with time is measured by allowing a small amount of light to leak through one of the mirrors to impinge on a fast photodetector. The decrease in intensity with time is called the 'ringdown time', and contributions to the ringdown time by species absorption of light may be readily separated from other causes of loss of intensity (scattering, mirror imperfection, etc.). A scan of ringdown time versus wavelength can yield the absorption spectrum of a species present in extremely low concentrations. This is shown schematically in Fig. 4.

Busch et al. [23] have reviewed the technique and applications of CRDS to trace sensing to 1999. Two reviews [24,25] have provided a review of applications to explosives. Dagdigian [26] has written a review of optical methods, including CRDS, employed for

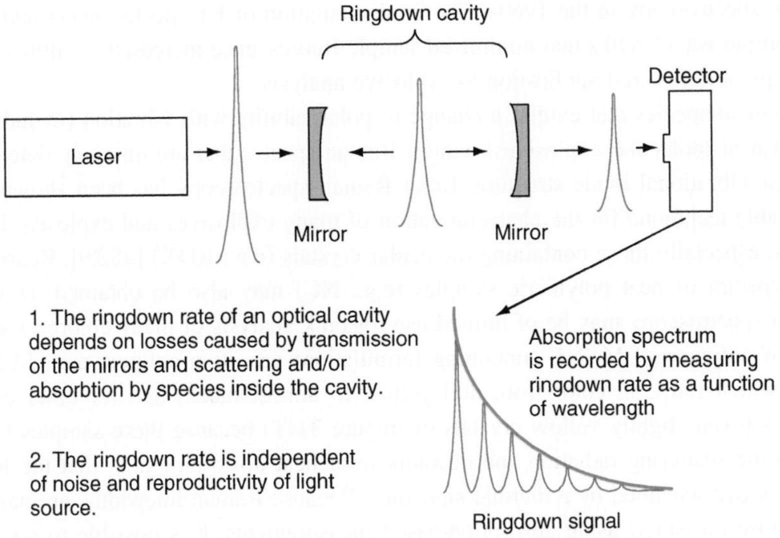

Fig. 4. A description of a cavity ringdown spectroscopy (CRDS) apparatus.

detection of decomposition of explosives and energetic materials (EMs). Todd et al. [18] have used a mid-IR OPO to measure vapor-phase mid-IR spectra of common explosives (TNT, TATP, RDX, PETN, and Tetryl) using the CRDS technique. Parts per billion concentration levels were detected with no sample pre-concentration. A collection/flash-heating sequence was implemented to enhance detection limits for ambient air sampling. Detection limits were not determined but were expected to approach 75 ppt for TNT, with similar concentration levels for the other explosives.

2.2. *Laser Raman spectroscopy*

Analytical techniques based on Raman spectroscopy have been widely used for explosive detection and characterization. Following theoretical prediction of inelastic light scattering in transparent media [27], the effect was experimentally verified in liquids by Raman in 1928, and the phenomenon is known as the Raman effect. The first demonstration of the Raman effect in gases was demonstrated by R. W. Wood (HCl gas) and F. Rasetti (CO and CO_2). The frequencies observed in Raman scattering correspond to the frequency of the incident light shifted by some characteristic frequency of the scattering molecule. The difference in energy between the incident and scattered photons (the Raman spectrum of the molecule) is typically a function of the vibrational energy levels within a molecule.

Before invention of the laser, Raman spectroscopy relied on arc lamps to provide incident light, and long periods of exposure were necessary to record a spectrum. The advent of high intensity, monochromatic laser radiation generated renewed interest in

Raman spectroscopy in the 1960s. The implementation of FT spectrometers and lasers with output wavelengths that minimized sample fluorescence increased the utility of the technique and fostered application to explosive analysis.

Chemical species that exhibit a change in polarizability with vibration (including all known neat molecular explosives) exhibit Raman spectra that are uniquely determined by their vibrational mode structure. Laser Raman spectroscopy has been shown to be a valuable technique for the characterization of many explosives and explosive formulations, especially those containing molecular crystals (e.g., RDX) [28,29]. Reasonably good spectra of neat polymeric samples (e.g., NC) may also be obtained. However, Raman spectroscopy may be of limited use for bulk analysis of many colored formulations of polymeric EM (NC containing formulations are often colored, e.g., JA2, M9, M30, which range in color from dull yellow to almost black) and for other colored samples (even slightly yellow crystals of impure TNT) because these samples tend to absorb the scattering radiation and decompose or heat up to an extent that the Raman signal is overwhelmed by a thermal signature. Because Raman linewidths are narrower than those measured using absorption-based measurements, it is possible to see slight impurities in samples and in mixtures of many explosive materials. Laser Raman spectra of the explosive formulation C-4, its main ingredient RDX, and samples of RDX of different origin are shown in Fig. 5 [10].

The first review of laser Raman spectroscopy of explosives appeared in the late 1960s [30]. A patent application for laser Raman applied to the remote identification of hazardous gases from explosives decomposition was filed a few years later [31].

Because many explosive samples fluoresce when exposed to visible laser radiation, the use of Raman spectroscopy for explosive analysis was accelerated by the development of long-wavelength laser sources and FT-Raman techniques. Beginning in the late 1970s, FT-Raman spectroscopy began to see use for analysis of propellants and energetics characterization [32]. Also around this time, coherent anti-Stokes Raman spectroscopy (CARS) began to be developed for explosive analysis [33]. The use of laser Raman spectroscopy for the trace identification of EM was first reported by Carver, with limits of detection of 1 ng or less for RDX, PETN, and TNT [34]. Trott et al. [35] reported single-pulse Raman studies of the solid explosive triaminotrinitrobenzene (TATB).

As laser FT-Raman spectroscopy found wider use, Raman spectroscopy in general found wider application in the study of EM. Survey FT-Raman spectra of most common explosives and propellants were reported in the early 1990s [28,29,36]. CARS for real-time diagnostics of explosions was first reported in 1991 [37]. Hare et al. [38] used picosecond Raman spectroscopy to study energy transfer in shock-initiated explosives. Gupta [39] used fast time-resolved Raman spectroscopy to study the flow of vibrational energy behind a shock wave in nitromethane. Also around this time, the use of the Renishaw Raman microscope for explosive detection and analysis was first reported [40]. Limits of detection for most explosives studied were in the picogram range.

During this time, there was an increasing effort to employ Raman spectroscopy as a tool in a field deployable explosives detector [41]. Demonstration of enhancement of the Raman signal as the wavelength of the incident light approaches the wavelength

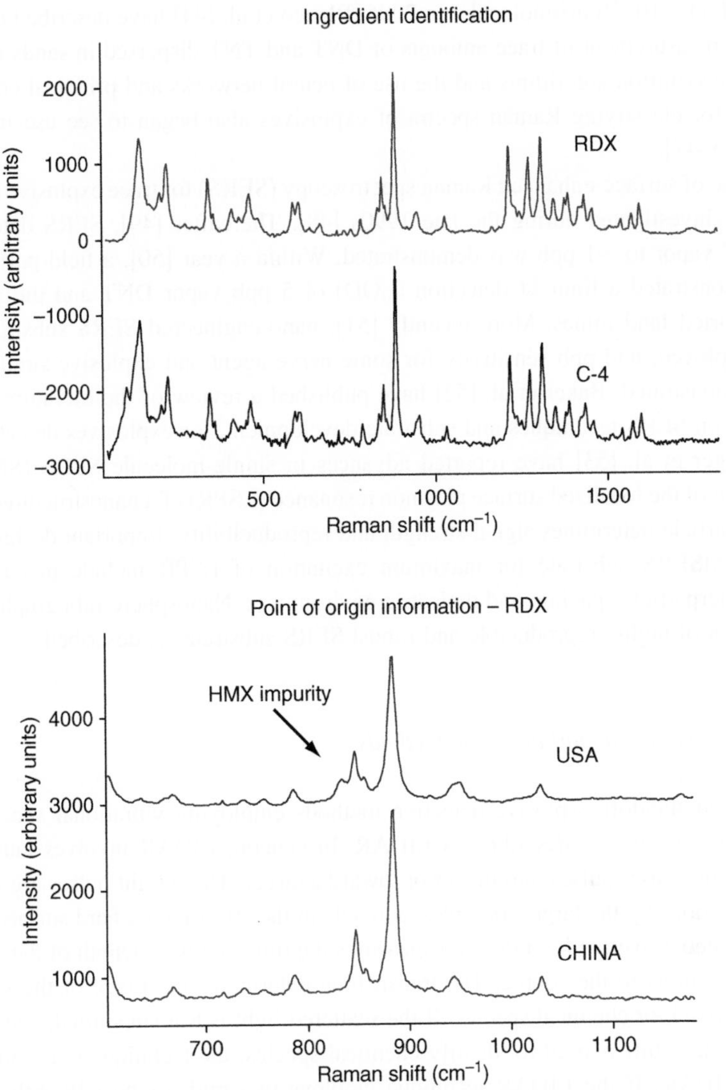

Fig. 5. Laser Fourier transform (FT)-Raman spectra of the explosive formulation C-4, its main ingredient RDX, and samples of RDX of different origin. The exciting laser wavelength was 1064 nm [10].

of an allowed transition (resonance Raman spectroscopy, RRS), and the elimination of fluorescence when using incident radiation near 244 nm was first reported for explosives in 1997–1998 [42,43]. Raman measurements on dilute TNT and DNT solutions in acetonitrile with 248 nm laser excitation have shown that for the $-NO_2$ stretching Raman modes, there is significant enhancement [7]. For TNT, the Raman cross-section of the 1351.6 cm^{-1} $-NO_2$ stretching mode is 5.18 $(\pm 0.4) \times 10^{-26}$ cm^2/molecule, and for 2,4-DNT the Raman cross-section of the 1351.3 cm^{-1} $-NO_2$ stretching mode is

$7.25\ (\pm 1.1) \times 10^{-26}\,cm^2$/molecule. In 2006, Blanco et al. [44] have described the use of RRS for measurement of trace amounts of DNT and TNT dispersed in sands and soils. Pattern recognition algorithms and the use of neural networks and principal component analysis for classifying Raman spectra of explosives also began to see use in the late 1990s [45–47].

The use of surface-enhanced Raman spectroscopy (SERS) for trace explosive detection was first investigated during the late 1990s [48]. Thereafter [49], SERS detection of 2,4-DNT vapor to ~1 ppb was demonstrated. Within a year [50], a field-portable unit had demonstrated a limit of detection (LOD) of 5 ppb vapor DNT and the ability to locate buried land mines. More recently [51], nano-engineered SERS substrates have been employed, and ppb sensitivity for some nerve agent and explosive simulants has been demonstrated. Baker et al. [52] have published a review of the literature in 2005, focusing on SERS techniques and substrate development, for explosives detection.

Dieringer et al. [53] have reported advances in single-molecule SERS (SMSERS). Excitation of the localized surface plasmon resonance (LSPR) of a nanostructured surface or nanoparticle determines signal strength and reproducibility. Important design criteria of the SMSERS substrate for maximum excitation of LSPR include material, size, shape, interparticle spacing, and dielectric environment. Nanosphere lithography for the fabrication of highly reproducible and robust SERS substrates is described.

2.3. *Laser-based standoff detection methods*

Laser-based standoff explosive detection methods employing vibrational spectroscopy are usually of the type described as LIDAR. In general, LIDAR involves launching a series of short laser pulses into the air or toward a target. These light pulses are scattered in all directions by the target (particles, aerosols in the atmosphere, hard surface targets, etc.). A gated, sensitive light detector measures the time and wavelength of the scattered light that returns to the source. The transit time determines the range of the scattering object, aerosol, or chemical species. If the scattered light is Raman shifted, and analysis of the Raman shift is used to identify chemical species, the technique is usually called Raman LIDAR. If the LIDAR technique employs two mid- or near-IR pulsed lasers and determines the identity of chemical species by measuring the differential absorption between two pulses with similar transit times, the technique is usually called DIAL.

LIDAR analysis grew out of NASA programs on remote winds and aerosols measurements of the late 1960s. The potential for performing remote Raman analysis of species with visible laser excitation in the atmosphere was explored as early as 1973 [54,55] and of bulk and surface materials starting in the 1990s [56,57]. Sharma et al. [58] have measured Raman spectra of TATB and HMX at a distance of 10 m (Fig. 6). Standoff Raman spectroscopy has been demonstrated for detection of organic materials at ranges of 100 m using 532 nm [59] and 500 m using a UV laser and large collection optics [60].

As mentioned previously, in addition to the $(1/\lambda^4)$-increase of the Raman scattering cross-section with UV excitation, further enhancement can also occur when the

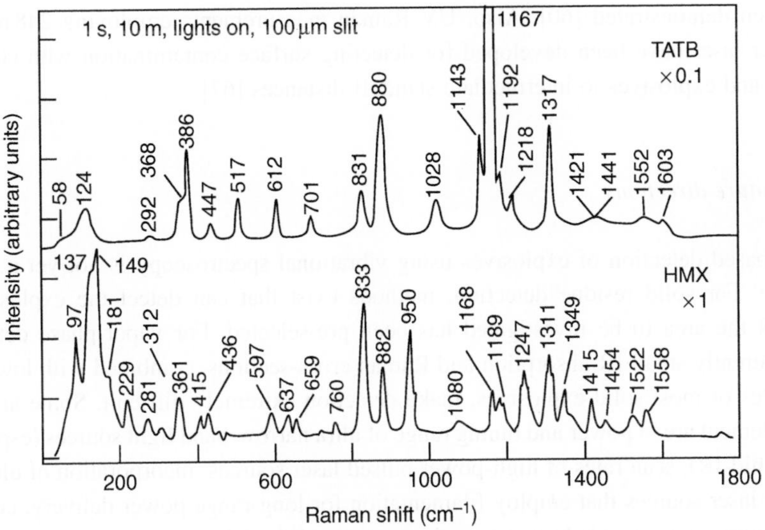

Fig. 6. Remote Raman spectra of explosives triaminotrinitrobenzene (TATB) and HMX at a standoff distance of 10 m (used with permission from [58]).

excitation frequency approaches an electronically excited state of the molecule [7,61]. This resonance Raman enhancement can range from factors of 2–3 orders of magnitude. Fluorescence is suppressed because the onset typically occurs at excitation wavelengths above ~270 nm [62,63]. Resonance Raman detection is in the solar-blind region; so, RR UV LIDAR can be operated both during day and nighttime. In the case of Raman LIDAR, the signal will be attenuated because of large absorption due to ozone and higher molecular scattering at 248 nm [64].

Carter et al. [65] have used a standoff Raman system for detecting explosive materials at distances up to 50 m in ambient light conditions. In the system, light is collected using an 8-in. Schmidt–Cassegrain telescope fiber-coupled to an $f/1.8$ spectrograph with a gated intensified charge-coupled device detector. A frequency-doubled Nd:YAG (532 nm) pulsed (10 Hz) laser is used as the excitation source for measuring remote spectra of samples containing 8% explosive materials. The explosives RDX, TNT, and PETN as well as nitrate- and chlorate-containing materials were used to evaluate the performance of the system. Laser power and detector gate width studies were performed to determine the effects of laser heating and photodegradation (significant for TNT residues) and to evaluate performance in high levels of ambient light (e.g., sunlight).

Schultz et al. [12] have investigated monostatic and bistatic frequency modulated (FM) DIAL at standoff distances up to 2.5 km. This group is also evaluating miniature QC laser transmitters for multiplexed chemical sensing. Vaicikauskas [19] has used IRDIAL capable of sensing pollutant gases at distances up to several kilometers.

Ultraviolet mini-Raman LIDAR for standoff, in situ identification of chemical surface contaminants has been reported [66]. Using semi-portable equipment, UV Raman spectroscopic identification of bulk organic compounds at distances of over half a kilometer has been demonstrated [60]. Also, UV Raman measurements employing 248 nm KrF excimer laser have been developed for detecting surface contamination with chemical agents and explosives to intermediate standoff distances [67].

2.4. Future directions

Laser-based detection of explosives using vibrational spectroscopy is not yet a mature science. For solid residue detection, methods exist that can detect the explosive, as long as the area to be investigated has been pre-selected. For vapor-phase detection, the inherently small IR absorption and Raman cross-sections, combined with low vapor pressures of most solid explosives, make detection extremely difficult. Some areas for improvement are in power and tuning range of ultra narrow-band light sources (especially in the mid-IR), scan rates of high-power pulsed laser sources, incorporation of ultra-fast (fs, as) laser sources that employ filamentation for long-range power delivery, compact high-power sources to help move equipment out of the laboratory, quantum-control of laser pulses for enhancement of absorption cross sections, new modulation techniques, low power consumption fast detectors, improved spectrographs, and rapid pre-selection of areas with a high probability of explosives contamination.

3. Laser-induced breakdown spectroscopy (LIBS)

A relatively new method for optically detecting explosives is LIBS. LIBS is an atomic emission spectroscopy technique used for the real-time, nondestructive determination of elemental composition and requires no sample preparation. The technique relies on a microplasma created by a focused laser pulse, typically several nanoseconds in length, to dissociate molecules and particulates within the plasma volume. The subsequent emission can be resolved spectrally and temporally to generate a spectrum containing emission lines from the atomic, ionic, and molecular fragments created by the plasma. A single laser shot and subsequent data analysis can take place in under a second. The basic LIBS experimental setup is shown in Fig. 7. Typically, a Nd:YAG laser is used to produce a pulse width of a few nanoseconds. The optical system is used to focus the laser pulse. For a typical setup, the laser power at the focal point is $>1\,GW/cm^2$ to produce a microplasma. Emission from the microplasma is then collected by a series of lenses and delivered to a spectrometer to resolve the collected light. Finally, the spectrally resolved light arrives at a detector to generate a LIBS spectrum. An example is shown in Fig. 8.

In the past, LIBS has been primarily used to analyze one or a few elements, mostly metals [68–70]. More recently, with the advent of high-resolution, broadband spectrometers, the capability of LIBS to identify compounds could be realized. Every element on the periodic Table has atomic emission lines that emit in the visible spectrum. A broadband spectrometer allows one to capture all of the elements in the sample interrogated by the laser-generated plasma, provided they are present in sufficient abundance. Instead

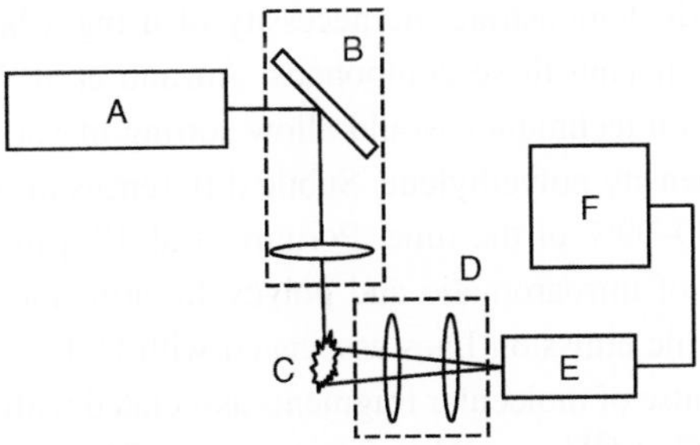

Fig. 7. Block diagram of laser-induced breakdown spectroscopy experimental setup (A) pulsed laser, (B) focusing optics, (C) microplasma, (D) collection optics, (E) spectrometer, and (F) data analyzer.

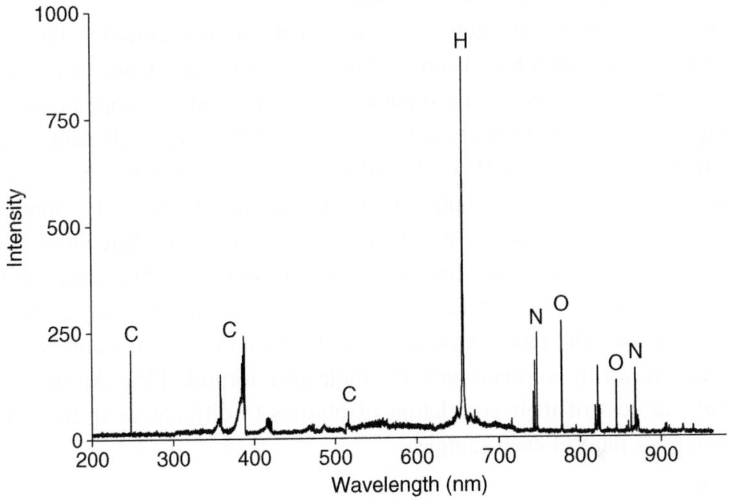

Fig. 8. Single-shot laser-induced breakdown spectroscopy (LIBS) spectrum of type I plastic collected from a 320 mJ, 8 ns laser pulse.

of concentrating on a small portion of a LIBS spectrum, all the emitting elements in a sample can be observed in a full broadband LIBS spectrum. Thus, the presence of atomic emission lines and the relative intensity of the atomic emission lines to one another can be used to identify the sample. Now, LIBS can more readily be applied to various materials beyond metals, including plastics and other organic compounds, biological materials, and other hazardous compounds [71–77]. The carbon, hydrogen, oxygen, and nitrogen atomic emission lines are commonly used to identify organic compounds. Figure 8 is an example of a LIBS spectrum of plastic. A prominent carbon atomic emission line is located at 247 nm and a hydrogen atomic emission line, the H_{α} line, at 656 nm. Oxygen atomic emission lines observed due to neutral oxygen (O I) can be observed at 777 and 845 nm. Nitrogen atomic emission lines observed due to neutral nitrogen (N I) can be observed at 744, 746, and 868 nm. The atomic emission lines of elements associated with organic compounds demonstrate the necessity of using a broadband spectrometer (200–1000 nm) for identifying these compounds. Anzano et al. [71] used LIBS to see whether linear correlation techniques would allow sorting of various plastics, including polystyrene and high-density polyethylene. Subtle differences in intensities allowed successful identification 90–99% of the time. Portnov et al. [75] used LIBS to investigate the spectral signatures of nitroaromatic and polycyclic aromatic hydrocarbon samples. They observed the atomic emission lines associated with C, H, N, and O, but they also observed emission because of molecular fragments associated with the CN ($B^2\Sigma - X^2\Sigma^+$) violet system and the C_2 ($d^3\Pi g - a^3\Pi u$) Swan system. These fragments were used to successfully show differences between the compounds studied. Ferioli et al. [73] have used LIBS to study hydrocarbon air mixtures (C_3H_8, CH_4, and CO_2 in air). The strength of the C, N, and O atomic emission lines is investigated in relation to the concentration of carbon and hydrogen present in the samples.

In addition to using the absolute intensities of the atomic emission lines, the peak intensity ratios of these lines have been used to analyze samples. Tran et al. [77] analyzed the atomic intensity ratios of several organic compounds with the hope to determine the empirical formula of a compound based on the ratios from several elements. Calibration curves were built based on C:H, C:O, and C:N atomic emission ratios from various compounds that covered a wide range of stoichiometries. Then, four compounds with known stoichiometries were tested against the calibration curves. The ratios determined from the calibration curves were compared with the actual stoichiometries and showed accuracy of 3% on average. In the study of nitroaromatic and polycyclic aromatic hydrocarbon samples, the ratios between C_2 and CN and between O and N of different samples were shown to correlate with the molecular formula [75]. Anzano et al. [71] also attribute success of their correlation of plastics to differences in the C/H atomic emission intensity ratio of each sample.

3.1. LIBS of explosives

LIBS has been shown to be a successful avenue for detecting organic material and in some cases determining the type of organic material. Applying LIBS to energetic organic material detection and identification is of interest for various applications; including force protection, security concerns, forensic analysis, etc. The success of LIBS for identifying organic compounds based on atomic emission intensity ratios lead researchers at the US Army Research Laboratory (ARL) to investigate characteristics of LIBS spectra of explosive compounds. LIBS spectra were collected from various explosive materials, including highly purified RDX, HMX, TNT, PETN, and NC as well as operational explosive and propellants; C-4, A-5, M-43, LX-14, and JA2 [78]. All of the atomic lines expected – carbon, hydrogen, nitrogen, and oxygen – are present. They also discuss how an energetic organic compound might be discriminated from other materials by using the oxygen to nitrogen atomic emission intensity ratio. The scheme to identify energetic compounds versus nonenergetic organic compounds is rooted in the observation that EM have larger amounts of oxygen and nitrogen relative to carbon and hydrogen. For example, the chemical formulas of RDX and TNT are $C_3H_6N_6O_6$ and $C_7H_5N_3O_6$, respectively. Some potential interferents include super glue, $C_{10}H_{10}O_4N_2$, plastics, repeating hydrocarbon chains of some sort, or nylon $(C_{12}H_{22}O_2N_2)_n$. As can be seen, oxygen and nitrogen are in much greater abundance in the explosives relative to the carbon and hydrogen. To identify EM, in addition to identifying the carbon, hydrogen, oxygen, and nitrogen atomic emission lines, it is necessary to track the oxygen and nitrogen atomic emission intensity lines relative to the carbon and hydrogen atomic emission intensity lines.

The following example illustrates how LIBS spectra may be applied to explosive detection. RDX was dissolved in acetone and then applied to an aluminum substrate. A thin residue of RDX was left after the acetone evaporated. LIBS spectra collected of RDX on aluminum and plain aluminum are shown in Fig. 9. No carbon or hydrogen atomic emission line is present in the aluminum sample. LIBS spectra were also collected with the sample under argon to eliminate the nitrogen and oxygen contribution from atmosphere. The composition of atmosphere is ~80% nitrogen and ~20% oxygen, a 1:4 oxygen to nitrogen ratio. In RDX ($C_3H_6N_6O_6$), the oxygen to nitrogen stoichiometric ratio is 1:1. In the LIBS spectra in Fig. 9, the intensity ratio of oxygen atomic emission line intensity to nitrogen atomic emission line intensity from RDX in air is ~2. The intensity ratio of oxygen to nitrogen from RDX in argon is ~5. The increase in the oxygen to nitrogen intensity ratio demonstrates the ability of LIBS to track relative amounts of elements in a sample. In this case, the increase of oxygen relative to nitrogen

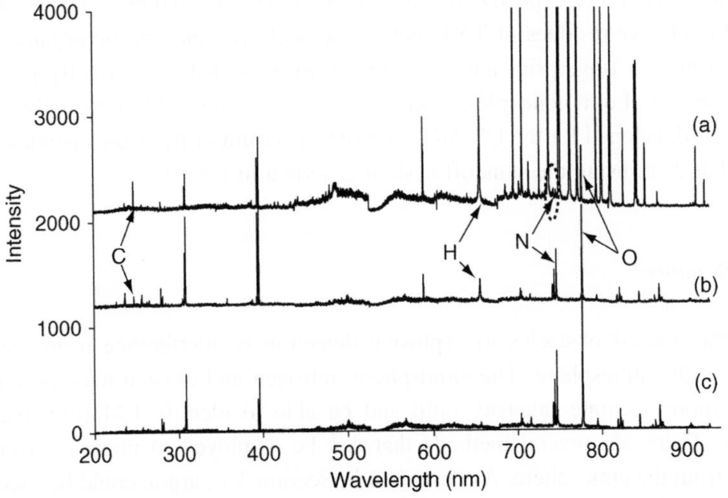

Fig. 9. Single-shot laser-induced breakdown spectroscopy (LIBS) spectra collected with man-portable LIBS system of (a) RDX residue on aluminum under argon, (b) RDX on aluminum in ambient atmosphere, and (c) plain aluminum in ambient atmosphere. Carbon (C), Hydrogen (H), Nitrogen (N), and Oxygen (O) are shown for the RDX samples.

when air is displaced by argon shows the oxygen and nitrogen atomic emission line intensities are entirely due to RDX and not from atmospheric oxygen and nitrogen.

3.2. *LIBS systems*

The standard laboratory LIBS system has been described in Fig. 7. Owing to the simplicity of the setup, different types of configurations can be used depending on the application. In particular, systems can be developed for field use [72,79]. Small rugged lasers and spectrometers can be used for portable instruments. LIBS biggest impact may be in the field. At the US ARL, a man-portable instrument was recently developed [72]. A small Nd:YAG laser is contained in a hand-held wand. The focusing optics and collection optics are also in the wand. The collected light is delivered to a backpack spectrometer through a fiber optic cable. Then, the data are analyzed by an onboard computer and displayed in a heads up display. Various LIBS spectra have been collected using the unit ranging from plastic land mine casings to liquid chemical warfare simulants [72]. In Fig. 9, the LIBS spectrum of the RDX residue on aluminum was collected by the man-portable LIBS system. The carbon, hydrogen, oxygen, and nitrogen are all from the RDX residue because the sample is taken under argon.

Another configuration for LIBS in the field is a standoff system capable of sampling tens of meters away from the instrument. In this case, a higher-power laser is used in conjunction with telescopic optics that focus the laser and collect the light from the plasma. Several optical configurations to achieve telescopic focusing and collecting for standoff operation have been used [80–83]. Standoff LIBS has been used for various applications, most typically elemental analysis. Recently, a standoff system was used to collect LIBS spectra of explosive residues at 30 m [81]. A blind test was performed that included explosive residues of TNT and C-4 as well as nonenergetic organic materials such as a human fingerprint and car paint. Post-data analysis correctly predicted an explosive material or a nonexplosive material six of six times. More recently, a standoff system was developed for the US ARL. A LIBS spectrum of RDX on a painted substrate collected at 20 m using the standoff system is shown in Fig. 10.

3.3. *LIBS future*

One of the biggest obstacles to explosive detection is interference from oxygen and nitrogen in the atmosphere. The atmospheric nitrogen and oxygen must be diminished to get a more accurate intensity ratio and be able to identify EM from background materials. There are several methods that can be employed to minimize oxygen and nitrogen from the atmosphere. As described in Section 3.1., argon could be used to blow across the surface and displace atmospheric oxygen and nitrogen. While this method is effective for close contact studies, it cannot be used with standoff applications. Two methods that may have promise for explosive detection at a standoff distance are double-pulse LIBS and femtosecond LIBS. Double-pulse LIBS involves two pulses aligned

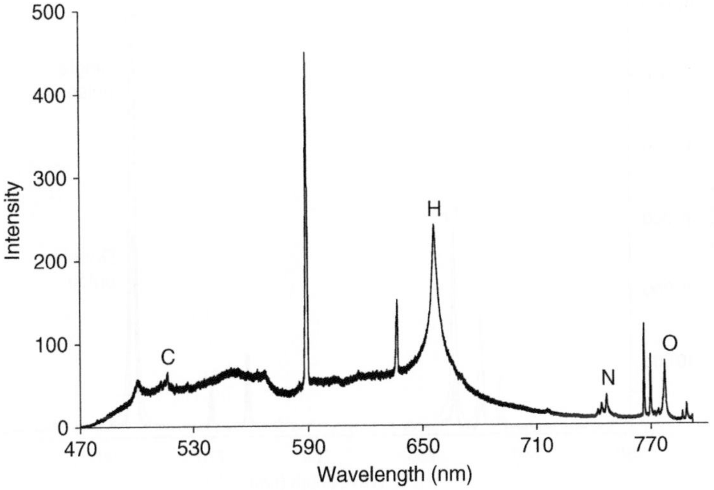

Fig. 10. Laser-induced breakdown spectroscopy (LIBS) spectrum of RDX collected at 20 m with ARL standoff LIBS system. The elements present due to RDX are labeled.

collinearly, separated by a few microseconds, interacting with the sample. Advantages of double-pulse LIBS include an increased signal and better reproducibility from shot to shot [84–87]. The reason for the increased signal has been attributed to several factors, including greater mass ablation, a wider region of high temperature in the plasma; so, more atoms are excited and less laser shielding because of a decrease of gas density [84]. It is the last reason for the increase in signal that double-pulse LIBS may be applied to explosive detection with great effect. When the first pulse hits, it impacts the sample and displaces the surrounding gas. The second pulse arrives and interacts with the material within the first plasma. Therefore, the influence from atmospheric oxygen and nitrogen will be decreased. In the case of RDX, the oxygen to nitrogen ratio should increase going from single-pulse to double-pulse LIBS. In Fig. 11, the oxygen and nitrogen region (~725–800 nm) from a double-pulse LIBS spectrum of RDX and a single-pulse LIBS spectrum of RDX is compared. The oxygen to nitrogen peak intensity ratio is larger for the double-pulse spectrum. The double-pulse method may be effective for utilizing standoff LIBS for explosive detection. The standoff system developed for ARL has double-pulse capability; the LIBS spectrum in Fig. 10 is collected from a plasma generated by a double-pulse laser at a standoff distance of 20 m.

The other method that could be employed for standoff LIBS is using femtosecond lasers. Traditionally, LIBS is performed with nanosecond laser pulses. Femtosecond pulses have been shown to have advantages over nanosecond pulses for LIBS applications [88–90]. Most importantly, because of the shorter time ($\sim 10^{-15}$ versus 10^{-9} s), all of the energy is deposited into the sample and not the surrounding atmosphere. With the longer pulse, the tail end of the laser reheats the plasma, thus leading to the air entrainment within the plasma and contributing to oxygen and nitrogen signal. The femtosecond pulse

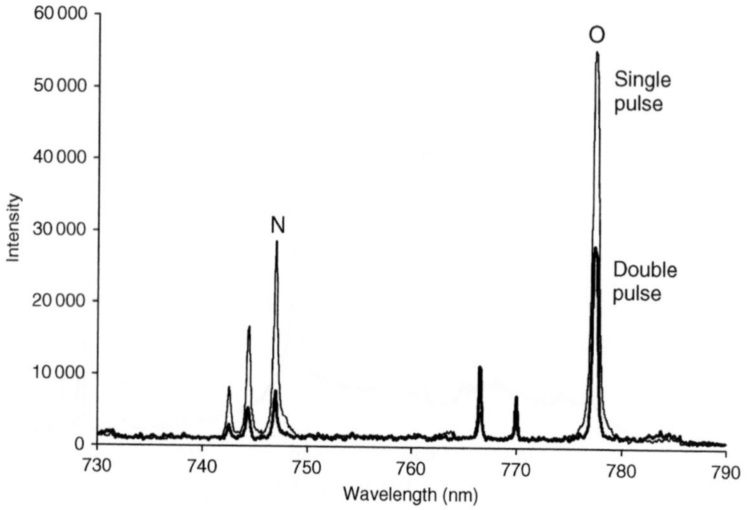

Fig. 11. Comparison of laser-induced breakdown spectroscopy (LIBS) spectra of RDX using the double-pulse configuration and the single-pulse configuration. Two 160-mJ pulses were separated by 2 μs for the double-pulse spectrum. One 320-mJ pulse was used to collect the single-pulse spectrum. The O to N intensity ratio for the double pulse and the single pulse is 4 and 2, respectively.

only deposits energy into the sample, thus minimizing the oxygen and nitrogen from atmosphere. Femtosecond LIBS for explosive detection is in the beginning stages [91]. As femtosecond lasers become more mature, performing femtosecond LIBS in the field becomes more plausible. More fundamental studies will be needed in the laboratory to achieve better understanding of the femtosecond pulse material interaction for LIBS.

3.4. LIBS summary

There are concerns with LIBS that still need to be overcome to make explosive detection more reliable. As mentioned previously, interference from atmospheric oxygen and nitrogen is a problem when trying to detect explosives. In addition, the plasma also samples everything around the explosive, such as the surface material an explosive rests on as well as any environmental contamination, i.e., dust, oil. Designing systems that can preferentially sample a residue can help eliminate surrounding interference. The double-pulse LIBS minimizes the contribution from the surrounding atmosphere. A different wavelength may allow more absorption of the laser energy into the residue as apposed to the underlying surface. For the standoff application, eye safety is a concern. One way to alleviate this problem is to use eye-safe laser wavelengths, such as 1.5 μm. The laser power at the focus can still cause eye damage, but the majority of the laser path length and any reflections of the beam will be in the eye-safe region. The LIBS technique also suffers from poor reproducibility and sensitivity. The poor reproducibility can be caused by a number of things, including pulse-to-pulse variability, differences in the laser coupling with the sample, and sample heterogeneity. Some of these issues can be solved by using lasers with better pulse-to-pulse uniformity or even move to femtosecond pulses that have been shown to increase reproducibility [88–90]. While LIBS has been shown to be able to detect explosive residues, the LOD has not yet been established. In general, most LODs are around ppm. However, this is highly dependent on the surrounding environment. There are several things that can be done to enhance the LIBS signal. The double-pulse method and/or using femtosecond pulses have been shown to increase signal intensity. Also more efficient spectrometers and detectors are being developed to maximize light collection. Even with these issues still in need of resolution, LIBS has many advantages as an explosive detection technique. The ability to design field-ready instruments is a major advantage. LIBS requires no sample preparation and can perform analysis on solid samples, making it extremely attractive as an explosive detector in the field. Most explosive detection techniques require the explosive to be in vapor phase, limiting sampling capability because most military-grade explosives have low vapor pressures (RDX is ~ppt). As there is no sample preparation, there is no waste or cost generated by consumables. In addition, a spectrum can be collected in real time allowing instantaneous detection of a potential threat. The simple components allow different types of system configurations, leading to development of man-portable systems for point detection or larger standoff systems for detecting explosives at a distance. These attributes make the use of LIBS for explosive detection a promising technique.

4. Other laser-based methods for explosive detection

In addition to the IR, Raman and LIBS methods previously discussed, a number of other laser-based methods for explosives detection have been developed over the years. The following section briefly describes the ultraviolet and visible (UV/vis) absorption spectra of EM and discusses the techniques of laser desorption (LD), PF with detection through resonance-enhanced multiphoton ionization (REMPI) or laser-induced fluorescence (LIF), photoacoustic spectroscopy (PAS), variations on the light ranging and detecting (LIDAR) method, and photoluminescence. Table 2 summarizes the LODs of several explosive-related compounds (ERC) and EM obtained by the techniques described in this section.

4.1. UV/vis absorption

Typically, the absorption spectra of EM are obtained to determine the optimal wavelength for laser ignition with the minimal pulse energy. When attempting to selectively ablate an explosive residue on the surface of a substrate for LIBS detection, the ablation laser should ideally be at a wavelength that is efficiently absorbed by the explosive – otherwise the laser energy passes through the residue and is absorbed by the substrate. To determine the most desirable laser wavelength for trace residue detection,

Table 2. Detection limits for explosive-related compounds (ERC) and energetic materials (EM)

EM/ERC	Technique	Reference	Phase	LOD
Nitroglycerin (NG)	PAS, 6 μm, 9 μm, and 11 μm	[142]	Vapor	0.28 ppb
	PAS, 9.6 μm	[145]	Vapor	0.23 ppb
Nitrobenzene (NB)	REMPI/TOF, 226.3 nm	[101]	Vapor	Sub-attomole
	REMPI/TOF, 226 nm (100 °C)	[108]	Vapor	2.4 ppm
	PF-REMPI, 193 nm	[113]	Vapor	0.49 ppm
	LP-LIF, 222–272 nm (10-100 Torr air)	[123]	Vapor	~500 ppb
	SPI-TOF-MS, 118.2 nm	[107]	Vapor	17–24 ppb
Dinitrobenzene (DNB)	LP-LIF, 248 nm (100 Torr, 500 Torr air)	[124]	Vapor	11–13 ppb
Ethylene glycol	PAS, 6 μm, 9 μm, and 11 μm	[142]	Vapor	1.5 ppm
Dinitrate (EGDN)	PAS, 9.6 μm	[145]	Vapor	8.26 ppb
Cyclotrimethylene	IRMPD-LIF, CO_2 laser + 280 nm (150 °C)	[119]	Vapor	–
Trinitramine (RDX)	REMPI/TOF, 226 nm (100 °C)	[108]	Vapor	8 ppb
	PF-REMPI (with electrodes), 227 nm (1 atm, air)	[103]	Vapor	7 ppb
	SPF-REMPI, 248 + 226 nm (1 atm, 298 K)	[104]	Solid	~14 ng cm^{-2}
	PF-LIF, 227 nm (1 atm, air)	[103]	Vapor	ND
	Pyrolysis-LIF, 227 nm (1 atm, air)	[103]	Vapor	~1.6 ppm
	PAS, 5.8–6.7 μm	[141]	Vapor	~ppb
	PAS, 9.6 μm, 10.6 μm	[147]	Solid	–
Cyclotetramethylene tetranitramine (HMX)	IRMPD-LIF, CO_2 laser + 280 nm (180 °C)	[119]	Vapor	–
Nitrotoluene (NT)	R2PI, 266 nm (1 atm, He)	[112]	Vapor	–
	R2PI, 213 nm (1 atm, He)	[112]	Vapor	–
	PF-REMPI, 193 nm	[113]	Vapor	0.10–0.12 ppm
	PF-REMPI, 226 nm	[113]	Vapor	15–36 ppm
Dinitrotoluene (DNT)	UV CRDS	[94]	Vapor	<1 ppb
	R2PI, 266 nm (1 atm, He)	[112]	Vapor	ND
	R2PI, 213 nm (1 atm, He)	[112]	Vapor	–
	LP-LIF, 248 nm (100 Torr, 500 Torr air)	[125]	Vapor	2.7–3.7 ppb
	SPI-TOF-MS, 118.2 nm	[107]	Vapor	~40 ppb
	PAS, 5.8–6.7 μm	[141]	Vapor	~ppb
	PAS, 6 μm, 9 μm, and 11 μm	[142]	Vapor	16 ppm
	PAS, 9.6 μm	[145]	Vapor	0.50 ppb

Table 2. (Continued)

EM/ERC	Technique	Reference	Phase	LOD
Trinitrotoluene (TNT)	UV CRDS	[93]	Vapor	<1 ppb
	R2PI, 266 nm (1 atm, He)	[112]	Vapor	ND
	REMPI/TOF, 226 nm (100 °C)	[108]	Vapor	24 ppb
	PF-REMPI, 193 nm	[113]	Vapor	0.21 ppm
	PF-REMPI, 226 nm	[113]	Vapor	1.7 ppm
	PF-REMPI (with electrodes), 227 nm (1 atm, air)	[103]	Vapor	70 ppb
	LIF, 226 nm	[121]	Vapor	~40 ppb
	LIF (2 μJ, 473 K), 226 nm	[122]	Vapor	4 ppm
	PF-LIF (1 atm, air)	[103]	Vapor	37 ppm
	PF-LIF, 248 nm (24 °C, 1 atm air, 15 cm)	[127]	Vapor	<8 ppb
	PF-LIF, 248 nm (28 °C, 1 atm air, 2.5 m)	[128]	Vapor	<15 ppb
	PAS, 5.8–6.7 μm	[141]	Vapor	~ppb
	PAS, 9.6 μm, 10.6 μm	[147]	Solid	–
	Fluorescence LIDAR (close-contact)	[148]	Aqueous	1 ppm
	Fluorescence LIDAR (500 m)	[148]	Aqueous	100 ppm
Pentaerythritol	PF-REMPI (with electrodes), 227 nm (1 atm, air)	[103]	Vapor	2 ppb
tetranitrate (PETN)	PF-REMPI (with electrodes), 454 nm (1 atm, air)	[103]	Vapor	16 ppm
	PF-LIF, 227 nm (1 atm, air)	[103]	Vapor	ND
	PF-LIF (with pyrolysis), 227 nm (1 atm, air)	[103]	Vapor	2.2 ppm
	PF-LIF (with pyrolysis), 454 nm (1 atm, air)	[103]	Vapor	140 ppm
	PAS, 5.8–6.7 μm	[141]	Vapor	~ppb
Triacetone triperoxide	LP-TOF-MS, 795 nm (130 fs, 840 μJ, 298 K)	[135]	Vapor	–
(TATP)	LP-TOF-MS, 266 nm (5 ns, 30 mJ, 298 K)	[135]	Vapor	–

LOD, limit of detection; ND, nondetection.
LODs for several studies were not determined.

therefore, the broadband absorption spectra of explosives are needed. In addition to the IR absorption spectra discussed previously (Section 2.), the UV/vis spectra of explosives have been recorded. Although most EM do not absorb strongly in the visible and near-IR regions, they do possess strong UV absorption spectra.

Smit [92] obtained the absorption spectrum from 190–550 nm for 2,2′,4,4′,6,6′-hexanitrostilbene (HNS), RDX, HMX, and PETN in both solution and KCl disks. The strong absorption maxima for HE in the UV suggest the use of frequency quadrupled Nd:YAG lasers for LIBS detection of trace residue explosives, rather than the visible or IR wavelengths previously used (Section 3.). Furthermore, laser ablation/excitation in this wavelength region reduces the laser pulse energy necessary for breakdown of the explosive sample. Not only does a lower pulse energy reduce the damage to the substrate (while still sampling all or most of the explosive residue), it becomes especially important for standoff detection, where maintaining high laser pulse energies at long distances (>100 m) could be an issue.

More recently, Usachev et al. [93] recorded the absorption spectrum of gaseous TNT by conventional absorption spectroscopy using a Xe arc lamp as the source of UV emission (195–300 nm) as well as CRDS with a pulsed dye laser (225–235 nm). After obtaining the broadband UV absorption spectrum, CRDS was applied to determine the real-time behavior of the TNT vapor number density at different temperatures (5–110 °C). They determined that the LOD of TNT vapor by CRDS is less than 1 ppb. Ramos and Dagdigian [94] recently presented a comprehensive study of the detection of vapors of DNBs and DNTs by UV CRDS. Their work showed that UV CRDS can detect ERCs at sub-ppb levels without any pre-concentration. However, unlike IR CRDS, measurements at atmospheric pressure result in a slight loss of detection sensitivity (5–10%) because of Rayleigh scattering. In addition, because the UV absorption of nitro-compounds is broad and relatively structure-less (Fig. 12), UV/vis absorption spectroscopy cannot provide positive identification of specific nitro-compounds. Owing

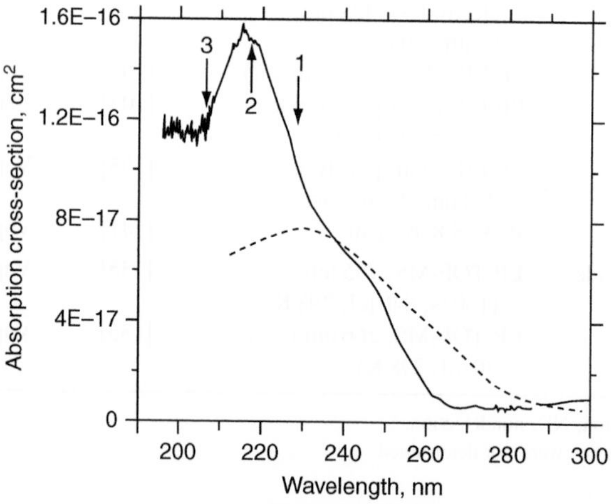

Fig. 12. The ultraviolet absorption spectrum of 2,4,6-trinitrotoluene, reproduced with permission from [93]. The solid line is the spectrum of gas-phase TNT measured at 92 °C; the dotted line is the spectrum of the TNT ethanol solution measured in reference 10 of [93]. Arrows labeled as 1, 2, and 3 show the peaks corresponding to NO A $^2\Sigma^+(v=0,1,2) \leftarrow$ X $^2\Pi(v=0)$ absorption transitions, respectively.

to the high sensitivity of the technique, UV CRDS may be useful as a screening tool supported by more selective detection techniques.

4.2. Laser desorption

Laser desorption (LD) is the formation of gas-phase neutral species by the interaction of a pulsed laser with a solid or liquid material. The goal in LD is to vaporize the sample material with minimal fragmentation, often as a means of sample introduction for mass spectrometry (MS). Fragmentation reduces the selectivity of the detection technique because the vaporized fragments can no longer be uniquely identified as belonging to the parent molecule. Sample fragmentation for large, labile species such as explosives can be minimized through the use of a low-power laser or shorter pulse width.

Huang et al. [95] used a pulsed laser to volatilize explosive materials for ion mobility/MS measurements. By using relatively low laser power (532 nm, $<10^7\,W/cm^2$), neutrals characteristic of the molecule were produced primarily. Ionization of the neutral species was achieved in a second step, by means of either a ^{63}Ni-β-foil atmospheric pressure ionization source or an additional laser pulse. Morgan et al. [96] suggested the use of laser thermal desorption where the IR laser pulse serves as a source of thermal energy to increase the vapor pressure of the explosive, significantly enhancing the detection capabilities of vapor-based sensing methods. This technique provides a faster, easier alternative to heating the sample in an oven or similar device.

4.3. Photofragmentation (PF)-fragment detection

Because large, fragile molecules are susceptible to nonradiative relaxation processes such as fragmentation, optical detection of EM is challenging. EMs also possess relatively weak and featureless absorptions. PF is an alternative to the direct spectroscopic approach. Characteristic fragments produced through laser photolysis of the parent molecule often prove more amenable to spectroscopic detection through their relatively sharp and strong features. Because the degree of fragmentation can be controlled by the laser fluence, different species can be identified by their fragmentation patterns [97]. Laser PF-fragment detection (FD) spectrometry has received considerable attention in recent years for its ability to detect labile species such as EMs and ERCs [98–100]. Most EMs contain one or more NO_2 functional groups, which UV excitation readily dissociates into NO [92,101,102]. In PF-FD, one or more lasers are used to photofragment the EM and facilitate the detection of the characteristic NO photofragment through excitation [103]. The NO photofragments can be monitored utilizing prompt emission excited by the photolysis, LIF, or REMPI. Detection limits are typically in the low ppb to ppm [104].

The first reported work on laser ionization of explosives by Marshall et al. [105,106] proposed the generally accepted fragmentation mechanism of the nitro-containing explosives following excitation of the parent molecule using a nanosecond laser pulse [107]. In the first step, the parent molecule fragments ~100 fs after initial nanosecond laser excitation, resulting in the release of NO_2 and other byproducts. NO_2 predissociates at wavelengths of less than 400 nm, making ionization difficult, and its weak oscillator strength and large radiative lifetime (~50–120 μs) make it difficult to detect [108]. The NO_2 fragment absorbs additional photons and is excited to a predissociative state, resulting in the formation of NO (radiative lifetime ~200 ns) and O. Next, the NO fragment undergoes subsequent photon absorption enabling detection of the ions resulting from a REMPI process through the A–X (0–0, 0–1, or 0–2) electronic transition from the ground state $X\,^2\Pi$ to the first excited state $A\,^2\Sigma^+$, or through stimulated fluorescence emission. Subsequent studies of UV laser (210–270 nm) induced dissociation of nitrotoluene isomers in the gas phase concluded that both dissociation ionization (dissociation followed by ionization) and ionization dissociation (ionization followed by dissociation) mechanisms are possible [109].

NO_2 absorbs in the visible region, whereas NO absorbs in the UV; both molecules absorb in the IR, however, few lasers can be tuned to both, and H_2O interference in the IR can reduce the detection sensitivity. Therefore, detection of these characteristic fragments of explosives is usually accomplished using UV or visible wavelength lasers. Figure 13 shows a partial energy level diagram of NO_2 and NO, along with some of the detection schemes described in Section 4.3.1. through Section 4.4. inclusive. The following sections describe PF followed by REMPI detection (PF-REMPI), PF followed by LIF (PF-LIF), and several variations of the PF-FD technique.

4.3.1. PF-REMPI

Resonance-enhanced multiphoton ionization is a relatively efficient soft ionization technique for producing molecular ions. REMPI is based on the enhancement of the ionization process when there are electronic states resonant with the energy of one or more of the incident photons. REMPI is generally performed in a two-photon process, $(1+1)$ REMPI, where the first photon excites the molecule to an intermediate electronic state and the absorption of a second photon results in ionization. The energy of the photons can be chosen such that ionization is induced without fragmentation. As ionization depends on absorption of the first photon, the technique selectively ionizes the target molecules. The resulting ions can be detected with a mass spectrometer or a pair of miniature electrodes [110]. Efficient REMPI pathways are known for a number of species and lead to high instrumental sensitivities and low detection limits [111]. In PF-REMPI of EM, PF of the EM produces NO, as described in Section 4.3. The maximum NO^+ yield from REMPI occurs at 226.3 nm, corresponding to a $(1+1)$ process through the $A\,^2\Sigma\ (\nu=0) \leftarrow X\,^2\Pi_{1/2}(\nu=0)$ transition [97].

In 1990, Zhu et al. [112] performed resonant two-photon ionization (R2PI) at 266 and 213 nm of substituted nitrobenzenes (NBs). Typically, in REMPI, the molecular ion itself is not observed for large parent molecules. They found that although extensive fragmentation occurs under vacuum, R2PI under atmospheric pressure (1 atm, He) has

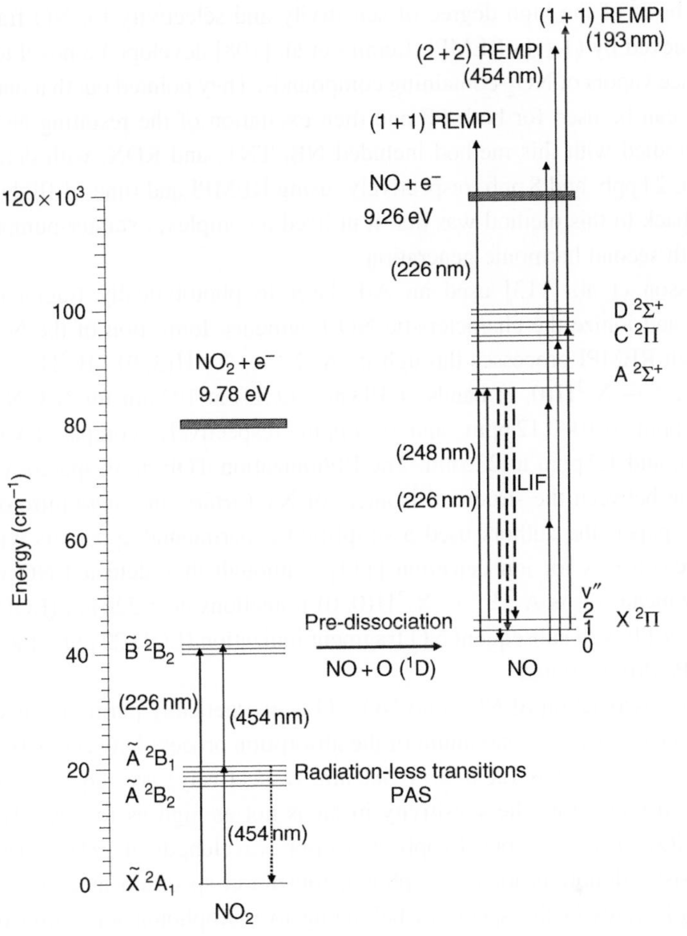

Fig. 13. Partial energy level diagram of NO_2 and NO. Several detection schemes including REMPI, LIF, and PAS are shown using laser wavelengths of 193, 226, and 454 nm.

the potential of being a soft ionization source for even very fragile molecules. From their study of *p*-nitroaniline, *o*-nitroaniline, nitroanisole, nitrophenol, nitrotoluene, 2,6-DNT, 2,4-DNT, and 2,4,6-TNT, they showed that unique ions are produced for the different isomers.

Clark et al. [101] developed a very sensitive and selective procedure for the detection of nitroaromatic vapors. Not only were they able to distinguish NB and *o*-nitrotoluene, two of the simplest explosive compounds, they could also distinguish NO^+ from NO/NO_2 gases and nitroaromatic compounds by their laser-induced mass spectra and wavelength dependence (245–250 nm) of the fragment ions. In addition, they demonstrated subattomole sensitivity for NB through generation of intense NO^+ ion signals using REMPI at 226.3 nm.

On the basis of the high degree of sensitivity and selectivity for NO fragments that can be achieved by (1 + 1) REMPI, Lemire et al. [108] developed a novel technique for sensing trace vapors of NO_2-containing compounds. They pointed out that one laser tuned to 226 nm can be used for both PF and then excitation of the resulting NO fragments. Samples studied with this method included NB, TNT, and RDX, with detection limits of 2.4 ppm, 24 ppb, and 8 ppb, respectively, using REMPI and time-of-flight (TOF) MS. One drawback to this method was that it utilized a complex, excimer-pumped dye laser system with second harmonic generation.

Simeonsson et al. [113] used an ArF laser to photolytically fragment the target molecules and ionize the characteristic NO fragments. Ionization of the NO molecules was through REMPI processes through its A $^2\Sigma^+ \leftarrow X\ ^2\Pi(3,0)$, B $^2\Pi \leftarrow X\ ^2\Pi(7,0)$, and/or D $^2\Sigma^+ \leftarrow X\ ^2\Pi(0,1)$ bands at 193 nm. LODs at 193 nm for NB, NT, and TNT were 0.49 ppm, 0.10–0.12 ppm, and 0.21 ppm, respectively, compared with 2.4 ppm, 15–36 ppm, and 1.7 ppm at 226 nm. The PF/ionization TOF mass spectra were used to discriminate between the structural isomers of NT (*ortho*- and *meta*-nitrotoluene). In a subsequent paper, the authors used a simplified experimental apparatus with a pair of miniature electrodes for ion detection [114]. Although they detected NO with (1 + 1) REMPI by means of its A $^2\Sigma^+ \leftarrow X\ ^2\Pi(0,0)$ transitions near 226 nm (LOD 1 ppb) and NO_2 by laser PF with subsequent NO fragment ionization (LOD 22 ppb), the two species could not be differentiated.

The photodissociation of NO_2 into NO + O is energetically permitted at wavelengths less than 400 nm, with the maximum of the absorption process between 380 and 400 nm. As three or four photons (at 226 or 380 nm, respectively) are required for photodissociation and ionization, the sensitivity in air is not as high as for NO. The detection limit for NO_2 in air is about 10 ppb at a laser wavelength of 380 nm [97]. NO has a characteristic double-headed, two-photon ionization spectrum at 226.4 and 227 nm, but nearby features in the spectrum belonging to multiphoton ionization processes in molecular oxygen reduce the sensitivity [97]. Guizard [115] suggested a (2 + 1) ionization scheme through the C $^2\Pi(\nu = 0) \leftarrow X\ ^2\Pi(\nu = 0)$ transition at 380 nm, which avoids the problems associated with the molecular peaks of oxygen. Ledingham [97] later demonstrated a LOD for NO of about 100 ppt using this scheme.

Simeonsson and Sausa [99] detected NO near 452 nm by (2 + 2) REMPI through its A $^2\Sigma^+ - X\ ^2\Pi(0,0)$ and (1,1) transitions, while NO_2 was detected by laser PF to NO followed by ionization. They found that spectral differentiation was possible because the internal energy of the NO photofragments differed from the 'ambient' NO. Measurement of the vibrationally excited NO through its A $^2\Sigma^+ - X\ ^2\Pi(0,3)$ band at 517 nm was also demonstrated. The LODs for NO and NO_2 were under 100 ppb (20–40 ppb_ν at 449–452 nm, 75 ppb_ν at 517.5 nm). Discrimination between NO and NO_2 photofragment detection at 226 nm was also demonstrated [116]. A 2001 ARL Tech Report summarized the extensive work done using PF-REMPI (and PF-LIF) of explosives at ARL by Sausa et al. [117].

Schmidt et al. [111] developed an atmospheric pressure laser ionization (APLI) source based on REMPI in pulsed gas expansions close to the inlet nozzle orifice (at high molecule densities). Their approach was to shift the ionization volume to the high-pressure continuous flow regime of expansion where high particle densities predominate to significantly enhance the REMPI detection limit in mass spectrometric applications (with TOF-MS). The sampling stage allows for measurements of reactive or thermally unstable species. This arrangement gave an LOD for NO of 0.9 ppt, an improvement of a factor of roughly 400 over a conventional skimmed molecular beam setup.

4.3.2. PF-LIF

A second technique frequently used to detect photofragmented species is LIF. Unlike UV/vis or other absorbance spectroscopy, fluorescence is a zero-background method and provides high sensitivity and selectivity when a laser is used as the excitation source. In this technique, the laser is tuned, so that its frequency matches that of an absorption line of some atom or molecule of interest. The absorption of the photons by this species produces an electronically excited state, which then radiates producing fluorescence emission characteristic of the species. Rodgers et al. [118] were the first to suggest a new method for the in situ detection of nonfluorescing molecular species by combining PF and PF-LIF. The species is first photolyzed at wavelength λ_1, producing one or more vibrationally excited fragments. Before vibrational relaxation, one the photofragments is pumped into a bonding excited state by a second laser pulse at wavelength λ_2. The fluorescence is sampled at wavelength λ_3, where $\lambda_3 < \lambda_2$ and λ_1. They demonstrated this technique on NO_2 (among other atmospheric gases) and estimated a detection limit of 3 ppt.

Zuckermann et al. [119] studied IR multiphoton dissociation (IRMPD) of RDX and HMX in a supersonic jet. A CO_2 laser was used for dissociation, with a pulsed frequency-doubled dye laser tuned to 280 nm to excite the A $^2\Sigma(\nu = 1) \leftarrow X^2\Pi(\nu = 0)$ transition of OH. OH radicals were observed by LIF, indicating that OH loss is a primary process in the unimolecular dissociation of nitramines such as RDX and HDX. Guo et al. [120] employed UV excitation to study the decomposition of RDX and HMX from their first excited electronic states. NO was observed as one of the initial dissociation products using both TOF-MS and LIF. Because the LIF of OH is well known and quite intense, they decided to look for it as well. Despite a calculated transition intensity 1.5 times that of NO, however, the OH radical was not observed as a UV dissociation product using LIF.

Wu et al. [121] developed a technique based on PF-LIF of NO to measure the concentration of EM in soil and other media. Laser radiation near 226 nm was used to photodissociate the EM to NO_2, which pre-dissociated into NO. The ground state NO then absorbed a second 226-nm photon to undergo a resonant transition A $^2\Sigma^+(\nu' = 0) \leftarrow$ X $^2\Pi(\nu'' = 0)$, subsequently producing NO A $^2\Sigma^+(\nu' = 0) \rightarrow$ X $^2\Pi(\nu'' = n)$ fluorescence. The concentration of the EM was inferred from the intensity of the NO fluorescence. They found that the PF-LIF signal intensity significantly increases when the sample (TNT) was heated above 343 K, but that heating causes physical and chemical changes in the sample. The LOD of TNT in soil was estimated to be 40 ppb. Factors influencing the PF-LIF signal such as sample temperature, laser power, and heating time were investigated in a subsequent paper [122]. The LOD for 2-μJ laser power and a TNT sample temperature of 473 K was 4 ppm. The suitability of the technique for field implementation was discussed along with plans to develop a PF-LIF laser-based sensor for use with the US Army Corps of Engineers' Waterways Experiment Station's cone penetrometer to measure the concentration of sub-surface TNT in situ.

In 1999, Swayambunathan et al. [103] detected trace concentrations of TNT, PETN, and RDX by laser PF-FD spectrometry using both PF-REMPI (with miniature electrodes) and PF-LIF in air with LODs in the low ppb–ppm range. They determined that collisional quenching of NO (A $^2\Sigma^+$) by N_2 and O_2 and reactions with O_2 were more pronounced in the LIF experiments than in the REMPI experiments because the radiative lifetime of the A $^2\Sigma^+$ intermediate state is relatively long, ~215 ns. In contrast, ionization from the intermediate A $^2\Sigma^+$ state is instantaneous (within the 6 ns laser pulse) in REMPI, allowing little time for quenching and other reactions to occur. In addition, nearly all the ions produced by REMPI are collected by the electrodes, while only part of the signal in the LIF experiments is collected (the fluorescence was viewed through a small cone in a direction perpendicular to the excitation beam). Consequently, the LIF signals for RDX and PETN were very weak for room temperature samples. However, the LIF intensity increases with increasing temperatures above the melting point (~413 K for PETN, 476–477 K for RDX). The pyrolysis-LIF technique involves pyrolysis of the EM with subsequent detection of the pyrolysis products NO and NO_2 by LIF (227 nm) and PF-LIF (454 nm), respectively [117]. PF-FD experiments on EMs were also performed with a visible laser because of the potential advantages compared with a UV laser, namely, that the laser can be easily transmitted through optical fibers over 10–30 m distances. Their studies showed that using 454 nm results in a lower PF efficiency and a higher LOD for PF-REMPI. Using 454 nm also results in lower NO excitation efficiency for pyrolysis-LIF and therefore a higher LOD.

Daugey et al. [123] demonstrated one-color PF-LIF detection of NB using 222–272 nm photodissociation and detection of NO through the A $(\nu' = 0) - X(\nu'' = 0\text{–}4)$ transitions. They found that in addition to the vibration-less ground state NO, a significant amount of vibrationally excited NO is also produced and can be detected free from interference by atmospheric NO. An LOD ~500 ppb in air (10 and 100 Torr) was achieved by monitoring NO through excitation of A $^2\Sigma^+(\nu' = 0) \leftarrow X\ ^2\Pi(\nu'' = 2)$ transition and detection through A $^2\Sigma^+(\nu' = 0) \rightarrow X\ ^2\Pi(\nu'' = 1)$ at ~236 nm. Shu et al. [124] then detected trace concentrations of dinitrobenzene (DNB) with a LOD of 13 ppb in 100 Torr of air and 11 ppb in 500 Torr of air using 248 nm dissociation and excitation. As with the previous study on NB, the fluorescence was collected at shorter wavelengths than the exciting radiation, precluding background fluorescence or ambient ground-state NO interference. This technique was subsequently used to detect DNT with a LOD of 3.7 ppb in 100 Torr of air and 2.7 ppb in 500 Torr of air [125,126].

Arusi-Parpar et al. [127] developed a unique scheme to remotely detect trace amounts of TNT vapor at atmospheric pressure and 24°C using a single 248-nm laser beam. Detection is based on the photodissociation of TNT vapor followed by LIF of the A $^2\Sigma^+(\nu' = 0) \leftarrow X\ ^2\Pi(\nu'' = 2)$ transition of the NO photofragments (LOD <8 ppb).

The authors estimated that at least 30% of the photodissociated TNT molecules produce NO with a $\nu'' = 0$, 1, 2 ratio of 1:0.5:0.1. As demonstrated by Shu et al. [123–126] at low pressures, there are two important advantages in detecting vibrationally excited NO radicals versus the ground-state NO produced by photodissociation: (i) the collected fluorescence is at lower wavelengths than the exciting laser (in contrast with the fluorescence from either TNT molecule or other molecules that exist in air) and (ii) there is no background fluorescence from the ground-state NO that is present naturally in air (which reduces the probability of a false alarm). Ambient conditions dramatically shorten the NO fluorescence lifetime, mainly because of quenching by oxygen. Using relatively high laser energy (~5 mJ), a large interaction volume (unfocused laser), an improved detection system (with tailor-made spectral filters), and a background-free scheme, they were able to develop a more sensitive detection system than in references [121] or [103]. This technique was later demonstrated at 2.5 m under near-ambient conditions (1 atm and 28°C) with a LOD <15 ppb [128].

4.3.3. Femtosecond ionization/dissociation

Photounstable molecules such as nitro-compounds tend to quickly dissociate after photoexcitation by a nanosecond laser, which greatly reduces the probability for the absorption of the additional photons necessary for ionization. The resulting photofragments are not specific enough to unambiguously identify the parent molecule. Ultrashort-pulse laser radiation provides both high intensities (10^{12}–10^{17} W cm^{-2}) and pulse durations shorter than the rotational timescales of molecules. It has been demonstrated that using ultrashort pulses decreases the interaction time such that the multiphoton ionization process is finished before the intermediate energy level can be depleted through fast relaxation processes (e.g., intramolecular energy redistribution because of internal conversion, intersystem crossing, or fast dissociation processes). A review by Ledingham and Singhal [129] discussed the use of ultrafast lasers for PF/ionization.

Kosmidis et al. [130] compared 90 fs photodissociation of NB and the nitrotoluene isomers at 375 nm to 10 ns photodissociation at the same wavelength. They demonstrated that only the femtosecond laser produces parent and heavy mass peaks (Fig. 14). A nonresonant multiphoton process results in molecular ionization. The mass spectra of the three NT isomers have analytically differentiable spectra. Using multiphoton excitation of nitrotoluene compounds with 170-fs laser pulses at either 412 or 206 nm, Tonnies et al. [131] demonstrated that the mass spectra of NT, DNT, and TNT exhibit a clear molecular ion or OH loss signal despite intense fragmentation. The two-photon absorption at 206 nm and four-photon absorption at 412 nm were not resonantly enhanced. They found that the intensity of the molecular ion signal decreased with increasing substitution and that mass spectra obtained at 412 nm show a higher degree of fragmentation than those at 206 nm. Isomer-specific detection after multiphoton ionization was demonstrated at both wavelengths for two isomers of NT and only at 412 nm for two of the DNT isomers. In a continuation of the work of Tonnies et al. [131], Weickhardt and Tonnies [132] obtained mass spectra of all isomers of NT and four isomers of DNT

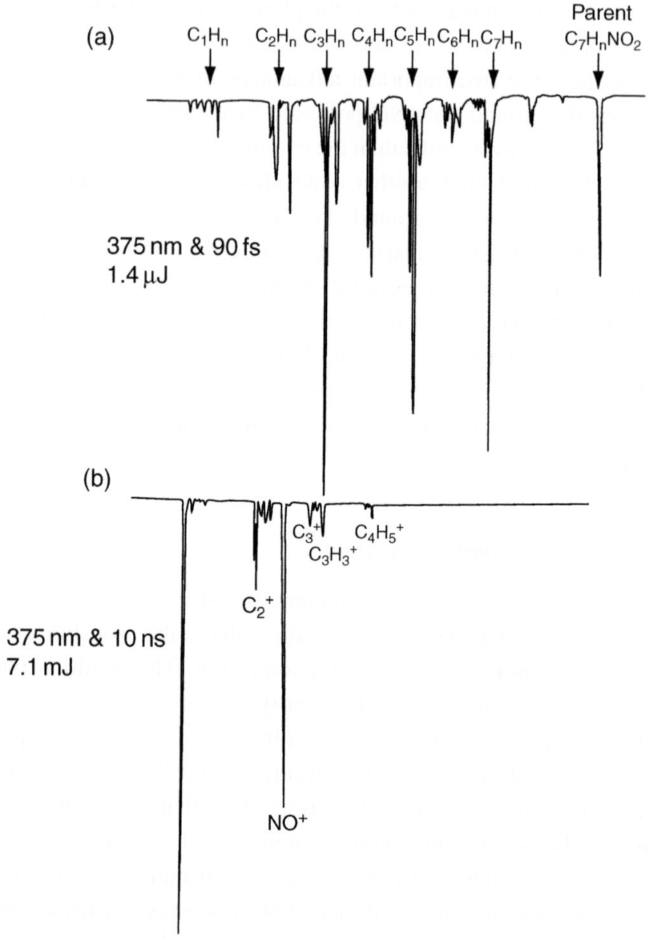

Fig. 14. Comparison of femtosecond and nanosecond fragmentation spectra of *m*-nitrotoluene at 375 nm, reproduced with permission from [130]. Only the 90-fs laser pulse shows parent and heavy mass peaks, while the 10-ns laser pulse results in much smaller photofragments.

using ultrashort laser pulses at 206 or 412 nm. Compounds with a nitro group in position 2 and/or 6 showed a pronounced ortho effect leading to the loss of OH. The 412 nm light resulted in a broader variety of fragments and additional signals in the higher mass range that could be used for isomer identification.

Hankin et al. [133] demonstrated femtosecond ionization following 266 nm desorption of solid samples of trinitrobenzene (TNB), TNT, and trinitrophenol (TNP). They confirmed the advantages of ultrafast ionization, namely, the formation of characteristic precursor and structure-specific fragment ions. The optimum intensities for efficient LD without ionization were determined for the compounds studied. Differences between femtosecond ionization of vapor samples of explosives [131,132] and laser desorbed molecules were also discussed.

Osorio et al. [134] performed TOF-MS measurements of TNT and RDX on soil surfaces. They used tunable UV radiation from a 130 fs laser to monitor the kinetic energy distribution of NO/NO_2 photofragments released by the dissociation of TNT and RDX. Analysis of the kinetic energy distribution of the photofragments revealed differences in the processes for NO and NO_2 ejection in different substrates. Mullen et al. [135] detected triacetone triperoxide (TATP) by laser photoionization. Mass spectra in two time regimes were acquired using nanosecond (5 ns) laser pulses at 266 and 355 nm and femtosecond (130 fs) laser pulses at 795, 500, and 325 nm. The major difference observed between the two time regimes was the detection of the parent molecular ion when femtosecond laser pulses were employed.

4.3.4. SPF-FD

Cabalo and Sausa [104] introduced a novel technique for detection of explosives with low vapor pressure called surface PF-FD (SPF-FD). Although techniques such as CRDS, PF-LIF, and PF-REMPI are ideal for TNT in the gas phase, the vapor pressure of most explosives is too low to detect using those methods, especially at room temperature. In SPF-FD, a UV laser (248, 266, or 355 nm) is used to photofragment RDX on a surface, and a second time-delayed laser (226 nm) ionizes the characteristic NO fragment by means of its $A\ ^2\Sigma^+ - X\ ^2\Pi(0,0)$ transitions by (1 + 1) REMPI (SPF-REMPI). The maximum signal was observed with a photolysis wavelength of 248 nm, where the absorption coefficient of RDX is the strongest. A detection limit of $\sim$14 ng/cm^2 at 1 atm and 298 K was demonstrated.

4.3.5. SPI-TOF-MS

Mullen et al. [107] demonstrated the detection of EM using single photon laser ionization (SPI). Although ultrafast ionization has been successfully applied to the selective detection of explosives, the complex, nonruggedized instrumentation required makes the technique unsuitable for field scenarios. SPI is a more robust ionization method that does not involve resonant excitation of an intermediate state (unlike REMPI) – the parent molecule is directly ionized using a vacuum ultraviolet (VUV) photon. By frequency tripling, the third harmonic output (354.6 nm) of a Nd:YAG laser in xenon, 118.2 nm (10.49 eV) photons can be generated. At these energies, bulk gases such as nitrogen, oxygen, and water do not have sufficient energy to ionize directly but most organic compounds do ionize. Ions of EM produced by two or more VUV photons are detected using TOF-MS (SPI-TOF-MS). Vapors from the samples (NB, 1,3-DNB, *o*-NT, 2,4-DNT, 2,4,6-TNT, and TATP) were introduced into the instrument using a capillary GC column. They found that SPI of the nitro-containing explosives yields mass spectra dominated by the parent molecular ion, although TATP undergoes extensive fragmentation. LODs were determined for NB (17–24 ppb) and DNT ($\sim$40 ppb).

4.4. Photoacoustic spectroscopy

Photoacoustic (or optoacoustic) spectroscopy is an indirect detection method that measures thermal energy imparted to a gas in close contact with a photo-excited material (solid, liquid, or gas). The absorbing sample is typically in close contact with a monatomic gas. Following sample photoexcitation, energy is rapidly transferred to the surrounding gas. If the photoexcitation is modulated at an acoustic frequency, the resulting thermal pulse in the surrounding gas may be detected by a microphone [136]. High-resolution photoacoustic spectra are measured using either a pulsed laser or a wavelength- or amplitude-modulated laser (e.g., with a mechanical chopper). PAS is an attractive technique because of its relative simplicity, ruggedness, and overall sensitivity. In addition to obtaining absorption spectra, PAS can be used to measure collisional relaxation rates, determine substance compositions, and monitor reactions [137].

Because optoacoustic detectors respond only to absorbed radiation, much weaker absorptions can be detected than with traditional absorption spectroscopy methods, which depend on detecting small differences between large signals [138]. It is particularly effective at high pressures for weak fluorescers or species that pre-dissociate with laser absorption [110]. The magnitude of the measured photoacoustic signal is given by $S = S_m PC_\alpha$, where C is a cell-specific constant (units of Pascal centimeters per watt), P is the power of the incident laser radiation (watts), α is the absorption coefficient of the transition that is being interrogated (cm^{-1}), and S_m is the sensitivity of the microphone (volts per Pascal) [139]. Because the signal is proportional to the incident laser power, the detection sensitivity of trace gases benefits from the use of higher laser powers. This technique is also well suited to diode laser spectroscopy of explosives, where the laser bandwidth is much narrower than the absorber bandwidth (Section 2.1.1.).

In 1972, Kreuzer et al. [140] demonstrated the optoacoustic detection of 10 pollutant gases using CO and CO_2 lasers, including 0.1-ppb detection of NO_2 at 6.22 μm. Claspy et al. [141] were the first to detect explosive vapors using PAS. Periodic heating with a chopped IR beam (5.8–6.7 μm) of a cell containing the vapor sample resulted in pressure fluctuations that were detected with a sensitive microphone. They identified EGDN, DNT, TNT, TOVEZ, dynamite, RDX, and PETN in the atmosphere at partial pressures of 10^{-6} Torr or less (on the order of ppb) with their system. In 1975, Angus et al. [138] reported the first use of a tunable visible laser with an optoacoustic cell, estimating a 10 ppb NO_2 detection limit from low-resolution CW dye laser excitation scans over the range 575–625 nm. Subsequent studies by Claspy and colleagues investigated atmospheric interferents (NO, NO_2, CH_4, C_4H_{10}, and H_2O) for explosives detection at 6, 9, and 11 μm [142] and used a pulsed dye laser (480–625 nm) to detect NO_2 with 4 ppb sensitivity at 600 nm [143].

Terhune and Anderson [144] measured NO_2 sensitivities of better than 0.1 ppb during in situ measurements of aerosols using an acoustically resonant spectrophone and a 514.5-nm Ar^+-ion laser. By constructing a nonmetal cell to minimize surface adsorption, Crane [145] was able to use a CO_2 laser (9.6 μm) to obtain optoacoustic absorption spectra of EGDN, DNT, and NG vapors with LODs of 8.26, 0.23, and 0.50 ppb, respectively.

Water vapor, which occurs in high concentrations in air and absorbs throughout the IR, was found to be the limiting interferent. Freid [146] measured the 488 nm PAS detection limit of NO_2 in NO, N_2, H_2O, and O_2 matrices with identical sensitivities of 5 ppb in all matrices except O_2, which decreased the NO_2 signal. A possible explanation given is that the energy deposited into the NO_2 molecules was transferred to O_2 instead of being released as heat, thus degrading the optoacoustic signal.

In 1995, Hasue et al. [136] reported low-resolution photoacoustic spectra of 18 powdered EM. A 500 W xenon lamp was used as a light source for coverage from 400–800 nm and a 300 W halogen lamp for 800–1600 nm. In general, EMs show peaks in the 600–800 and 1400–1600 nm ranges. Spectra of picric acid, TNT, PETN, ammonium perchlorate (AP), tetryl, composition-B, dinitromethyloxamide, RDX, HMX, black powder, potassium nitrate, ammonium nitrate fuel oil (ANFO), composition C-4, composition A-3, Hexal, nitroguanidine, and EDNA were obtained. The energy required to initiate explosives with a ruby laser at 694.3 nm was also correlated with their photoacoustic signals at that wavelength. The acquired spectra for Hexal and PETN demonstrated that both aluminum and active carbon improve the absorption of laser light for EMs at 820 nm.

Pastel and Sausa [110] detected NO_2 (LOD 400 ppb at 1 atm) with a one-photon absorption photoacoustic process (Fig. 13) by means of $\tilde{A}\ \tilde{X}^2B_1(0,8,0) - \tilde{X}\ ^2A_1(0,0,0)$. This work, which employed a dye laser operating near 454 nm, was the first report of a high-resolution visible NO_2 photoacoustic spectrum. They found that low laser intensities favor NO_2 photoacoustic detection, whereas high laser intensities favor NO detection through REMPI (LOD 160 ppb). Fig. 15 shows the high-resolution photoacoustic spectrum of NO_2 and the NO REMPI spectrum.

Prasad et al. [147] used vibrational PAS in conjunction with published conventional IR data and *ab initio* calculations to assign the normal modes of RDX and TNT. The IR

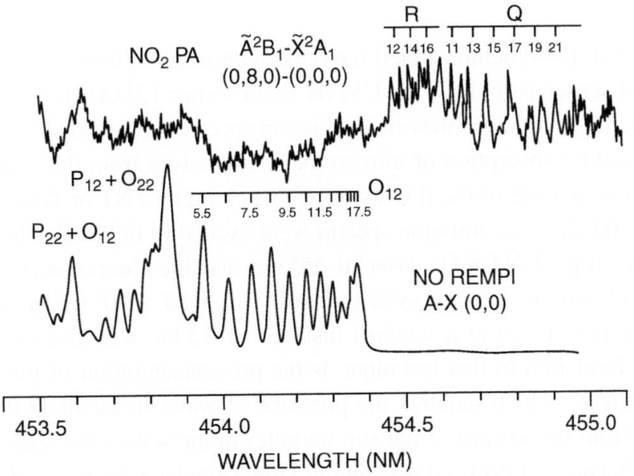

Fig. 15. High-resolution photoacoustic spectrum of NO_2 (top) and REMPI spectrum of NO (bottom), reproduced with permission from [110].

PAS measurements on powder samples were made using a line tunable CW CO_2 laser source. They found that the most suitable bands of TNT for spectrochemical analysis are at 978 cm^{-1} (C–H out of plane bending) and 946 cm^{-1} (C–N stretching motions); for RDX, the persistent bands are at 1045 cm^{-1} (N–N stretching) and 941 cm^{-1} (O–N–O bending motions).

More than 30 years of research has shown that explosive vapor detection using PAS is limited by surface adsorption and decomposition at elevated temperatures. The technique is also limited by interferents, especially in the IR (e.g., water vapor in air). Several promising new methods for PAS have recently been suggested, however. Webber et al. [139] incorporated near-IR diode lasers and optical fiber amplifiers to enhance sensitivity. This new approach to wavelength modulation PAS is applicable to all species that fall within the gain curves of optical fiber amplifiers. By wavelength-modulating the laser, the acoustic signals from wavelength-independent sources such as window absorption and continuum spectra from broadband absorbers are eliminated by demodulation. They demonstrated the technique, which is the first use of fiber amplifiers to enhance PAS, with ammonia detection at 1532 nm (LOD <6 ppb). The same group recently evaluated the performance of a field deployable tunable CO_2 laser photoacoustic spectrometer, demonstrating detection of chemical warfare agents at 4 ppb with a probability of false positives less than $1{:}10^6$. Because CO_2 lasers cannot provide continuous tuning, they also discuss the use of QC lasers for PAS.

4.5. *LIDAR variations*

Several systems based on LIDAR (Section 2.3.) have been developed for explosives and explosive device detection. While not fitting the conventional LIDAR experiment definition, these systems apply LIDAR principles to the detection of explosive sources.

Simonson et al. [148] demonstrated remote detection of explosives in soil by combining distributed sensor particles with UV/vis fluorescence LIDAR technology. The key to this approach is that the fluorescence emission spectrum of the distributed particles is strongly affected by absorption of nitroaromatic explosives from the surrounding environment. Remote sensing of the fluorescence quenching by TNT or DNT is achieved by fluorescence LIDAR – the emission spectra were excited in field LIDAR measurements by a frequency-tripled Nd:YAG laser at 355 nm and the fluorescence collected with a telescope and various detector systems housed in a $10' \times 50'$ trailer. TNT has been detected in the ppm range at a standoff distance of 0.5 km with this system (Fig. 16). An important limitation to this technique is the pre-concentration of the explosives on the sensor particles, which requires the presence of water to facilitate the transport of the explosive from the surface of the soil particles to the sensor particles.

Xiang and Sabatier [149] used a scanning laser Doppler vibrometer (LDV) to detect acoustic-to-seismic surface motion. This technique exploits airborne acoustic waves penetrating the ground and causing seismic motion. When an anti-personnel landmine

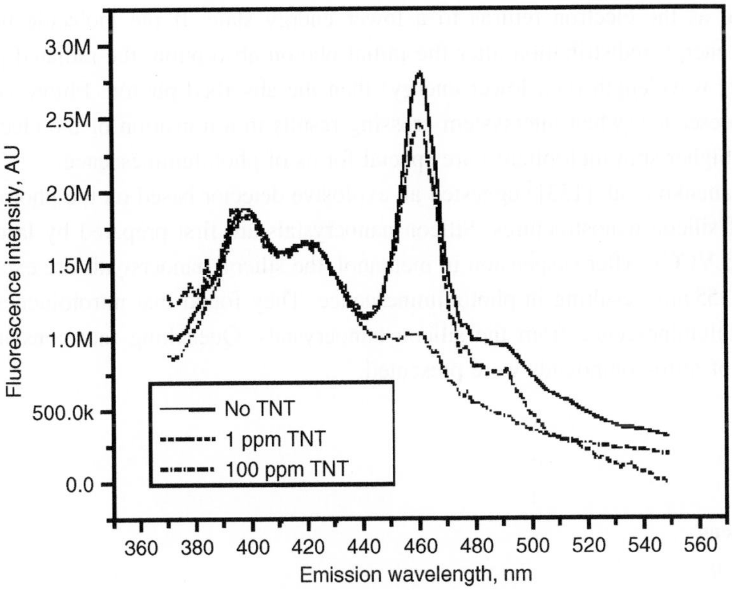

Fig. 16. Laboratory-collected fluorescence emission from sensor particles on soil contaminated with TNT, reproduced with permission from [148].

is buried in the soil, the acoustically coupled ground motion is altered. The LDV measures the motion of the ground surface because the surface vibration causes a Doppler frequency shift of the reflected laser light. Color maps can be formed which image the mine location, size, and shape. Although mines can be identified with a high probability of detection with this technique, ground clutter can cause high false alarm rates.

Finally, LIDAR has been used to detect honeybees trained to search for explosives [150–152]. Honeybees are faster to train than dogs, do not need to be leashed, and will not setoff landmines. They can also detect explosives with sensitivity comparable to, or better than, dogs (low ppt or ppq) and a range of 1–2 km. When the honeybees detect the explosive, they hover over the sample location before returning to the hive. Direct LIDAR has been used to map the areas of highest bee concentration, but a clear line of sight with no vegetation or other obstructions is needed [152]. Recently, it was demonstrated that a frequency-modulated signal resulting from the light scattered off honeybee wings (170–270 Hz) could be used to discriminate the return signal of a honeybee from the cluttered background [153].

4.6. Photoluminescence

Photoluminescence is a process in which a molecule absorbs a photon in the visible region, exciting one of its electrons to a higher electronic excited state, and then radiates a photon as the electron returns to a lower energy state. If the molecule undergoes internal energy redistribution after the initial photon absorption, the radiated photon is of longer wavelength (i.e., lower energy) than the absorbed photon. Fluorescence and phosphorescence (when intersystem crossing results in a transition of the electron to a state of higher spin multiplicity) are special forms of photoluminescence.

Germanenko et al. [153] suggested an explosive detector based on the photoluminescence of silicon nanostructures. Silicon nanocrystals are first prepared by laser vaporization (LVCC). After suspension in methanol, the silicon nanocrstyals are excited by a laser at 355 nm, resulting in photoluminescence. They found that nitrotoluenes quench the photoluminescence from the silicon nanocrystals. Quenching rate constants for a number of nitro-compounds were presented.

References

[1] J. I. Steinfeld and J. Wormhoudt, Annu. Rev. Phys. Chem., 49 (1998) 203.
[2] D. O. Henderson, Y. S. Tung, R. Mu, A. Ueda, W. Curby, A. G. Mercado and X. Li, Trends Appl. Spectrosc., 2 (1998) 139.
[3] D. S. Moore, Rev. Sci. Instrum., 75 (2004) 2499.
[4] B. C. Dionne, D. P. Rounbehler, E. K. Achter, J. R. Hobbs and D. H. Fine, J. Energ. Mat., 4 (1986) 447.
[5] P. A. Pella, J. Chem. Thermodyn., 9 (1977) 301.
[6] T. F. Jenkins, D. C. Leggett and T. A. Ranney, Vapor Signatures from Military Explosives Part 1. Vapor Transport from Buried Military-Grade TNT, Special Report 99–21, Office of the Chief Engineers, US Army Corps of Engineers, Cold Regions Research & Engineering Laboratory (1999).
[7] A. J. Sedlacek III, S. D. Christesen, T. Chyba and P. Ponsardin, Proc. SPIE-Int. Soc. Opt. Eng., 5269 (2004) 23.
[8] G. Herzberg. Vol I. Spectra of Diatomic Molecules, and Vol II. Infrared and Raman Spectra, New York: Van Nostrand Reinhold, 1945.
[9] K. L. McNesby and R. A. Pesce-Rodriguez, 'Applications of Vibrational Spectroscopy in the Study of Explosives', in Handbook of Vibrational Spectroscopy, J. M. Chalmers and P. R. Griffiths (Eds.), West Sussex, UK: John Wiley and Sons (2002) 3152.
[10] N. F. Fell, J. W. Widder, S. V. Medlin, J. B. Morris, R. A. Pesce-Rodriguez and K. L. McNesby, J. Raman Spectrosc., 27 (1996) 97.
[11] J. A. Silver, Appl. Opt., 31 (1992) 707.
[12] J. F. Schultz, M. S. Taubman, W. W. Harper, R. M. Williams, T. L. Myers, B. D. Cannon, D. M. Sheen, N. C. Anheier Jr., P. J. Allen, S. K. Sundaram, B. R. Johnson, P. M. Acker, M. C. Wu and E. K. Lau, Proc. SPIE-Int. Soc. Opt. Eng., 4999 (2003) 1.
[13] M. G. Allen, D. J. Cook, B. K. Decker, J. M. Hensley, D. I. Rosen, M. L. Silva, D. M. Sonnenfroh and R. T. Wainner, Proc. SPIE-Int. Soc. Opt. Eng., 5732 (2005) 134.
[14] J. Wormhoudt, J. H. Shorter and C. E. Kolb, Mater. Res. Soc. Symp. Proc., 418 (1996) 143.
[15] H. Riris, C. B. Carlisle, D. F. McMillen and D. E. Cooper, Appl. Opt., 35 (1996) 4694.
[16] C. Bauer, P. Geiser, J. Burgmeier, G. Holl and W. Schade, Appl. Phys. B, 85 (2006) 251.
[17] F. J. Effenberger and A. G. Mercado, Proc. SPIE-Int. Soc. Opt. Eng., 3384 (1998) 104.

[18] M. W. Todd, R. A. Provencal, T. G. Owano, B. A. Paldus, A. Kachanov, K. L. Vodopyanov, M. Hunter, S. L. Coy, J. I. Steinfeld and J. T. Arnold, Appl. Phys. B, 75 (2002) 367.
[19] V. Vaicikauskas, V. Kabelka, Z. Kuprionis and M. Kaucikas, Proc. SPIE-Int. Soc. Opt. Eng., 5958 (2005) 59581K:1.
[20] A. K. Chaudhary, G. C. Bhar and S. Das, J. Appl. Spectrosc., 73 (2006) 123.
[21] S. W. McKnight, C. A. DiMarzio, W. Li and R. A. Roy, Proc. SPIE-Int. Soc. Opt. Eng., 4038 (2000) 734.
[22] H. Kim, A. Lagutchev and D. D. Dlott, Propell. Explos. Pyrot., 31 (2006) 116.
[23] W. W. Busch and M. A. Busch (Eds.), An Ultratrace-Absorption Measurement, ACS Symposium Series, Vol. 720, Washington, DC: American Chemical Society, 1999.
[24] J. B. Spicer, P. J. Dagdigian, R. Osiander, J. A. Miragliotta, X.-C. Zhang, R. Kersting, D. R. Crosley, R. K. Hanson and J. Jeffries, Proc. SPIE-Int. Soc. Opt. Eng., 5089 (2003) 1088.
[25] J. I. Steinfeld, R. W. Field, M. Gardner, M. Canagarantna, S. Yang, A. Gonzalez-Casielles, S. Witonsky, P. Bhatia, B. Gibbs, B. Wilkie, S. L. Coy and A. Kachanov, Proc. SPIE-Int. Soc. Opt. Eng., 3853 (1999) 28.
[26] P. J. Dagdigian, Adv. Ser. Phys. Chem., 16 (2005) 129.
[27] A. Smekal, Naturwiss, 11 (1923) 873.
[28] J. Akhavan, Spectrochim. Acta, Part A, 47A (1991) 1247.
[29] K. L. McNesby, J. E. Wolfe, J. B. Morris and R. A. Pesce-Rodriguez, J. Raman Spectrosc., 25 (1994) 75.
[30] P. J. Hendra, 'Laser-Raman Spectroscopy Applied to Some Chemical Problems', presented at Mol. Spectrosc., Proc. Conf. 4th (Univ. Southampton, Southampton, UK).
[31] G. R. Abell and C. E. Gillespie, Remote Sensing and Analyzing of Gaseous Materials Using Raman Radiation, US Patent # 3625613 19680628 (1971).
[32] J. M. Schnur, Application of Picosecond and Light Scattering Spectroscopies to the Study of Energetic Materials, NRL-MR-4324, N. Res. Lab., Washington, DC, USA. (1980).
[33] D. R. Crosley and M. A. Schroeder, Development of Inverse Raman Spectroscopy for Probing Rapidly Decomposing Explosives and Propellants, ARBRL-TR-02345, Ballistic Res. Lab., Army Armament Res. Dev. Command, Aberdeen Proving Ground, MD, USA (1981).
[34] F. W. S. Carver and T. J. Sinclair, J. Raman Spectrosc., 14 (1983) 410.
[35] W. M. Trott and A. M. Renlund, Appl. Opt., 24 (1985) 1520.
[36] C. M. Hodges and J. Akhavan, Spectrochim. Acta, Part A, 46A (1990) 303.
[37] F. Grisch, M. Pealat, P. Bouchardy, J. P. Taran, I. Bar, D. Heflinger and S. Rosenwaks, Appl. Phys. Lett., 59 (1991) 3516.
[38] D. E. Hare, I.-Y. S. Lee, J. R. Hill, J. Franken, H. Suzuki, B. J. Baer, E. L. Chronister and D. D. Dlott, MRS Proceedings, 418 (1995) 337.
[39] Y. M. Gupta, G. I. Pangilinan, J. M. Winey and C. P. Constantinou, Chem. Phys. Lett., 232 (1995) 341.
[40] C. Cheng, T. E. Kirkbride, D. N. Batchelder, R. J. Lacey and T. G. Sheldon, J. Forensic. Sci., 40 (1995) 31.
[41] I. R. Lewis, N. W. Daniel Jr., N. C. Chaffin, P. R. Griffiths and M. W. Tungol, Spectrochim. Acta, Part A, 51A (1995) 1985.
[42] R. J. Lacey, I. P. Hayward, H. S. Sands and D. N. Batchelder, Proc. SPIE-Int. Soc. Opt. Eng., 2937 (1997) 100.
[43] H. S. Sands, I. P. Hayward, T. E. Kirkbride, R. Bennett, R. J. Lacey and D. N. Batchelder, J. Forensic. Sci., 43 (1998) 509.
[44] A. Blanco, L. C. Pacheco-Londono, A. J. Pena-Quevedo and S. P. Hernandez-Rivera, Proc. SPIE-Int. Soc. Opt. Eng., 6217 (2006) 621737/1.
[45] N. W. Daniel Jr., I. R. Lewis and P. R. Griffiths, Mikrochim. Acta, Suppl., 14 (1997) 281.
[46] I. R. Lewis, N. W. Daniel Jr. and P. R. Griffiths, Appl. Spectrosc., 51 (1997) 1854.

[47] D. S. Moore, Fresen. J. Anal. Chem., 369 (2001) 393.
[48] J. W. Hass III, J. M. Sylvia, K. M. Spencer, T. W. Johnston and S. L. Clauson, Proc. SPIE-Int. Soc. Opt. Eng., 3392 (1998) 469.
[49] K. M. Spencer, J. M. Sylvia, J. A. Janni and J. D. Klein, Proc. SPIE-Int. Soc. Opt. Eng., 3710 (1999) 373.
[50] J. M. Sylvia, J. A. Janni, J. D. Klein and K. M. Spencer, Anal. Chem., 72 (2000) 5834.
[51] J. F. Bertone, K. L. Cordeiro, J. M. Sylvia and K. M. Spencer, Proc. SPIE-Int. Soc. Opt. Eng., 5403 (2004) 387.
[52] G. A. Baker and D. S. Moore, Anal. Bioanal. Chem., 382 (2005) 1751.
[53] J. A. Dieringer, A. D. McFarland, N. C. Shah, D. A. Stuart, A. V. Whitney, C. R. Yonzon, M. A. Young, X. Zhang and R. P. Van Duyne, Faraday Discuss., 132 (2006) 9.
[54] T. Hirschfeld, Appl. Opt., 13 (1974) 1435.
[55] T. Hirschfeld, E. R. Schildkraut, H. Tannenbaum and D. Tanenbaum, Appl. Phys. Lett., 22 (1973) 38.
[56] S. M. Angel, T. J. Kulp and T. M. Vess, Appl. Spectrosc., 46 (1992) 1085.
[57] P. G. Lucey, T. F. Cooney and S. K. Sharma, P. Lunar Planet Sci. C., 29 (1998) 1354.
[58] S. K. Sharma, A. K. Misara and B. Sharma, Spectrochim. Acta, Part A, 61A (2005) 2404.
[59] S. K. Sharma, A. K. Misara, P. G. Lucey, S. M. Angel and C. P. McKay, Appl. Spectrosc., 60 (2006) 871.
[60] M. Wu, M. Ray, K. H. Fung, M. W. Ruckman, D. Harder and A. J. Sedlacek III, Appl. Spectrosc., 54 (2000) 800.
[61] D. A. Long. Raman Spectroscopy, New York: McGraw-Hill, 1977.
[62] S. A. Asher and C. R. Johnson, Science, 225 (1984) 311.
[63] J. M. Dudik, C. R. Johnson and S. A. Asher, J. Chem. Phys., 82 (1985) 1732.
[64] G. P. Anderson, A. Berk, P. K. Achary, M. W. Matthew, L. S. Bernstein, J. H. Chetwynd, H. Dothe, S. M. Adler-Golden, A. J. Ratkowski, G. W. Felde, J. A. Gardner, M. L. Hoke, S. C. Richtsmeier, B. Pukall, J. Mello and L. S. Jeong, Proc. SPIE-Int. Soc. Opt. Eng., 4049 (2000) 176.
[65] J. C. Carter, S. M. Angel, M. Lawrence-Snyder, J. Scaffidi, R. E. Whipple and J. G. Reynolds, Appl. Spectrosc., 59 (2005) 769.
[66] M. D. Ray, A. J. Sedlacek III and M. Wu, Rev. Sci. Instrum., 71 (2000) 3485.
[67] N. S. Higdon, T. H. Chyba, D. A. Richter, P. L. Ponsardin, W. T. Armstrong, C. T. Lobb, B. T. Kelly, R. D. Babnick and A. J. Sedlacek III, Proc. SPIE-Int. Soc. Opt. Eng., 4722 (2002) 50.
[68] D. A. Rusak, B. C. Castle, B. W. Smith and J. D. Winefordner, Crit. Rev. Anal. Chem., 27 (1997) 257.
[69] D. A. Rusak, B. C. Castle, B. W. Smith and J. D. Winefordner, TrAC, 17 (1998) 453.
[70] I. Schechter, Rev. Anal. Chem., 16 (1997) 173.
[71] J. M. Anzano, I. B. Gornushkin, B. W. Smith and J. D. Winefordner, Polym. Eng. Sci., 40 (2000) 2423.
[72] F. C. DeLucia Jr., A. C. Samuels, R. S. Harmon, R. A. Walters, K. L. McNesby, A. LaPointe, R. J. Winkel Jr. and A. W. Miziolek, IEEE Sensors Journal, 5 (2005) 681.
[73] F. Ferioli and S. G. Buckley, Combust. Flame, 144 (2006) 435.
[74] J. D. Hybl, G. A. Lithgow and S. G. Buckley, Appl. Spectrosc., 57 (2003) 1207.
[75] A. Portnov, S. Rosenwaks and I. Bar, Appl. Opt., 42 (2003) 2835.
[76] A. C. Samuels, F. C. DeLucia Jr., K. L. McNesby and A. W. Miziolek, Appl. Opt., 42 (2003) 6205.
[77] M. Tran, S. Sun, B. W. Smith and J. D. Winefordner, J. Anal. At. Spectrom., 16 (2001) 628.
[78] F. C. DeLucia Jr., R. S. Harmon, K. L. McNesby, R. J. Winkel Jr. and A. W. Miziolek, Appl. Opt., 42 (2003) 6148.

[79] K. Y. Yamamoto, D. A. Cremers, M. J. Ferris and L. E. Foster, Appl. Spectrosc., 50 (1996) 222.
[80] R. Grönlund, M. Lundqvist and S. Svanberg, Opt. Lett., 30 (2005) 2882.
[81] C. Lopez-Moreno, S. Palanco, J. Javier Laserna, F. De Lucia, Jr., A. W. Miziolek, J. Rose, R. A. Walters and A. I. Whitehouse, J. Anal. At. Spectrom., 21 (2006) 55.
[82] S. Palanco, C. Lopez-Moreno and J. J. Laserna, Spectrochim. Acta, Part B, 61 (2006) 88.
[83] B. Salle, J. L. Lacour, E. Vors, P. Fichet, S. Maurice, D. A. Cremers and R. C. Wiens, Spectrochim. Acta, Part B, 59 (2004) 1413.
[84] M. Corsi, G. Cristoforetti, M. Giuffrida, M. Hidalgo, S. Legnaioli, V. Palleschi, A. Salvetti, E. Tognoni and C. Vallebona, Spectrochim. Acta, Part B, 59 (2004) 723.
[85] C. Gautier, P. Fichet, D. Menut, J.-L. Lacour, D. L'Hermite and J. Dubessy, Spectrochim. Acta, Part B, 60 (2005) 792.
[86] J. Scaffidi, S. M. Angel and D. A. Cremers, Anal. Chem., 78 (2006) 24.
[87] D. N. Stratis, K. L. Eland and S. M. Angel, Appl. Spectrosc., 54 (2000) 1270.
[88] K. L. Eland, D. N. Stratis, D. M. Gold, S. R. Goode and S. M. Angel, Appl. Spectrosc., 55 (2001) 286.
[89] B. Le Drogoff, J. Margot, M. Chaker, M. Sabsabi, O. Barthelemy, T. W. Johnston, S. Laville, F. Vidal and Y. von Kaenel, Spectrochim. Acta Part B, 56 (2001) 987.
[90] X. Liu, D. Du and G. Mourou, IEEE J. Quant. Electron., 33 (1997) 1706.
[91] Y. Dikmelik and J. B. Spicer, Proc. SPIE-Int. Soc. Opt. Eng., 5794 (2005) 757.
[92] K. J. Smit, J. Energ. Mat., 9 (1991) 81.
[93] A. D. Usachev, T. S. Miller, J. P. Singh, F. Y. Yueh, P. R. Jang and D. L. Monts, Appl. Spectrosc., 55 (2001) 125.
[94] C. Ramos and P. J. Dagdigian, Appl. Opt. 46 (2007) 620.
[95] S. D. Huang, L. Kolaitis and D. M. Lubman, Appl. Spectrosc., 41 (1987) 1371.
[96] J. S. Morgan, W. A. Bryden, J. A. Miragliotta and L. C. Aamodt, Johns Hopkins APL Technical Digest, 20 (1999) 389.
[97] K. W. D. Ledingham, Physica Scripta, T58 (1995) 100.
[98] J. B. Simeonsson and R. C. Sausa, Appl. Spectrosc. Rev., 31 (1996) 1.
[99] J. B. Simeonsson and R. C. Sausa, Appl. Spectrosc., 50 (1996) 1277.
[100] J. B. Simeonsson and R. C. Sausa, TrAC, 17 (1998) 550.
[101] A. Clark, K. W. D. Ledingham, A. Marshall, J. Sander and R. P. Singhal, Analyst, 118 (1993) 601.
[102] A. Marshall, A. Clark, K. W. D. Ledingham, J. Sander and R. P. Singhal, Int. J. Mass Spectrom. Ion Processes, 125 (1993) R21.
[103] V. Swayambunathan, G. Singh and R. C. Sausa, Appl. Opt., 38 (1999) 6447.
[104] J. Cabalo and R. Sausa, Appl. Spectrosc., 57 (2003) 1196.
[105] A. Marshall, A. Clark, R. Jennings, K. W. D. Ledingham, J. Sander and R. P. Singhal, Int. J. Mass Spectrom. Ion Processes, 116 (1992) 143.
[106] A. Marshall, A. Clark, R. Jennings, K. W. D. Ledingham and R. P. Singhal, Int. J. Mass Spectrom. Ion Processes, 112 (1992) 273.
[107] C. Mullen, A. Irwin, B. V. Pond, D. L. Huestis, M. J. Coggiola and H. Oser, Anal. Chem., 78 (2006) 3807.
[108] G. W. Lemire, J. B. Simeonsson and R. C. Sausa, Anal. Chem., 65 (1993) 529.
[109] C. Kosmidis, A. Marshall, A. Clark, R. M. Deas, K. W. D. Ledingham and R. P. Singhal, Rapid Commun. Mass Spectrom., 8 (1994) 607.
[110] R. L. Pastel and R. C. Sausa, Appl. Opt., 35 (1996) 4046.
[111] S. Schmidt, M. F. Appel, R. M. Garnica, R. N. Schindler and T. Benter, Anal. Chem., 71 (1999) 3721.
[112] J. Zhu, D. Lustig, I. Sofer and D. M. Lubman, Anal. Chem., 62 (1990) 2225.
[113] J. B. Simeonsson, G. W. Lemire and R. C. Sausa, Appl. Spectrosc., 47 (1993) 1907.

[114] J. B. Simeonsson, G. W. Lemire and R. C. Sausa, Anal. Chem., 66 (1994) 2272.
[115] S. Guizard, D. Chapoulard, M. Horani and D. Gauyacq, Appl. Phys. B, 48 (1989) 471.
[116] R. L. Pastel and R. C. Sausa, Appl. Opt., 39 (2000) 2487.
[117] R. C. Sausa, V. Swayambunathan and G. Singh, Detection of Energetic Materials by Laser Photofragmentation/Fragment Detection and Pyrolysis/Laser-Induced Fluorescence, ARL-TR-2387, U.S. Army Research Laboratory (2001).
[118] M. O. Rodgers, K. Asai and D. D. Davis, Appl. Opt., 19 (1980) 3597.
[119] H. Zuckermann, G. D. Greenblatt and Y. Haas, J. Phys. Chem., 91 (1987) 5159.
[120] Y. Q. Guo, M. Greenfield and E. R. Bernstein, J. Chem. Phys., 122 (2005) 244310.
[121] D. D. Wu, J. P. Singh, F. Y. Yueh and D. L. Monts, Appl. Opt., 35 (1996) 3998.
[122] G. M. Boudreaux, T. S. Miller, A. J. Kunefke, J. P. Singh, F.-Y. Yueh and D. L. Monts, Appl. Opt., 38 (1999) 1411.
[123] N. Daugey, J. Shu, I. Bar and S. Rosenwaks, Appl. Spectrosc., 53 (1999) 57.
[124] J. Shu, I. Bar and S. Rosenwaks, Appl. Opt., 38 (1999) 4705.
[125] J. Shu, I. Bar and S. Rosenwaks, Appl. Phys. B, 70 (2000) 621.
[126] J. Shu, I. Bar and S. Rosenwaks, Appl. Phys. B, 71 (2000) 665.
[127] T. Arusi-Parpar, D. Heflinger and R. Lavi, Appl. Opt., 40 (2001) 6677.
[128] D. Heflinger, T. Arusi-Parpar, Y. Ron and R. Lavi, Opt. Comm., 204 (2002) 327.
[129] K. W. D. Ledingham and R. P. Singhal, Int. J. Mass Spectrom. Ion Processes, 163 (1997) 149.
[130] C. Kosmidis, K. W. D. Ledingham, H. S. Kilic, T. McCanny, R. P. Singhal, A. J. Langley and W. Shaikh, J. Phys. Chem. A, 101 (1997) 2264.
[131] K. Tonnies, R. P. Schmid, C. Weickhardt, J. Reif and J. Grotemeyer, Int. J. Mass Spectrom., 206 (2001) 245.
[132] C. Weickhardt and K. Tonnies, Rapid Commun. Mass Spectrom., 16 (2002) 442.
[133] S. M. Hankin, A. D. Tasker, L. Robson, K. W. D. Ledingham, X. Fang, P. McKenna, T. McCanny, R. P. Singhal, C. Kosmidis, P. Tzallas, D. A. Jaroszynski, D. R. Jones, R. C. Issac and S. Jamison, Rapid Commun. Mass Spectrom., 16 (2002) 111.
[134] C. Osorio, L. M. Gomez, S. P. Hernandez and M. E. Castro, Proc. SPIE-Int. Soc. Opt. Eng., 5794 (2005) 803.
[135] C. Mullen, D. Huestis, M. Coggiola and H. Oser, Int. J. Mass Spectrom., 252 (2006) 69.
[136] K. Hasue, S. Nakahara, J. Morimoto, T. Yamagami, Y. Okamoto and T. Miyakawa, Propell. Explos. Pyrot., 20 (1995) 187.
[137] L.-G. Rosengren, Appl. Opt., 14 (1975) 1960.
[138] A. M. Angus, E. E. Marinero and M. J. Colles, Opt. Comm., 14 (1975) 223.
[139] M. E. Webber, M. Pushkarsky and C. K. N. Patel, Appl. Opt., 42 (2003) 2119.
[140] L. B. Kreuzer, N. D. Kenyon and C. K. N. Patel, Science, 177 (1972) 347.
[141] P. C. Claspy, Y. H. Pao, S. Kwong and E. Nodov, IEEE J. Quant. Electron., 11 (1975) D37.
[142] P. C. Claspy, Y.-H. Pao, S. Kwong and E. Nodov, Appl. Opt., 15 (1976) 1506.
[143] P. C. Claspy, C. Ha and Y.-H. Pao, Appl. Opt., 16 (1977) 2972.
[144] R. W. Terhune and J. E. Anderson, Opt. Lett., 1 (1977) 70.
[145] R. A. Crane, Appl. Opt., 17 (1978) 2097.
[146] A. Fried, Appl. Spectrosc., 36 (1982) 562.
[147] R. L. Prasad, R. Prasad, G. C. Bhar and S. N. Thakur, Spectrochim. Acta, Part A, 58 (2002) 3093.
[148] R. J. Simonson, B. G. Hance, R. L. Schmitt, M. S. Johnson and P. J. Hargis Jr., Proc. SPIE-Int. Soc. Opt. Eng., 4394 (2001) 879.
[149] N. Xiang and J. M. Sabatier, J. Acoust. Soc. Am., 113 (2003) 1333.
[150] J. J. Bromenshenk, C. B. Henderson and G. C. Smith, 'Appendix S: Biological Systems (Paper II)', in Alternatives for Landmine Detection, J. MacDonald, J. R. Lockwood,

J. McFee, T. Altshuler, T. Broach, L. Carin, R. Harmon, C. Rappaport, W. Scott and R. Weaver (Eds.), Santa Monica, CA: RAND Corp. (2003) 273.

[151] K. S. Repasky, J. A. Shaw, R. Scheppele, C. Melton, J. L. Carsten and L. H. Spangler, Appl. Opt., 45 (2006) 1839.

[152] J. A. Shaw, N. L. Seldomridge, D. L. Dunkle, P. W. Nugent, L. H. Spangler, J. J. Bromenshenk, C. B. Henderson, J. H. Churnside and J. J. Wilson, Opt. Express, 13 (2005) 5853.

[153] I. N. Germanenko, S. T. Li and M. S. El-Shall, J. Phys. Chem. B, 105 (2001) 59.

Chapter 11

Detection of Explosives by Terahertz Imaging

John F. Federici[a], Dale Gary[a], Robert Barat[b] and Zoi-Heleni Michalopoulou[c]

[a]*Department of Physics, New Jersey Institute of Technology, Newark, NJ 07102, USA*
[b]*Otto York Department of Chemical Engineering, New Jersey Institute of Technology, Newark, NJ 07102, USA*
[c]*Department of Mathematical Sciences, New Jersey Institute of Technology, Newark, NJ 07102, USA*

Counterterrorist Detection Techniques of Explosives
Jehuda Yinon (Editor)

Contents

1. Introduction

Since 2001, there has been an increased interest in the potential of terahertz (THz) detection for imaging of concealed weapons, explosives, and chemical and biological agents [1]. As suggested in Table 1, there are three major factors contributing to this interest:

(1) Terahertz radiation is readily transmitted through most non-metallic and non-polar mediums, thus enabling THz systems to 'see through' concealing barriers such as packaging, corrugated cardboard, clothing, shoes, book bags, etc. to probe the potentially dangerous materials contained within.
(2) Many materials of interest for security applications including explosives, chemical agents, and biological agents have characteristic THz spectra that can be used to fingerprint and thereby identify these concealed materials (Fig. 1).
(3) Terahertz radiation poses either no or minimal health risk [2–6] to either a suspect being scanned by a THz system or the system's operator.

Table 1. Comparison of terahertz radiation with other radiation for explosive detection

Competing Technologies	Transmission through metals	Transmission through common dielectrics	Material signature	Non-ionizing
Terahertz	No	Yes	Yes	Yes
X-Ray	Yes	Yes	No	No
Infrared	No	No	Yes	Yes
Millimeter-Wave	No	Yes	No	Yes

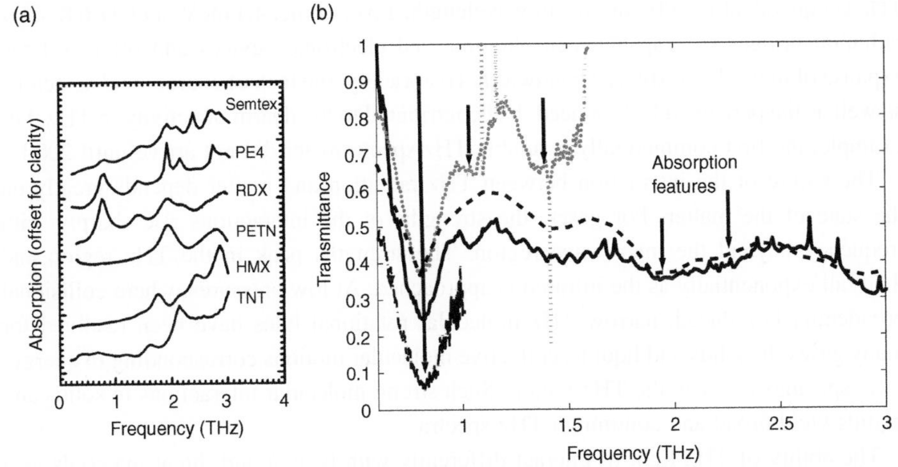

Fig. 1. (a) THz absorption spectra from Kemp et al. [38]. (b) Comparison study of transmission spectra of RDX using THz-TDS and Fourier Transform Infrared (FTIR) spectroscopy with data from Kemp et al. and Huang et al [39]. The dashed line is RDX from Kemp et al. The other curves are from Huang et al., where the bottom curve is RDX measured by THz-TDS, the solid curve is RDX measured by FTIR, and the top curve is C-4 measured by THz-TDS.

As plastic explosives, fertilizer bombs, and chemical and biological agents increasingly become weapons of war and terrorism, and the trafficking of illegal drugs increasingly develops as a systemic threat, effective means for rapid detection and identification of these threats are required. One proposed solution for locating, detecting, and characterizing concealed threats is to use THz electromagnetic waves to spectroscopically detect and identify concealed materials through their characteristic transmission or reflectivity spectra in the range of 0.5–10 THz. For example, many explosives (e.g., C-4, HMX, RDX, and TNT) and illegal drugs (e.g., methamphetamine) have characteristic transmission/reflection spectra in the THz range that could be distinguishable from other materials such as clothing, coins, and human skin. In essence, these materials should appear as different 'colors' to the THz detector as compared with non-hazardous items. Using THz spectroscopy, it should be possible to detect explosives or drugs even if they are concealed because the THz radiation is readily transmitted through plastics, clothing, luggage, paper products, and other non-conductive (non-metallic) materials. Metals completely block or reflect THz waves. By comparing measured reflectivity THz spectra with known calibration spectra, one may identify the presence of these agents and distinguish them from benign objects.

2. Terahertz radiation

2.1. Spectral range

Terahertz radiation lies between the microwave and the infrared regions of the electromagnetic spectrum. 'Terahertz' typically ranges from 0.1×10^{12} to 10×10^{12} Hz. One THz is equivalent to 300 microns in wavelength, 1 ps in time, 4.1 meV, and 47.6 K. THz radiation bridges the gap between photonic and electronic devices and offers a large expanse of unused, unexplored bandwidth. Historically, the lack of sources and detectors, as well as the perceived lack of need, had contributed to the dearth of activity in THz. For example, the first commercially available THz spectrometer did not arrive until 2000.

The nature of the interaction between THz radiation and matter depends greatly on the state of the matter. For gases, the strengths of the interactions rise sharply with frequency beyond the microwave region. The strengths peak in the THz region and then fall exponentially as the infrared is approached. At low pressures, where collisional broadening is reduced, narrow THz molecular rotational lines have been resolved for many gases. In solids and liquids, collective molecular motions corresponding to energy-level spacings occur in the THz region. Such strong molecular interactions in solids and liquids yield broad and continuum THz spectra.

The ability of THz light to interact differently with benign and threat materials as a function of THz frequency yields a highly flexible foundation for THz imaging security screening based on spectroscopy. In general, non-polar, non-metallic solids such as plastics and ceramics are at least partially transparent and reflective in the 0.2–5 THz range. Non-polar liquids are transparent as well, whereas polar liquids, such as water, are highly

Table 2. Collection of representative absorbance peak positions of some explosives. Conversion to units of wave-numbers requires multiplication by $33\,cm^{-1} = 1\,THz$

Material explosive	**Feature band center position frequency (THz)**	**Ref**
Semtex-H	0.72, 1.29, 1.73, 1.88, 2.15, 2.45, 2.57	35
PE-4	0.72, 1.29, 1.73, 1.94, 2.21, 2.48, 2.69	35
RDX/C-4	0.72–0.8, 1.26, 1.73	13, 35, 41, 43, 45
PETN	1.73, 2.01, 2.51	13, 35
HMX	1.58, 1.84, 1.91, 2.21, 2.57	13, 35
TNT	1.44, 1.7, 1.91, 5.6, 8.2, 9.1, 9.9	13, 35, 42, 47
NH_4NO_3	4, 7	1, 13, 28

absorptive. This is because absorption in the THz range of the electromagnetic spectrum is generally due to the rotational motions of dipoles within a material. Crystals formed from polar liquids are substantially more transparent because the dipolar rotations have been frozen out; however, these crystals may exhibit phonon resonances in the THz range. THz time domain spectroscopy (TDS), Fourier Transform Infrared (FTIR) experiments, and computer simulation of the far-infrared spectra of organic molecules show vibrational features associated with intermolecular hydrogen bond relative motions [7–9]. Gases can have distinctive spectroscopic fingerprints in the THz range [10].

The THz imaging technique is based on the use of THz electromagnetic waves to spectroscopically detect and identify concealed explosives through their characteristic transmission or reflectivity spectra in the THz range. Experimental data that have appeared in the literature suggest that many materials that are relevant to security applications have characteristic THz reflection or transmission spectra (see references listed in Table 2). Typical clothing items and paper and plastic packaging should appear transparent in the THz regime. Metals completely block or reflect THz waves. Ceramic guns and knives would partially reflect the THz light. Images of concealed objects such as concealed metallic or plastic knives are sharp and can be identified when imaged with THz [11,12]. Skin, because of its high water content, would absorb nearly all T-Rays. The energy would be harmlessly dissipated as heat in the first 100 microns of skin tissue. A THz reflection image of a person would show the outline of clothing and the reflection of objects beneath (such as weapons or key chains), but the person's skin would appear substantially dark.

2.2. *Sample preparation*

Before discussing the THz spectra of explosives, some discussion should be made regarding explosive sample preparation method. Ideally, for security applications, the THz spectra should be independent of sample preparation. However, structural changes of the material and impurities or fillers introduced during sample preparation can affect

THz spectra. As an example, there are discrepancies in the reported THz spectra of RDX [38,39,90]. In addition, the experimental method itself can introduce spectral artifacts that can mask the true THz spectral fingerprint. These include multiple reflections within the sample, multiple reflections between the sample and its holder, and multiple reflections from the system, etc. [13].

To uniquely identify the intrinsic feature of the material, one method of sample preparation is to pelletize the explosive powders or crystals [14]. It is standard practice in far-infrared (THz) spectroscopy to press samples into pellet form to measure the THz transmission spectra. When the sample is a powder with a grain size comparable to the THz wavelength (about 300 microns), the powder strongly scatters the THz radiation. Another method of sample preparation is to mix the material (e.g., RDX) with an inert matrix or filler material to create a pellet. The filler is typically a material that is transparent in the THz such as polyethylene. This allows dilute concentrations of a highly absorbing agent to be measured.

The presence of a matrix material and/or a pressed pellet, however, can influence the measured THz spectra. For example, the use of a matrix material with an index of refraction that closely matches with that of the powder can minimize the scattering by reducing the effect of the dielectric mismatch in the THz range. In general, one would need to analyze the spectroscopic data that treat the composite sample as one dielectric material (powder) imbedded in another dielectric host material (Garnett theory) [15]. The proximity of grains of powder to each other can lead to interaction between neighboring grains that modifies effective dielectric of composite material and consequently the THz spectra [16,17]. It has been suggested [39] that several of the THz peaks in RDX are due to crystalline nature of the powder and long-range oscillations of the crystal structure. To conduct a thorough experiment with pellets, one should try different fillers, packing fractions, etc. to ensure that the spectral features of interest are not artifacts resulting from a particular form of sample preparation.

While it is good science to pelletize samples to try to understand the scientific origin of various THz peaks, terrorists or smugglers are not going to prepare their samples, so that they are readily detectable! A realistic field-ready THz detection system should be able to measure or detect the threat in a realistic form (e.g., improvised explosive devices) or in a realistic filler (e.g., RDX in a plastic matrix to make C-4). Despite varying preparation methods, many spectral features of explosives, for example, are reproducible and not sensitive to sample preparation. While there are some slight variations in the THz spectra of C-4 and RDX that can be attributed to the matrix or sample preparation methods, there clearly are peaks/features in the THz spectra that are independent of whether one is measuring an RDX grains, powder, or RDX mixed in a matrix (e.g., C-4).

2.3. Granular solids

For granular solids, apart from concentration (or particle density), moisture content, binder materials and other factors, the grain sizes of the solid itself play a major role in determining the THz spectra. Especially for solids having grain sizes comparable to THz wavelengths, the extinction spectra are greatly influenced by scattering losses that partially obscure the characteristic phonon resonances leading to complications in the quantitative analysis and subsequent material identification. The situation gets particularly critical when the materials of interest do not have any sharp intrinsic material absorption peaks in THz range. For these non-absorbing solid materials, therefore, one is required to consider the effects of scattering because of variation in sizes and shapes of the grains while analyzing their THz spectra. The effect of scattering because of variation of grain sizes is an important consideration for THz security screening imaging systems, as it directly affects the ability of a system to successfully classify different lethal agents such as explosives based on spectroscopy.

Almost all of the practical experimental arrangements that measure the attenuation of an electromagnetic wave in terms of either the reduction of the original field amplitude or intensity or energy of the incident wave upon the propagation through a finite distance through a material medium, essentially estimate the sum effects of two fundamental processes: absorption and scattering. Moreover, as these two processes are not mutually independent, it is rather a very difficult task to isolate the effects of these two processes analytically. Nevertheless, careful experimental observation [18,19] and theoretical analysis [20–22] can estimate the average attenuation of the electromagnetic wave caused by the scattering and absorption processes independently. In most of the cases, the procedure involved to extract the scattering contribution from the total attenuation of the electromagnetic wave is computationally challenging [23], especially whenever multiple scattering events take place. Material properties such as grain size and shape with respect to the frequency of the incident electromagnetic radiation set the regime of particular type of scattering processes such as Rayleigh scattering, Mie scattering, and others [22].

Even though THz spectroscopy has been extensively used during the past decade to characterize numerous materials in the far-infrared region of the spectrum, only recently, granular solids are being inspected using THz radiation and experimentally observed absorption peaks are being reported [24–32]. The submillimeter wavelength scale of THz radiation implies that scattering will have significant effect on THz signals while passing through granular solids and will partially obscure the characteristic phonon resonances of the solid material leading to complications in the quantitative analysis. The radiation will either experience Mie scattering (proportional to ν^m where m could be 1.1, 1.2, 1.5, 2.0 etc. depending on the material) or much stronger Rayleigh scattering (proportional to ν^4) depending on the size distribution of the grains which could be either comparable or much smaller in scale to the THz wavelength, respectively. Despite this significant influence of scattering on the propagation of THz radiation through solids, experimental and theoretical work on scattering studies has been very limited in the THz range [33–35].

As an example of the importance of grain size on THz spectroscopic identification of explosives, the effect of morphology of granular solids on their THz spectra in the region from 0.2 to 1.2 THz using samples of ammonium nitrate (NH_4NO_3) of different grain sizes was studied. It was found (Fig. 2) that NH_4NO_3 has monotonically

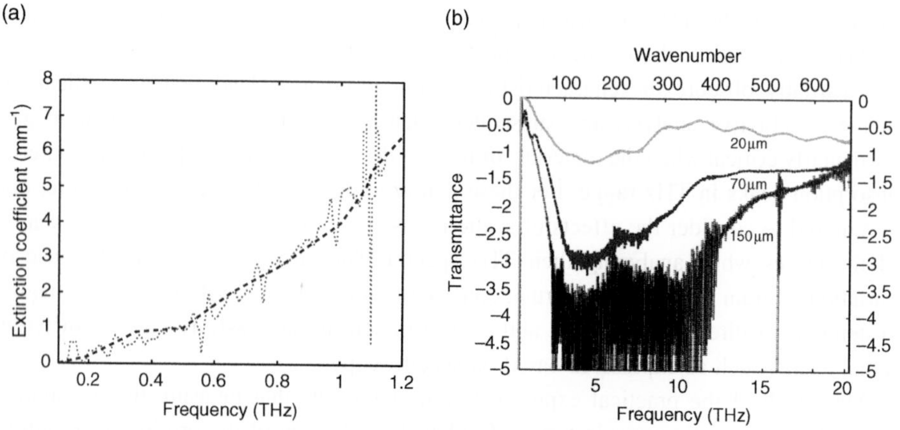

Fig. 2. (a) THz spectra of ammonia nitrate (1-mm thick sample). The dotted line corresponds to measured data of Federici et al. [1]. The dashed line is from Cook et al. [14]. (b) Extended transmission spectra to 21 THz with nominal sample thickness listed from Federici et al. [1].

increasing absorption spectra lacking any characteristic phonon resonance in the far IR range up to 3 THz [1,14]. Therefore, it was expected that by changing the grain sizes of the sample, the effect of the scattering could be studied. It was found that the transmission and extinction spectra of samples of different grain sizes of NH_4NO_3 show a characteristic trend with variation of grain sizes indicating scattering contribution to the total attenuation. The transmission spectra was later fitted to an $\sim\nu^2$ dependence, and therefore, Mie scattering was suspected to play a major role in this case [30,31].

It has been shown [36] that the experimentally obtained THz extinction spectra of NH_4NO_3 and as well as granular salt, chalk, sugar, and flour of known grain sizes can be predicted on the basis of the Mie Scattering model at lower THz frequencies for smaller grain sizes. These materials exhibit an increase in THz extinction with frequency that matches with the prediction of Mie theory in the limit of weak scattering. Assuming that this trend in the spectral shape is characteristic of the material, previous reports have suggested that this might offer a possibility for spectral identification in the lower THz frequencies [24].

Therefore, to test the possibility, a generic curve based on Mie theory [36] can be plotted which predicts the total extinction coefficient (absorption plus scattering) for M different materials, with refractive indices n_M, and a range of values of size parameter, x, to account for any variation of their size and/or wavelength. The size parameter is defined as the ratio between the size of the grain and the center wavelength of the probing radiation. Figure 3 shows the expected extinction coefficient as a function of the size parameter for materials with different real indices of refraction. In the lower frequency region, this plot provides an incisive estimation of the total extinction for materials with a particular refractive index value and grain size. As shown in part (b) [where the graph of part (a) has been enlarged to show only up to the first undulation in the data], these power law frequency dependences of the extinction coefficient vary with the real index of refraction for a constant grain size of 100 μm.

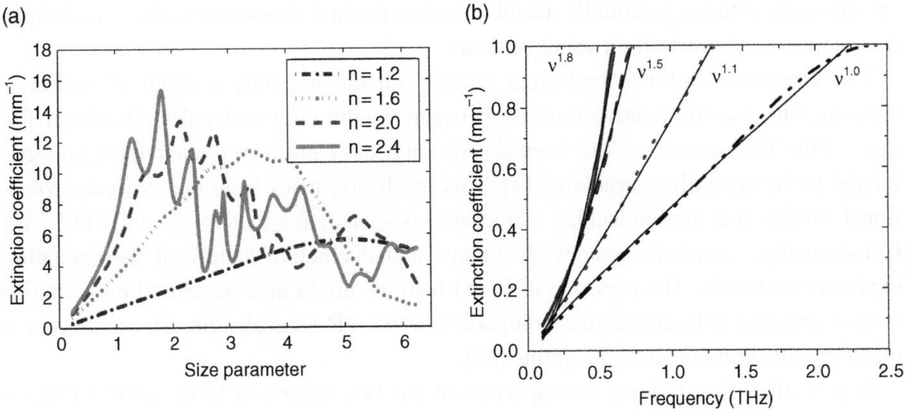

Fig. 3. (a) Theoretically predicted extinction coefficients for materials having refractive indices, n, and size parameters, x, as predicted from Mie scattering theory and (b) the normalized extinction coefficient for a grain size of 100 μm as a function of frequency.

Herrmann et al. [24] reported experimental findings on absorption spectra of solid grains where they have shown the frequency dependence to be approximately ν^1 and ν^2 respectively, for flour ($n \sim 1.4$) and sugar ($n \sim 2.0$). On a direct comparison of these frequency trends with the theoretical plot of Fig. 3(b), one may find that the predicted extinction spectra according to Mie scattering theory are to be indeed $\nu^{1.0}$ and $\nu^{1.8}$ for those materials with those particular grain sizes. This effectively suggests that their observed spectra were primarily due to single event scattering contribution of the grains under study (flour ~50 μm and sugar ~300 μm). Moreover, from Fig. 3(a), one can readily argue that these frequency trends are not unique or representative of the materials used, as extinction is a function of the refractive indices, n, and as well as the size parameters, x of the samples. In fact, for a specific material with different grain sizes, one would obtain varying frequency dependence of extinction. Likewise, the grain sizes of different materials could be chosen in such a way that the extinction spectra of all those different materials with varying grain sizes would show a particular frequency trend. For example, 200 μm grains of flour and 450 μm grains of sugar would yield the same frequency trend of $\nu^{1.1}$ in the extinction spectra [36]. Therefore, the frequency dependence of extinction spectra alone cannot be used for identifying non-absorbing spherical granular solids of unknown grain sizes in the weak scattering region.

2.4. *Explosives*

Table 2 contains an abbreviated summary of materials with their reported absorbance peak positions. Most of these explosives show distinct THz fingerprints in the 0.1–3 THz range (Fig. 1). As many of these explosives typically have low vapor pressure [14], THz spectroscopy offers a potentially useful detection method alternative to those techniques that require a relatively high vapor pressure.

The vibrations within a molecular crystal cell are not only a result of molecular motions, but also the relative motions between neighboring molecules. Dominant features of the THz spectra are the sharp absorption peaks caused by phonon modes directly related to the crystalline structure [14]. This result originates from the molecular vibrational modes and intramolecular vibrations associated, for example, with RDX [39]. Consequently, vibrational modes are unique and distinctive feature of the crystalline explosive materials. The presence of broad features might also be caused by scattering from a structure with dimensions comparable to the THz wavelength. This can occur in materials that contain fibers or grains [37].

In particular, Fig. 1 shows a comparison of the THz spectra of RDX and C-4 [38,39]. Note that the absorption features at 0.8, 1.5, 2.0, 2.2, and 3 THz are exhibited in both the data sets. It should be noted that the main absorption features of RDX are still present when RDX is mixed with a matrix material to make C-4. The spectral feature at 1.1 THz from Huang et al. [39] is not present in the data from Kemp et al. [39]. The sharp features in C-4 (THz-TDS) are remnants of water. The spectral feature at 1.1 THz is not present in both the data sets.

As an example of the THz spectral assignment of peaks in crystalline explosives, the THz spectra of RDX [39] and TNT [40] have been compared with a SPARTAN ab initio molecular simulations [41] study using density functional analysis. The TNT molecular vibrational modes >5 THz agree well with the measured THz absorption. Major spectral features should not exist below 1 THz for known molecular conformations of RDX. Ab initio calculation of the vibration modes of one single RDX molecule compared well with the experimental FTIR data beyond 600 cm^{-1} [41]. The observed 0.8 THz feature of RDX (Fig. 1) is interesting because it does not appear to correspond to an inherent molecular vibrational mode. It is therefore reasonable to attribute this significant absorption feature at ~0.8 THz to interactions between the RDX molecules. Possible explanations include molecular conformations [e.g., partial rotations of the nitro (-NO_2) groups], the acoustical branch of such interaction, a weak hydrogen bond between two RDX molecules, or a partial rotation of the RDX molecule.

First principles solid-state density functional analyses have also been performed on the explosive pentaerythritol tetranitrate (PETN) [42] to further understand the relationships between the choice of computational parameters and the predictions of molecular and solid-state properties, such as intermolecular interactions within the crystal cell, in the THz region. This study concluded that the Becke-Perdew functional has the best overall performance and that the choice of basis set is most critical.

2.5. Explosive detection in a transmission versus reflective geometry

There are two likely implementations for explosive detection: a transmission mode and a reflection mode. In both the cases, THz sources illuminate a person or pallet under study. In the transmission mode, the THz radiation propagates through the material to be inspected to the receiving detector. As many explosive materials strongly absorb THz radiation, a transmission mode is effective for fairly small quantities of explosives (e.g., about 0.1-mm thick slab of RDX or about 20 mg of illicit drugs). Much thicker samples become virtually opaque to the THz. In this case, a reflection mode is preferred. In this scenario, a bright THz source illuminates the person under study, for example. The THz radiation propagates through the clothing, reflects from the hidden explosive, propagates back through the clothing, and is finally detected.

While many transmission spectra of explosives have been measured, only recently have reflection spectra [43–49] been reported. Characteristic spectral peaks appear in transmission and reflection spectra at roughly the same wavelength. A simple Kramers–Kronig argument suggests that this is reasonable: Rapid changes in the THz transmission as a function of frequency can be attributed to rapid changes in the imaginary index of refraction (absorption coefficient). By Kramers–Kronig transformations, rapid changes in the imaginary index of refraction are exhibited in the real index of refraction. As the reflectance of a material is determined by the real and imaginary indices of refraction, rapid changes in the imaginary index of refraction should be exhibited in both the transmission and the reflection measurements. However, the reflectivity of an object depends not only on the indices of refraction but also on the surface roughness, polarization of the incident THz radiation, and other geometric factors.

2.6. *Barrier materials*

Under realistic circumstances, a target threat material will be concealed in a package, under clothing, etc. Such 'barrier' materials that conceal the threat material have their own characteristic THz transmission spectra that must be taken into account. For example, a barrier material might absorb THz radiation and therefore limit the maximum thickness of material that the THz can 'see through'. It is also possible that the barrier material could exhibit similar THz absorption peaks ('colors') that could mask the spectral characteristics of the explosives or illicit drugs. However, most barrier materials such as cloth, paper, and plastic show a THz absorption spectrum that is gradually increasing with frequency, particularly below 3 THz [1,37,50].

The detection of land mines using THz spectroscopic imaging [40,51,52] has some unique considerations in terms of barrier materials. Currently, there are an estimated 100 million landmines buried worldwide. Anti-personnel mines are small devices and contain minimal amounts of metal. One potential advantage of THz technology is improved spatial resolution ($\sim$1mm) that should provide better discrimination of mines from harmless objects.

Detection of landmines requires THz to propagate through sand and soil. It is anticipated that imaging through such grainy and moist material will significantly affect the detectability of non-metallic objects hidden beneath it. Sand is measured to have a loss of 8.2 dB/cm% where % denotes the water weight percentage. The non-metallic landmine detection proof of concept was done by imaging a Neoprene grommet (OD = 25.4 mm; ID = 12.7 mm) under 1.2 cm of dry sand [40]. An image is formed by scanning every 300 μm for 100 × 100 pixels. The grommet can clearly be seen in the image.

2.7. *Atmosphere*

Both close proximity (<3m) and standoff detections (>3m) of concealed threat materials require consideration of THz transmission through barrier materials. A standoff THz detection system must be capable of propagating short (3–100 m) distances through the atmosphere. THz propagation through the humid atmosphere has been discussed by several groups [7,10,53]. Atmospheric absorption, which is dominated by water and oxygen as one moves from the millimeter to the THz range [10], absorbs THz radiation at specific frequencies. However, there are numerous transmission bands throughout the THz frequency range [7]. While the THz attenuation in the transmission bands is frequency dependant, a typical number is 50 dB/km at 0.8 THz. A standoff sensing system should be tuned to these transmission bands (e.g., centered at 0.5, 0.65, 0.87, 1.02, 1.29, 1.36, and 1.52 THz) to minimize the absorptive losses of THz radiation by the atmosphere. Although atmosphere attenuation is predominately due to absorption of the THz radiation by water vapor, other attenuation effects such as dust, smoke, fog, and rain could effect the THz transmission. It is thought that rain would increase the effective attenuation because of increased water absorption, whereas typical micron-scale particles such as dust, pollens, and smoke, would not greatly scatter the THz radiation [53] because the THz wavelength is much longer than typical micron-scale particles. However, there are little or no experimental data on atmospheric scattering of THz radiation [53].

2.8. *Safety considerations*

One of the appeals of THz radiation for security screening is that, at present and anticipated power levels, it appears to be safe for living tissue. Some evidence is presented here.

Berry et al. [54] employed guidelines for skin exposure to THz radiation (15 GHz to 115 THz) drawn from American National Standard for the Safe Use of Lasers (ANSI Z136.1) and from the IEEE Standard for Safety Levels with Respect to Human Exposure to Radio Frequency Electromagnetic Fields (C95.1). They concluded that the maximum permissible average beam power was 3 mW, suggesting that typical THz imaging systems are safe.

Human skin cells (keratinocytes) were tested [5] in vitro for their ability to differentiate after exposure to 0.1–3 THz radiation generated by two different pulsed Ti:Sapphire laser systems. The first generated THz by directing 20 ps laser pulses onto an electro-optic photoconverter to produce a 1 μW average output. The second impacted 0.25 ps laser pulses onto a gallium arsenide (GaAs) photoconductive antenna. The average THz power was ~1 mW with an exposure spot size of 130 micron to 3.7 mm. Compared to normal (unexposed cells), the radiated cells showed no ill effects. No statistically significant changes in ability to differentiate were observed.

Refractive indices and absorption coefficients were determined in vitro for human skin and other tissues [3]. The successful measurements, while limited to samples from only one person, do suggest that in vivo THz imaging is feasible.

In vitro tests exposing human blood samples from nine different donors to 0.12–0.13 THz radiation were performed [6]. Samples were irradiated at 0.6–1 mW average power for 20 min. Pulse trains (4 μs) of THz at a 2 Hz repetition rate consisting of 50 ps micropulse 'bunches' were delivered from a free-electron laser. Cell activity was not altered by the exposure to THz, suggesting that no chromosomal or other damage.

Pulsed THz radiation is typically non-ionizing; hence, it is suitable for medical imaging [2]. Various biomedical samples were imaged using 0.15 ps THz pulses generated from 775 nm laser pulses directed onto either a ZnTe crystal or a GaAs antenna. The samples were also successfully classified.

Terahertz radiation is actually employed for medical imaging. THz pulses applied in reflection were used to distinguish cancerous cells from normal tissue both in vitro and in vivo [55]. This technique is based on the different hydration levels in the normal and abnormal skin tissue and the strong interaction of THz with water molecules. One mW average power pulses over a 0.1–2.7 THz bandwidth were generated from 800 nm excitation of a GaAs antenna. For example, studies of basal cell carcinoma using THz pulsed imaging have shown a real difference between healthy tissue and tumors. Whether in reflection or transmission modes, these differences, because of changes in the refractive index and absorption coefficient, provide the contrast needed to distinguish cancerous from healthy tissue [56].

The information presented above suggests that THz radiation expected to be used for scanning of human subjects in a practical security application will not be harmful. In fact, THz radiation shows great promise as a medical imaging technique.

3. Terahertz imaging techniques

In this section, we discuss various methods of THz imaging and several important issues that must be considered in developing a THz imaging detection system for security applications.

3.1. THz imaging modalities

THz imaging has been rapidly evolving due to advances in THz sources, detectors, and device fabrication methods. A high-speed THz imaging device that can scan and positively identify harmful materials at a rapid (video) frame rate is in high demand for security applications. THz images typically have diffraction limited spatial resolution in the ~1 mm range. For security screening applications, this is a sufficient resolution to resolve the shapes of guns, knives, and other concealed weapons with comparable resolution to simple visible video images [92].

THz imagers are at the forefront of R&D for new imaging and sensing modalities for defense and security-related applications. The importance and technological impact of THz imaging array systems for homeland and defense security needs have prompted the US Defense Advanced Research Projects Agency (DARPA) to fund the THz Imaging Focal-plane Technology (TIFT) program. The goal of this program is to demonstrate a large, multi-element detector receiver focal plane array that is sensitive to radiation in the THz band above 0.557 THz. The THz imaging system to be developed will be able to operate effectively at standoff range with a diffraction-limited, two-dimensional (2-D) video-rate imaging.

It is widely recognized that any imaging or detection modality, be it THz imaging or something else, cannot be the 'silver bullet' for security screening. Each modality has its own strengths and weaknesses. When various imaging modalities are used as part of a sensor suite, the collective strengths of the security screening methods are integrated into an overall solution. As an example, consider the issue of screening of trucks at a port of entry or checkpoint. There are detection systems for monitoring of nuclear radiation. X-ray systems could penetrate through the thin metallic walls of the truck to probe the contents but may have difficulty in identifying some types of materials inside. How would the driver be screened? The driver might be carrying concealed weapons, a suicide bomb, a trigger mechanism for a weapon hidden in the truck, or other threats that would be overlooked if only the vehicle were screened. A potential solution is the security 'car wash' wherein the vehicle passes through one set of screening systems that can probe through the metallic structure of the truck or storage containers. At the same time, the driver walks through a hallway or portal that contains its own sequence of imaging sensors (human-safe) to detect concealed weapons.

The image contrast mechanism for detection of weapons depends on their material composition. At THz frequencies, metals appear dark in transmission images and bright in reflection. For high-energy explosives such as RDX and C-4, the THz spectra are characterized by molecular and intra-molecular vibrational modes. Theoretical analysis indicates that the THz spectrum for anthrax is a result of a polariton mode of the anthrax spore. In this mode, the electric field of the THz radiation couples to the dielectric charge displacement in the shell of the spore [57]. Plastic knives are discernable in THz images not because of any spectral features (most plastics are either transparent or semi-transparent to THz radiation) but rather by the scattering of THz radiation from the edges enabling the THz image to show the outline of the object.

There are three modes of THz imaging. The simplest is a single transmitter and detector, i.e., line-of-sight detection. An image is obtained on a point-by-point basis by scanning the transmitter/detector pair over the sample under test and recording the THz phase and amplitude at each point. Using this method, THz images of macroscopic objects have been obtained [58]. Using this method, THz imaging has been extended to THz tomography [59] and sub-wavelength spatial resolution near-field microscopy [60–62].

An alternative THz imaging method is to up-shift the THz spatial (and spectral) information to an alternative electromagnetic frequency range for which image acquisition is more easily implemented [63]. For example, in the electro-optic THz imaging method [64], THz images are shifted into the visible range where conventional CCD cameras acquire images at a very rapid rate.

A third class of THz imaging has emerged. Borrowed relatively recently from their extensive use in astronomy and radar ranging [65], THz synthetic aperture imaging uses the measured THz phase and amplitude from multiple positions or from multiple beam paths to reconstruct a THz image. Examples of synthetic aperture imaging include THz impulse imaging [66], which shares many features with optical holography. By measuring the phase and amplitude of the scattered THz radiation as a function of scattering angle and rotation of the target, the geometrical shape of the target can be reconstructed.

Other methods of image reconstruction measure scattered THz radiation and use an appropriate wave propagation equation to back-propagate the scattered radiation to reconstruct the scattering object [67,68]. While the previous two examples of synthetic aperture imaging use a finite number of detectors at specific positions to reconstruct THz images, synthetic phased array THz imaging [69–71] utilizes arrayed optical mirrors to reconstruct field amplitude or energy density, diffraction-limited, THz images. Many individual images can be recorded and superimposed to produce a higher-resolution image.

The interferometric imaging method of synthetic aperture image reconstruction, described in detail in Section 4, detects the THz electric field at multiple locations and then uses the correlated phase and amplitude of the electric field from selected pairs of detectors [72].

3.2. Close range versus standoff detection

Several of the THz imaging methods described above have shown promise for THz security screening. Imaging of concealed weapons can be roughly classified by the effective range of the imaging. Close range (<3m) imaging offers solutions to a wide variety of applications. Over short distances, atmospheric attenuation and scattering is minimal. Screening of mail, packages, and baggage is a close range application, whereas detection of explosives, weapons, and illicit drugs on people approaching a checkpoint or portal are a standoff application. Standoff (>3m) THz imaging allows for long-range screening and detection and potentially offers a solution for scanning persons in open areas such as airports, stations, etc. As discussed below, the range requirements partially determine the architectural choices of the detection system.

3.3. Pulsed versus continuous wave sensing

Terahertz imaging approaches have typically used either short-pulsed laser or continuous wave (CW) THz generation and detection. The short-pulsed method usually involves the generation and detection of sub-picosecond THz pulses using either photoconductive antenna structures or optical rectification in a non-linear crystal. Pulsed sources seem to be more favorable (in particular for close proximity applications) because they can be used for acquiring depth information. Spectral information is retrieved by a Fourier transform of the time-domain data to the frequency domain.

Continuous wave generation of THz radiation by photo-mixing (beating) of two infrared laser sources commenced in the 1990s through the seminal work of Brown, McIntosh, and Verghese [73,74]. The technology for growth, design, and characterization of the low-temperature (LT) grown GaAs photo-mixers has improved dramatically [73,74], enabling the use of CW THz systems for sensing, spectroscopy [27,53], and imaging applications [75–79]. The key material component of LT-GaAs and higher power ErAs:GaAs photo-mixers is the presence of nanoparticulates that reduce the charge carrier lifetime in the material to the sub-picosecond level thereby enabling optical mixing to the THz range.

While there have been improvements in photo-mixer performance with ErAs:GaAs materials [80], the photo-mixer approach has been limited by the achievable output power and device reliability [81]. To circumvent the carrier lifetime limits of the GaAs system, several groups utilized p-i-n photodiode structures [81]. THz photo-mixing by resonant excitation of plasma oscillations in quantum well structures has also been considered [82,83]. While the GaAs systems use infrared lasers at ~800 nm wavelengths, the InAlAs/InGaAs system can utilize inexpensive telecommunication lasers operating near 1.5 μm [84]. In addition to semiconductor-based systems, CW THz generation and detection have been demonstrated using non-linear optical crystals [85].

An example of quasi CW THz detection [86] uses a THz wave parametric oscillator (TPO) consisting of a Q-switched Nd:YAG laser and parametric oscillator [87,88]. In this technique, $MgO:LiNb_3$ is employed as a non-linear material to generate CW THz. Silicon prisms couple the THz radiation from the non-linear crystal where it is detected using a pyroelectric detector. THz images are collected at discrete THz frequencies and then spectroscopically analyzed using a component spatial pattern analysis method to determine sample composition.

The short-pulsed laser method usually involves the generation and detection of THz pulses using either photoconductive antenna structures or optical rectification in a non-linear crystal. In one version of the short-pulsed laser technique [89], the entire waveform of a single THz pulse is recorded by encoding the temporal profile of the THz pulse on a chirped ultra-fast laser pulse. Each wavelength of the infrared laser pulse corresponds to a different time window of the THz pulse. The temporal profile of the THz pulse is measured by analyzing the spectral components of the infrared pulse. In the CW generation method, two narrowband infrared laser sources are miss-tuned by roughly 1 THz. CW THz radiation is generated by the mixing of the two laser sources in a non-linear crystal. One limitation in utilizing these techniques is that coherent CW or short-pulsed laser sources are required. Moreover, the laser sources that generate and detect the THz radiation must retain a coherent phase relationship with each other. Using these methods, the imaging of an incoherent THz source is not possible.

Pulsed sources seem to be more favorable (in particular for close proximity applications) because they can be used for acquiring depth information. For example, a pulsed THz imaging method was used [11,12,38,90] to demonstrate a proof of principle for security screening. Hidden objects (e.g., scalpel, alumina ceramic, acrylic plastic, and SX2 sheet explosive) were imaged through barrier materials (e.g., inside a parcel post box or under layers of clothing or shoes). The advantages of pulsed THz TDS is that broad spectral information (0.1–3 THz) can be acquired from a single picosecond THz pulse as well as the depth information from the difference in arrival times of the short pulses. CW imaging systems have the advantage of higher THz power at a distinct THz frequency. A further advantage of CW THz spectroscopy over pulse THz TDS is that narrow spectral features are easier to measure using CW techniques because of the spectrally narrow CW THz radiation and because of the lack of a long scanning delay line that would be required for high spectral resolution using a time-domain system.

3.4. Standoff sensing modalities

For standoff detection, a dominating issue in choosing a pulsed or CW source is the need to propagate through the atmosphere. As discussed in Section 2.7., only certain bands of THz frequencies are appropriate transmission windows for remote detection applications. For THz time-domain pulses that are generated by short-pulse laser systems (e.g., mode-locked Ti:Sapphire lasers), the THz spectrum of the pulse spreads over several transmission bands. However, the THz power that is *outside* of the transmission bands is highly absorbed. Consequently, the amount of usable THz power in the pulse is drastically reduced. In addition, the time duration of the picosecond THz pulses is considerably lengthened due to the water vapor absorption. Thus, pulsed time-of-flight THz detection is effectively impossible. For propagation of THz pulses through 2.4 m of a humid atmosphere, a 1-ps pulse is broadened to a time duration in excess of 30 ps. After 100 m, the pulse is well beyond 100 ps in duration [7]. To use a time-domain THz pulse for standoff detection, the entire waveform (e.g., 100 ps for 100 m standoff) would need to be digitized.

Despite the difficulty imposed by the attenuation and dispersion of the atmosphere on subpicosecond THz pulses, there has been some effort to extract spectroscopic signatures of explosives, such as RDX, from standoff ranges up to 30 m [91]. However, the measured spectral absorption peak of RDX artificially broadens, due to the absorption and dispersion in the atmosphere, as the distance to the target increases. This broadening at standoff distances suggests that spectroscopic identification of explosives such as RDX might be problematic because the apparent spectral shape cannot be directly compared to a spectral standard curve. To circumvent the atmospheric attenuation and dispersion for standoff detection, CW imaging systems can be tuned to atmospheric transmission windows. Of course, because CW THz outputs are spectrally narrow, these systems will require tuning to probe materials at several wavelengths.

One complication with imaging at different THz frequencies is that the overall spatial resolution of the THz imaging depends on the wavelength of the lowest THz frequency. This is essentially a diffraction effect: the larger THz wavelengths cannot spatially resolve as small an object as the smaller wavelengths. Consequently, when THz images are analyzed (e.g., by artificial neural networks (ANNs) for interferometric THz images, as discussed in Section 4) for the spectral presence of an explosive, the edges or boundaries of the explosive might be subject to uncertainty.

3.5. THz imaging systems

State of the art time-domain THz imaging security systems are capable of raster scanning a single transmitter/receiver over an object with up to 100 pixels per second acquisition speed [92]. THz images are generated by analyzing the reflected THz time-domain wave at each transmitter/receiver position corresponding to one pixel in the image. The waveforms could be analyzed according to various parameters such as time-of-flight, total THz power, THz power in spectral bands (for spectroscopic imaging), peak-to-peak waveform amplitude, etc. The images of Fig. 4 are generated using a gray scale pixel intensity that is logarithmically proportion to the integrated THz power (0.2–2 THz). The briefcase image took 30 min to scan with a pixel size of 1.5 mm. Raster-scanned (20 pixels per second) spectroscopic reflection images of RDX, lactose, and sucrose pellets have also been demonstrated [43].

Rapid raster-scanned CW imaging systems have also been developed. In one such implementation, a 0.2 THz Gunn diode source and Schottky diode detector [93] could raster-scan an object at a 512 pixels per second acquisition rate. The advantages of pulsed THz TDS compared to CW imaging is that broad spectral information (0.1–3 THz) can

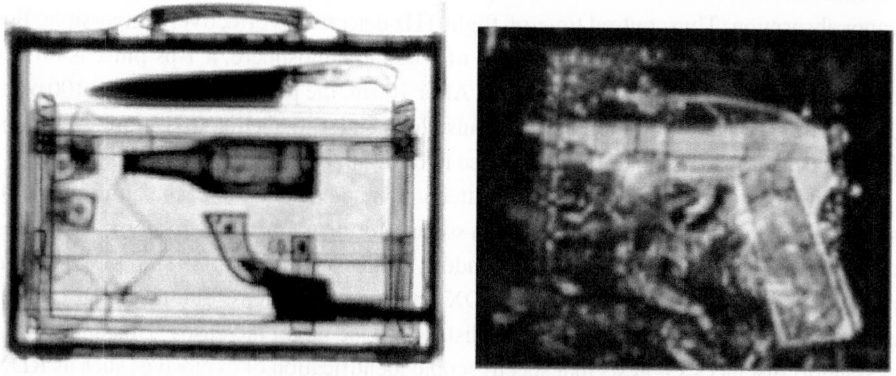

Fig. 4. Terahertz time-domain images from Zimdars et al. [92]: (a) transmission through a briefcase, (b) a gun hidden on a person beneath a jacket.

be acquired from a single picosecond THz pulse as well as the depth information from the difference in arrival times of the short pulses. CW systems require either frequency tuning or multiple sources to generate spectral information. Moreover, depth information through time-of-flight measurements is not possible with CW systems. However, as has been previously discussed in Section 3.2., short pulse methods are not practical for standoff detection imaging.

Clearly one solution and a challenge to high-speed imaging are to fabricate an imaging array to replace the point-by-point raster-scanning of an object. Ideally, one would like to have the equivalent of a THz digital camera. The issue of real-time THz imaging requires advances on three fronts: the development of sensitive THz detectors, the integration of the detectors into an imaging array, and the development of bright THz sources with sufficient power to illuminate a large area ($>1\,m^2$) while penetrating barriers (e.g., clothing and briefcases) and exhibiting a tunability enabling spectral identification of target materials.

Several array approaches are being explored. In one approach, a commercial mid-infrared micro-bolometer focal plane array is used in conjunction with a bright quantum cascade laser (QCL) THz source to demonstrate video-rate THz imaging at 2.52 THz [94]. While the room-temperature micro-bolometers are not optimized for detection in the THz range, there is sufficient sensitivity for ~13 dB S/N for a single video frame. To improve the sensitivity of the detector elements, superconducting NbN micro-bolometer detectors are under development [95]. When these detectors are used for direct detection of THz radiation, it is desirable to optimize the superconducting transition temperature (T_c) to a low value (~9 K) because the noise-equivalent power (NEP) of the detector is proportional to $T_c^{3/2}$. Using these micro-bolometers, passive indoor (i.e., no active THz illumination source) THz imaging with an equivalent background fluctuation of 0.2–0.5 K has been achieved. A 64-element NbN micro-bolometer array is currently under development.

While the direct detection micro-bolometers have superior sensitivity, they exhibit a fairly slow response time (~0.5 ms). If one optimizes the speed of response at the expense of sensitivity, i.e., by making the bolometric thin film, the high speed of response (~1 ns) allows heterodyne detection of THz radiation. Imaging arrays of only a few hot electron bolometrix (HEB) heterodyne receivers have been fabricated [96]. While the superconducting NbN HEB mixers offer high sensitivity and require low local oscillator power ($<1\,\mu W$), one of the major drawbacks, as with the direction detection micro-bolometers, is the cryogenic cooling requirements. As of yet, imaging arrays of only a handful of elements have been fabricated. For a detailed analysis of heterodyne versus homodyne THz detection, the reader is referred to Brown [53].

Why are there currently no THz imaging arrays available for security screening? First, while imaging array technology has been developed for the visible and infrared, relatively few development resources have been focused on a THz imaging array. Second, there are several technical challenges in realizing a THz focal plane array for security screening including room-temperature operation with high sensitivity, integration of individual THz detectors into an array structure, and distribution of 'power' to the individual detectors (i.e., laser power to gate time-domain THz detectors or local oscillator power for THz heterodyne mixers).

In the absence of a kilopixel THz camera, one needs to play 'tricks' to increase the imaging speed and pixel count of images. One method is to convert THz to another frequency range (e.g., visible) for which imaging cameras are readily available. Another method is a synthetic aperture imaging/interferometric imaging approach [72].

4. Interferometric imaging with terahertz

Interferometric imaging has been suggested as a novel imaging modality for standoff detection of explosives, weapons, and other threats. To apply the interferometric synthetic aperture imaging method to the THz range, the basic techniques of radio interferometry [65] are employed. Signals at two or more points in space (i.e., the aperture plane at which the detectors are located) are brought together with the proper delay and correlated both in phase and in quadrature to produce cosine and sine components of the brightness distribution. This technique thus measures both amplitude and phase of the incoming signals. If measured from a sufficient number of points in the aperture plane, the original brightness distribution can be synthesized (imaged) through standard Fourier inversion. From the phase delay in wavefront arrival at the sensor positions, the direction and location of the source can be determined [97]. The instantaneous response of an interferometer to point sources can be analyzed by knowing the signal paths.

Theoretically, the THz interferometric imaging array approach should have sufficient spatial resolution to detect centimeter-sized concealed explosives from standoff distances. A longer-term advantage is that interferometric imaging may produce more information than a single line-of-sight system. With repeated measurements, it should be possible to apply image processing/computational techniques to multiple images and THz sources to aid in noise and false alarm reduction. In the millimeter wave range, it is anticipated that synthetic aperture techniques could enable imaging systems (1-m aperture) that are lightweight and only a few wavelengths in thickness [98].

4.1. Theory

The imaging interferometer consists of an array of individual detectors or sensors. Each pair of THz detectors measures the amplitude and phase of incoming THz radiation. As a wavefront of THz radiation encounters the array, each pair of detectors measures one spatial Fourier component of the incoming THz radiation as determined by the separation (baseline) of the detector pair. Each spatial Fourier component is represented as a point in the Fourier transform (u-v) plane. To image the source, additional measurements from other baselines must be carried out. As every baseline determines a singe spatial frequency, it is best to have a zero-redundant [98] (i.e., non-periodic) detector array. For a given number of detectors N, there are $N(N-1)/2$ possible baseline combinations.

An image is generated from the spatial Fourier components of all the different pair combinations. The quality of an image depends on the coverage of the u-v plane, i.e., the number of different points generated in the u-v plane. This in turn depends on the arrangement of the detecting elements of the imaging interferometer array. The primary concern in designing the configuration of detectors is to obtain an efficient coverage of the u-v plane over a range determined by the required angular resolution.

For interferometric detection, the correlation of the electric fields at the various pairs of detectors is calculated. It can be shown that the mutual coherence function of the electric fields at detector placements $r_1 = (x_1, y_1)$ and $r_2 = (x_2, y_2)$ can be written as [65]

$$C_{1,2} = \int_S \frac{\sigma_E(r') \exp[ik(r_1 - r_2)]}{r_1 r_2} \mathrm{d}S' \tag{1}$$

where $\sigma_E(\mathbf{r}')$ is the time-averaged intensity of the surface at $\mathrm{d}S'$, and the integral is over the surface S of the radiating source (Fig. 5). Assuming that $|\mathbf{r}' - \mathbf{r}|/Z_0 \ll 1, r_1 - r_2$ becomes

$$r_1 - r_2 = \frac{x_1^2 - x_2^2 + y_1^2 - y_2^2}{2Z_0} + \frac{(x_2 - x_1)x' + (y_2 - y_1)y'}{Z_0} \tag{2}$$

where Z_o is the distance along the z axis between the object and imaging array, and x', y', z' locate the spatial coordinates of the object. Using the following definitions, $u = k(x_1 - x_2)/2\pi$, $v = k(y_1 - y_2)/2\pi$, $\xi = x'/Z_0$, and $\eta = y'/Z_0$, the coherence function of Eq. 1 can be cast into the form [65]

$$C_{1,2}(u, v) = [\exp(i\delta)] \int_{-\infty}^{\infty} \int_{-\infty}^{\infty} \sigma_E(\xi, \eta) \exp[-i2\pi(u\xi + v\eta)] \mathrm{d}\xi \mathrm{d}\eta \tag{3}$$

where the denominator of Eq. 3 has been approximated as $1/r_1 r_2 \approx 1/Z_0^2$. If an object is in the far-field ($\delta \ll 1$), the phase shift $\delta = k(x_1^2 - x_2^2 + y_1^2 - y_2^2)/2Z_0$ can be neglected, and Eq. 3 will relate the coherence function in the antenna plane to the brightness distribution

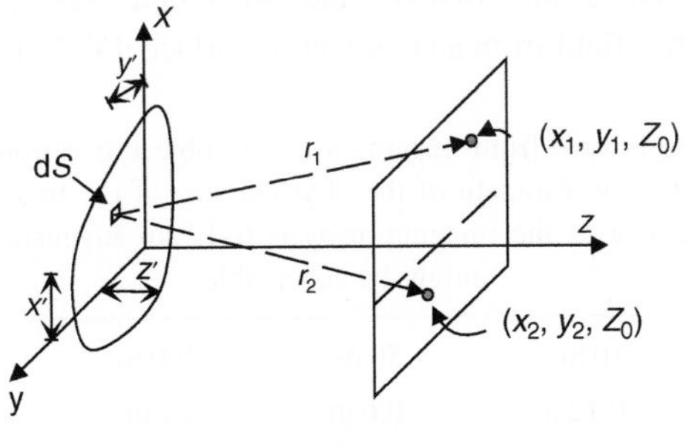

Fig. 5. Source $\mathrm{d}S'$ on surface of $z = z'$ plane irradiating a pair of detectors on the imaging array plane located at (x_1, y_1, Z_o) and (x_2, y_1, Z_o).

of the source(s). By an inverse Fourier transform, the brightness distribution (image) can be reconstructed by measuring the coherence function for a given arrangement of detectors in the sensor array. Imaging with a significant phase shift results in a blurred image quality.

4.2. *Near-field versus far-field*

With radio wave interferometric imaging of astronomical sources, it is typically assumed that the incoming radiation consists of planar waves. In this limit, any curvature of the incoming wavefronts is negligible. This 'far-field' assumption is used for synthetic imaging and simplifies the imaging reconstruction process. However, most standoff THz applications do not fall in the 'far-field' limit, and the simplified inverse Fourier transform of the electric field correlation – the far-field image reconstruction [99] – must be modified [100] to account for the curvature of the wavefronts in the near-field.

The condition for negligible phase shift δ can be approximated as $Z_0 \gg b^2/\lambda$, where b is the largest baseline length of the imaging array and λ is the THz wavelength. As an example, assume that a 2.5-cm object needs to be imaged at various distances. The angular resolution of a planar array can be approximated as $\theta_{\min} = \lambda/b$. At a distance Z_0 away, the lateral spatial resolution is $\Delta L_{\text{lat}} \approx \theta_{\min} Z_0 \approx \lambda Z_0/b$. Using $\delta \approx b^2/Z_0\lambda$ as an estimate for the far-field limit of a planar array, the limit can now be estimated as $\delta \approx Z_0\lambda/\Delta L_{\text{lat}}^2$. To maintain a 2.5-cm lateral resolution at various distances, the maximum baseline for a planar imaging array can be estimated as $b \approx \lambda Z_0/\Delta L_{\text{lat}}$. Using $\delta \sim b^2/Z_0\lambda$ as an estimate of the far-field limit for a planar array, the limit can be estimated as $\delta \sim Z_0\lambda/\Delta L_{\text{lat}}^2$. Table 3 summarizes the corresponding maximum baseline required and phase error. Note that, for this application, the far-field criteria of $\delta \ll 1$ is never satisfied. THz interferometric imaging for standoff applications *must* include contributions from the near-field.

The 'blurring' of the THz image for standoff applications can be greatly reduced by changing the arrangement of the detectors in the imaging array. In essence, the arrangement of the detectors in the array must be modified to 'focus' the imaging array at standoff distances. In the analysis, we match the curvature of the imaging array to that of the wavefront. We assume that two individual detectors of the spherical imaging array measure an electric field from an element of surface dS'. For simplicity, we assume

Table 3. Estimated phase error δ from imaging a 2.5 cm object at various distances using 1 THz radiation. The baseline b is an estimate of the physical size of the imaging array. Note that for distances of 5–10m, the size of the imaging array is 6–12 cm suggesting that a hand-held unit might be acheivable

Z_0	5 m	10 m	50 m	100 m	500 m	1000 m
b	0.06 m	0.12 m	0.6 m	1.2 m	6 m	12 m
δ	2.4	4.8	24	48	240	480

that the detectors lay on a spherical surface whose radius of curvature R_o is centered on the origin. The correlation between the two wavefronts at the two detectors can be calculated from Eq. 1 with r_j expressed in spherical coordinates. One can show that the coherence function becomes [72,100]

$$C_{1,2}(\mathrm{u},\mathrm{v}) = [\exp(i\bar{\delta})] \int\int \sigma_E(\xi,\eta)\exp[-i2\pi(u\xi+v\eta)]\,d\xi d\eta \tag{4}$$

where $\xi = \bar{x}'/R_o$, $\eta = \bar{y}'/R_o$, $v = k(\bar{y}_1 - \bar{y}_2)/2\pi$ and $u = \mathrm{k}(\bar{x}_1 - \bar{x}_2)/2\pi$. In these expressions, $\bar{x}_2 - \bar{x}_1 = R_o(\sin\theta_2\cos\phi_2 = \sin\theta_1\cos\phi_1)$ and $\bar{y}_2 - \bar{y}_1 = R_o(\sin\theta_2\sin\phi_2 - \sin\theta_1\sin\phi_1)$ are the spherical coordinates of the detectors and the phase error is $\bar{\delta} = kz'(\cos\theta_2 - \cos\theta_1)$.

If the near-field phase error $\bar{\delta}$ can be neglected, Eq. 4 would then relate the coherence function to the brightness distribution of the source. Assuming that the azimuthal angles of the detectors are not equal, the magnitude of the phase error can be calculated, assuming that the largest azimuthal angle is determined by the distance to the object and the largest baseline distance. For the baseline and distances summarized in Table 3, the largest possible value of $\cos\theta_2 - \cos\theta_1 \approx 7.2 \times 10^{-5}$ independent of the distance to the object. If $\bar{\delta} = 1$, the approximate depth of focus (range of z') can be estimated as that for which the phase errors are small and the object is in focus in the near-field. For a frequency of 1 THz, the depth of focus would be ± 0.7 m.

An alternative configuration that eliminates the phase error is to arrange the detectors in a circle. In this configuration, all detectors have the same azimuthal angle, thereby enforcing $\bar{\delta} = kz'(\cos\theta_2 - \cos\theta_1) = 0$. For this circular arrangement, the first-order phase error vanishes, implying a much larger depth of focus.

The form of the coherence function of a near-field imaging array approaches the far-field planar array limit when $R_o \to \infty$ because at very large radii of curvature, the surface of the spherical array now becomes flat. When $R_o \to \infty$, $\exp(i\delta) = \exp[ikz'(\cos\phi_2 - \cos\phi_1)] \to 1$ because the azimuthal angles are virtually zero. As $Z_0 \to \infty$, $\exp(i\delta) = \exp(ikb^2/Z_0) \to 1$.

4.3. Simulated interferometric images

As an example of the potential resolution of a standoff interferometric imaging array, Fig. 6 illustrates a simulated 35° field of view image of the NJIT logo from a distance of 25 m. The number of detectors in the circular array varies from 49, 100, to 200. Note that, as the number of detectors in the array is increased, the image becomes clearer. This is expected because increasing the number of detectors increases the number of Fourier components that can be measured, increases the resolution, and reduces the distortion in a reconstructed image. Standard image processing techniques can be used to 'clean' the reconstructed images to reduce the distortion in the reconstructed images [101]. Note that the 200-element image shows minimal lateral distortions. The circular array architecture is capable of a depth of focus larger than ±10% of the distance to the target and a large field of view (35°).

Fig. 6. A THz emitting object in the shape of the NJIT logo. The number of detectors in the circular imaging array increases from $N = 49$, $N = 100$, and $N = 200$ from left to right. The distance to the object is 25 m distance with a ~35° field of view. The images are unprocessed (uncleaned) images.

4.4. *Interferometric versus focal plane array*

Unlike a focal plane imaging array, synthetic aperture/interferometric imaging uses intensity and phase information correlated from pairs of detectors. Whereas detectors are spaced periodically in a focal plane array (e.g., a rectangular arrangement), the interferometric approach requires an aperiodic detector spacing. For N detectors, there are $N(N-1)/2$ detector pairs corresponding to $N(N-1)/2$ pixels in a reconstructed image. As the number of pixels scales as the N^2 rather than N, fewer detectors are needed and faster frame rates possible for the interferometric approach as compared with either a raster-scan or an N detector focal plane array. However, $N(N-1)/2$ correlation evaluations are needed for the interferometric imaging array.

As an example, consider the two types of arrays for imaging with a set resolution at a given distance using the same number of detectors and a fixed area corresponding to 1596 pixels. The number of detectors is chosen to be $N = 57$, so that the number of pixels in the THz image is the same for both the focal plane array image (with scanning) and the interferometric imaging method. The focal plane array, at a rate of 57 pixels per scan, would need 28 scans to cover the entire area. In this estimate, it is assumed that the time to digitize or acquire the data from each detector is the same for both the interferometric and the focal plane array approach. This is a reasonable assumption, for example, if the same or comparable detectors are used for both the approaches and the correlation calculations are not a rate-limiting step. The interferometric imaging array can record the entire THz image with one scan. However, there is a tradeoff – 1596 correlations are needed to reconstruct the interferometric image. This correlation hardware requirement is not necessary with a focal plane array. In essence, one is trading imaging speed for backend image processing.

If there is no time restriction to generate an image, one can reduce the required number of detectors for interferometric imaging accordingly. For example, if the aperiodic interferometric array is rotated, the equivalent number of pixels in the reconstructed image is $M^*N(N-1)/2$ where M is the number of unique rotational positions of the array. Assuming 28 different rotation positions for the interferometric array, the corresponding time to acquire an image would be the same for both the techniques. However, to maintain the number of pixels at 1596, only 11 detectors are required in the interferometric approach compared with 57 for the focal plane array. This reduction in the number of detectors reduces the required number of correlations from 1596 to 55. If there are technical limitations or difficulty in fabricating focal plane THz detector arrays with a large number of detectors, synthetic aperture approaches, such as interferometric imaging, can reduce the number of required detectors while maintaining the same number of image pixels.

5. Experimental terahertz imaging

5.1. Terahertz photo-mixing system

To experimentally demonstrate THz image reconstruction using interferometric imaging, a CW THz generation and detection method is employed. A conceptual diagram of the receiving array is shown in Fig. 7. Two infrared lasers are tuned to have a difference frequency of ~1 THz to power the fiber optically coupled THz photo-mixer receivers. Owing to the relatively low-power requirements of the receiver, the optical power from a single set of lasers can be distributed using a fiber optic splitter to power each element of the imaging array. The receiving array is tunable by simply adjusting the difference frequency of the infrared lasers.

The experimental system that is used to demonstrate [72] the concept of interferometric imaging in the THz range is shown in Fig. 8. In advance of the availability of multiple THz detectors for a complete N element detector array, we use a single homodyne detector at multiple positions to sample the phase and amplitude of the wavefront from a point-like THz source. The electric field correlation can then be calculated for each 'pair' of detector positions, thereby mimicking the performance of an N element detector array. That information is used to test the near-field imaging performance and techniques expected for an interferometric array. The relative reference phase for each 'detector' remains fixed because the infrared signals are delivered to the receiver through a fixed-length fiber pigtail.

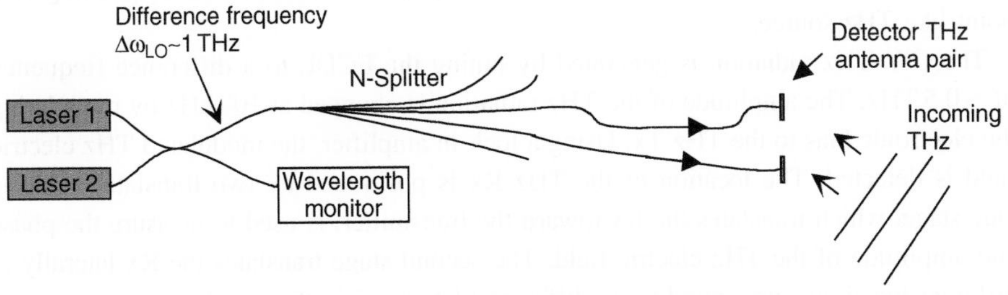

Fig. 7. Schematic representation of interferometric imaging array receiver.

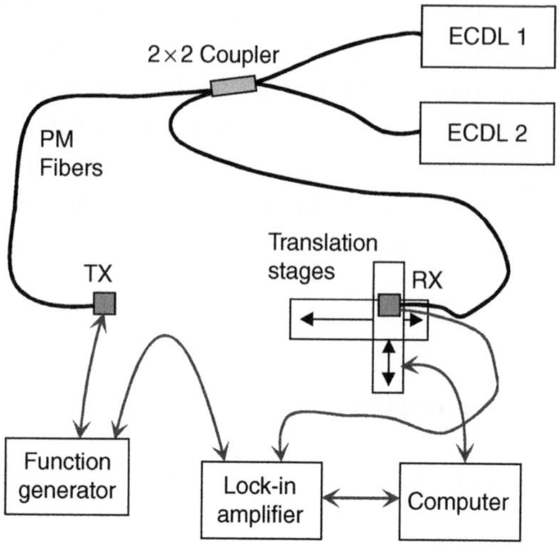

Fig. 8. Schematic of continuous wave (CW) photo-mixing apparatus used for demonstration of interferometric imaging.

Two external cavity diode lasers (ECDL) in the Littrow configuration are used as the infrared laser sources. The ECDL can be electronically tuned from 778 to 782 nm. The line-width of each laser is ~2 MHz with an output power of ~100 mW. The lasers are very sensitive to feedback. Consequently, an optical isolator is utilized to prevent back-reflections into the laser cavity. In addition, the fiber launch assembly is designed such that the back reflection from the fiber front surface is not reflected back into the laser cavity. A 2×2 fiber coupler is used to combine the infrared laser beams into fibers for delivery to the THz transmitter (Tx) and receiver (Rx) modules. As the photo-mixing process requires that the electric fields of the optical sources be collinear, polarization-maintaining fiber is used to deliver the optical power to the fiber pigtailed THz transmitter and receiver. The THz modules are LT grown GaAs bowtie-type photo-mixers. The Tx emits radiation when a voltage bias is applied to the LT GaAs in the presence of the two infrared beams. The Rx generates a voltage in the presence of the infrared beams which is proportional to the THz electric field. The THz radiation is emitted within 30° [102] using a hyper-hemispheric silicon lens that is attached to the THz module. On the basis of previous measurements [102], one would expect the THz radiation at ~0.5 THz to be emitted as a spherical wavefront corresponding to a point-like THz source.

The CW THz radiation is generated by tuning the ECDL to a difference frequency of ~0.5 THz. The amplitude of the THz radiation is chopped at 100 kHz by modulating the electronic bias to the THz Tx. Using a lock-in amplifier, the modulated THz electric field is detected. The location of the THz Rx is positioned by two translation stages. One stage, which translates the Rx toward the transmitter, is used to measure the phase and amplitude of the THz electric field. The second stage translates the Rx laterally at different locations corresponding to different 'detectors' in the interferometric imaging array. The initial spacing between the Rx and the Tx is ~14 cm.

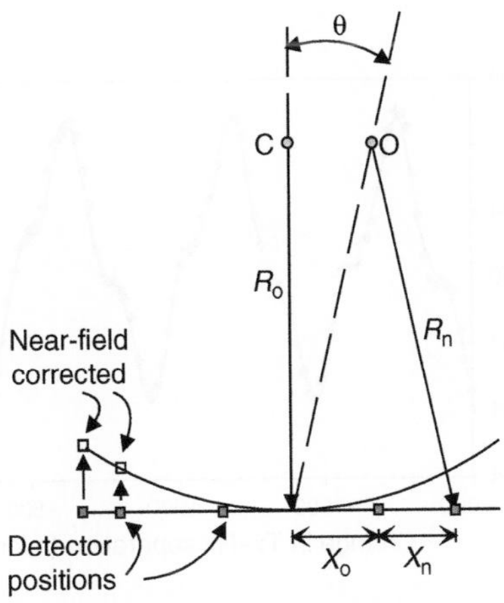

Fig. 9. Geometry and location of detector array positions (solid squares) and source location (point O at an angle of θ) for near-field correction.

The geometry and location of the Tx relative to the array detector positions is shown in Fig. 9. For the near-field correction, the appropriate phase is added to the measured THz phase such that the position of the detectors is moved from the line of solid squares to positions (open squares) along the curved path. The radius of curvature of the curved path is R_0. The distance from the source to each detector position (x_n) is given by R_n. The value of x_0 represents the distance from the $x_n = 0$ detector to the origin. Point O is the location of the THz source that makes an angle θ relative to the normal to the linear array of detectors. The position of the N detectors is given by x_n while x_0 represents the offset of the $x_n = 0$ detector position from the origin. For the values of R_0 investigated, the variation in the measured THz amplitude in a linear (solid squares) compared to curved geometry (open squares) is rather small. The predominant effect on the THz electric field is the additional change in phase from positioning the detectors on the linear compared with curved geometry. For small angles θ, the phase change is given by

$$\phi = k\left[R - \sqrt{(R_o \tan\theta)^2 + R_o^2}\right] + \phi_{off} \tag{5}$$

where $R = \sqrt{(x_o + x_n - R_o \tan\theta)^2 + R_o^2}$ and ϕ_{off} represents a constant offset phase.

5.2. Generation of homodyne waveform

The configuration of Fig. 8 is commonly referred to as a homodyne configuration. The homodyne waveform is determined by recording the THz electric field as a function of separation between the Tx and the Rx. (A negative value for the separation corresponds to a decreasing distance between the Tx and the Rx.) Displacement of the

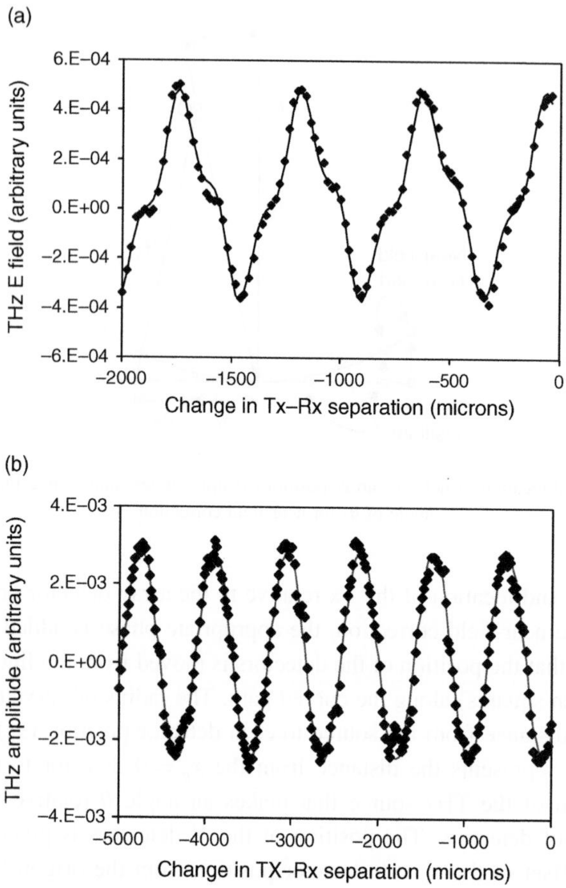

Fig. 10. Homodyne waveforms as acquired by changing the separation between the Tx and Rx in Fig. 8 – fundamental at 0.535 THz (a) and third harmonic at 1.605 THz (b).

detector to N different positions in the transverse direction will simulate an N element detector array.

Figure 10 shows a typical measured homodyne waveform and the corresponding numerical fit (solid lines). The measured THz waveform exhibits both the fundamental ECDL difference frequency (Fig. 10(a)) and higher harmonics – predominantly the third harmonic (Fig. 10(b)). Multiple harmonic generation in THz photo-mixers has been previously reported [103]. By fitting the observed waveform to a sum of harmonic sinusoidal functions, the amplitude and phase of the THz electric field can be determined separately for the fundamental and third harmonic. The solid line shows a numerical fit to the data. The fundamental extracted frequency, 0.535 THz, compares well to the expected frequency based on the frequency difference of the two ECDL. The extracted E field amplitudes and phases are 3.37×10^{-4} and 2.17 radians for 0.535 THz (Fig. 10(a)) and 5.61×10^{-5} and 3.94 radians for the 1.605 THz third harmonic, respectively (Fig. 10(b)).

5.3. Generation and near-field corrections of 1-D images

Unlike focal plane array imaging, for which the individual detectors are typically spaced at regular intervals, interferometric imaging is most effective for an aperiodic detector spacing that is non-redundant. At each detector position, a homodyne waveform is recorded (by scanning the separation distance between the Tx and Rx) and the THz amplitude, phase, and frequency extracted from a numerical fit to the data. As the homodyne waveform is periodic over 2π, multiples of 2π can be added to the measured phase to 'unwrap' its value. The lateral detector positions that are used in the experiment are given by the formula $x_n = 0.0125\cdot(1.7^n - 1.7^6)$ where x is in centimeters and $n = 0, 1\ldots 7$ is the corresponding label for each detector position. The $n = 6$ detector position, which is roughly in the geometric middle of the array, is chosen to be at $x = 0$. The effective distance between the Tx and Rx is $Z_o \sim 14\,\text{cm}$.

Figure 11 shows the measured THz phase and amplitude as functions of lateral Rx position compared to theoretical predictions. Figure 11(a) compares the measured THz amplitude to that expected for a point source at $R_0 = 14.4\,\text{cm}$. The measured THz amplitudes fall just below this line as is expected because the Tx is known to emit a peak THz electric field along the optical axis [102]. Figure 11(b) compares the 'unwrapped' measured phase to the expected phase for a point source emitting spherical wavefronts a distance R_0 from the Rx. The solid line is a theoretical fit to the measured phase using Eq. 5. The incoming wavefronts, based in particular on the measured phase, are curved rather than planar (for which the phase would be constant). Therefore, one might expect some near-field distortions to the reconstructed image, particularly if the detector positions exceed an $\sim$5 mm baseline. The extracted parameters from Eq. 5 for R_0, x_0, θ, and ϕ_{off} are 14.4 cm, 0.275 cm, 0.29°, and 3.38 radians, respectively.

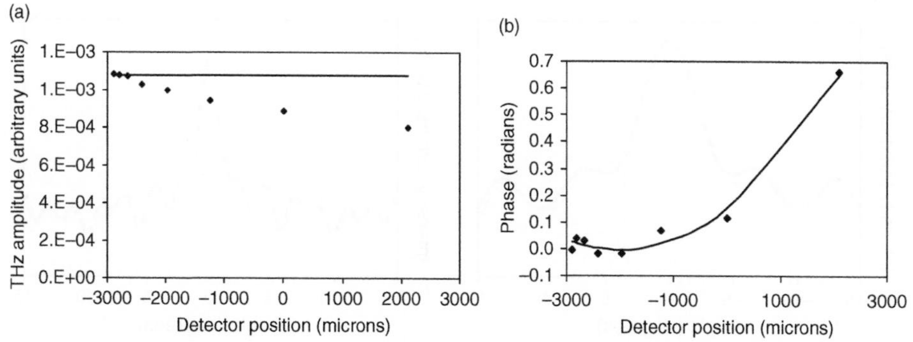

Fig. 11. (a) The measured THz amplitude as a function of lateral detector position. (b) The measured phase as a function of lateral detector position. The symbols represent the data that are extracted from the homodyne waveforms (e.g., Fig. 10). The solid lines are theoretical values, assuming that the source is a point source located at R_o (i.e., incoming spherical waves of constant amplitude).

Ignoring any near-field correction, the THz line image is reconstructed [1,72] from Eq. 4 using the following formulas

$$\sigma_E(\xi) = \sum_{i=1}^{N(N-1)/2} \{Re[C(u_i)]\cos(2\pi u_i \xi) - Im[C(u_i)]\sin(2\pi u_i \xi)\} \tag{6}$$

$$u_i = \frac{(x_n - x_m)}{\lambda} \tag{7}$$

where σ_E is the brightness distribution (image), $\xi = x/Z_o$, x_n and x_m are the detector positions, and the iteration of n and m is chosen such that each baseline combination [from the $N(N-1)/2$ possible] is included. $C(u_i)$ is the correlation function given by

$$C(u_i) = A_i e^{i\Delta\phi_i} \tag{8}$$

where $A_i = E_n E_m$ and $\Delta\phi_i = \phi_n - \phi_m$ represent the product of the electric field amplitudes and change in phase for each baseline pair combination. The summation over i includes in the reconstructed image the contribution from each spatial Fourier component corresponding to each unique baseline separation. The above equations represent the simplest reconstruction of the image: the correlation function for each possible unmeasured baseline pair combination is assumed to be zero.

The reconstructed 0.535 THz image of the Tx point source is shown in Fig. 12(a). In essence, the THz image is a measure of the point-spread function of the detection array. Note that the reconstructed image closely resembles that of a point source. As the largest baseline is small compared with the distance to the source, the phase of the wavefront only varies slightly across the detector array. Consequently, near-field corrections to the reconstructed image would be small. From theoretical considerations [1,72], the angular resolution of a planar array can be approximated as $\theta_{\min} = \lambda/b$. At a distance R_o away,

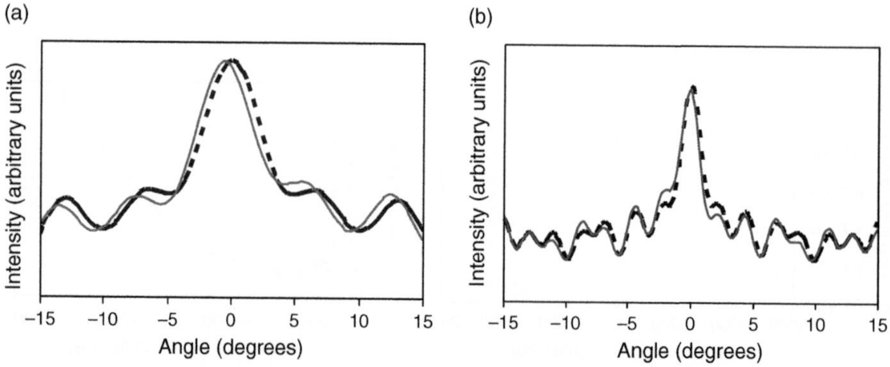

Fig. 12. (a) Reconstructed image of 0.535 THz source located directly in front of the detector a distance of 14.4 cm away. The dashed line is the expected image reconstruction assuming a point source at infinity (FWHM = 4.8°). (b) Similar reconstructed image for 1.602 THz. Note that the resolution of the central peak FWHM = 1.6° is improved by a factor of three corresponding to a 1/3 reduction in the THz wavelength.

the lateral spatial resolution is $\Delta L_{\text{lat}} \approx \theta_{\min} R_{\text{o}} \approx \lambda R_{\text{o}}/b$. For the maximum baseline value $b = 5$ mm used in this experiment, the angular resolution is estimated to be 6.4°. Note that the full-width at half maximum (FWHM) for the reconstructed image (4.8°) is close to this value, indicating a good correspondence between the reconstructed image and theoretical predictions. A reconstruction of the (third harmonic) 1.6 THz image, shown in Fig. 12(b), reveals a narrowing of the central peak to a FWHM of 1.6° corresponding to a factor of 3 reduction compared with the fundamental peak (Fig. 12(a)). This is consistent with the expected 1/3 reduction because of the third harmonic of the THz wavelength.

While the experimental data match well with what one would expect for a point THz source, the 4.8° angular resolution is not adequate for standoff applications. Previous analyses of THz interferometric images show that the wavelength-dependant spatial resolution may lead to spectral misidentification of the boundaries of the object [1]. To improve angular resolution, one must increase the maximum baseline of the imaging array. However, one would expect larger near-field distortions in the reconstructed THz image because (a) wavefront curvature will be more evident, and (b) the value of the THz E field (Fig. 10(a)) is not nearly constant along a spherical wavefront but rather exhibits the 30° divergence angle of the Tx with peak intensity along the optical axis [102].

To demonstrate improved spatial resolution, the detector positions in the imaging 'array' are increased by a factor of 10:

$$x_n = 0.125(1.7^n - 1.7^6) \tag{9}$$

The corresponding minimum and maximum baselines are ~875 μm and ~5 cm, respectively. The initial distance to the point source is $R_{\text{o}} = 40.9$ cm. The resulting extracted parameters from Eq. 5 for R_{o}, x_{o}, θ, and ϕ_{off} are 40.9 cm, 0.520 cm, 0.01°, and 1.55 radians, respectively. These data are used to reconstruct a THz line image. Figure 13(a) is a comparison of the reconstructed line image using Eq. 6 (no near-field

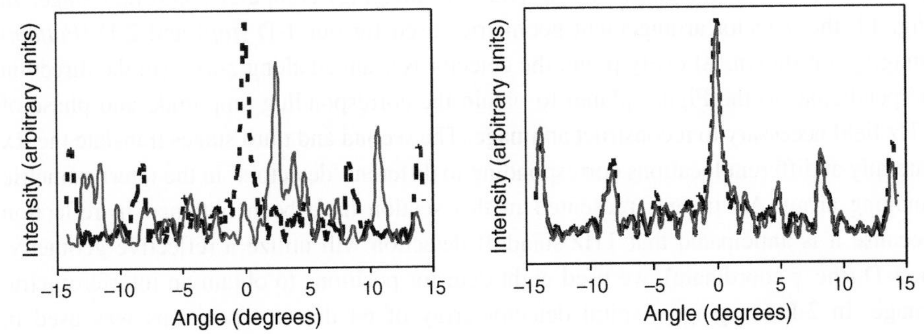

Fig. 13. (a) Comparison of a reconstructed line image (solid line) without near-field correction and theoretical image (dashed) for a point source at infinity. (b) Comparison of reconstructed line image (solid line) including a near-field correction.

correction) with theoretical predictions for a point at infinity (i.e., planar wavefronts). When the image is not corrected for near-field distortions (i.e., the spherical wavefronts), there is very poor agreement between the reconstructed image and the theoretical prediction. However, when the near-field corrections are included in the image reconstruction, (Fig. 13(b)) the agreement is greatly improved. The measured FWHM of 0.6° is about an order of magnitude smaller than Fig. 12(a) corresponding to the factor of 10 increase in the largest baseline. The smaller peaks at $\sim\pm 8°$ and $\sim\pm 14°$ are called side lobes. They are an artifact of the image reconstruction process and the finite number of sampled baselines.

The near-field correction is calculated conceptually by repositioning the detector positions from a linear arrangement to a spherical arrangement that matches the curvature of the incoming wavefront. The theoretical phase delay from a point source at normal incidence is subtracted from the measured phase. As shown in those Figures, the curvature of the phase (indicating a curved wave front) is removed by the near-field correction, yielding a linear dependence of phase on detector position. The slopes of the near-field-corrected phase versus detector position plots indicate the direction to the source.

5.4. Generation of 2-D images

Our experimental setup for 2-D CW THz interferometric imaging in reflection is a variation of Fig. 8. The infrared outputs of two Littman ECDL (Sacher Lion MLD1000) each operating at 0.78 μm are joined in a 2×2 fiber coupler. The lasers are detuned to a beat frequency of 0.3 THz. In the output of the coupler, this mixed CW radiation is used to power both THz emitter and detector both LT grown GaAs bowtie-type photoconductive dipole antennae. The power in both channels is $\sim$25 mW. The THz emitter is also driven by a 50-kHz oscillating electric field with an amplitude of 10 V from a function generator, which also supplies a reference signal for the lock-in amplifier used to detect the THz signal.

In our current configuration, only one THz detector is used. This detector is mounted on a 2-D (for 1-D imaging) or 3-D (for 2-D imaging) computer-controlled stage. In Fig. 14, the detector arrangement geometries used for our 1-D (*top*) and 2-D (*bottom*) imaging are shown. At every point, the detector is scanned along z-axis (in the direction perpendicular to the Figure plane) to obtain the corresponding amplitude and phase of THz field necessary to reconstruct an image. The second and third stages translate the Rx laterally at different locations corresponding to different ‘detectors’ in the interferometric imaging array. All images presented in this section have been obtained in reflection because it is anticipated that THz standoff detection will utilize a reflective geometry. In 1-D (the x-coordinate), we used eight detector positions to obtain an interferometric image. In 2-D imaging, a spiral detector array of 64 detector positions was used in XY plane.

With amplitudes and phases of the THz field, correlations of all detector pairs are performed, and the interferometric image is reconstructed using Eq. 8. In Fig. 15, 2-D

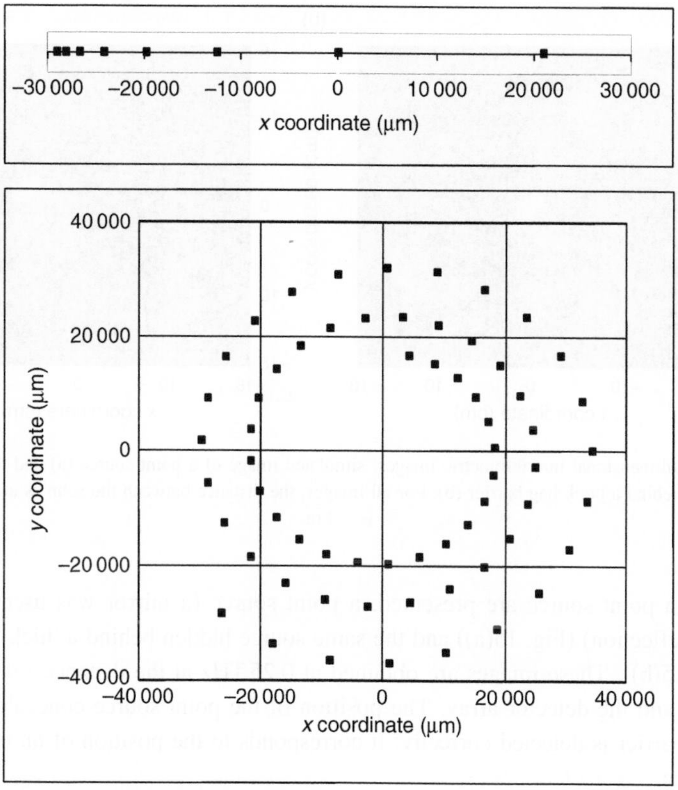

Fig. 14. Non-regular detector array arrangements: 1-D (top), 2-D (bottom).

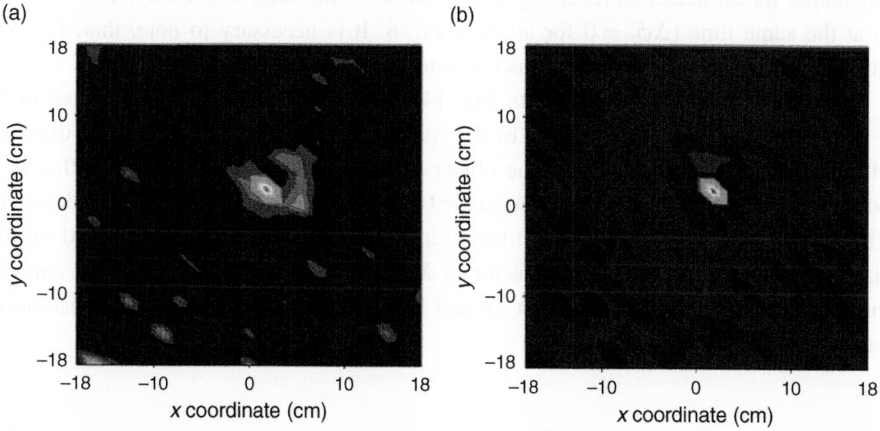

Fig. 15. Two-dimensional interferometric images with an ~1-m distance between the source and the detector: an open point source (a), a point source behind a book bag barrier (b). Red corresponds to maximum THz intensity, whereas blue corresponds to minimum.

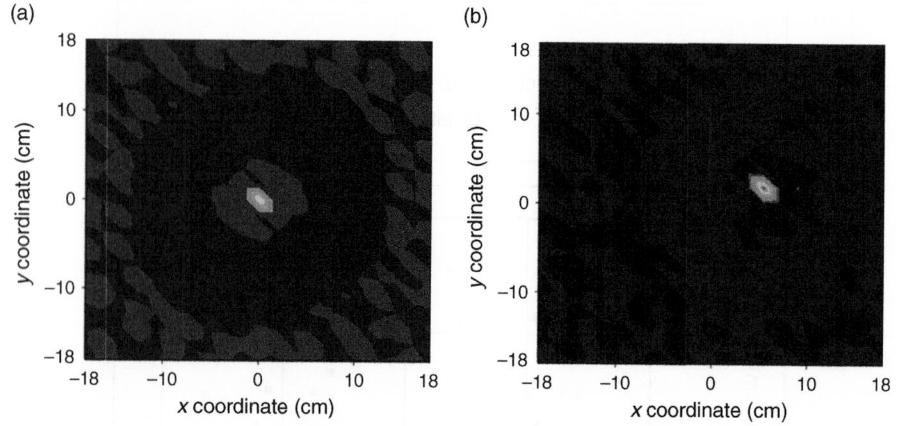

Fig. 16. Two-dimensional interferometric images: simulated image of a point source (a) and the image of a metal object behind a book bag barrier (b). For all images, the distance between the sources and the detector is ~1 m.

images of a point source are presented: a point source (a mirror was used as a point source in reflection) (Fig. 15(a)) and the same source hidden behind a thick nylon book bag (Fig. 15(b)). These images are obtained at 0.25 THz at the distance ~1 m between the source and the detector array. The position of the point source concealed behind a book bag barrier is detected correctly: it corresponds to the position of an unconcealed point source.

In Fig. 16(b), the image of a metal object concealed behind the book bag barrier is shown. The size of the object is ~3×3 cm. Also, the image of the metal object in reflection is in a good agreement with the corresponding simulation (Fig. 16(a)) calculated for all detectors receiving a THz wave of the same amplitude ($A_l = 1$ for all l) at the same time ($\Delta\phi_l = 0$ for all l) in Eq. 8. It is necessary to note, that, for 2-D images, the phase of THz waves was corrected according to Eq. 7.

For the spiral detector geometry of Fig. 14, the angular resolution is calculated to be ~0.015 radians which corresponds to the spatial resolution of ~1.5 cm at the distance of 1 m. Owing to low resolution, the object image is not very different from that of a point source. However, the object concealed behind the barrier is confidently detected. The maximum angular (and spatial) resolution in our experiment is determined by the size of stages used to move the detector in X and Y directions. In Fig. 16, we can see that the maximum baseline does not exceed 7 cm. With different stages, this resolution can be significantly increased.

6. Interferometric image analysis

Terahertz images by themselves are generally not sufficient to identify hidden explosives because of the complicating effects of barrier materials, noise, and potential image reconstruction artifacts. THz imaging at multiple frequencies, however, provides two features that can be combined and exploited for accurate identification with low probability of false alarm: (i) spatial and (ii) spectral variability. Spatial analysis admits the location of object boundaries; image intensities for objects so identified can then be used to classify high or low reflectance materials. For the spectral analysis, intensity patterns from multifrequency images of an object are matched to known spectral response signatures of specific explosive agents to determine whether an agent is present.

Each of the aforementioned features, spatial and spectral, can be independently employed for explosive detection. No detection method is foolproof, but the combination of the two reduces identification uncertainty and matching error. Two approaches to THz image analysis are discussed in Sections 6.1. and 6.2.

6.1. Component pattern analysis

The first method [87,88] uses a matrix form of the Beer–Lambert absorption law in a transmission mode. A 2-D material sample is raster-scanned by a pulsed, tunable THz source to generate an N rows $\times$ L columns image matrix $[I]$ where N is the number of THz frequencies used and L is the number of pixels. The values of $[I]$ are the measured total absorbance at each pixel. Separate absorption experiments with known materials of interest establish the THz spectra both graphically and through the $N \times M$ spectra matrix $[S]$, where M is the number of components. The collection of species can include non-agent materials that can affect the absorptions; for example, barrier materials with spectra that are typically weakly frequency dependent. The spatial patterns of the agents are contained in the $M \times L$ matrix $[P]$. The values of $[P]$ effectively contain agent concentration information.

Based on the Beer–Lambert law, the matrices are related by

$$[I] = [S][P] \tag{10}$$

The total absorbance entries of $[I]$ are, in effect, weighted sums of the absorbance of each material at a specific location. For example, if two different agents are present at one location behind a barrier material, assuming spectra for all components exist in $[S]$, then matrix $[P]$ will reveal how much of each component is present and in what spatial pattern. The matrix $[P]$ is determined, for cases where $N > M$, from

$$[P] = ([S]^t[S])^{-1}[S]^t[I] \tag{11}$$

where superscript t indicates matrix transpose, and -1 indicates matrix inverse. This approach has been successfully used to identify pellets of aspirin and palatinose (a food sweetener), as well as the illegal drugs methamphetamine and MDMA hidden in small plastic bags [87,88].

6.2. Artificial neural networks

The second method uses ANNs [104] to analyze THz images, obtained in either reflection or transmission mode, and classify pixels or groups of pixels within the image as to their component material (e.g., explosive, barrier material, metal, and skin). ANNs are collections of interconnected computational cells that simulate biological neural networks. Many ANN architectures have been proposed and have been applied successfully to estimation, optimization, and classification. In summary, ANNs are multivariate non-linear models, in which inputs are mapped to outputs through a sequence of transformations. Components of the networks are adaptively determined, often in a learning stage, during which ANNs are trained to detect certain patterns by mapping known and controlled input sequences to specific outputs. In the current application, ANNs are trained to recognize the THz intensity 'signatures' corresponding to explosives, paper, metal, skin, barrier materials, etc. The inputs can be single pixels or neighborhoods of pixels at multiple frequencies. ANN outputs can be a number uniquely representing a material of interest.

The Multilayer Perceptron (MLP) [104] is the most widely employed ANN architecture in classification tasks. As shown in Fig. 17, it contains a layer of input nodes accepting the input patterns (spectral signatures, here), one or more hidden layers of nodes, and an output layer of nodes. Weighted connections lead from one layer to

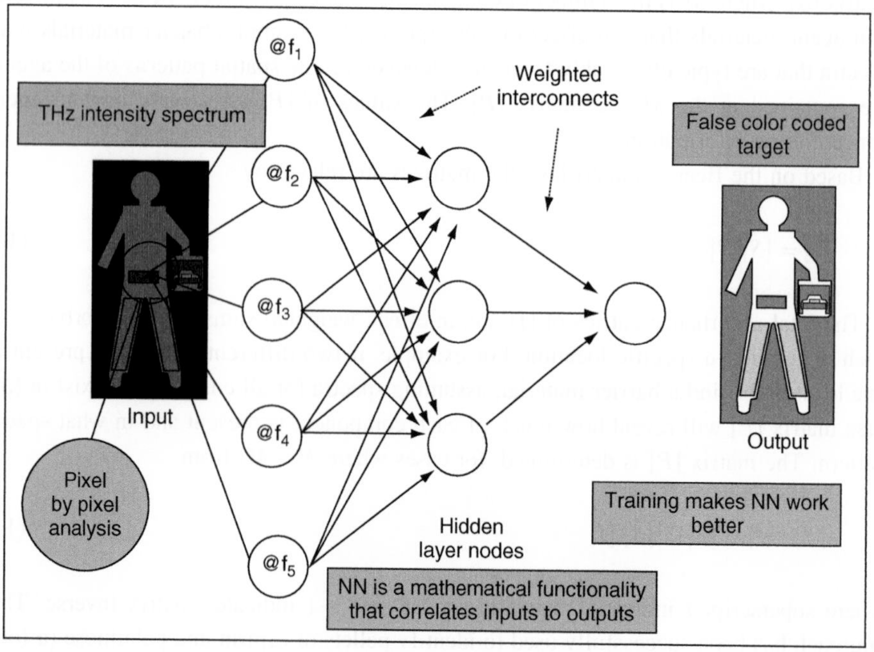

Fig. 17. Use of a multilayer perceptron-type artificial neural network to analyze an interferometric image of a suicide bomber.

the next. These weighted connections are summed at each node and filtered through a non-linear function, such as sigmoids and hyperbolic tangents. The weight values are optimized during the training stage, when a set of input patterns and desired output values for each input pattern are presented to the MLP. Weights are originally randomized and are then iteratively updated to make the output of the MLP as close to the desired output as possible according to a specified penalty function. A straightforward training strategy for MLPs is back-propagation, minimizing the mean squared error between desired and network calculated outputs. The magnitude of the modifications to the weights is controlled by a learning rate parameter. The numbers of hidden layers and nodes within layers, the number and exact nature of training patterns, the learning rate, and the type of non-linearity used at layers are parameters selected based on empirical knowledge and experimentation and can affect the quality and efficiency of the training process.

To demonstrate the applicability of MLPs to explosive detection, we generated synthetic reconstructed interferometric images of two objects, one metal and one consisting of RDX, both embedded in background. The images, each consisting of 1×500 pixels, were generated for frequencies of 0.7 and 0.9 THz. The metal object was imaged with the same intensity at both frequencies because it is assumed to be highly reflective. The RDX has a spectral frequency-dependent intensity. This difference in imaging at different frequencies is a 'fingerprint' we use to identify materials as belonging in one or the other category.

An MLP was constructed with two hidden layers and hyperbolic tangent non-linearities. Input patterns used for training were synthetic interferometric images containing metal or RDX at many different locations within an image. Some input patterns contained no object and corresponded to background. The MLP was trained with back-propagation to match correct outputs to corresponding inputs, i.e., to identify the input patterns as metal, RDX, or background on the basis of their spectral fingerprint. When images of metal were used as inputs, the MLP was trained to produce an output of 1, whereas RDX required an output of 2. When only background was present, the network was trained to produce 0 at the output. The completion of training was indicated by a small error between desired and obtained outputs.

Once training was complete, the MLP was tested on images not previously used for training. Results for the classifications of pixels in the test images fell into one of three classes: metal, RDX, and background. Most pixels were correctly identified, except for a few pixels at the boundaries between metal and background and background and RDX. Such boundary problems have also been reported elsewhere [86–88] and can be attributed to reconstruction artifacts in the interferometric images. Misclassification at boundaries can be reduced by applying homogeneity rules to pixel neighborhoods or through interferometric image pre-processing.

Figure 17 shows a potential application of the MLP to the analysis of a multi-frequency, reflective interferometric THz image of a suicide bomber. Each pixel, or group of pixels such as a row, is analyzed using a trained MLP. The resulting false-color image would indicate to an operator the presence of an explosive.

Although MLPs can provide good results identifying correctly the explosive agent, they have limitations and have not always been successful. The simulated case of a person with a rectangle of RDX explosive material attached to their chest under a shirt was considered [105]. In this simulation, the shirt is assumed transparent to THz. The person also carries a candy bar and wears a belt with a metal buckle. An MLP was trained by processing training vectors representing spectral signatures of RDX, skin, and sugar (representing candy). During the training stage, the network learned to associate different outputs with distinct (on the basis of spectral signature) materials. For demonstration purposes, the output levels are here mapped to colors. Interferometric images of the explosive-carrying person were then generated at six frequencies in the THz range. Each 6-D pixel (corresponding to the same spatial element for the six frequencies) was presented to the MLP. The results, shown in the top three plots of Fig. 18, demonstrate that the MLP correctly detected RDX on the chest. The resolution, however, is poor. When all six frequencies are used in training and testing, large regions of the person are erroneously mapped to RDX and candy; the actual candy bar and buckle do not appear in the output.

Resolution improved when the network was trained and tested on data at four and five frequencies. This improvement is due to the removal of the lowest THz frequencies that correspond to the poorest spatial resolution. Omitting information at the two lowest frequencies enabled the identification of both the RDX rectangle and the metal buckle. The colors, however, are widely wrong.

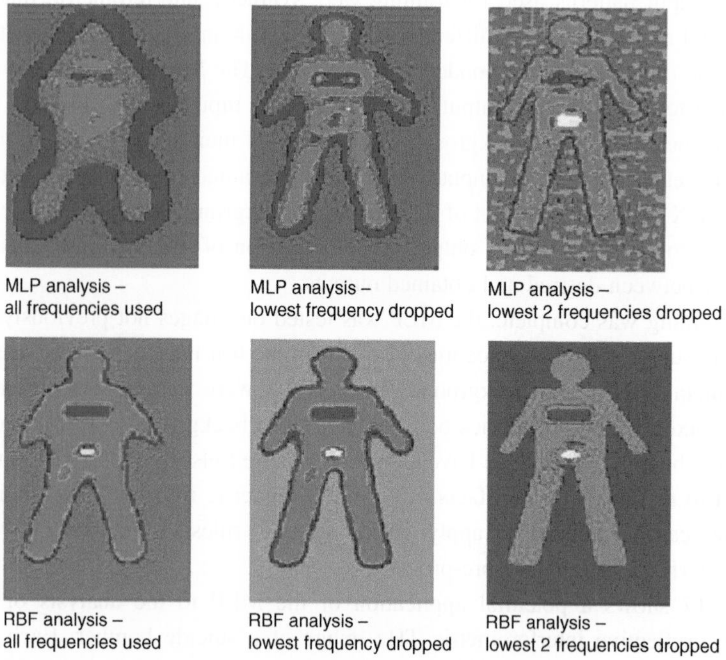

Fig. 18. Bitmap of MLP and RBF analyses of bomber problem – (white = metal, orange = candy, red = RDX, green = skin, grey = off-person, blue = unknown).

To determine whether alternative ANN architectures can lead to improved resolution and successful agent detection, Radial Basis Function (RBF) networks [106] were considered for the same problem. RBFs are networks with one hidden layer associated with a specific, analytically known function. Each hidden layer node corresponds to a numerical evaluation of the chosen function at a set of parameters; Gaussian waveforms are often the functions of choice in RBFs. The outputs of the nodes are multiplied by weights, summed, and added to a linear combination of the inputs, yielding the network outputs. The unknown parameters (multiplicative weights, means and spreads for the Gaussians, and coefficients for the linear combination of the inputs) are determined by training the RBF network to produce desired outputs for specific inputs.

An RBF network was trained to identify RDX, metal, candy, and skin similarly to the MLP and was then applied to pixels from the image of the person with the RDX rectangle. The RBF results, shown in the bottom row of Fig. 18, show a much-improved spatial resolution in comparison with the MLP results. The evidence of 'unknown' results is effectively gone. The improved performance of the RBF is consistent with its preferred capability with 'classification' problems, especially with limited data. Objects are correctly identified, with slight color inaccuracies in border transition zones (i.e., edges). Each object – correctly identified in each case – is visible as a central major-sized color generally surrounded by other color(s) – an artifact corresponding to transition zones between the object and the background. Best results were obtained when the lowest frequency was dropped. When two frequencies were dropped, the RBF incorrectly identifies the person as being covered in or made of candy, although the RDX is still correctly revealed.

7. Conclusions and challenges

Terahertz detection and imaging are showing great potential for security applications, especially because THz radiation poses no health risk for scanning of people. However, key hardware challenges remain. Most likely, the first imaging systems to be deployed will be single THz frequency or will detect the total integrated THz power, with true spectroscopic imaging arrays to follow. Proximity (<3 m) imaging systems are approaching commercialization while standoff detection remains an area of active research and development. The key issue remaining for proximity detection is to increase the frame rate for real-time imaging. Other issues that need to be addressed as THz technology moves out of the research laboratory and into the field include issues of portability, hand-held size, cost, and power requirements.

While many non-metallic, non-polar materials are transparent to THz radiation, many target compounds have characteristic THz spectra that can be used as a 'fingerprint' for identification. While crystalline, high energy explosives have characteristic THz fingerprint spectra that can be used to identify these threats, home-made NH_4NO_3 bombs could pose a change to THz security applications because these materials can have featureless THz spectra below 3 THz.

The biggest challenge to THz security applications is standoff detection. As the standoff distance increases, one must consider the effect of the humid atmosphere, dust, smoke, etc. as well as possible barrier materials. At standoff distances, picosecond pulsed measurements become problematic. To overcome the attenuation losses of barrier materials and the atmosphere, higher-power sources need to be developed. In conjunction, compatible low-noise THz receivers need further development. As the need to penetrate through barriers or the need to increase the standoff distance increases, higher-power THz sources are required. However, all of the high-power sources typically still have long coherence lengths. The long coherence lengths lead to (i) standing wave effects among the various layers or surfaces that are probed and (ii) THz speckle that results from the interference of THz waves that are coherently reflected from different locations on an object. Both of these effects can distort or mask the spectral image.

To date, most work on THz standoff detection assumes that the THz power reflected from barrier materials (i.e., backscatter) can be ignored. This is certainly a valid assumption in the transmission mode or using a close proximity time-domain technique for which the reflections do not appear in the detected time-window. In the reflection mode, one must detect the reflected power from the threat material and differentiate that contribution from the reflected light from barrier materials. For a fairly transparent material (e.g., corrugated paper), the amount of diffusively reflected radiation from the barrier material should be fairly small and easily distinguished from the reflection from the explosive. For highly scattering barrier material, the diffuse reflection from the front surface is much more problematic because the ‘real’ reflection from the explosive is highly attenuated – because it passes through the attenuating barrier – as compared with the reflection from the barrier itself. This problem might be circumvented by using a differential reflection technique. This is accomplished by measuring the difference in reflection of two THz frequencies. This differential technique effectively removes the contribution of the reflections from the barrier materials and amplifies the THz spectral presence of the threat material.

Acknowledgments

The authors thank the Technical Support Working Group (TSWG) Explosives Directorate, the US Army, and the Foundation at NJIT for supporting our work in THz imaging. Discussions with Prof. J. M. Joseph are gratefully acknowledged.

References

[1] J. Federici, B. Schulkin, F. Huang, D. Gary, R. Barat, F. Oliveira, and D. Zimdars, Semicond. Sci. Technol., 20 (2005) S266.
[2] T. Loffler, K. Siebert, S. Czasch, T. Bauer, and H. Roskos, Phys. Med. Biol., 47 (2002) 3847.
[3] G. C. Walker, E. Berry, N. N. Zinov’ev, A. J Fitzgerald, R. E. Miles, C. Martyn, and M. A. Smith, www.comp.leeds.ac.uk/comir/research/terahertz/GRN39678.html (2003).

[4] E. Berry, J. Biol. Phys., 29 (2003) 263.
[5] R. H. Clothier and N. Bourne, J. Biol. Phys., 29 (2003) 179.
[6] M. R. Scarfi, M. Romano, R. Di Pietro, O. Zeni, A Doria, G. P. Gallerano, E. Giovenale, G. Messina, A. Lai, G. Campurra, D. Coniglio, and M. D'Arienzo, J. Biol. Phys., 29 (2003) 171.
[7] T. Yuan, H. Liu, J. Xu, F. Al-Douseri, Y. Hu, and X. Zhang, Proc. SPIE, 5070 (2003) 28.
[8] T. A. Heimer and E. J. Heilweil, Bull. Chem. Soc. Japan, 75 (2002) 899.
[9] D. L. Woolard, T. Koscica, D. L. Rhodes, H. L. Cui, R. A. Pastore, J. O. Jensen, J. L. Jensen, W. R. Loerop, R. H. Jacobsen, D. M. Mittleman, and M. C. Nuss, J. Appl. Toxic., 17 (1997) 243.
[10] F. C. De Lucia, Sensing with Terahertz Radiation, D. Mittleman (Ed.), Springer, New York (2003).
[11] D. A. Zimdars and J. S. White, Proc. SPIE, 5411, (2004)
[12] D. A. Zimdars, Proc. SPIE, 5070 (2003) 108.
[13] D. Woolard, T. Globus, E. Brown, L. Werbos, B.Gelmont, and A. Samuels, Proceedings of the 5th Joint Conference on Standoff Detection for Chemical and Biological Defense, Williamsburg, VA (2001).
[14] D. J. Cook, B. K. Decker, G. Maislin, and M. G. Allen, Proc. SPIE, 5354 (2004) 55.
[15] R. W. Cohen, G. D. Cody, M. D. Coutts, and B. Abeles, Phys. Rev. B, 8 (1973) 3689.
[16] J. Federici and H. Grebel, Sensing Science and Technology at THz Frequencies, Vol. II. Emerging Scientific Applications & Novel Device Concepts, D. Woodard, M. Shur, W. Loerop (Eds), World Scientific, Hackensack, NJ, USA (2003).
[17] H. Altan, F. Huang, J. Federici, A. Lan, and H. Grebel, J. Appl. Phys., 96 (2004) 6685.
[18] D. Bhanti, S. Manickavasagam, and M. P. Mengüç, J. Quant. Spectrosc. Radiat. Transf., 56 (1996) 591.
[19] X. H. Wu, A. Yamilov, H. Noh, H. Cao, E. W. Seelig, and R. P. H. Chang, J. Opt. Soc. Am. B, 21 (2004) 159.
[20] F. Borghese, P. Denti, G. Toscano and O. I. Sindoni, Appl. Opt., 18 (1979) 116.
[21] Y. L. Xu, Appl. Opt., 36, (1997) 9496.
[22] L. Kai and P. Massoli, Appl. Opt., 33 (1994) 501.
[23] T. Wriedt and U. Comberg, J. Quant. Spectrosc. Radiat. Transfer, 60 (1998) 411.
[24] M. Herrmann, M. Tani, M. Watanabe, and K. Sakai, IEEE Proc. Optoelectron., 149 (2002) 116.
[25] M. R. Kutteruf, C. M. Brown, L. K. Iwaki, M. B. Campbell, T. M. Korter, and E. J. Heilweil, Chem. Phys. Lett., 375 (2003) 337.
[26] B. Ferguson, S. Wang, H. Zhong, D. Abbott and X. C. Zhang, Terahertz for Military and Security Applications, R. Jennifer Hwu (Ed.), Proc. SPIE, 5070 (2003) 7.
[27] T. L. J. Chan, J. E. Bjarnason, A. W. M. Lee, M. A. Celis, and E. R. Brown, Appl. Phys. Lett., 85 (2004) 2523.
[28] P. C. Upadhya, Y. C. Shen, A. G. Davies, and E. H. Linfield, Vibr. Spec., 35 (2004) 139.
[29] H. S. Chua, J. Obradovic, A. D. Haigh, P. C. Upadhya, O. Hirsch, D. Crawley, A. A. P. Gibson, L. F. Gladden, and E. H. Linfield, IEEE Joint 29th International Conference on Infrared and Millimeter Waves, and 12th International Conference on Terahertz Electronics, 29th IRMMW-2004/12th THz-2004, Germany (2004) 399.
[30] A. Sengupta, PhD Dissertation, Characterization of Novel Materials Using Terahertz Spectroscopic Techniques, New Jersey Institute of Technology, USA (2006).
[31] A. Sengupta, A. Bandyopadhyay, R. B. Barat, D. E. Gary, and J. F. Federici, Terahertz Science and Technology Topical Meeting, Florida, USA, The Optical Society of America, Washington D.C. (2005) ME6.
[32] T. M. Korter, R. Balu, M. B. Campbell, M. C. Beard, S. K. Gregurick, and E. J. Heilweil, Chem. Phys. Lett., 418 (2006) 65.
[33] J. Pearce and D. M. Mittleman, Physica B, 338 (2003) 92.

[34] Z. Jian, J. Pearce, and D. M. Mittleman, Phys. Rev. Lett., 91 (2003) 039903.
[35] S. Mujumdar, K. J. Chau, and A. Y. Elezzabi, Appl. Phys. Lett., 85 (2004) 6284.
[36] A. Bandyopadhyay, A. Sengupta, R. B. Barat, D. E. Gary, and J. F. Federici, Proc. SPIE, 6120 (2006) 61200H.
[37] M. B. Campbell and E. J. Heilweil, Proc. SPIE, 5070 (2003) 38.
[38] M. C. Kemp, P. F. Taday, B. E. Cole, J. A. Cluff, A. J. Fitzgerald, W. R. Tribe, Proc. SPIE, 5070, 44 (2003).
[39] F. Huang, B. Schulkin, H. Altan, J. Federici, D. Gary, R. Barat, D. Zimdars, M. Chen, and D. Tanner, Appl. Phys. Lett., 85 (2004) 5535.
[40] M. J. Fitch, D. Schauki, C. A. Kelly, and R. Osiander, Proc. SPIE, 5354 (2004) 45.
[41] B. M. Rice and G. F. Chabalowski, J. Phys. Chem. A, 101 (1997) 8720.
[42] D. G. Allis, D. A. Prokhorova, A. M. Fedor, T. M. Korter. Proc. SPIE, 6212 (2006) 62120F.
[43] Y. C. Shen, T. Lo, P. F. Taday, B. E. Cole, W. R. Tribe, and M. C. Kemp, Appl. Phys. Lett., 86, (2005) 241116.
[44] A. Sengupta, A. Bandyopadhyay, R. Barat, D. Gary, and J. Federici, Proc. SPIE, 6120 (2006) 61200A.
[45] M. Fitch, C. Dodson, Y. Chen, H. Liu, X.-C. Zhang, and R. Osiander, Proc. SPIE, 5790 (2005) 281.
[46] Y. Chen, H. Liu, M. Fitch, R. Osiander, J. Spicer, M. Shur, and X.-C. Zhang, Proc. SPIE, 5790 (2005) 19.
[47] K. Yamamoto, M. Yamaguchi, F. Miyamaru, M. Tani, M. Hangyo, T. Ikeda, A. Matsushita, K. Koide, M. Tatsuno, and Y. Minami, Jpn. J. Appl. Phys., Part 2, 43 (2004) L414.
[48] D. W. van der Weide, Sensing with Terahertz Radiation, D. Mittleman (Ed.), Springer (2003).
[49] Y. Shen, P. F. Taday, and M. C. Kemp, Proc. SPIE, 5619 (2004) 82.
[50] J. Bjarnason, T. Chan, A. Lee, M. Celis, and E. Brown, Appl. Phys. Lett. 85 (2004) 519.
[51] H. Zhong, N. Karpowicz, J. Partridge, X. Xie, J. Xu, and X.-C Zhang, Proc. SPIE, 5411 (2004) 33.
[52] R. Osiander, J. A. Miragliotta, Z. Jiang, J. Xu, and X.-C. Zhang, Proc. SPIE, 5070 (2003) 1.
[53] E. R. Brown, Terahertz Sensing Technology, Vol. 2, Emerging Scientific Applications & Novel Device Concepts, Woolard, Loerop, Shur (Eds), World Scientific (2003).
[54] E. Berry, G. Walker, A. Fitzgerald, N. Zino'ev, M. Chamberlain, S. Smye, R. Miles, and M. Smith, J. Laser Appl., 15, 3 (2003) 192.
[55] R. Woodward, B. Cole, V. Wallace, R. Pye, D. Arnone, E. Linfield, and M. Pepper, Phys. Med. Biol. 47 (2002) 3853.
[56] E. Pickwell, B. Cole, A. Fitzgerald, M. Pepper, and V. Wallace, Phys. Med. Biol. 49 (2004) 1595.
[57] A. Majewski, P. Miller, R. Abreu, J. Grotts, T. Globus, and E. Brown, Proc. SPIE, 5790 (2005) 74.
[58] B. B. Hu and M. C. Nuss, Opt. Lett., 20 (1995) 1716.
[59] D. M. Mittleman, S. Hunsche, L. Boivin, and M. C. Nuss, Opt. Lett. 22 (1997) 904.
[60] F. Buersgens, R. Kersting, and H.-T. Chen, Appl. Phys. Lett. 88 (2006) 112115.
[61] H-T Chen, R. Kersting, and G. Cho, Appl. Phys. Lett., 83 (2003) 3009.
[62] J. Federici and O. Mitrofanov, Phys. Med. Biol., 47 (2002) 3727.
[63] C. A. Schuetz and D. W. Prather, Proc. SPIE, 5619 (2004) 166.
[64] Z. Jiang and X.-C. Zhang, Opt. Lett., 23 (1998) 1114.
[65] A. Thompson, J. Moran, and G. Swenson, Interferometry and Synthesis in Radio Astronomy, 2nd Edition, Wiley Interscience, New York (2001).
[66] K. McKlatchy, M. Reiten, and R. Cheville, Appl. Phys. Lett., 79 (2001) 4485.
[67] A. Ruffin, J. Decker, L. Sanchez-Palencia, L. Le Hors, J. Whitaker, T. Norris, and J. Rudd, Opt. Lett., 26, (2001) 681.

[68] T. D. Dorney, J. L. Johnson, J. Van Rudd, R. G. Baraniuk, W. W. Symes, and D. M. Mittleman, Opt. Lett. 26 (2001) 1513.
[69] J. O'Hara and D. Grischkowsky, Opt. Lett., 26 (2001) 1918.
[70] J. O'Hara and D. Grischkowsky, Opt. Lett., 27, (2002) 1070.
[71] J. O'Hara and D. Grischkowsky, J. Opt. Soc. Am. B, 21 (2004) 1178.
[72] A. Bandyopadhyay, A. Stepanov, B. Schulkin, M. Federici, A. Sengupta, D. Gary, J. Federici, R. Barat, Z.-H. Michalopoulou, and D. Zimdars, J. Opt. Soc. Am. A, 23 (2006) 1168.
[73] K. McIntosh, E. Brown, K. Nichols, O. McMahon, W. DiNatale, and T. Lyszczarz, Appl. Phys. Lett., 67 (1995) 3844.
[74] S. Duffy, S. Verghese, and K. McIntosh, Sensing with Terahertz Radiation, D. Mittleman (Ed.), Springer, New York (2002).
[75] T. Loffler, K. J. Siebert, H. Quast, N. Hasegawa, G. Lota, R. Wipe, T. Hahn, M. Thomson, R. Leonhardt, and H. G. Roskos, Phil Trans. R. Soc. Lond. A, 362 (2004) 263.
[76] K. J. Siebert, T. Loffler, H. Quast, M. Thomason, T. Bauer, R. Leonhardt, S. Czasch, and H. G. Roskos, Phys. Med. Biol,. 47 (2002) 2743.
[77] K. J. Siebert, H. Quast, R. Leonhardt, T. Loffler, M. Thomson, T. Bauer, H. G. Roskos, and S. Czasch, Appl. Phys. Lett., 80 (2002) 3003.
[78] T. Kleine-Ostmann, P. Knobloch, M. Koch, S. Hoffman, M. Breede, M. Hofmann, G. Hain, K. Pierz, M. Sperling, and K Donhuijsen, Electr. Lett., 37 (2001) 1461.
[79] I. S. Gregory, W. R. Tribe, B. E. Cole, C. Baker, M. J. Evans, I. V. Bradley, E. H. Linfiled, A. G. Davies and M. Missous, Electr. Lett., 40 (2004) 143.
[80] J. E. Bjarnason, T. L. J. Chan, A. W. M. Lee, E. R. Brown, D. C. Driscoll, M. Hanson, A. C. Gassard, and R. E. Muller, Appl. Phys. Lett., 85 (2004) 3983.
[81] S. Verghese, K. A. McIntosh, and E. R. Brown, Appl. Phys. Lett., 71 (1997) 2743.
[82] V. Ryzhii, I. Khmyrova, and M. Shur, J. Appl. Phys., 91 (2002) 1875.
[83] V. Ryzhii, I. Khmyrova, A. Satou, P. Vaccaro, T. Aida, and M. Shur, J. Appl. Phys., 92 (2002) 5756.
[84] A. Malcoci, A. Stohr, A. Sauerwald, S. Schulz, and D. Jager, Proc. SPIE, 5466 (2004) 202.
[85] A. Nahata, J. T. Yardley, and T. Heinz, Appl. Phys. Lett,. 81 (2002) 963.
[86] K. Kawase, Y. Ogawa, and Y. Watanabe, Opt. Express, 11 (2003) 2549.
[87] Y. Watanabe, K. Kawase, and T. Ikari, Appl. Phys. Lett., 83 (2003) 800.
[88] K. Kawase, Y. Ogawa, and Y. Watanabe, Proc. SPIE, 5354 (2004) 63.
[89] Q. Wu, T. D. Hewitt, and X.-C. Zhang, Appl. Phys. Lett., 69 (1996) 1026.
[90] W. R. Tribe, D. A. Newnham, P. F. Taday, and M. C. Kemp, Proc. SPIE, 5354 (2004) 168.
[91] H. Zhong, A. Redo, Y. Chen, and X. C. Zhang, Proc. SPIE, 6212 (2006) 62120L.
[92] D. Zimdars, J. White, G. Stuk, A. Chernovsky, G. Fichter, and S. L. Williamson, Proc. SPIE, 6212 (2006) 62160O.
[93] N. Karpowicz, H. Zhong, C. Zhang, K.-I Lin, J.-S. Hwang, J. Xu, and X.-C. Zhang, Appl. Phys. Lett., 86 (2005) 054105.
[94] A. Wei Min Lee and Q. Hu, Opt. Lett., 30 (2005) 2563.
[95] A. Luukanen, L. Gronberg, P. Helisto, J. S. Penttila, H. Seepa, H. Sipola, C. R. Cietlein, and E. N. Grossman, Proc. SPIE, 6212 (2006) 62120Y.
[96] E. Gerecht, D. Gu, S. Yngvesson, F. Rodriguez-Morales, R. Zannoni, and J. Nicholson, Proc. SPIE, 5790 (2005) 149.
[97] D. H. Johnson and D. E. Dungeon, Array Signal Processing, Prentice Hall (1993)115.
[98] N. A. Salmon, S. Hayward, R. L. Walke, R. Appleby, Proc. SPIE, 5077 (2003) 71.
[99] J. Federici, D. Gary, B. Schulkin, F. Huang, H. Altan, R. Barat, and D. Zimdars, Appl. Phys. Lett., 83 (2003) 2477.
[100] K. Walsh, B. Schulkin, D. Gary, J. Federici, R. Barat, and D. Zimdars, Proc. SPIE, 5411 (2004) 9.

[101] J. A. Hogbom, Astron. Astrophs. Suppl., 15 (1974) 417.
[102] J. Van Rudd and D. Mittleman, J. Opt. Soc. Amer. B, 19 (2002) 319.
[103] M. Tani, P. Gu, M. Hyodo, K. Saki, and T. Hidaka, Opt. Quant. Electron., 32 (2000) 503.
[104] B. Widrow, D. E. Rumelhart, and M. A. Lehr, Communication of the ACM, 37 (1994) 93.
[105] F. Oliveira, R. Barat, B. Schulkin, F. Huang, J. Federici, D. Gary, and D. Zimdars, Proc. SPIE, 45 (2004) 5411.
[106] J. C. Principe, N. R. Euliano, and W. C. Lefebvre, Neural and Adaptive Systems, John Wiley & Sons, New York (2000).

Chapter 12

Explosives Detection Personnel Portals

Kevin L. Linker

Contraband Detection Technologies, Sandia National Laboratories, Albuquerque, NM 87185, USA

Counterterrorist Detection Techniques of Explosives
Jehuda Yinon (Editor)

Contents

1. Introduction

The presence of an explosive on a person ('person' is defined as anywhere on the body, including fingers, clothing, hair, etc.) can be characterized in two different forms: bulk and/or trace. In the bulk form, the actual explosive of some macroscopic mass quantity is located on the person in one or more location(s). An explosive in the trace form consists of residue (vapor or particulate) that remains on the person when he or she has been near an explosive or has some mass of explosive hidden on his or her person. To detect an explosive on a person, there are two principal detection categories. The first detection category is locating the actual bulk explosive using methods such as a physical search or with techniques such as X-ray backscatter, millimeter-wave, or low-power microwave. The second detection category provides notification that traces of explosive residue exist on the person. This information can be provided to personnel who can question the person and perform a physical search to determine the location of the actual explosive. Figure 1 shows the general categorization of explosives on people.

Some detection methods, such as a physical search, offer a high probability of identifying the presence of an explosive but suffer from being invasive and time-consuming techniques that can also place personnel in harm's way if the explosive is detonated. Other methods result in a somewhat lower probability of detection but are non-invasive and less time-consuming and offer some standoff protection for personnel. The choice of methods is a decision for the end-user and depends on a variety of considerations. These considerations include:

(1) the explosive threat, i.e., the size of explosive impact and where and how the explosive might be expected to be placed on the person;

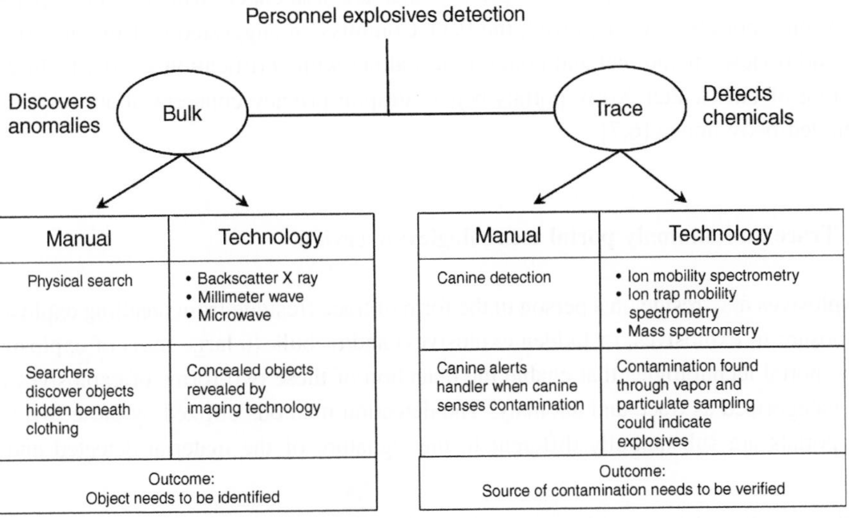

Fig. 1. Explosive detection on people.

(2) the vulnerability that needs protection from the explosive threat of people or infrastructure;
(3) the location where the detection will be made, such as an airport, nuclear power plant, or public venue;
(4) initial and recurring costs; for example, one technology may have a low purchase cost compared with another device, but higher maintenance costs in the long term;
(5) installation requirements such as the availability of electrical power, any other needed utilities, and floor space;
(6) effectiveness, which includes the probability that the explosive threat will be detected.

Explosives detection personnel portals are prime candidates for applications where (i) a human-borne explosive threat is a concern, (ii) the flow of pedestrian traffic can be directed through a defined single point, and (iii) the facility requires a medium to high throughput (tens to hundreds) of people. High throughput rates imply a need for technology with non-invasive methods of detection. The portals described in this chapter utilize trace and bulk detection technologies with very sensitive detection capabilities with relatively low false alarm levels.

Personnel portal research and development accelerated in the 1990s as a result of the bombing of PanAm Flight 103 over Lockerbie, Scotland in 1988. In 2004, a report from the 9/11 Commission highly recommended that the US Congress and the Transportation Security Administration (TSA) consider screening airline passengers for explosives [1]. During 2004 and 2005, the TSA began field-testing these newly developed trace detection portals at 15 different airports [2,3]. The portals under test were developed through funding and technical oversight provided by the Transportation Security Laboratory (TSL)[1] [4,5]. Part of the purpose of the field tests was to address factors such as reliability, high throughput, and privacy that are associated with use in airports. If personnel portals were deployed, the 9/11 Commission suggested that the government should review operational and policy issues along with certification standards. In 2004, testing of backscatter X-ray portals began, despite privacy concerns about revealing a detailed body image [6,7].

2. Trace and anomaly portal technologies: overview

Explosives may reside on a person in the form of trace (residue from handling explosives, exposure to explosives, or hidden explosives) and/or bulk (a large mass of explosives). The portal technologies that enable the detection of these two forms of explosives may be categorized as trace and anomaly. The detection methods utilized by these two types of portals are substantially different in the signature of the material detected and the

[1] The former Federal Aviation Administration Aviation Security Research and Development effort is now the Transportation Security Laboratory (TSL).

information provided to operators. For both types of portals, a physical search may be required to resolve the information supplied by the portal.

Trace portals sample the air surrounding the person and any dislodged particulate matter and then analyze the sample for explosive residue. If explosives are present, the portal's detector identifies its chemical signature and the portal alerts the operator, noting the particular explosive detected. The location and amount of explosives are not identified.

Anomaly portals detect an irregularity caused by the bulk explosive (or other contraband) and its location on the person and then provide visual information to the operator. Anomaly portals detect masses of objects (i.e., guns and knives), including explosives, concealed beneath clothing but external to the body. To detect these masses, some anomaly portals utilize imaging techniques such as X-rays or millimeter waves to provide visual information. Another anomaly portal uses dielectric changes to detect masses. Currently, anomaly portals do not identify the particular explosive (a chemical characteristic) or other type of contraband. Operators interpret the data provided by the portal to make a judgment about the source of the alarm.

Anomaly portals create an image of the body that provides a representative data display. The representation may be a direct image of the person or an image of a 'generic human' with the anomalies projected onto that image, thereby avoiding privacy concerns. Whereas some models require operator interpretation of the display, other vendors have algorithms that can highlight anomalous (or potential problem) areas on the body and alert the operator to those areas. This energetic technology is able to penetrate clothing and the resulting display can highlight concealed items. Images of the person's body are also part of the information presented to the screener, but it is possible to conceal this image and present it only to higher levels of authority as needed. Algorithms have been developed to mask the body image while retaining the anomalous images. Metal detectors also detect anomalies, such as those created by the presence of guns or knives; so while they are not suitable for explosives, they can detect some bombs if the bombs have metal components.

Trace portals could also be used for bulk detection because of the likelihood that a mass of explosive concealed on the person would present an adequate chemical signature. The combination of a trace and anomaly portal would provide a powerful multi-sensor platform that would offset the limitations of the individual technologies. Currently, a commercially available multi-sensor explosives detection personnel portal that combines trace and anomaly methods does not exist.

3. Trace detection portal: background

3.1. Subsystems that contribute to better detection

The principal function of the trace detection portal is to analyze explosive residue extracted from an air sample taken from around the person's body. The typical sequence of operation is as follows:

- The person enters the portal and stands in a defined sample collection space.
- The system collects an air sample for a predetermined amount of time, which varies from several seconds to tens of seconds, depending on the particular trace portal design.
- Using the on-board detector, the system analyzes the sample for explosive residue.
- The system presents the result to the operator either visually and/or audibly in another few seconds. A trace explosives detection portal system will provide a signal (e.g., red light/green light) output depending on the concentration of explosives reaching the detector. If needed, the particular explosive detected can be relayed to the screener.

The total cycle time, from the time the person enters the portal until the operator has an answer, is on the order of tens of seconds, and usually less than a minute.

The trace detection of explosives on a person is not a trivial task. Current trace portals are sophisticated analytical systems designed to detect a wide variety of explosives. These portals incorporate many subsystems to accomplish explosive detection. Three subsystems (sample collection, preconcentration, and detection) have significant influence on the ability of the portal to detect explosives and are as follows:

- *Sample collection* – Removes the explosive (vapor and/or particulate) from the person and transports the sample to the preconcentration subsystem. Current trace portals use non-contacting means of pulsating air jets, commonly referred to as 'puffers', and sampling airflow (tens to hundreds of liters of air per second) to accomplish the removal and transport of particles and vapor. The sample collection area needs to be designed properly to remove, contain, and transport explosives from the entire body, from head to feet. Sample collection is important in trace explosives detection because of the limited amount of explosive residue available to extract from a person [8].
- *Preconcentration* – Concentrates the explosive delivered from the sample collection subsystem. The preconcentration subsystem is a mechanical system that extracts the explosive vapor and/or particulate from the airflow of the sample collection subsystem. The preconcentration subsystem also serves as an impedance-matching device between the sample collection airflow and the detector airflow (which moves tens to hundreds of cubic centimeters of air per minute). Preconcentration in trace portals increases the concentration ratio (10–10^5 times) of the explosives delivered from the sample collection subsystem to the detector [8].
- *Detection* – Accepts a sample from the preconcentrator and makes an analysis. Several types of detectors are currently incorporated in trace portals. These detectors include ion mobility spectrometers or mass spectrometers. The detection subsystem needs to be reliable and rugged to perform hundreds to thousands of analyses per day with the necessary sensitivity and specificity required in the trace portal system. Most detectors are concentration-sensitive devices.

3.2. Using mathematical models to describe subsystems and performance

Trace detection portal systems can be mathematically modeled [8]. The model provides the basis for describing trace portal subsystems and is independent of the specific sampling technique, preconcentration technology, or detector. Through the model, a relative measure of a trace portal system can be determined, which provides a means for comparing trace detection portals.

Equation 1 shows the relationship of sample collection, preconcentration, and detection in the portal. This equation contains measurable parameters such as (i) sample removal and transport from a person, (ii) collection (adsorption) and release (desorption) of the explosive in the preconcentrator, (iii) additional preconcentration, if applicable, (iv) a term that accounts for flow mismatches between the detector inlet and the preconcentrator subsystem outlet, and (v) a proportionality term relating signal strength to concentration at the detector.

$$S = M_0 \eta_r \eta_c C_1 \cdots C_n \left(\frac{Q_d}{Q_n} \right) k \qquad (1)$$

where

S = signal output from the detector (red light/green light depending on alarm thresholds)
M_0 = initial mass of explosives residue or traces on a person (vapor and/or particles)

- Sample collection factors
  η_r = fraction of explosive that is removed from the person by sample collection subsystem
  η_c = fraction of explosive removed that is collected in the preconcentration subsystem
- Preconcentration factors
  C_1 = concentration gain of the first preconcentration stage
  C_n = concentration gain of the nth preconcentration stage if applicable, with each preconcentration stage included, from 1 to n
- Detection factors
  Q_d = inlet (sampling) flow rate of the particular detector
  Q_n = output flow rate from the final preconcentration stage
  k = proportionality term relating signal strength to concentration at the detector.

The efficiencies for sample removal and collection are between zero and one and can be represented as:

$$0 < \eta_r = 1.0$$

$$0 < \eta_c = 1.0$$

Although it may appear that η_c is part of the preconcentration term and not sample collection, this sample collection is a function of both the sample collection subsystem and preconcentrator. The sample collection subsystem is designed to efficiently remove explosives from the person and transport the explosives to the preconcentrator. Thus, η_r accounts for both the removal and the transport. The preconcentrator then traps or collects the explosives (η_c) for subsequent detection. Sample collection and preconcentration must be integrated for optimal performance in a trace portal detection system. In other words, the preconcentrator should be designed to accept the entire volumetric airflow of the sampling system. Any mismatch in these subsystems will decrease the ability of the portal to detect explosives.

$C_1 \ldots C_n$ define the preconcentrator performance by combining the efficiencies of adsorption and desorption and the ratio of the volume of flow into each stage during adsorption and out of each stage during desorption. The terms $C_1 \ldots C_n$ are further defined as follows:

$$C = \eta_p V_a / V_p \tag{2}$$

where

η_p = preconcentrator efficiency, $0 < \eta_p \leq 1.0$
V_a = volume of airflow that passed through the preconcentrator during the adsorption (collection) phase
V_p = internal volume of the preconcentrator or volume into which the explosives sample is desorbed.

The preconcentrator efficiency η_p accounts for several factors associated with the preconcentrator performance. These factors include explosive passing through the preconcentrator during the adsorption (collection) phase, explosive loss due to adsorption on cool internal surfaces, and molecular decomposition associated with the desorption phase in the preconcentrator. The value for η_p ranges from 0 (no explosive delivered to the detector) to 1 (entire explosive sample delivered to the detector). Ideally, η_p would be 1.

Figure 2 indicates how the initial mass, M_0, is fractionally distributed through a trace detection portal system. The mass and location of M_0 on a person is a complex parameter to determine. Considerable research has been performed to understand the amount and distribution of trace explosives material on the body [9–11]. Even with a better understanding of M_0, there is still a wide variability in M_0 in a real-world environment. The mass and location of explosive on the person is never exactly known. Establishing a mass for M_0 as a standard is convenient for comparative testing between portal systems, but the convenience comes at some risk [12]. The risk in setting a standard for M_0 is that it sets a reference that may not be a realistic representation of the explosives on people. If the reference mass for M_0 is set to a particular amount and placed in particular locations on the person, then a certain number of people with

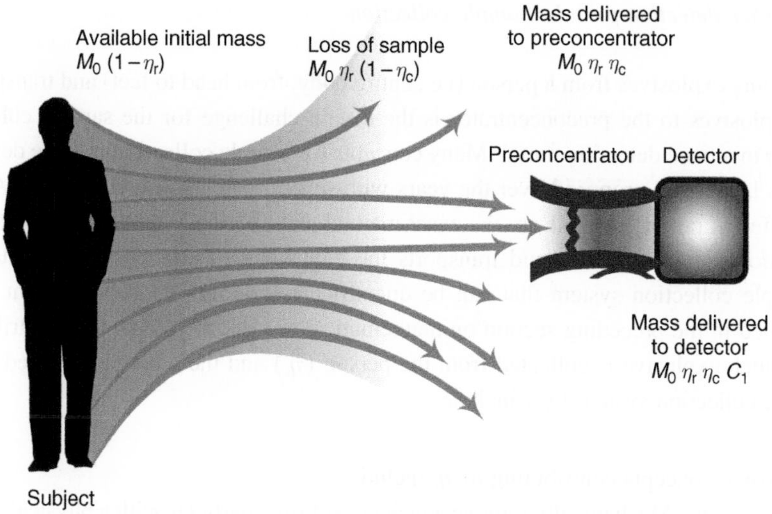

Fig. 2. Explosive mass distribution in trace portal detection system.

explosives will pass through the portal with less M_0 and different locations. If possible, an appropriate portal test should include both standardized testing with a known M_0 and real-world scenarios using actual bulk explosives concealed on people to characterize the limits of the trace detection portal.

The mathematical model demonstrates the importance of sample collection (η_r), preconcentration (η_c), concentration (C), and detection (k) in a complete trace portal detection system. Of these three subsystems, the detector is the most understood. Considerable information is available that quantifies the sensitivity, specificity, and limits of detection (LOD) for a particular detection method when used for trace explosives detection [13–15]. For trace detection portals, the selection of the detection method is based on performance, initial cost, and maintenance issues. The remaining subsystems (sample collection and preconcentration) are the most variable and least understood for their contribution to trace portal performance. Optimizing the explosive removal and transport in sample collection along with preconcentration will enhance the performance of the entire trace detection system. The sensitivity of the detector will help determine the performance needed from the sample collection and preconcentration.

In general, trace detection portal systems are recognized more by the type of the detector rather than the method of sample collection and preconcentration. Emphasis on the detector occurs because the detection method tends to be the 'high-tech' part of the portal system. But, as has been shown above, the detector is no more important than the collection and preconcentration subsystems. All subsystems must be appropriately considered and optimized in a trace detection portal system. A very sensitive detector is of no consequence if the explosive is not removed, transported, and delivered.

3.3. Trace detection portals: sample collection

Removing explosives from a person (i.e., entire body, from head to feet) and transporting the explosives to the preconcentrator is the design challenge for the sample collection system in a trace detection portal. Many concepts for sample collection in trace detection portals have been proposed over the years with several patents issued [16–25]. Among the different portal concepts are the exact method and efficiency with which the sample collection system removes and transports the explosives. These portal concepts have a sample collection system that can be quantified by η_r and η_c in the system model described in the preceding section on mathematical models. Concepts that contribute to how much explosive is collected from the person (η_r) and the amount collected by the sample collection system (η_c) include

- Removal concepts contributing to η_r include
  - *Abrasion* – Mechanically remove attached explosive particles with a physical contact method, i.e., rubbing or wiping
  - *Agitation* – Mechanically remove attached explosive particles with an external non-contacting method, i.e., high velocity air jets [26]
  - *Ionization* – Statically remove attached explosive particles with an external neutralizing charge, i.e., ionizing bars and nozzles
  - *Airflow* – Remove explosive vapors with an airflow, i.e., externally induced air movement with a fan or air jets
  - *Heat* – Elevate the temperature of the explosive to enhance explosive vapor production with an external heat source, i.e., radiant heating.
- Transport concepts contributing to η_c include
  - *Airflow* – Transport explosive particles and vapors, i.e., externally induced airflow with a fan or air jets
  - *Barrier* – Control the dispersion of removed and transported explosives with a physical barrier, i.e., a door.

All of the currently available commercial trace portals, which will be described later, utilize the same underlying theme in the sample collection subsystem for explosive removal and transport. Specifically, the most frequently implemented sample collection concept can be categorized as ‘non-contacting’ or ‘non-invasive’. This definition implies that no part of the portal physically contacts the person to collect an explosive sample. All of these non-contacting trace detection portals use air jets to dislodge explosive particles from the person and use airflow to transport explosives away from the person.

The idea of using airflows to collect a substance similar to explosive vapors and/or particles is not unique. From an engineering systems design perspective, collecting an explosive from a person can be directly compared to local exhaust ventilation systems used in industry for contamination control, e.g., welding hoods, open vapor tanks, and fume hoods [8,27,28]. Local exhaust ventilation systems operate on the principle of capturing a contaminant (vapors or particles) at or near the source using a fan to induce airflow into a filtration device. In a trace portal system, the contaminant is the explosives sample (vapor or particulate), the source is the person, and the filtration device is the preconcentrator. Allowances in local exhaust ventilation design account for the target itself if the contamination is being expelled at high velocities or elevated temperatures. The local exhaust method is preferred in industrial applications because it is more efficient in contaminant collection and requires less horsepower in the fan. The difference between explosive collection and typical contamination collection results from chemical properties that are specific to explosives. Specifically, explosive molecules are 'sticky' and tend to adhere to a surface easily [29,30]. Second, a limited amount of explosive can be collected because of the very low vapor pressures of many common explosives [4,13,14,31]. Effectively removing the explosive particles from the subject can be complex [26,32]. In the real world, the sample collection subsystem must enable explosives removal and transport without offending the person. For example, using excessively strong airflows and/or jets of air that strike people inappropriately or uncomfortably (for example, a strong blast of air to the face) would be an unacceptable portal design.

As mentioned earlier, all subsystems of a trace portal must be integrated to facilitate explosives detection. After removal and transport of the explosive, the explosive must be collected for eventual delivery to the detector. The preconcentrator receives the explosive from the sample collection subsystem. A well-designed preconcentrator will enhance a marginal sample collection system.

3.4. *Trace detection portals: preconcentrator*

The preconcentrator is a mechanical system used to concentrate the limited mass of explosive delivered from the sample collection subsystem [8]. From the sample collection subsystem airflow, the preconcentrator adsorbs explosives (vapor or particulate). The adsorbing surface is then heated to desorb the explosives into the airflow stream for delivery either to a detector or another preconcentration stage. Concentration of the explosives sample occurs because the explosive contained in the sample collection airflow prior to adsorption is now contained in a smaller volume after desorption. Equation 2 shows the relationship between concentration and volumes related to the preconcentrator.

Figure 3 shows the operation of the preconcentrator. In Step 1, the preconcentrator processes the airflow from the sample collection subsystem. Minute amounts of explosive particles and vapors are entrained in this airflow. In Step 2, the preconcentrator adsorption surface is heated to vaporize the collected and concentrated explosives sample into the airflow of the detector inlet.

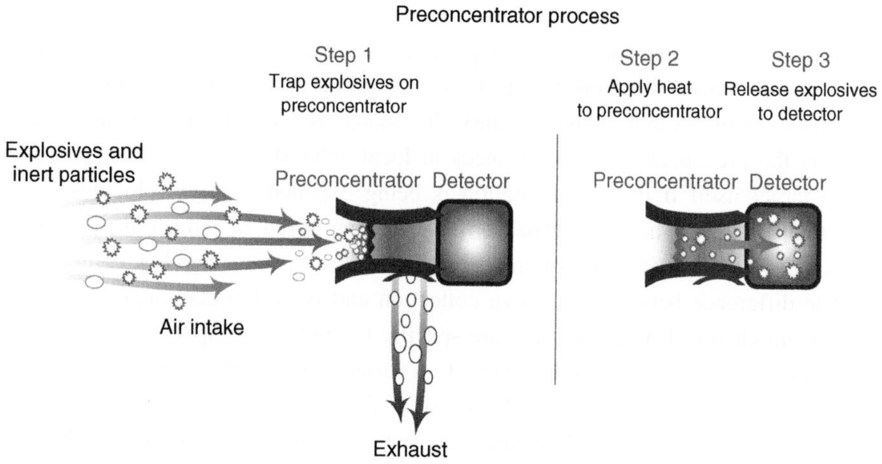

Fig. 3. Preconcentrator operation.

The preconcentrator is a critical subsystem in a trace detection system because of its ability to collect minute amounts of explosive material. Without preconcentration, the amount of material available to detect would be well below the LOD for many detectors. In addition, the preconcentrator serves as an impedance-matching device between the airflow of the sample collection subsystem and the airflow to the detector. The characteristics of a well-designed preconcentrator include

- Inlet airflow to the preconcentrator matches the airflow of the sample collection subsystem
- Minimal pressure drop during adsorption, which reduces fan horsepower requirements
- Minimal pressure drop during desorption, which reduces the restriction to the detector inlet
- Minimal internal volume, which maximizes concentration, C, in Eq. 2
- Efficient at adsorbing explosive (vapor and particle) and desorbing explosive, which maximizes η_p in Eq. 2, i.e., high probability of adsorption in the collection phases and low probability of decomposition in the desorption phase
- Rapid desorption, which maximizes the output concentration pulse
- Quick cycle between adsorption and desorption, which maximizes system throughput
- Cost-effective, easily maintained, and robust for real-world applications.

This section described the importance of sampling and preconcentration at a systems level and was presented using a mathematical model. Background material for local exhaust ventilation systems was provided, as well as its application in trace explosives detection systems. For trace portals, large-volume sampling is necessary for the collection and transport of sample. Finally, when large-volume sampling and preconcentration are integrated, the trace detection of explosives is greatly enhanced.

3.5. Current trace detection portals

Several trace explosives detection portals are currently manufactured and marketed by various companies. These trace portals are an outgrowth of funded research and development programs through the TSL and the Department of Energy (DOE) [4,5,33]. Some initial trace portal concepts have been dropped or changed in design due to lack of performance or practicality in real-world environments. The portals that have continued from research and development are able to detect extremely low levels of explosives with high accuracy.

3.5.1. General Electric Security: EntryScan[3]

The trace detection portal manufactured and marketed by General Electric (GE) Security is called the EntryScan[3] [34], shown in Fig. 4. This portal was the result of a collaborative development effort between Ion Track Instruments (ITI), which was later acquired by GE, and Pennsylvania State University (Penn State), funded by the TSL [4,5,23,32]. The detection method used in this portal is ion mobility spectrometry (IMS) [35]. GE utilizes their patented Ion Trap Mobility Spectrometer (ITMS) for detection [36].

To begin the cycle, a person enters the portal and stands in the sampling space defined by the floor, ceiling, and three walls, which is similar to the two-wall metal detection portals found at airports. The sampling space can be defined as a semi-open-hood design

Fig. 4. General Electric Security, EntryScan[3] (Image provided courtesy of GE Security).

because there is some communication between the sampling space and the external environment during sampling [8]. The design is considered 'semi-open' instead of 'open' because three defined walls are used during sampling. One of the walls, however, is a full-height, clear-view door on the exit side of the portal, which is closed during sampling. This exit side door is used to retain some of the sample within the portal as well as provide a barrier for process control. The direction of the sampling airflow over the person is vertically upward toward a single-stage $10 \times 10\,cm^2$ preconcentrator located in the ceiling [32]. The sampling airflow passes from the sampling space through the preconcentrator and back into the environment, which is a 'single-pass' design. Removal and transport of explosives is accomplished with ~20 air-jet nozzles in the portal side walls that puff and a fan that draws air through the preconcentrator at 50 l/s. The air jets remove explosive from the person and drive the explosive overhead to be captured in the preconcentrator. An on-board air compressor powers the air jets.

According to GE and Penn State, an additional component of the sample collection is associated with the temperature of the human body generating a 'human convection plume' [32,34]. The plume is said to promote a natural upward convection of air along the entire body at a rate of 0.3–0.5 m/s, with a volumetric flow equal to the fan flow of 50 l/s. With this portal design, when a pure explosive vapor is released in the portal sampling space ~25% of the vapor is captured and detected [32]. After the person has been screened, the operator is notified through audile and/or visual indicators on an interface computer screen on the exit side of the portal. If the person is cleared, the door swings open to allow exit. A complete cycle for this portal, defined from the time the person stops inside the portal to a notification to the operator, is ~15–20 s.

3.5.2. Smiths Detection: Sentinel II

Smiths Detection manufactures and markets the Sentinel II[2] trace detection personnel portal [37], shown in Fig. 5. Sandia National Laboratories performed the original concept and development of the Sentinel II through funding from the TSL and the DOE [22,24,38–40]. Smiths Detection incorporates their own IMS design called the Ionscan® [35]. The person enters the portal's sampling space in a walk-through direction similar to the design of metal detectors used in most airports.

Two walls on either side of the person, the floor, ceiling, and a single-arm, waist-high swing gate on the exit side define the sampling area. This sampling space is defined as an open-hood design because of the interaction between the sampling space and the external environment during sampling [8]. The design is 'open' because two defined walls are needed during sampling. The sample collection subsystem takes advantage of the characteristics of air movement with directional blowing of air and unidirectional exhausting of air [8]. The airflow in this portal is modeled after the laminar flow clean room routinely used in microelectronic fabrication or hospitals and invented at Sandia

[2] Sentinel II was originated by Barringer Instruments prior to acquisition by Smiths Aerospace and renamed Smiths Detection.

Fig. 5. Smiths Detection, Sentinel II (Image provided courtesy of Smiths Detection).

in 1960 [41–43]. Air curtain fans in the ceiling blow air directionally downward over the body at ~2 m/s, taking advantage of gravity to collect larger particles, hair, lint, and other debris that may have attached explosive trace material. An on-board air compressor drives 30+ air jets in the sidewalls, thereby providing agitation to remove explosives from the person, and the explosive vapor and particles are entrained in the downward airflow and carried to the unidirectional exhaust slot near the floor and then through a two-stage preconcentrator [44–46]. The airflow design in this portal is a single-pass design. The volumetric airflow through the first stage of its two-stage preconcentrator is about 300 l/s. The downward airflow is designed to cover the complete body from head to feet. Laboratory testing has demonstrated that sample collection efficiencies between the head, body, and feet with this design are nearly identical [47]. In this portal design, with an explosive vapor source introduced into the sampling space, ~50% of the source is collected and detected [47]. Once the person has been screened, a message on a monitor notifies the operator of the result. If the person is cleared, the single-arm gate swings open to allow exit. For this portal, the cycle time is about 15–20 s. A doubling of throughput is possible with the two-stage preconcentrator design of this portal. This doubling is accomplished by alternating a person through the portal while simultaneously shuttling a sample between the first and second stages of the two-stage preconcentrator and the detector. This doubling of throughput could be coordinated with other security measures such as metal detectors or X-ray devices, making the 'corralling' of people possible while increasing throughput in the portal and the overall checkpoint.

3.5.3. Syagen Technology: Guardian™

The Guardian™ Explosives Detection Security Portal (Fig. 6) is manufactured and marketed by Syagen Technology, Inc., of Tustin, CA [48]. This portal was the result of two research and development efforts that included collaboration between Syagen and Sandia National Laboratories, funded by TSL, and a project at Sandia, funded by DOE [33,49]. The detector utilized in the portal is a quadrupole ion trap, time-of-flight (QitTof™) mass analyzer [50,51]. The mass analyzer identifies explosives by time-of-flight and mass-to-charge (m/z) ratio, which increases the accuracy of the identification of the explosive. The foundation of the sample collection method in this portal is an air shower [52].

The person enters the portal's sampling area, which is defined by four walls, the floor, and the ceiling, similar to an elevator with two sets of sliding doors. Two of the portal's walls are doors across the entrance and exit planes that actuate between 'open' and 'closed' to process people through the portal. To initiate screening, the entrance side door is opened and the exit side door is closed, which stops the person's progress through the portal. Once the person is inside the portal, the entrance door is closed. One of the walls, both doors, and the ceiling are transparent to enable external visual contact while allowing light into the sampling area. The sampling area of this portal is considered a 'closed-hood' design because communication between the sampling space and the external environment during sampling does not exist [8]. The sample collection subsystem utilizes directional blowing air through more than 20 air-jet nozzles located in the walls and a single unidirectional exhaust near the floor. Airflow from the nozzles

Fig. 6. Syagen Technology, Guardian (Image provided courtesy of Syagen Technology).

accomplishes both explosive removal and transport. The exhausted air is processed through the first stage of a two-stage preconcentrator at the rate of 500 l/s [44–46]. Air exiting the first-stage preconcentrator enters a centrifugal blower where its pressure is increased. The pressurized air is channeled back to the nozzles for another cycle through the sampling space, preconcentrator, and blower. This portal concept is a 'multiple-pass' design because the air used in sampling is returned continuously to the sampling space. This portal design has a vapor removal and collection efficiency of at least 60% [53]. Because of the high explosive removal efficiency within the first-stage preconcentrator, the air returned to the sampling space is free of any explosive material. Therefore, contamination of the sampling space or person from a prior occurrence is negated. The cycle time for this portal is ~15–20 s. After a person is screened for explosives and cleared, the results are reported to the operator through a display monitor. If the person is cleared, the exit door automatically opens, while the entrance door remains closed, allowing the person to exit. If an explosive is detected, the exit door remains closed and the entrance door is re-opened, allowing the person to retrace their entrance for off-line screening. With integrated entrance and exit doors, the Guardian™ can be used in applications that require physical barriers and process control of people through a checkpoint where explosives detection is needed between non-secured and secured areas. Finally, the two-stage preconcentrator design enables the portal's throughput rate to be halved if needed.

4. Anomaly detection portals

4.1. Background

Anomaly detection portals gather information regarding bulk items such as explosives or other concealed items, such as knives and guns, hidden under clothing but external to the body. These portals are characterized as 'active' systems, because these systems incorporate a source or transmitter and a detector or receiver to interrogate a person. Examples of active systems use X-rays, millimeter waves, and microwaves. X-ray and millimeter-wave portals irradiate the body and analyze the backscattered radiation. Objects concealed on the body reflect the radiation differently than the body itself, resulting in an image of the object. Using the image, a trained operator can make a judgment whether an explosive is present. Another personnel portal technique uses a microwave field to measure a change in dielectric constant to detect material within the portal. The location of suspected threat objects are then visually presented to the screener [4]. Portals using these methods are commercially available and some US prisons use backscatter systems (specifically X-ray systems) for detecting contraband such as drugs and weapons.

Anomaly portals have two factors that have affected public acceptance of the associated technologies: radiation (both ionizing and non-ionizing) exposure and privacy concerns, despite verification of safe levels of radiation and advances in technology to address privacy concerns. The radiation levels emitted by these portals are well documented and accepted by several recognized standards, cited in the following section. Both issues will be addressed in the following sections.

4.2. X-ray portals

Anomaly portals with X-rays as the active system use low-energy collimated X-rays to scan the body in two dimensions: X and Y. From the two-dimensional scanning, an image can be obtained based on the principle of Compton scattering (also called backscatter) interactions and imaging software. Compton scattering depends on two physical properties of the material: density and atomic number. The density and atomic number (Z) of a material will dictate how the X-ray is absorbed, scattered, and transmitted. Material with high atomic numbers (high Z) such as metals absorb the X-rays efficiently while low atomic number (low Z) materials that contain elements such as the nitrogen in explosives are better at scattering X-rays. Lower density items such as clothing and hair are transparent to X-rays. Both organic (e.g., explosives, narcotics, and ceramics) and inorganic (e.g., metals) materials are detected using X-rays. From the interactions of material characteristics and X-rays, an image of the person and objects on the person is generated. The image will appear dark for high-Z material, light for low-Z material, and shades of gray for other materials. Trained operators can assess the displayed image for potential contraband. To obtain an image, the subject to be screened stands in front of a small vertical wall, behind which is located the X-ray source and detector. A scan of one side of the body requires 10 s or less. If needed, the person turns 180°, placing the back toward the wall for a second screen.

Total screening of a person is ~20 s or less with a total, non-penetrating radiation exposure (front and back) of 20 microREMs or less. In comparison, a medical X-ray procedure produces 50 000 times more radiation (1 000 000 microREMs) than a portal exposure. A 2-h airplane flight exposes the subject to 1000 microREMs or 50 times more than a portal dose. In December 2003, the National Council on Radiation Protection and Measurements (NRCP) issued a formal commentary at the request of the Food and Drug Administration (FDA) 'on radiation protection issues concerning exposure to ionizing radiation from radiation-producing devices used for non-medical security purposes', particularly for screening humans [54]. The American National Standards Institute (ANSI) [55] recommends that the effective dose per scan should be 0.1 microSieverts (μSv) or less per scan. NCRP states that an individual could receive ~2500 scans per year and would not exceed the recommended limit.

4.2.1. AS&E: SmartCheck

The SmartCheck system is manufactured and marketed by American Science and Engineering, Inc. (AS&E) [56]. Figure 7 shows the SmartCheck portal. AS&E utilizes their Z® Backscatter image system to provide an enhanced display for the screener to review.

Fig. 7. SmartCheck, manufactured by AS&E (Image provided courtesy of AS&E, Inc.).

Figure 8 is an X-ray image from the portal. The system reveals a variety of anomalies on the body such as explosives, narcotics, metal guns and knives, and weapons made of plastic or ceramic. AS&E offers privacy software to address the subject's concerns about images that are too revealing.

4.2.2. Rapiscan®: Secure 1000

Rapiscan® Systems manufactures and markets the Secure 1000 X-ray system [57]. Figure 9 shows the Secure 1000 system. Typical scan time (for one side of the subject) for this portal is 8 s or less with a radiation level of 10 microREMs. The Secure 1000 provides images of concealed items such as explosives, narcotics, weapons (including plastic, ceramic, and composite), and metals. A color image is presented to an operator's monitor for review. Figure 10 shows an image from the Secure 1000.

4.3. Dielectric portals

To detect a change in the dielectric constant, a non-ionizing, standing microwave field is established in the screening space during a scan of a person. Low-energy microwaves are transmitted at a frequency of about 5.5 gigahertz (GHz) to penetrate clothing and indicate an anomaly [58,59]. The microwaves interact with material within the screening space, causing the waves to reflect from or to penetrate into a surface.

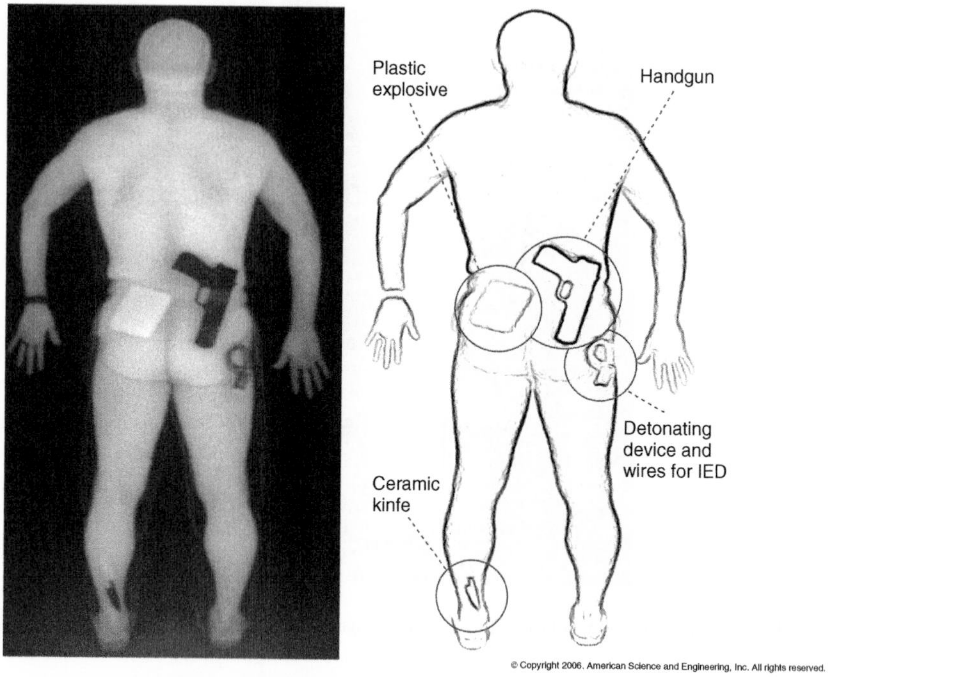

Fig. 8. SmartCheck Image from AS&E (Image provided courtesy of AS&E, Inc.).

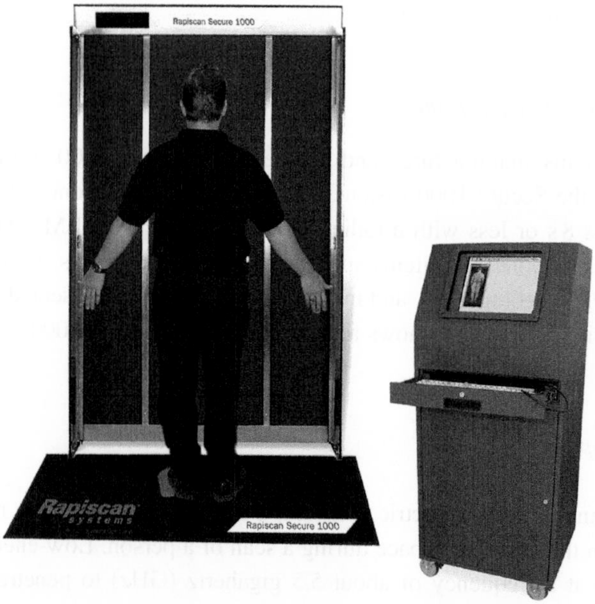

Fig. 9. Rapiscan Secure 1000 (Image provided courtesy of Rapiscan Systems).

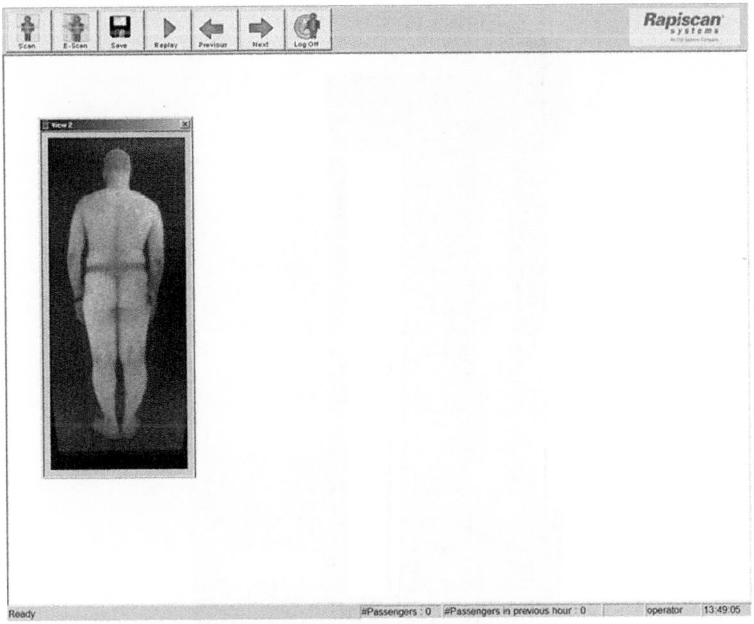

Fig. 10. Rapiscan Secure 1000 Image (Image provided courtesy of Rapiscan Systems).

4.3.1. EMIT Technologies: People Portal II

The People Portal II (PPII), manufactured and marketed by EMIT Technologies LLC, identifies anomalies on a person's body by detecting changes in the dielectric constant within a defined space [60]. Figure 11 shows the People Portal II, which is based on a proprietary microwave technology platform called Electro-Magnetic Impedance Translation (EMIT).

Because of the reflection and penetration, the microwaves' phase and amplitude change from the originally transmitted microwaves. Phase changes are associated with the velocity of the microwave traveling through material while an amplitude change is a result of the dielectric constant (electrical resistivity) of the material in which the microwave is traveling. To detect the change in dielectric constant, a proprietary, bi-focusing microwave lens is used as an antenna. Using an internal database, computer software compares the known dielectric constant of the human alone in the screening space with the detected change in dielectric constant resulting from an anomalous material in the same space. The anomalies are highlighted on the display and their locations are provided to the operator through visual notification. The PPII does not identify the specific anomalous material but was designed to detect metallic and non-metallic weapons, explosives, and narcotics. Figure 12 shows a typical detection and notification. The person enters the portal, stands on the feet locators on the floor, and holds two vertical handles that allow images to be taken of the entire body, thus eliminating 'blind' spots under the arms. Total screening time of a person is ~5 s while the system completes a 360° scan of the subject. To provide privacy, the image of

Fig. 11. People Portal II from EMIT Technologies (Image provided courtesy of EMIT Technologies LLC).

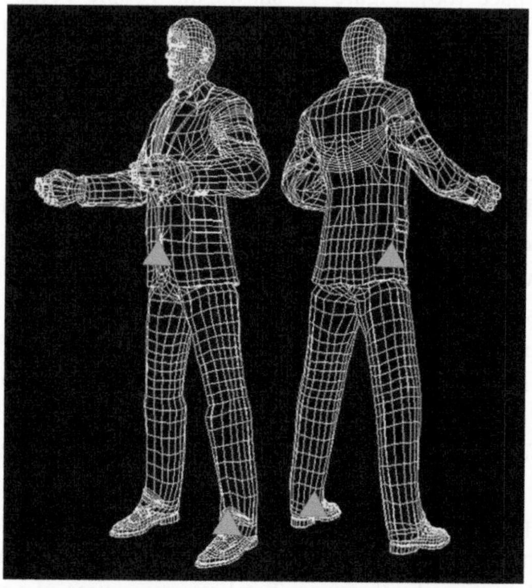

Fig. 12. EMIT Technologies: People Portal II Image.

the body is represented by a wire-frame Figure with the anomaly location(s) indicated. During the body scan, dielectrometers in the floor also scan the person's shoes for material. The manufacturer states that the microwave emission from the portal complies with national and international standards for screening personnel [61].

4.4. *Millimeter-wave portal*

Another approach to detect anomalies on the body is the use of millimeter-wave technology, which is non-ionizing low-power radiation, enabling its use with people for detecting explosives, drugs, plastics, ceramics, wood, paper, metals, and other anomalies concealed under clothing.

4.4.1. L3 communications millimeter-wave portals

L3 Communications manufactures and markets two portals that use low-power, active millimeter-wave technology: the SafeScout100™ Security Portal [62], shown in Fig. 13, and the ProVision™ Body Screening System portal. The SafeScout100™ was originally developed by SafeView, Inc., and was developed at the Pacific Northwest National

Fig. 13. L3 SafeView: SafeScout™ Portal (Image provided courtesy of L3 Communications, Inc.).

Laboratory with additional support from the Federal Aviation Administration (FAA). The ProVision™ Body Screening System portal is a more recent addition [63]. In real time, both portals provide a holographic image of the person along with materials concealed under the clothing. An operator can then view the image and determine if any contraband is present. The holographic image is generated by a high-speed image algorithm, which receives information from a millimeter-wave array/transceiver. While the person stands in the cylindrical chamber with arms raised, the array/transceiver illuminates the body with millimeter waves over a wide frequency bandwidth (10–40 GHz) and collects the reflected signal from the person's body [64–68]. From the reflected millimeter wave's amplitude and phase, a wideband, three-dimensional, image reconstruction algorithm creates an image that can be viewed. The image generated of the person is near 360°. The portals do not automatically identify the type of anomaly on the person. A person enters the portal screening area and stands while the portal takes a scan. A complete scan of the body requires ~2–5 s. Privacy concerns are addressed through software solutions that restrict viewing of the actual image. The company is developing software to identify anomalies on the body to help operators with data interpretation.

5. Future development of personnel portals

Both types of portal (trace and anomaly) have benefits that far outweigh the perceived shortcomings. While the trace portals can detect and identify minute amounts of explosive, the portals cannot distinguish between mere contamination and bulk quantities or specify the location on the body. The anomaly detectors can indicate that some type of material exists beneath the clothing and can indicate a location on the body, but they cannot identify the type of explosive. To take advantage of both portal types' strengths and to offset their complementary weaknesses, the next step in personnel portal development could be the creation of a dual technology portal. This orthogonal device would be capable of detecting multiple types of contraband (both weapons and explosives), showing the location on the body, identifying the type of explosive, and indicating trace contamination that could identify persons who had been handling explosives, in addition to bulk carriers.

Combining trace and anomaly technologies will provide a powerful detection system that would be very difficult to defeat. In the recent past, the portals highlighted in this chapter existed only as an idea, one-of-a-kind, or laboratory novelty, and now all are commercially available portals. Researchers will develop new portal systems as detectors, methods, and threats change.

References

[1] D.A. Shea and D. Morgan, Detection of Explosives on Airline Passengers: Recommendation of the 9/11 Commission and Related Issues, CRS Report for Congress, February 7, 2005.

[2] 15 Airports and 15 Months, 'Puffer' Explosives Detection Testing Finishes Up, National Safe Skies Alliance, 6 (2005) 8.

[3] C. Fotos, Beyond Metal Detectors, Homeland Security for Government and Business Leaders Worldwide, 2(2) (2005) pp. 32–35.

[4] S.F. Hallowell, Screening people for illicit substances: a survey of current portal technology, *Talanta* 54 (2001) 447.

[5] C. Woodyard, Labs work to formulate bomb detectors, *USA Today*, page 5B, February 20, 2001.

[6] A. Schatz, New airport body X-rays reveal all, *The Wall Street Journal*, 244(59) (2004) p. D3.

[7] T. Frank, Air travelers stripped bare with x-ray machine, *USA Today*, page 1A, May 15, 2005.

[8] K.L. Linker, Large-Volume Sampling and Preconcentration. Third Explosives Detection Technology Symposium and Aviation Security Technology Conference, Atlantic City, NJ, 2001.

[9] C.J. Miller, D.F. Glenn, and S.D. Hartenstein, Mapping of Explosives Contamination Resulting from the Manufacture and Transport of Improvised Explosive Devices, INEEL/INT-99-612, December 1999.

[10] C.J. Miller, D.F. Glenn, and S.D. Hartenstein, Detection of Explosives Contamination Using a Portal Monitor and Subsequent Mapping of Explosives Contamination Resulting from the Manufacture and Transport of Improvised Explosive Devices, INEEL/INT-2000-47, January 2000.

[11] C.J. Miller, G. Elias, N.L. Schmitt, C. Rae, R.C. Regalado, and R.A. Jordan, Quantification of Explosive Contamination from Improvised Explosive Device Manufacturers and Couriers, INEEL/EXT-02-00621, May 2002.

[12] R.T. Chamberlain, Dry Transfer Method for the Preparation of Explosives Test Samples, US Patent No. 6,470,730 (2002).

[13] M. Nambayah and T.I. Quickenden, A quantitative assessment of chemical techniques for detecting traces of explosives at counter-terrorist portals, *Talanta* 63 (2004) 461.

[14] D.S. Moore, Instrumentation for trace detection of high explosives, *American Institute of Physics*, 75 (2004) 2499.

[15] J. Yinon, Field detection and monitoring explosives, *Trends in Analytical Chemistry*, 21 (2002) 4.

[16] L.C. Showalter et al., Air Curtain Device, US Patent No. 4,045,997 (1977).

[17] A.H. Ellson, Apparatus or Detecting Explosive Substances, US Patent No. 4,202,200 (1980).

[18] M.D. Arney et al., Air-Sampling Apparatus with Easy Walk-In Access, US Patent No. 4,896,547 (1990).

[19] C.D. Corrigan et al., Explosive Detection Screening System, US Patent No. 4,987,767 (1991).

[20] A. Jenkins, Portal Vapor Detection System Patent, US Patent No. 4,964,309 (1990).

[21] C. Corrigan et al., Explosive Detection Screening System, US Patent No. 5,585,575 (1996).

[22] K.L. Linker et al., Vertical Flow Chemical Detection Portal, US Patent No. 5,915,268 (1999).

[23] G.S. Settles, Chemical Trace Detection Portal Based on the Natural Airflow and Heat Transfer of the Human Body, US Patent No. 6,073,499 (2000).

[24] K.L. Linker et al., Target Detection Portal, US Patent No. 6,334,365 (2002).

[25] J.H. Davies, Apparatus and Method for Screening People and Articles to Detect and/or to Decontaminate with Respect to Certain Substances, US Patent No. 6,375,697 (2002).

[26] R.C. Flagan, D.J. Phares, and G.T. Smedley, Aerodynamic Sampling of Particles from Surfaces, 2nd Explosives Detection Symposium & Aviation Security Conference, Atlantic City, NJ, 1996.

[27] *Industrial Ventilation, A Manual of Recommended Practice*, 22nd Edition, American Conference of Governmental Industrial Hygienists, 1995.

[28] *Air Pollution Engineering Manual* (2nd Edition), Los Angeles County Air Pollution Control District, Prepared For Environmental Protection Agency, May 1973.

[29] A. Fainberg, Explosives detection for aviation security, *Science* 225 (1992) 1531.

[30] F. Conrad, Explosives Detection: The Problem and Prospects, 25th Annual Meeting INMM, Columbus, OH, 1984.
[31] J. Yinon, *Forensic and Environmental Detection of Explosives*, John Wiley & Sons, Chichester, 1999.
[32] G.S. Settles, Sniffers: fluid-dynamic sampling for olfactory trace detection in nature and homeland security – the 2004 Freeman Scholar Lecture, *Journal of Fluids Engineering*, 127 (2005) 189.
[33] Integrated Personnel Checkpoint, Sandia National Laboratories Fact Sheet, Contraband Detection Department, Albuquerque NM, 09/28/2005, SAND2004-4799P.
[34] General Electric Security, http://www.geindustrial.com/geinterlogix/iontrack/prod_entryscan.html.
[35] G. Eiceman and Z. Karpas, *Ion Mobility Spectrometry*, 2nd Edition, CRC Press, Boca Raton, 2005.
[36] A. Jenkins, Ion Mobility Spectrometers, US Patent No. 5,200,614 (1993).
[37] Smiths Detection, http://trace.smithsdetection.com/products/Default.asp?Product = 24&section = Transportation
[38] J.E. Parmeter, K.L. Linker, C.L. Rhykerd, D.W. Hannum, F.A. Bouchier, and D.W. Hannum, Development of a Trace Explosives Detection Portal for Personnel Screening, IEEE International Carnahan Conference on Security Technology, Alexandria, VA, 1998.
[39] J.E. Parmeter, K.L. Linker, C.L. Rhykerd, D.W. Hannum, and F.A. Bouchier, Explosives detection portal for high-volume personnel screening, enforcement and security technologies, *Proceedings SPIE*, 3375 (1998) 384.
[40] G. Paula, Crime-fighting sensors, *Mechanical Engineering*, 120 (1998) 1.
[41] Clean room information web site, http://www.webopedia.com/TERM/C/clean_room.html.
[42] *Sandia Lab News*, 57 (2005) 2.
[43] S. Vorenberg, *The Albuquerque Tribune*, February 28, 2005.
[44] K.L. Linker et al., Particle Preconcentrator, US Patent No. 5,854,431 (1998).
[45] K.L. Linker et al., Particle Preconcentrator, US Patent No. RE38,797 E (2005).
[46] K.L. Linker et al., Two-Stage Preconcentrator for Vapor Particle Detection, US Patent No. 6,335,545 (2002).
[47] J.E. Parmeter, K.L. Linker, C.L. Rhykerd, and D.W. Hannum, Testing of a Walk-Through Portal for the Trace Detection of Contraband Explosives, 2nd Explosives Detection Symposium & Aviation Security Conference, Atlantic City, NJ, 1996.
[48] Syagen Technology, Inc., http://www.syagen.com/.
[49] K.A. Hanold, Mass Spectrometry Based Personnel Screening Portal, Proceedings of the 7th International Symposium on Analysis and Detection of Explosives, Edinburgh, Scotland, 2001.
[50] J.A. Syage, K.A. Hanold, and R.L. Woodfin (ed.), *Mass Spectrometry for Security Screening of Explosives, Trace Chemical Sensing of Explosives*, John Wiley & Sons, New York, 2006.
[51] J.A. Syage, K.A. Hanold, and M.A. Hanning-Lee, Mass Spectrometry Based Personnel Screening System, 42nd INMM Meeting, Phoenix, AZ, 2001.
[52] Liberty Industries, Inc., http://www.liberty-ind.com/airsh.htm.
[53] K. Hanold, Detection of Improvised Explosives and Pyrotechnic Compositions on Passengers, 54th ASMS Conference on Mass Spectrometry Proceedings, Seattle, WA, May 28 – June 1, 2006.
[54] NCRP Commentary No. 16, Screening of Humans for Security Purposes Using Ionizing Radiation Scanning Systems, National Council on Radiation Protection and Measurements, Bethesda, MD, December 15, 2003.
[55] American National Standards Institute, Radiation Safety for Personnel Security Screening systems Using X-Rays, ANSI/Health Physics Society N43.17-2002, Health Physics Society, McLean, Virginia.

[56] American Science and Engineering, Inc., http://www.as-e.com.
[57] Rapiscan® System, http://www.rapiscansystems.com.
[58] T. Yukl, Dielectric Personnel Scanning, US Patent No. 6,927,691 (2005).
[59] C. Seward and T. Yukl, Dielectric Portal for Screening People, Proceedings of the 2nd FAA Symposium on Explosives Detection Technology and Aviation Security Technology, Atlantic City, NJ, 1996.
[60] EMIT Technologies LLC, http://www.emittech.com.
[61] J.E. Reap, Aviation Security Research and Development Service Safety & Health Plan for Spatial Dynamics Portal, 1999.
[62] L3 Communications SafeView, http://www.safeviewinc.com.
[63] L3 Communications Security and Detection Systems http://www.dsxray.com/products/mmwave.htm.
[64] D.M. Sheen, D. L. McMakin, W. M. Lechelt, and J. W. Griffin, Circularly polarized millimeter-wave imaging for personnel screening, Passive Millimeter-Wave Imaging Technology VIII, Orlando, FL, 2005, Proceedings of the SPIE – The International Society for Optical Engineering, 5789(1) pp. 117–126, 2005.
[65] D.L. McMakin, D. M. Sheen, and T. E. Hall, Millimeter-wave imaging for concealed weapon detection, Nondestructive Detection and Measurement for Homeland Security, San Diego, CA, 2003, Proceedings of the SPIE – The International Society for Optical Engineering vol. 5048: 52–62, 2003.
[66] D.L. McMakin et al., Interrogation of an Object for Dimensional and Topographical Information, US Patent No. 6,507,309 (2003).
[67] D.L. McMakin et al., Interrogation of an Object for Dimensional and Topographical Information, US Patent No. 6,703,964 (2004).
[68] D.L. McMakin et al., Holographic Arrays for Threat Detection and Human Feature Removal, US Patent No. 7,034,746 (2006).

Chapter 13

Biological Detection of Explosives

Ross J. Harper[a] and Kenneth G. Furton[b]

[a]*Nomadics Inc., ICx Nomadics Inc., Stillwater, OK 74074, USA*
[b]*Department of Chemistry and Biochemistry, International Forensic Research Institute, Florida International University, Miami, FL 33199, USA*

Counterterrorist Detection Techniques of Explosives
Jehuda Yinon (Editor)

Contents

1. Introduction

Various biological detectors for explosives have been studied over the years with *Canis familiaris*, better known as the common dog, the most widely deployed detector to date. While this chapter will discuss the various biological detectors that have been studied, the focus is on detector dogs that are now used ubiquitously by law enforcement and private agencies for detection of many different items including explosives. The most common items of forensic interest for which canines have been deployed to locate are drugs, ignitable liquid residues, explosives, currency, human remains, and human scent (trailing, tracking, and lineups). The use of the canine as a detector is based on the well-established reliability and impressive selectivity and sensitivity associated with dog's sense of smell [1]. The canine olfactory system is a complicated biological detection system, which will be discussed in more detail in Section 2 and which makes an interesting contrast to the chemical detection of most field instruments. A recent report on standoff explosive detection techniques conducted by the National Academy of Sciences concluded that it is important to use multiple orthogonal detection methods (methods that measure the properties of explosives that are not closely related), as there is currently no single technique that solves the explosive detection problem. Canine studies that have been conducted include free-running and remote explosive scent tracing (REST) in which the volatile organic compounds (VOCs) are collected on a sorbent in the field and presented to the animal at a different location [2]. Biological explosive detectors, including detector dogs, can be considered orthogonal detectors to sensors under development, as they generally rely on different detection modalities.

2. Canine detection

The use of canines as a method of detection of explosives is well established worldwide, and those applying this technology range from police forces and military to humanitarian agencies in the developing world. Until recently, most data regarding optimal training protocols and the reliability of canine detection have been anecdotal, leading to successful challenges regarding the admissibility of evidence obtained with the assistance of canines and hampering the improvement in the performance of canines as biological explosive detectors [3]. Challenges facing the field of canine detection include the limited ability to evaluate canine performance with standardized calibration standards. Unlike instrumental methods, it is currently difficult to determine detection levels and perform a calibration of the canines' ability to locate scientifically valid quality-control checks. In addition, there are increasingly strict requirements being applied to the admissibility of the application of detector dogs in locating items of forensic interest, highlighting the need for a better scientific understanding of the process of canine detection.

2.1. *Explosive detection canines*

Although the sensitivity of some of the emerging instrumental technologies is on par with, if not beyond, that of the canines, the dogs still hold an advantage over the instruments on selectivity. Canines are renowned for their ability to locate target odorant molecules even in the presence of significant background chemicals. It is this selectivity combined with mobility and independent thinking that still ranks the canines as the current best method for real-time detection of explosives. Current research in the field includes the identification of odorants for canine detection of explosives and the development of what we are calling odor mimics, or training aids that contain the odorants that mimic the real substances. There are presently several theories about what is responsible for the canines' high selectivity and specificity to explosives: (i) canines alert to the parent explosives regardless of their volatility; (ii) canines alert to more volatile, non-explosive chemicals that are present in explosives and that are characteristic to explosives; or (iii) both parent explosives and characteristic volatiles are used to accurately locate explosives. To date, there are no definitive peer-reviewed studies to support any of these theories. A previous study reported results consistent with the theory of stimulus generalization, indicating that odor generalization is a function of the similarity of the vapor chemistry between trained and untrained target substances and the extent of training across multiple variants of the substances. Conclusions from current studies include the importance of identifying the variants of explosives that will yield optimal effectiveness [4].

2.2. *Olfaction theory*

Of all the senses, the sense of smell is the least understood and is a complicated biochemical process, which is beginning to be clarified. Chemoreception is the process of detecting chemical compounds by a living organism and involves molecular interactions with olfactory neurons by molecules that have moderate molecular weight, low polarity, a particular water solubility, high vapor pressure, and lipophilicity [5]. Numerous olfaction theories had been formulated including vibrational theory, membrane diffusion theory, Piezzo effect, complexation, polarization theory, chromatography analogy, hydrogen bond dispersion model, and tunneling vibrational theory. In 2004, the Nobel Prize in Physiology or Medicine was awarded to Richard Axel and Linda Buck for their discoveries of 'Odorant receptors and the organization of the olfactory system'. Although odor and odorant are often used interchangeably, odorants are molecules objectively definable in terms of their physicochemical characteristics and capable of being transposed by the olfactory system into odors. Odors therefore are the products of the nervous system and open to subjective interpretation of the mind. An analogy is vision where colors are the sensation and information processed by the central nervous system after light of different wavelengths falls on the retina [6]. Some of the processes thought to occur in transforming an odorant into a detected odor are illustrated in Fig. 1.

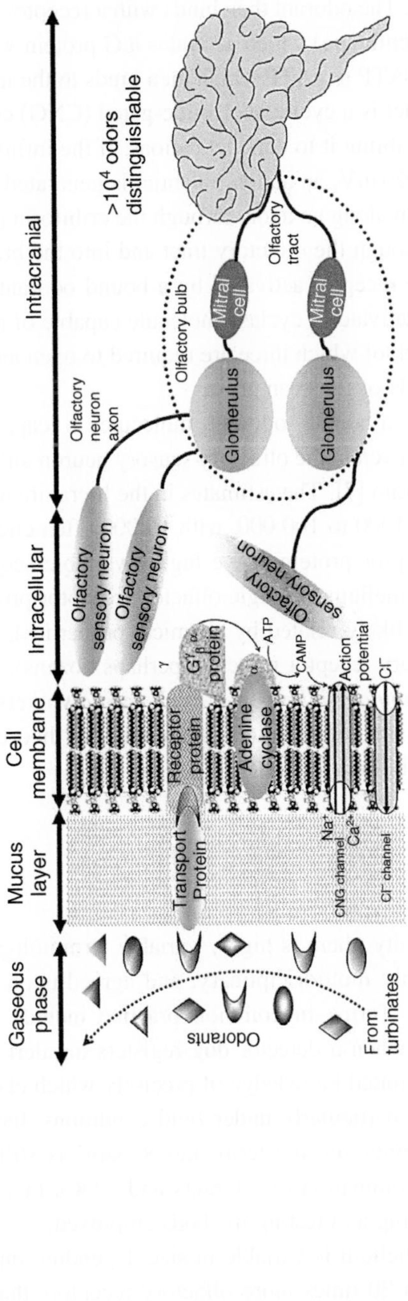

Fig. 1. Some of the processes transforming an odorant into a detected odor.

Odorants generally first enter the olfactory system by passing through the external nares of vertebrates where warming and humidification occur before passing through a set of cartilaginous flaps called turbinates that act to increase the surface area of the epithelium. Initially, odorants are thought to be associated with odorant-transport proteins that help transport the odorant from the inhaled air stream through the mucus to the cilia of the olfactory neurons [5]. The odorant then binds with a receptor protein (embedded in the phospholipid surface membrane) which activates a G protein which in turn activates adenine cyclase converting ATP to cATP which then binds to the intracellular face of an ion channel. This ion channel is a cyclic nucleotide-gated (CNG) channel similar to that found in photoreceptors, enabling it to conduct cations. If the influx of cations increases the membrane potential by 20 mV, an action potential is generated that propagates from this olfactory sensory neuron along an axon, through the cribiform plate to the glomerlus and mitral cells, then on through the olfactory tract and into the brain where the odor is interpreted. One membrane receptor activated by a bound odorant can activate tens of G proteins, each of which activates a cyclase molecule capable of producing a thousand cAMP molecules per second, of which three are required to open an ion channel through which hundreds of thousands of ions can cross.

Therefore, theoretically, a single odorant molecule is capable of producing a measurable electrical signal even in the olfactory sensory neuron although not necessarily a perceivable event in the brain [7]. The estimates in the literature as to how many odors can be detected range from 1000 to 100 000, with 10 000 often cited. Gene analysis has revealed that olfactory receptor proteins have highly variable sequences and are only encoded in the olfactory epithelium. A single olfactory receptor protein can be activated by multiple odorants (most likely related by chemical properties), and a single odorant can activate multiple olfactory receptor proteins (perhaps dozens). Olfactory perception in the end is therefore a *combinatorial* process. Buck and coworkers [8] estimated that if an odorant activated as few as three receptors, the number of theoretically discriminable odorants should be about a billion.

2.3. *Canine olfaction*

In the detector dog community, there is highly variable terminology utilized; scientific research in this area is highly multidisciplinary, and agreed-upon unified terminology is still under development. Terms in common practice include explosive 'odor' to describe the VOCs present when a detector dog registers an alert to a target material. Unfortunately, our current limited knowledge of precisely which chemicals are reaching a dog's olfactory receptors, particularly under field conditions, hampers the definitive determination of most odorants and the term 'target odor' is still in common use to describe what is likely a combination of odorants and VOCs that may or may not be odorants depending on training and testing methods employed.

The canine olfactory epithelium is variable in size depending on breed but has been shown to express more than 20 times more olfactory receptors than humans and have

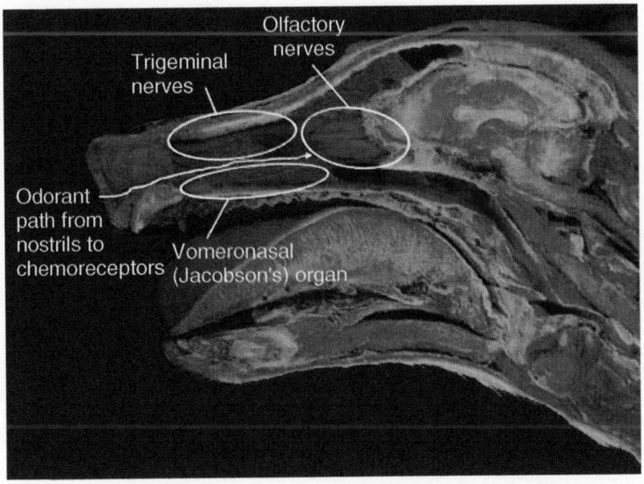

Fig. 2. Sagittal cross-section of canine skull identifying key olfaction components (portion of image from Virtual Canine Anatomy [43]).

an olfactory receptor repertoire around 30% larger than humans and a much lower percentage of pseudogenes [9]. The complete canine olfactory receptor repertoire is approximately 1300 genes compared with around 900 in humans. In addition, the olfactory receptors within a canine's nasal cavity are oriented to efficiently contact a high percentage of the inhaled molecules. Figure 2 shows a sagittal cross-section of a canine skull with the olfactory receptor region highlighted in addition to the Vomeronasal organ that is primarily responsible for detecting pheromones. The trigeminal nerves that are also present throughout the region and participate in the detection of chemical irritants may be present in the headspace of target materials.

The canine nose is a highly efficient sampling system. When searching, the dog inhales in short sharp breath at a frequency of 5–8 Hz, equivalent to 300–480 breaths per minute. In comparison, humans average 10–12 breaths per minute during normal activity. The volume of inhaled air is around 30 ml/s/nostril equating to approximately 3.6 l/min of sampled air [10]. The dynamics of airflow around the nostrils are such that the air is inhaled from the front and exhaled to the side [11]. Figure 3 illustrates possible inspiration and expiration pathways through the external nares of a canine nose with expired air directed by an alar fold toward the sides and the midlateral slit of the nare. The larger surface area of the canine olfactory epithelium, larger number of receptors present, larger variety of olfactory receptors, and highly efficient sampling system result in an overall increase in the sensitivity of a dog's nose compared with a human's nose. Estimates in the literature range from 10 times to millions of times more sensitive although there is no definitive number, and 100 times is a reasonable estimate although actual detection thresholds are highly variable depending on the odorant being studied and how the canine was trained.

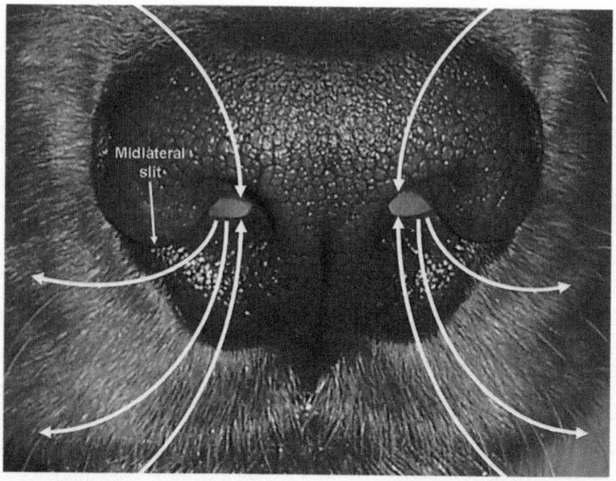

Fig. 3. The external nares of *Canis familiaris* illustrating possible inspiration and expiration pathways.

Scientific debate continues as to whether dogs detect odorants as vapors or whether they may be trapped and carried on particles inhaled. It has been shown with the aid of micron-sized graphite dust that inhaled particles are trapped by the hairs and mucus at the front of the nasal cavity, although a small fraction may pass through to the olfactory epithelium [12]. Accordingly, this would suggest that olfaction is primarily a vapor process, although particles may still play a role.

3. Instrumental options

Detector dogs still represent the fastest, most versatile, reliable, real-time explosive detection device available. Instrumental methods, while they continue to improve, generally suffer from a lack of efficient sampling systems, selectivity problems in the presence of interfering odor chemicals, and limited mobility/tracking ability [13].

3.1. Explosive detection technologies

There are a variety of detection technologies currently available and others under development.

Figure 4 illustrates some trace explosive technologies including separation techniques such as ranging from high-performance liquid chromatography (HPLC) and capillary electrophoresis (CE) commonly with fluorescence or electrochemical detection and gas chromatography (GC) combined with mass spectrometry (GC/MS) electron capture (GC/ECD) or luminescence detection. In addition, techniques based on mass spectrometry and ion mobility spectrometry (IMS) continue to improve [14]. Currently, the most widely deployed explosive screening technology deployed at airports is IMS that relies

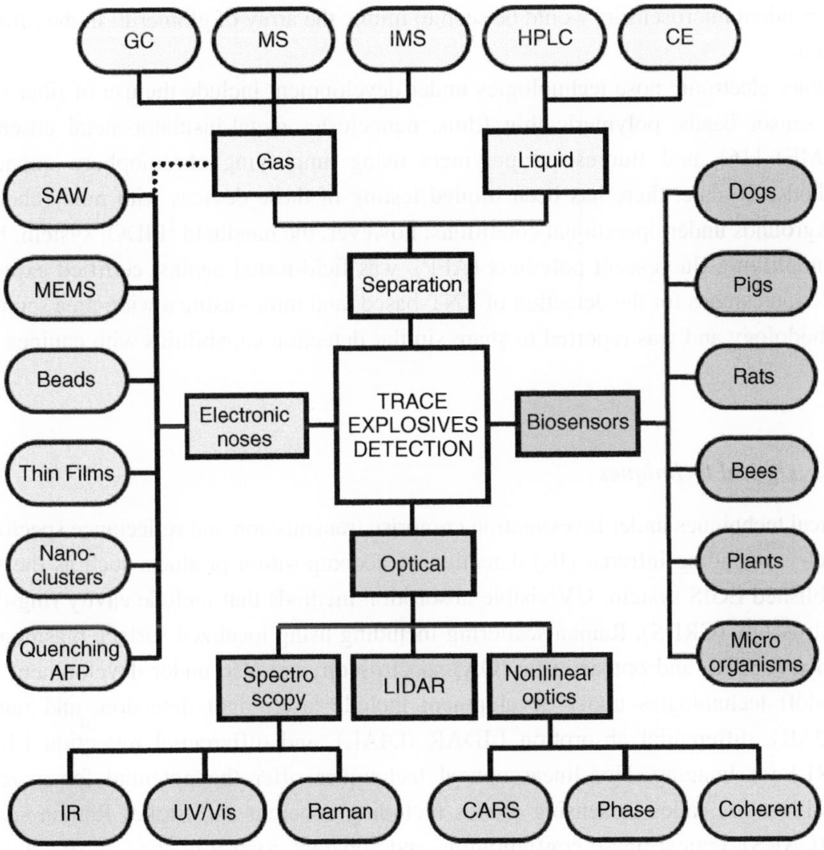

Fig. 4. Various techniques under investigation for detecting explosives.

primarily on the detection of particles contaminated on the outside of baggage or paper tickets. Recently, a new IMS inlet has been developed that allows for the detection of odor chemicals using solid-phase microextraction (SPME) sampling [15].

3.1.1. Electronic noses

Microsensors have the potential for selective GC detectors and also as remote sensors when combined in arrays often referred to as 'electronic noses'. Promising microsensors include surface acoustic wave (SAW) detectors normally coated with different semi-selective polymeric layers and microelectromechanical systems (MEMS) including microcantilever sensors. The hope is that, in the future, hundreds of such microcantilevers, coated with suitable coatings, may be able to achieve sufficient selectivity to provide a cost-effective platform for detecting explosives in the presence of potentially interfering compounds in real environments. This array of independent microsensors would be seen to mimic the array of glomeruli in the olfactory system.

Other electronic nose technologies under development include the use of fiber optics and sensor beads, polymeric thin films, nanocluster metal-insulator-metal ensembles (MIME) [16], and fluorescent polymers using amplifying chromophore quenching methods. To date, there has been limited testing of these devices with noisy chemical backgrounds under operational conditions; however, the handheld 'FIDO' system, based on amplifying fluorescent polymers (AFP), was field-tested against certified explosive detection canines for the detection of TNT-based land mines using a wide-area screening methodology and was reported to share similar detection capabilities with canines [17].

3.1.2. Optical techniques

Optical techniques under investigation comprise transmission and reflectance spectophotometry, including infrared (IR) detection of decomposition products such as the well-established EGIS system. UV/visible absorption methods that include cavity ring-down spectroscopy (CRDS), Raman scattering including using localized surface plasmon resonance (LSPR), and optoacoustic (OA) spectroscopy are also under development [13]. Standoff technologies under development include laser, light detection, and ranging (LIDAR), differential absorption LIDAR (DIAL), and differential reflection LIDAR (DIRL) for imaging. Non-linear optical techniques offer the potential for increased signal-to-noise ratios in sensing modes including coherent anti-Stokes Raman scattering (CARS), optical phase configuration, and coherent control of the specific states of molecules and optimize their luminescence [18].

4. Choosing between canine and instrumental detection

The wide variety of instrumental methods available makes the optimal choice and/or combination of detectors challenging. As mentioned previously, biological detectors and, specifically, detector dogs offer an orthogonal detection mechanism and should be included in an optimized layered approach to detecting explosive residues or trace odors. Table 1 highlights some of the advantages and disadvantages of canine and instrumental field detection, revealing how complementary operation rather than competitive operation should be considered. Canine detection generally has the advantage when it comes to selectivity, mobility, sampling system, standoff detection/scenting to source, remote/semi-autonomous guidance, cost, and speed. A major advantage of detector dogs is that they can be rapidly trained to alert to novel odorants and trained not to alert

Table 1. Comparison of canine detection and field instrument technologies

Aspect	Canine	Instrument
Selectivity (versus interferents)	**Very good**	*Sometimes problematic*
Mobility	**Very versatile**	*Limited at present*
Integrated sampling system	**Highly efficient**	*Problematic/often inefficient*
Scent to source/standoff potential	**Natural and quick**	*Difficult with present technology*
Turnaround time for training on novel odors	**Relatively rapid**	*Often time consuming*
Capable of remote guidance and mobile platform for other sensors	**Straightforward**	*Limited at present*
Initial cost	**Approximately $8000**	*Approximately 60 000*
Annual cost	**Approximately 3000 (vet and food bill)**	*Approximately $5000 (service contract)*
Intrusiveness	**Often innocuous (breed dependent)**	*Variable*
Overall speed of detection	**Generally faster**	*Generally slower*
Duty cycle	*Approximately Up to 1 h search duration (with conditioning) per dog before break*	**Approximately 24 h /day (theoretically but not practically)**
Calibration standards	*Not widely available, typically run individually*	**Widely available, can be run simultaneously**
I.D. of explosive	*Not trained to I.D. with different alerts*	**Presumptive I.D. possible**
Operator/handler influence	*A potential factor*	**Less of a factor**
Instrument lifetime	*Generally 6–8 years*	**Generally 10 years**
State of scientific knowledge	*Late emerging*	**Relatively mature**
Target chemical(s)	*Odorant chemicals/mostly unknown*	**Parent explosive(s)/well studied**
Courtroom acceptance	*Sometimes challenged*	**Generally unchallenged**
Degree of standardization	*Highly variable*	**Consistent standards**
Sensitivity	Very good	Very good
Environmental conditions	Potential affect (high temperatures)	Potential affect (high temperatures, dust)
Initial calibration	Generally performed by supplier	Generally performed by manufacturer
Operator training	Typically 40 h minimum	Typically a 40-h course
Certifications	Typically annual	Typically annual
Re-calibrations	Daily to weekly	Daily to weekly
Scientific foundations	Neurophysiology, behavioral psychology, analytical chemistry	Electronics, materials and computer science, analytical chemistry
Potential effects on performance	Disease conditions	Electronics/mechanical

to innocuous items that may have similar chemicals present in their headspace. This allows detector dogs to provide highly selective detection of the new targets in a matter of minutes or hours of training, whereas instruments may not be capable of detecting some novel targets without extensive changes in the hardware design and may require changes in sampling procedures and software systems that might take weeks to months to implement. The major disadvantages of detector dogs are their limited duty cycles (necessitating multiple teams for continuous coverage), possibility of operator influence (because the handler is dealing with an intelligent animal), generally shorter working lifetime, and the shortcomings of our scientific understanding of the odorants used by detector dogs resulting in often inadequate training materials/calibrators and certification protocols with accompanying courtroom challenges. In a variety of other aspects, canine and instrumental detectors are comparable including the timetable for operator training, (re-)calibration protocols, and potential effects on performance.

4.1. Standardization efforts

In recent years, there have been major efforts to standardize the procedures used with detector dogs and their optimal combination with instrumental detectors. Although there have been ongoing efforts by individual agencies and organizations to develop standards aimed at improving the performance and consistency of detector dog teams, it was not until the end of the twentieth century that major national and international efforts have been focused on this topic [1]. In 1998, the National Detector Dog Conference Series was established in the United States with biannual meetings addressing topics including optimal training and deployment strategies and certification standards. In Europe, the Interpol European Working Group on the Use of Police Dogs in Crime Investigation (IEWGP) was established in 1999 and was responsible for drawing up recommendations and regulations aimed at improving the efficiency of the use of police dogs in crime investigation and for promoting their practical application by specialized police units in Europe. The IEWGP completed their recommendations in 2003. In 2004, the Scientific Working Group on Dog and Orthogonal Detection Guidelines (SWGDOG) were established with biannual meetings beginning in 2005. SWGDOG best practice guidelines are in various stages of development and available at http://www.swgdog.org in nine major areas as follows: (i) unification of terminology; (ii) minimum training, certification, and maintenance standards; (iii) selection of serviceable dogs and replacement systems; (iv) kenneling, keeping, and health care; (v) selection and training of handlers and instructors; (vi) procedures on presenting evidence in court; (vii) research and technology; (viii) substance detector dogs (agriculture, arson, chemical/biological, drugs, explosives, human remains, other/miscellaneous); and (ix) scent dogs (scent identification, search and rescue, trailing dogs, and tracking dogs). It is expected that best practice guidelines will be published in all nine areas by 2007 with updates on these guidelines at least biannually.

5. Alternative biological detectors

In addition to canines, other animal and plant species and enzymatic technology have been proposed as alternative methods of biological explosive detectors. Most of the proposed alternatives are still in their infancy or have been shown to have sufficient shortcoming such that canine detection is still the only widely employed biological method for detecting explosives. Some of the proposed alternative biological detectors are discussed in this section.

5.1. Other mammalian species

In theory, almost any mammal has the potential to be utilized as an biological explosive detector with the primary limitations associated with the size and the mobility of the mammal as well as the trainability of the mammal. A research project in Tanzania, under the support of the Belgian research organization APOPO, trains African giant pouched rats to detect land mines. Reports indicate that rats may be capable of detecting similarly low levels of explosive odors compared with dogs with advantages including their small size and low cost, but with more challenging training and retrieval aspects [19]. The rats' small size permits an active alert at the site of the mine without the danger of detonation. At one time, truffle-detecting pigs were also considered for land mine detection; however, it was discovered that their weight was sufficient to trigger the devices.

5.2. Insects

Bees are also being studied as a biological explosive detection system. It has been demonstrated that bees are capable of detecting explosive odors at concentrations below those of most instruments and comparable to dogs [20]. The bees can be imaged or traced to the source or, more commonly, used to survey areas by examining chemical residues brought back to the hive. The advantages are that they can be trained quickly and will not set off any mines. Limitations are that bees do not fly at night, in heavy rain, or in cold weather (below 40°F).

5.3. Plants

Detecting environmental contaminants, including explosives, through vegetation sentinels has been a long-term goal of remote-sensing scientists including using the fluorescence signature of genetically modified (GM) vascular plants spectrally separable from controls [21]. Danish scientists at Aresa Biodetection have developed a GM cress that, when sown over a suspected mine field, will change its leaves from green to red on detection of buried explosives [22]. The GM thale cress is altered to change color should its roots contact NO_2 in the soil. Because this method relies on seepage of NO_2 from leaking mines, it has the potential to miss the more recent models that are specially sealed to obscure detection. Research conducted at Colorado State University by Medford et al. [23] has targeted the chlorophyll cycle in plants, leading to a 'de-greening' in the presence of target materials, where the foliage undergoes a reversible color change from green to white. The genetic coding responsible for the color change can be inserted into many different species, and it is foreseen that different target molecules may be learnt through 'directed evolution'. Although in its infancy, this technology appears to have a promising future.

5.4. Microorganisms

Microalgal biosensors are under development as specific, sensitive, rapid, and low-cost biosensors for contaminants including TNT. The method uses two different genotypes for detection: (i) a sensitive genotype to obtain sensitivity and (ii) a resistant mutant to obtain specificity [24]. A structure-based computational method has been proposed allowing redesign of protein ligand-binding specificities and was used to construct soluble receptors that bind TNT and were incorporated into synthetic bacterial signal transduction pathways, regulating gene expression in response to extracellular TNT. It is thought that such an approach might offer a cost-effective method based on spraying, say, a mine field with bacteria that respond to TNT by transcriptional activation of a fluorescent reporter gene [25].

5.5. Nanotechnology

Ongoing research at the University in Wales by the Kalaji group has led to the announcement of the 'nanodog'. Little is known yet about this technology; however, it is said to combine passive enzymatic detection of several explosives that will allow tentative identification of the explosive detected, down to part-per-trillion sensitivity [26]. Early testing has shown promise with rapid detection times, despite the youth of the technology.

6. The canines take on explosives

The canine community often observes, with disdain, that there is no such thing as waste at an explosive manufacturing plant; whatever is left over from one batch almost always ends up in the next. This introduces impurities and heterogeneity into the explosives, something which is welcomed by the trace analyst in the forensic laboratory but despised by experienced trainers wishing to imprint the optimum range of explosive odors on the dogs.

6.1. Explosive taggants and detection aids

The extremely low vapor pressures for many of the common explosives complicate the detection of these compounds directly. For this reason, the compound 2,3-dimethyl-2,3-dinitrobutane (DMNB) is one of four chemicals now added as a marker to plastic and sheet explosives, the others being 2-nitrotoluene (2-MNT) and 4-nitrotoluene (4-MNT) and ethylene glycol dinitrate (EGDN), although DMNB is by far the most commonly used. DMNB was chosen because of its high vapor pressure, high permeability through textiles, and uniqueness, and with no known industrial applications [27], its detection is highly indicative of the presence of explosives. The vapor pressure for DMNB is more than 1 million times greater than any of the nitramine explosives (table 2). The relatively involatile explosives that pose particular challenges to detection are highlighted (shaded),

Table 2. Organic explosive properties including vapor pressures

Explosive class	Explosive	Molecular weight (amu)	Vapor pressure at 25°C (torr)
Aliphatic nitro	Nitromethane	61.04	2.8×10^{1}
	2,3-Dimethyl-dinitrobutane (DMNB)	176.17	2.1×10^{-3}
Aromatic nitro	2,4-Dinitrotoluene (DNT)	182.14	2.1×10^{-4a}
	2,4,6-Trinitrotoluene (TNT)	227.13	5.8×10^{-6}
	2,4,6-Trinitrophenol (picric acid)	229.11	5.8×10^{-9}
Nitrate ester	Ethylene glycol dinitrate (EGDN)	152.06	7.0×10^{-2}
	Glycerol trinitrate (GTN)	227.09	3.1×10^{-4}
	Pentaerythritol tetranitrate (PETN)	314.14	1.4×10^{-8}
Nitramine	2,4,6-Trinitrophenyl methylnitramine (Tetryl)	287.15	5.7×10^{-9}
	Trinitro-triazacyclohexane (RDX)	222.12	4.6×10^{-9}
	Tetranitro-tetrazacyclooctane (HMX)	296.16	1.6×10^{-13a}
	Hexanitro-hexaazaisowurzitane (CL20)	438.19	Not available
Peroxide	Triacetone triperoxide (TATP)	138.08	3.7×10^{-1a}
	Hexamethylene triperoxide diamine (HMTD)	208.17	Not available
By-product	2-Ethyl-1-hexanol	130.23	1.4×10^{-1}
	Hydrogen peroxide	34.03	1.4×10^{0a}
	Acetone	58.08	2.3×10^{2}

[a] Extrapolated value.

as there may be limited to no available explosive molecules present in the headspace of these explosives, depending on the conditions encountered including how the explosives are packaged. In addition to possibly added taggants, in most cases, there are other VOCs that can be used to detect explosives and explosive formulations. Unfortunately, the current limited knowledge of the common VOCs in the wide variety of possible explosive combinations has hampered progress toward improved training of biological detectors and the use of target VOCs for instrumental detection as well.

6.1.1. Other VOCs available

Consideration should not only be paid to the vapor pressures of the explosive compounds but to the additives and by-products present in explosive mixtures. For example, a VOC observed in the headspace of many plasticized explosives is the chemical 2-ethyl-1-hexanol, which is produced through the hydrolysis breakdown of the plasticizer dioctyl sebacate and/or the binder dioctyl adipate. The vapor pressure of 2-ethyl-1-hexanol is 1.4×10^{-1} torr at room temperature, nearly 1 billion times more volatile than the parent explosives in the plastic matrix. It would be expected that a 2-ethyl-1-hexanol concentration greater than only 0.000001% by mass would still be observed as the dominant odor. Diphenylamine, present in smokeless powders in up to 2% by mass, has a vapor pressure several orders of magnitude higher than NG, DNT, or NC to be found in the powder. Utilizing these readily available chemicals to train detectors may therefore be useful. Even in the case of a volatile explosive such as triacetone triperoxide (TATP), the major starting materials and main decomposition products are acetone and hydrogen peroxide that have a vapor pressure higher than TATP itself [28].

6.2. Explosive products and combinations

Explosive compounds are rarely found in the pure state; they are frequently combined with other explosives, stabilizers, tagging agents, plasticizers, and other modifiers to increase the stability and efficiency of the explosive product in question. Over three dozen commercial products with the desired properties, as represented in Table 3, can be shown to share thress major target VOCs also listed [29,30].

Commercial explosives see application in demolitions and mining and include dynamites, water gels, gelatins, and slurries such as aquaspex or ANFO. These are available in stick (solid and gel) or liquid (slurry) form. Slurries see most application in open cast mining, where large quantities can be pumped into prepared wells. The commercial explosives are generally less stable than military charges because of the less extreme handling conditions expected. The military explosives include TNT, RDX, PETN, and combinations of these and related products. Military ordinance may be cast into warheads and mines, or plasticized to form explosives such as C-4 and Semtex.

Table 3. Examples of explosive mixtures

Explosive	Components	DNT	NG	P
Amatol	Ammonium nitrate + TNT	X		
Ammonal	Ammonium nitrate + TNT + *Al*	X		
ANFO (Amex or Amite)	Ammonium nitrate + fuel oil (Diesel)			
Black Powder	Potassium nitrate + *C* + *S*			
Composition A	RDX + wax			
Composition B	RDX + TNT	X		
Composition C-2	RDX + TNT + DNT + NC + MNT	X		X
Composition C-3	RDX + TNT + DNT + Tetryl + NC	X		X
Composition C-4	RDX + plasticizers			X
Composition D	RDX + TNT + Al + wax	X	X	
Cyclotol	RDX + TNT	X		
Datasheet (Flex-X)	RDX + plasticizers			X
DBX	TNT + RDX + ammonium nitrate + *Al*	X		
Demex 200	RDX + plasticizer			X
Detonation Cord (commercial)	PETN			
Detonation Cord (military)	RDX or HMX	X		
Dynamite (ammonia)	NG + NC + sodium nitrate		X	
Dynamite (gelatin)	NG + NC + ammonium nitrate		X	
Dynamite (military)	TNT			
HBX-1	RDX + TNT + *Al*	X		
Helhoffnite	NB + nitric Acid			
HTA	HMX + TNT + *Al*	X		
Nitropel	TNT	X		
Nonel Cord	HMX			
PE-4	RDX + plasticizer			X
Pentolite	PETN + TNT	X		
Picratol	TNT + ammonium picrate	X		
Primasheet 1000	PETN + plasticizers			X
Primasheet 2000	RDX + plasticizers			X
PTX-1	RDX + TNT + Tetryl	X		
PTX-2	RDX + TNT + PETN	X		
Red Diamond	NG + EGDN + sodium nitrate + ammonium		X	
SEMTEX A	PETN + plasticizers			X
SEMTEX H	RDX + PETN + plasticizers			X
Smokeless Powder (double based)	NC + NG		X	
Smokeless Powder (single based)	NC			
Smokeless Powder (triple based)	NC + NG + nitroguanidine/TNT	X	X	

(*Continued*)

Table 3. (Continued)

Explosive	Components	DNT	NG	P
Tetratol	TNT + Tetryl	X		
Time Fuse	Potassium nitrate + *C* + *S*			
Torpex	TNT + RDX + *Al*	X		
Tritonal	TNT + *Al*	X		
Water gel/slurry (aquaspex)	NG		X	
Water gel/slurry (hydromex)	Ammonium nitrate + TNT	X		
Water gel/slurry (powermex)	Ammonium nitrate + Sodium nitrate + EGMN			
Water gel/slurry (tovex)	Ammonium nitrate + Sodium nitrate + MMAN			

Al, aluminium; C, carbon; DNT, dinitrotoluene; EGMN, ethylene glycol mononitrate; MMAN, monomethylanine nitrate; NG, nitroglycerin; P, 2-ethyl-1-hexanol; S, sulphur.

6.3. Choosing the optimal training targets

Unlike narcotic detection canines, which are expected to face a predictable lineup of five or six drug odors, the explosive detection canine is expected to face dozens of different potential explosive products during its service. Narcotics detection canines are typically trained on cocaine (HCl and Base), heroin, and marijuana. They may be trained on additional drugs, including methamphetamine, MDMA, hashish, opium, mescaline, and LSD, depending on the training agency and the locations where they are deployed [31]. Although there are the six principal chemical categories of explosives, there are dozens of individual explosive chemicals that must be detected.

Depending on the training agency and deployment locations, explosive detection canines are currently trained on a wide variety of samples ranging from half a dozen samples to upward of 20. Using at least one representative sample from each explosive chemical class would require an acid salt such as ammonium nitrate, an aromatic nitro such as TNT, a nitrate ester such as PETN, a nitramine such as RDX, an aliphatic nitro such as DMNB, a peroxide such as TATP, and representative black and smokeless powders. Unfortunately, there is currently little scientific information available to aid in the optimal selection of training aids. Owing to the challenges in handling and storing of a wide variety of explosives, non-hazardous training aids are commercially available, but with limited types available and limited testing of their effectiveness under field conditions in double-blind studies.

6.4. Understanding explosive VOCs

The difficulty in choosing an optimal number of training aids lies in the multiple explosives within each category and is then confounded by the wealth of explosive products that employ various combinations of the explosives. One of the most important decisions a canine trainer has to make is in choosing which explosives to use as odor targets. Many trainers and canine programs choose to focus on one main explosive from each principal category. However, there are scant peer-reviewed data to demonstrate the utility of this practice. The difficulty in selecting the optimal number and combinations of training aids can be highlighted by looking at the choice of a smokeless powder training aid. Although bombs made from black and smokeless powder are generally relatively small, these devices are the ones most commonly used in criminal bombings in the United States and are readily available with millions of individuals purchasing these powders for sport use each year and hundreds of different formulations available from different manufacturers [32]. Finding one smokeless powder that adequately represents the hundreds of possibilities seems unlikely.

The Bureau of Alcohol, Tobacco, Firearms, and Explosives (ATF) has primary responsibility for control of firearms and explosives within the United States. Figures released documenting the use of explosive materials in domestic incidents over a 20-year period between 1976 and 1996 revealed that less than a minority of incidents featured a high explosive as the energetic material with black/smokeless powders by far the most often encountered as seen in Table 4 [33]. The most prevalent explosive fillers were black/smokeless powders, flammable liquids, and flash/pyrotechnic powders at 23.2, 22.9, and 9.4%, respectively. With high explosives accounting for such a small percentage of explosive incidents, it is clear that sufficient attention must be focused toward the detection of explosive devices from low explosives, a category that is often given less attention by canine trainers who favor traditional training on the high explosives.

6.5. Smokeless powder

Smokeless powder, as the name suggests, is a replacement propellant that does not produce the volumes of smoke associated with the combustion of black powder. Smokeless powders are classified into single, double, or triple based, according to the energetic materials included during production. The single-based propellants use only

Table 4. Explosive fillers 1976 through 1996 (high explosives in **bold**)

Filler	1976–1996
Unreported	9224 (24.3%)
Black /Smokeless Powders	**8810 (23.2%)**
Flammable liquids	8686 (22.9%)
Photo flash/fireworks powders	3545 (9.4%)
Chemicals	3475 (9.2%)
Other*	2117 (5.6%)
Commercial high explosives	1816 (4.8%)
Blasting agents	170 (0.4%)
C-4/TNT	56 (0.1%)
Total	**37 899 (100%)**

Table 5. Common additives in smokeless powder

Stabilizers	Plasticizers	Deterrents
Diphenylamine	Trinitroglycerin	Dinitrotoluene
Nitro-diphenylamine	Dinitrotoluene	Methyl centralite
Methyl centralite	Ethyl centralite	Ethyl centralite
Ethyl centralite	Dibutyl phthalate	Dibutyl phthalate

nitrocellulose as the energetic, whereas double-based propellants also include the addition of trinitroglycerin. Triple-based powders are rarely observed outside of high-caliber military applications and include nitroguanidine as the tertiary energetic. In addition to the energetics, other components are added during manufacture to provide the desired final product with some common examples summarized in Table 5. Stabilizers are used to increase shelf life, by removing nitric acid formed during decomposition of the nitrated energetics. Common stabilizers include diphenylamine and its nitrated derivatives, methyl centralite, and ethyl centralite. Plasticizers reduce the hygroscopicity of the powders, soften the powder granules, and reduce the need for solvents. Plasticizers used include dinitrotoluenes, phthalates, ethyl centralite, and trinitroglycerin itself.

Deterrents are used to coat the powder, reducing the initial burn rate, and lowering the burn temperature. Deterrents used include dinitrotoluenes, ethyl centralite, methyl centralite, and phthalates. Inorganic additives such as earth metal salts are added as flash suppressants, and opacifiers such as carbon black can be added to increase the reproducibility of the burn rate [34]. Note that several of the listed additive chemicals serve multiple purposes, as summarized in Table 5. Ethyl centralite can be applied as a stabilizer, a plasticizer, and a deterrent. Similarly, dinitrotoluenes can be used as both a plasticizer and a deterrent. Table 6 represents composition data from six powder

Table 6. Smokeless powder compositions from manufacturers' material safety data sheets

Manufacturer	Base	Nitro cellulose	Nitro glycerin	Dinitro toluene	Diphenyl amine	Ethyl centralite	Dibutyl phthalate
Accurate Arms	Single	<98%		<10%	<2%	<6%	<3%
Accurate Arms	Double	<85%	1–40%	<2%	<2%	<8.5%	<10%
Alliant	Double	% not listed	4–40%		0.5–1%		< 3.5%
Hodgdon	Single	>90%		1–10%	1%		
Hodgdon	Double	>85%	10%		1%		
IMR	Single	>90%		1–10%	1%		
IMR	Double	>85%	10%		1%		
Vihtavuori	Single	>90%			<3%	<6%	
Vihtavuori	Double	>65%	<25%		<3%	<6%	
Winchester	Double	40–70%	10–60%		0.5–1.5%	3–7%	1–5%

manufacturers, taken from their respective material data sheets. Different manufacturers often choose different additives, resulting in the potential individuality of VOC headspaces. Many of these are proven to be more volatile than the explosive itself, offering a characteristic odor signature for canine detection.

The wide variability of VOC signatures for smokeless powders and the variability in the smokeless powders used in the training of the canines tested may preclude the identification of odorants for low explosives with the most common VOC present, diphenylamine, not identified as an odorant for the detector dogs we have tested as discussed later. However, given the high occurrence of diphenylamine in smokeless powders, it may prove prudent to include a controlled source of diphenylamine during odor training, as it has the potential to be an explosive odorant.

6.6. Instrumental analysis of explosive VOCs

Presently, we show some highlights from ongoing research aimed at the identification of the most abundant VOCs above explosive samples used as canine training aids followed by field trials with certified law enforcement detection canines in an attempt to identify odorants. More details can be found elsewhere [35–39]. High explosives including cast and plasticized matrices are compared with smokeless powders and non-explosive training aids. Analysis employed SPME combined with GC/MS and GC/ECD. Figures 5 and 6 show typical headspace SPME–GC/MS total ion chromatograms demonstrating that explosives that contain TNT have 2,4-dinitrotoluene (DNT) and/or 2,4,6-trinitrotoluene (TNT) present as major components in the vapor phase and 2-ethyl-1-hexanol as a common primary component in addition to DMNB for tagged C-4. Cyclohexanone is also observed particularly in newer C-4 samples. Figures 7 and 8 compare some single- and double-based smokeless powders studied by SPME–GC/MS and SPME–GC/ECD, respectively. Compared with the high explosives, there was significantly more variation in the VOCs observed between brands of smokeless powder. Whereas diphenylamine was most commonly observed for single-based powders, nitroglycerin was seen in double-based powders. DNT was also contained in many of the brands regardless of whether they were single or double based sokeless powders. The brands containing DNT and those containing trinitroglycerin have the potential to mimic the high explosives containing TNT or NG, and careful selection of specific smokeless powder brands can potentially be used as an alternative to high explosives in certain training scenarios.

6.6.1. Non-explosive training aids

There is a significant demand for reliable inert training aids that can mimic the odor signature of explosives while being safe to handle. One leading line of such materials is the Non-Hazardous Explosives for Security Training and Testing (NESTT) range from Van Aken International (Rancho Cucamonga, CA). Purified explosive is diluted to 4–8% by silica granules or petrolatum jelly. Results from the petrolatum jelly aids gave

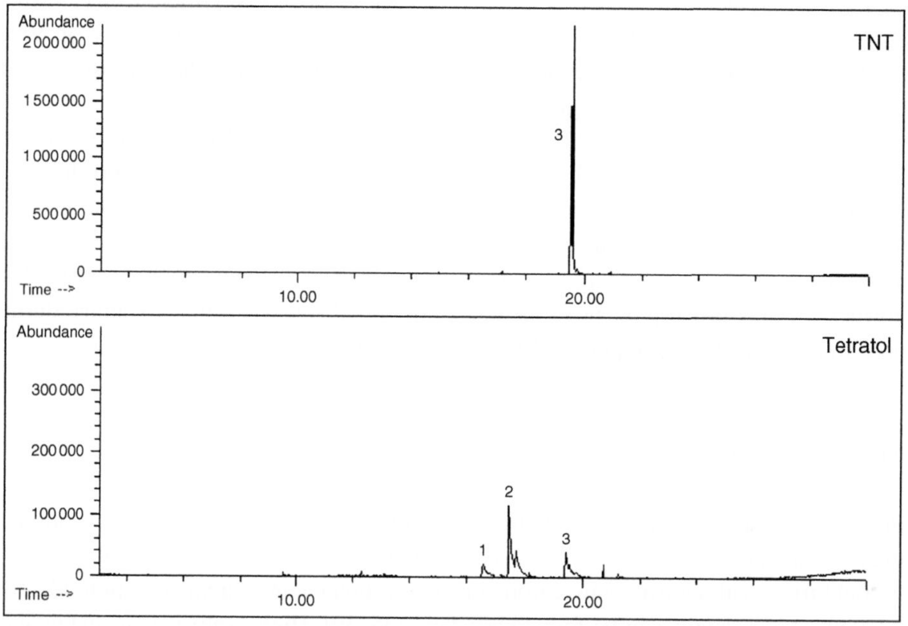

Fig. 5. Headspace SPME–GC/MS of selected cast explosives (1 = 1,3-dinitrobenzene, 2 = 2, 4-dinitrotoluene, 3 = 2,4,6-trinitrotoluene).

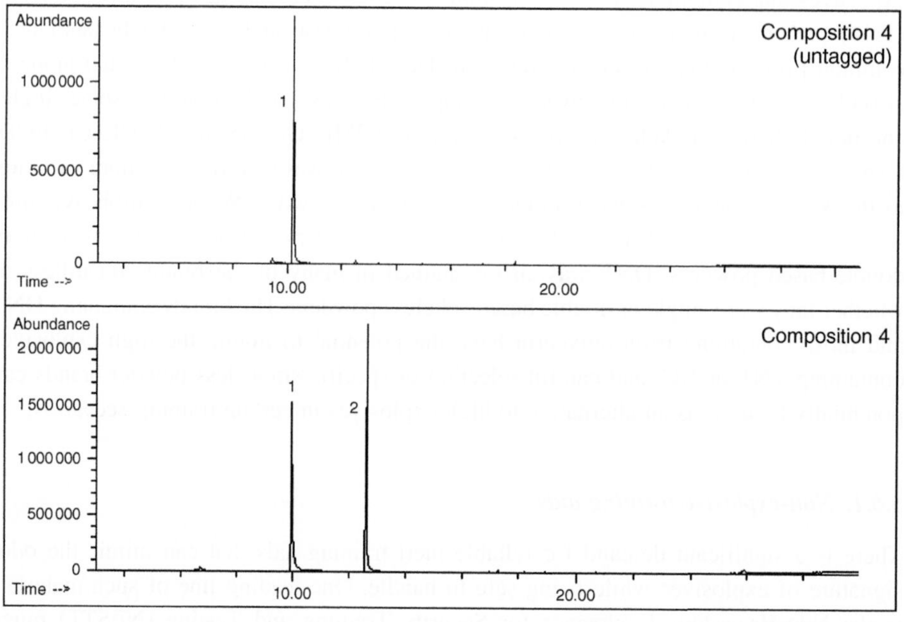

Fig. 6. Headspace SPME–GC/MS of Composition 4 (1 = 2-ethyl-1-hexanol,–2 = 2,3-dimethyl-2,3-dinitrobutane).

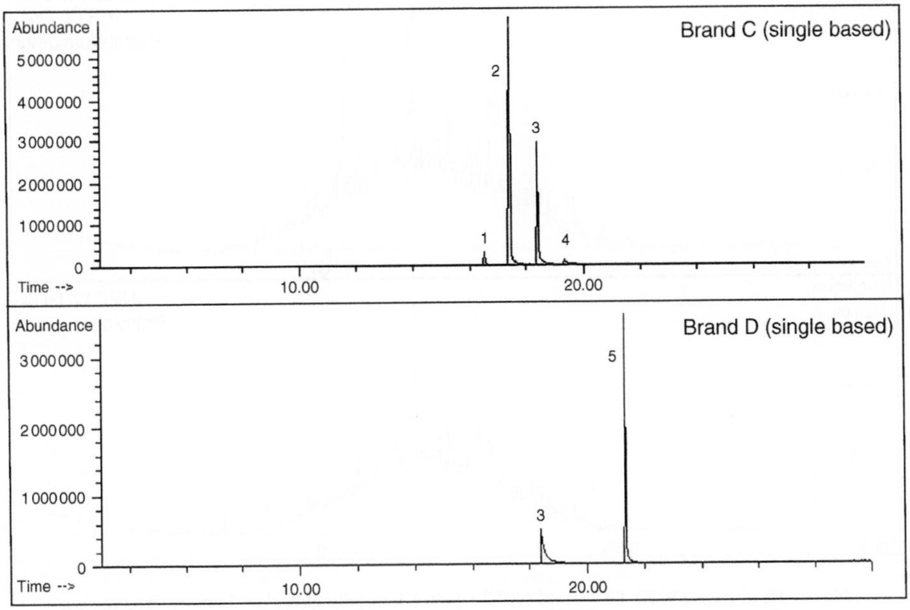

Fig. 7. Headspace SPME–GC/MS of selected single-based smokeless powders (1 = 2,6-dinitrotoluene, 2 = 2,4-dinitrotoluene, 3 = diphenylamine, 4 = 2,4,6-trinitrotoluene, 5 = ethyl centralite).

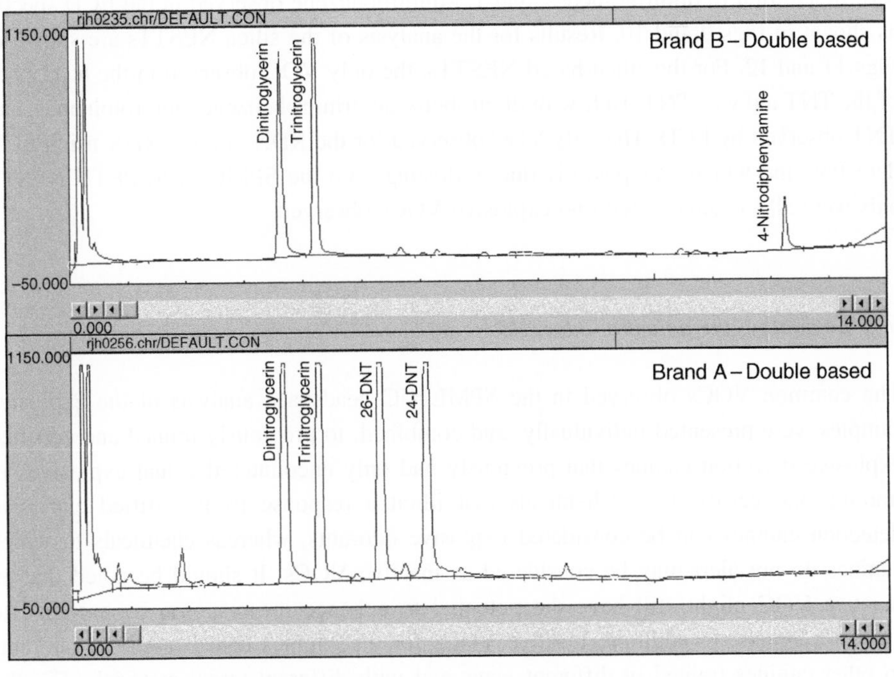

Fig. 8. Headspace SPME–GC/ECD of selected double-based smokeless powders.

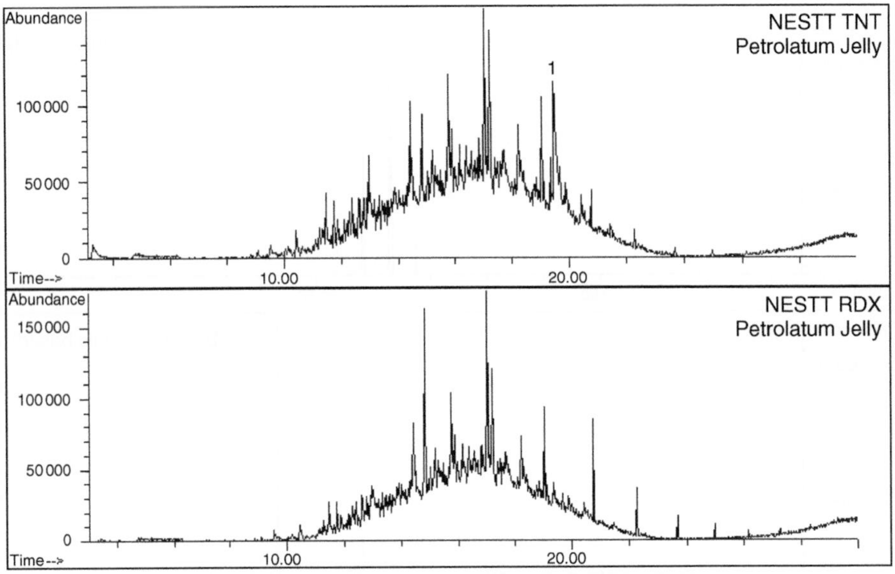

Fig. 9. Headspace SPME–GC/MS of NESTT petrolatum jelly aids (1 = 2,4,6-trinitrotoluene).

a large unresolved hydrocarbon baseline, with only TNT observed in the TNT aid by MS, in addition to dinitrotoluenes, and 1,3-dinitrobenzene observed in the ECD spectra as shown in Figs 9 and 10. Results for the analysis of the silica NESTTs are shown in Figs 11 and 12. For the silica-based NESTTs, the only VOC observed in the headspace of the TNT aid was TNT itself with dinitrobenzene, trinitrobenzene, dinitrotoluenes, and TNT observed by ECD. The only VOC observed for the RDX aid was RDX using ECD detection and which was possibly due to dusting onto the SPME fiber. PETN NESTT aids were also evaluated with no explosive VOCs observed.

6.7. Evaluating the potential odorants

The common VOCs observed in the SPME–GC headspace analysis of the explosive samples were presented individually, and combined, to previously trained and certified explosive detection canines that previously had only encountered actual explosives in training and certification. Chemicals that illicit a response from certified explosive detection canines can be considered explosive odorants, whereas chemicals to which canines do not alert may be considered as inactive VOCs. It should be noted that an inactive VOC might still have the potential to enhance the response by a canine to known odorants. In addition, inactive VOCs for the canines tested might be odorants to other canines trained in different ways and with different target materials. Finally, inactive VOCs might be useful target vapor chemicals for instrumental detectors.

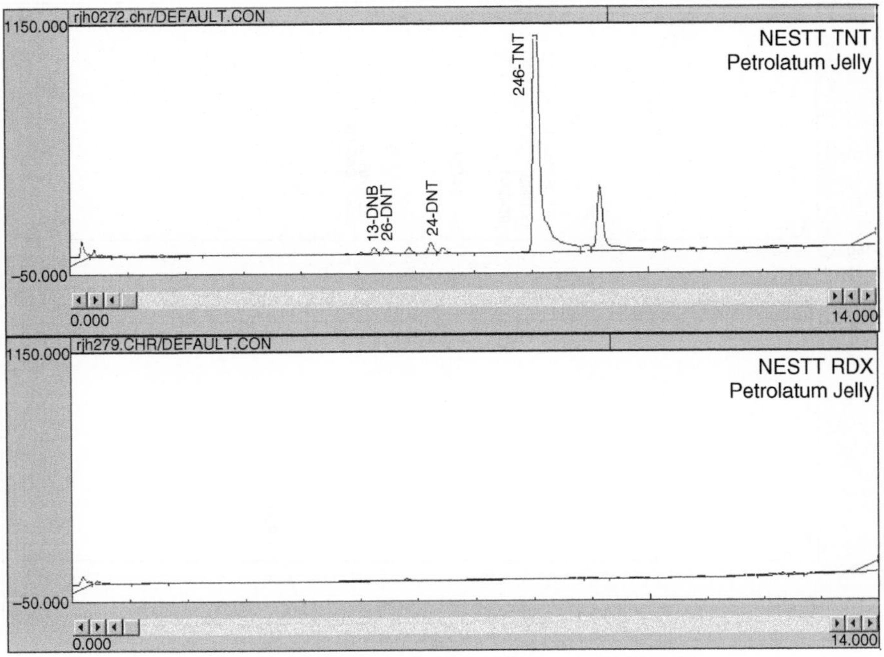

Fig. 10. Headspace SPME–GC/ECD of NESTT petrolatum jelly aids.

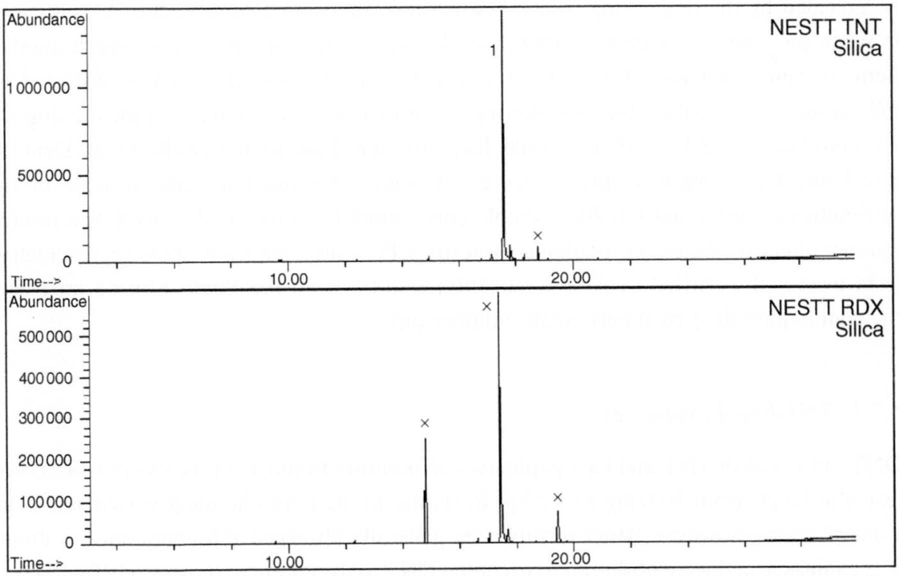

Fig. 11. Headspace SPME–GC/MS of NESTT silica aids (1 = 2,4-dinitrotoluene, x = siloxanes).

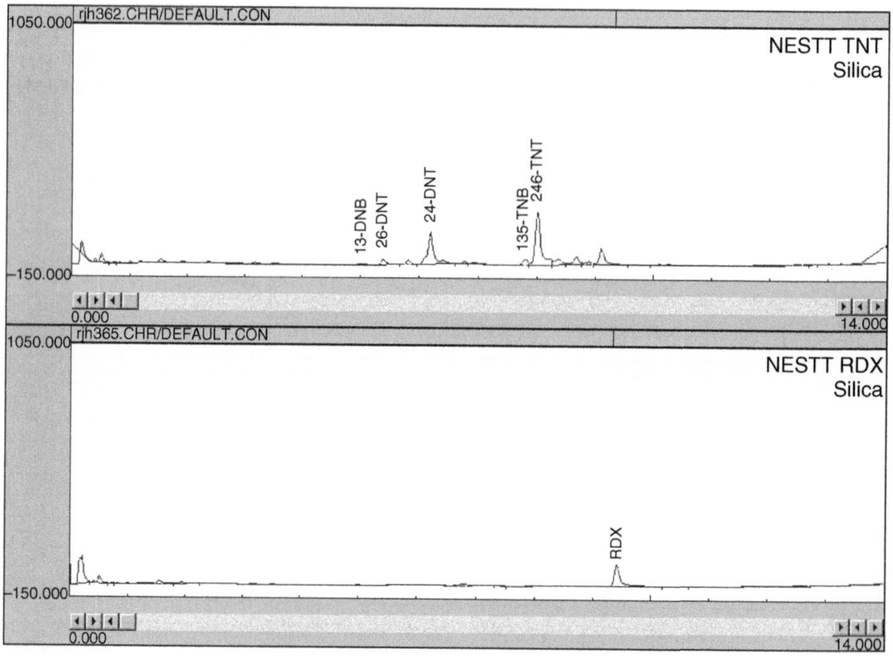

Fig. 12. Headspace SPME–GC/ECD of NESTT silica aids.

A canine 'alert' is a change in the behavior of the canine that can be recognized by the handler and ideally an observer as well. For most explosive detection canines, this is indicated by the dog sitting or lying down near the source of the odor. A 'non-alert' is when the canine is seen to sniff but walk away of its own accord to search another item. A canine 'interest' fills the gap between alert and non-alert and is indicated by the canines investigating but not alerting, such that the handler has to pull the dog off the item being searched, after dog has had sufficient time to alert/walk away. Detailed thresholds for odorants identified have not been determined to date because of the difficulties in delivering reliable variable concentrations in the field as well as a limited number of certified canines available for testing. Presently, we show some representative field tests with certified explosive detection canines from the South Florida area with the canine identified by a participant number only.

6.7.1. TNT-based explosives

DNT, observed in TNT and cast explosives in addition to most smokeless powders, was tested at levels from 10.0 mg to 1.25 g. Response to the odor chemical varied across all concentrations; however, better results were generally observed at higher concentrations. The canines tested were observed to favor the DNT to the parent explosive TNT, although whether this was a result of the chosen odorant during training or a vapor pressure effect remains unclear. Interestingly, 100 μg of analytical-grade TNT received fewer positive responses than 100 μg of military-grade TNT, suggesting that a contaminant, most likely residual DNT in the military product, increased the ease of detection.

6.7.2. Plasticized explosives

2-Ethyl-1-hexanol, present in the headspace of the majority of plasticized explosives, in addition to some smokeless powders, was tested on several occasions. Most success was observed at delivery levels of 5.0 mg, receiving an alert or interest from every canine. A detection level study was also performed for 2-ethyl-1-hexanol, with most dogs alerting to 1–10 mg of 2-ethyl-1-hexanol present. Present often only in fresh C-4, cyclohexanone was presented to the canines in various quantities. One representative canine field trial is summarized in Table 7 with 1.0 g of cyclohexanone presented and two of six canines alerting and a further three of six showing interest. In other tests with significantly lower amounts including 10.0 μl, no alerts were observed, indicating that this is a weak odorant for plasticized explosives. On the contrary, 2-ethyl-1-hexanol present in comparable amounts elicited alerts from the majority of the canines tested, indicating that this is a strong plastic explosive odorant. Table 7 summarizes some results for these two plastic explosive VOCs as well as tests for DNT and TNT. These field results are consistent with laboratory results reported by Waggoner et al. [40], where 2-ethyl-1-hexanol and cyclohenanone were proposed as odor signature chemicals for C-4 under behavioral laboratory conditions utilizing dilution olfactometry. The tagging agent

Table 7. Selected field-testing results of high-explosive odor compounds

Hide content	No alert	Interest	Alert	% Alert
Scratch box containing 100 μl acetonitrile on cotton in LDPE	106, 112, 113, 114, 115, 116	–	–	–
Scratch box containing 2-ethyl-1-hexanol	113	106	112, 114, 115, 116	66.7
Scratch box containing cyclohexanone	114	106, 112, 115	113, 116	33.3
Scratch box containing 1,3-dinitrobenzene	–	106, 113, 114	112, 115, 116	50.0
Scratch box containing 2,4-dinitrotoluene	–	106, 112, 114	113, 115, 116	50.0
Scratch box containing 2,4,6-trinitrotoluene	113, 114	106, 112	115, 116	33.3

DMNB was determined not to be an odorant by the majority of the canine population tested. On three separate field trials, there was not one alert to the DMNB source involving different canine teams and different amounts of DMNB ranging from 100 μg to 5 mg. Although this was not observed to be an odorant, it is recommended that DMNB be included in a training aid selection because of its sole application in the tagging of low-vapor-pressure explosives.

6.7.3. Smokeless powders

Although it is a common low-explosive odorant, DNT was tested previously for the high explosives, and thus, there was little benefit of testing it again by itself for the low explosives. Diphenylamine, ethyl centralite, and 2-nitrodiphenylamine were chosen to further study into possible smokeless powder odorants. As the most common low-explosive VOC, diphenylamine was tested at levels from 100 μg to 5 mg with no alerts seen as shown in Table 8. The Table also highlights the nearly 100% alerts by the canines tested to the TNT training sample as well as 50 μg of 2-ethyl-1-hexanol presented. Ethyl centralite was observed in both single- and double-based powders. Ethyl centralite present in smokeless powders at ~1–3% was also tested at up to 1.0 g samples to be representative of ~35–100 g of powder explosive with the vast majority of canines testing not alerting to this chemical as shown in Table 9. Present only in select double-based powders, 2-nitrodiphenylamine is less commonly observed than either diphenylamine or ethyl centralite. In two field tests, 2-nitrodiphenylamine failed to elicit an alert from the canines. It is highly probable therefore that (i) 2-dinitrodiphenylamine is too uncommon an odor to be chosen by the canines as an active odor, (ii) there are

Table 8. Selected field-testing results of assorted explosive odor compounds

Quart cans containing	No alert	Interest	Alert	% Alert
50 μl acetonitrile on cotton in open vial	101, 103, 106, 110, 113, 116, 117, 118, 119	–	–	–
Diphenylamine (50 μl at 10% w/v acetonitrile)	101, 103, 106, 110, 113, 116, 117, 118	119	–	–
2-Ethyl-1-hexanol (50 μl at 10% w/v acetonitrile)	–	117	110, 101, 103, 116, 106, 113, 119, 118	88.9
2,3-Dimethyldinitrobutane (50 μl at 10% w/v acetonitrile)	101, 103, 106, 110, 113, 116, 117, 118, 119	–	–	–
TNT Aid	–	–	110, 101, 103, 116, 106, 113, 119, 117, 118	100.0

Table 9. Selected field-testing results of double-based powder odor compounds

Hide content	No alert	Interest	Alert	% Alert
Empty electrical box	101, 108, 109, 110, 112, 115, 117, 119, 126	–	–	–
Electrical box containing 50 μl acetonitrile	101, 108, 109, 110, 112, 115, 117, 119, 126	–	–	–
Electrical box containing 2-nitrodiphenylamine (50 μl at 10% w/v acetonitrile)	101, 108, 109, 110, 112, 115, 117, 119, 126	–	–	–
Electrical box containing Ethyl centralite (50 μl at 10% w/v acetonitrile)	101, 109, 110, 115, 119, 126, 127, 128, 130	108	117, 112	16.7
Electrical box containing trinitroglycerin Tablets 25 × 0.4 mg	108, 109, 112, 119, 126, 127	117, 101, 115	110, 128, 130	25.0

other more common odors such as DNT and trinitroglycerin present in most powders that negate the requirement for the canines to 'learn' a new odor, or (iii) the canines used in this study had simply not been trained on any powder brands that contain 2-nitrodiphenylamine. The wide variability in observed odor chemical signatures for smokeless powders and the variability in the smokeless powders used in the training of the canines tested may preclude the identification of active odor chemical(s) for low explosives. Once again, however, although diphenylamine was not observed to be an odorant, it is recommended that diphenylamine be included in a training aid selection because of its nearly universal presence in single- as well as double-based smokeless powders.

6.7.4. NESTT aids

The complete range of NESTT aids (including the blank distracters) was obtained and placed in the same metal hide boxes used in the field trials earlier. Again, negative controls (the blank distracters) and positive controls (real explosives) were made available separately. The dogs had difficulty in locating the NESTT aids in several separate tests. In early field work, where NESTT Silica TNT and RDX aids had been used as positive controls during TNT and plasticized explosive odor chemical studies, and the handlers had familiarity with these aids, we observed some interest and alerts to NESTT samples. However, under double-blind conditions where canines had never been exposed to

Table 10. Selected field-testing results of silica-based NESTT aids

Hide Content	No Alert	Interest	Alert	% Alert
Quart Can containing NESTT Distracter 5 g silica in open vial	101, 103, 108, 109, 112, 113, 116, 117, 118	–	–	–
Quart Can containing NESTT Chlorate 5 g silica in open vial	101, 103, 108, 109, 112, 113, 116, 117, 118	–	–	–
Quart Can containing NESTT Nitrate 5 g silica in open vial	101, 103, 108, 109, 112, 113, 116, 117, 118	116	–	–
Quart Can containing ½ lb Deta Sheet (AID)	–	–	101, 103, 108, 109, 112, 113, 116, 117, 118	100
Quart Can containing NESTT PETN 5 g silica in open vial	101, 103, 108, 109, 112, 113, 116, 117, 118	–	–	–
Quart Can containing NESTT RDX 5 g silica in open vial	101, 103, 108, 109, 112, 113, 117, 118, 119	116	–	–
Quart Can containing NESTT TNT 5 g silica in open vial	101, 103, 108, 109, 112, 113, 116, 117, 118	–	–	–

NESTT samples, no alerts were observed as seen in Table 10 with only interest from one dog to the nitrate and RDX aid but 100% alerts to the actual explosive. Similar results were obtained using the petroleum jelly-based NESTTs demonstrating a lack of consistency in the results when using NESTT aids and that the mode of delivery may play an important role in the available odor of these aids. The significantly different headspace VOCs observed over NESTT samples and actual explosives and the poor results in canine field tests with NESTT samples appear to significantly limit their utility for training purposes. It is possible that a canine trained on the silica NESTT RDX training aid could alert to a C-4 sample containing RDX using RDX as an odorant; however, this chemical is present in a significantly lower quantity than the odorant 2-ethyl-1-hexanol and may have limited availability depending on the packaging of the explosive.

7. Conducting the certification

Assuring the reliability of the detection team requires a regular training regime and documentation of consistent imprinting with odorants as well as a regular completion of a certification program.

7.1. ATF recommendations

The ATF uses a food reward training methodology exposing canines to five basic explosive groups including chemicals used in an estimated 19 000 explosive formulations and has trained over 300 explosive detection canines, providing them both to domestic and to international agencies [41]. The ATF has an Odor Recognition Proficiency Standard (ORT) for explosive detection canines, which involves lines of one gallon paint cans and within each gallon can is a quart-sized can that may contain the explosive aid or distracting odor. The ATF states that to earn ATF certification, all dogs must pass a blind test and successfully detect 20 different explosive odors, two of which they were never exposed to during training. The ATF designates mandatory explosives, which include Black Powder, Dynamite, PETN, RDX, TNT, and double-based Smokeless Powder, and more than a dozen 'elective' explosives, which include Black Powder substitutes, Emulsions, SEMTEX, Slurries, single-based Smokeless Powder, Tetryl, and Water Gels.

It should be noted that the ORT gives a list of recommended explosives for a representative certification, and not an all-inclusive training kit, and much like a driving test does not cover all aspects of handling the vehicle, the student is expected to be fully capable of driving in all conditions prior to sitting the test. At first glance, this seems an extensive if not exhaustive list of explosive products; however, several of the elective explosives are duplicated in name. Many of the elective explosives therefore have the likelihood of duplicating each other and/or the mandatory explosives, and there is a significant potential for the choice of even 20 explosives to overlook the odor of some explosive combinations. The Federal Register list of explosive materials, administered by the ATF, contains 236 items [40] with scant published information about the major VOCs that might be used to optimize the selection of these targets for training and certification. Further confusion can arise from the use of common names in describing formulated explosives with similar names containing very different ingredients and those with very different name yet similar formulations. For example, most blasting agents are binary in nature, and water gel, emulsion, and slurry are three names often used to describe the same explosive product. Similarly, cast boosters may include Composition B and/or Tetryl.

Table 11 lists several formulations of Cast Primer, notably each one contains TNT, often in addition to other high explosives also on the ATF mandatory list, and thus, the use of those cast explosives would only duplicate odors at the expense of others.

Table 11. Different cast booster compositions

Cast booster/cast primer	Contents
Amatol	TNT + Ammonium nitrate
Composition B	TNT + RDX
Cyclotol	TNT + RDX
Pentolite	TNT + PETN
Tetratol	TNT + Tetryl
Torpex	TNT + RDX + Aluminum
Tritonal	TNT + Aluminum

7.2. Recommendations from scientific study

Examples of an optimized choice of explosive products to use in training regimes are given in Table 12 with 10 explosive products shaded that cover the majority of explosive odorants, both individually and in combination. Rather than performing regular training on all 28 targets that would be difficult to administer, careful selection of say 10 samples could be employed. Multiple smokeless powders should be included in a complete training regime. Further classification of the single- and double-based smokeless powders suggests the rotation of powders from various manufacturers, including Hodgdon, IMR, and Vihtavuori Lapua (particularly single-based Vihtavuori powders). Additional studies should allow for the grouping of the most important powders to be included in training. The remaining explosives (non-shaded) may also be included as additional non-essential training material for variability. Although HMX and Tetryl are not included in the 10 primary targets recommended here, there are few examples of those explosives being used in commercial or military products without the presence of TNT or any other explosive. It is not wise to use a combination explosive such as Composition B in place of two separate targets such as TNT and RDX. Given that TNT has a vapor pressure three orders of magnitude greater than RDX, it is highly probable that the canine will only imprint on the TNT, leaving them vulnerable to explosives containing only RDX. In the interests of maintaining accurate, concise, and scientifically sound records, it is important to record not only the explosives trained on but also the primary odor chemicals to which the canines have been presented. It is therefore recommended that initial imprinting and regular training include providing the major odorants individually, particularly with the less volatile nitramine explosives, as well as distracters that should include innocuous items which could contain some of the explosive odorants individually but not necessarily in the same proportions and/or may contain additional VOCs such that the signatures are unique and distinguishable.

7.3. Simplifying the selection further

Returning to Table 3 that lists common explosive products and whether or not they fall into one or more of three major groups [i.e., (i) all those containing DNT/TNT, (ii) all those containing NG, and (iii) all those containing plasticizers], it can be shown that 78% (35/45) of the known compositions may be represented by only three target odorants, leaving behind the inorganic salts and a few exceptions. This is not a recommendation to train on so few odorants but illustrates the issues to be considered in the selection of optimal target training odorants. It is also important to note that formulations of smokeless powders and high explosives change over time, and therefore, there should be ongoing research into the dominant VOCs and odorants of current explosives. The optimal number of targets is likely greater than 3 and less than 23, but the final number and identity of target explosives are yet to be agreed upon by the scientific

Table 12. Examples of optimized choice of explosives for canine training

Explosive	Nitro-methane	DMNB	DNT	TNT	EGDN	NG	PETN	NC	RDX	HMX	Tetryl	Ammonium nitrate	Potassium nitrate	Sodium
Amatol				X								X		
Black Powder													X	
Comp C-2			X	X				X	X					
Comp C-3			X	X				X	X		X			
Comp C-4 (untagged)									X					
Comp C-4 (tagged)		X								X				
Comp B				X					X					
Cyclotol				X					X					
Deta Sheet									X					
Det Cord (com)							X							
Det Cord (mil)									X					
Dynamite (com)					X	X								
Dynamite (mil)				X										
HTA 3 Sheet				X						X				
Kinepak (liquid)	X													
Kinepak (solid)												X		

(Continued)

Table 12. (Continued)

Explosive	**Nitro-methane**	**DMNB**	**DNT**	**TNT**	**EGDN**	**NG**	**PETN**	**NC**	**RDX**	**HMX**	**Tetryl**	**Ammonium nitrate**	**Potassium nitrate**	**Sodium**
Pentolite				X			X							
PTX 1				X					X		X			
Pyrodex														X
Semtex A							X							
Semtex H							X		X					
Smokeless (double) brand x						X		X						
Smokeless (double) brand y						X		X						
Smokeless (double) brand z						X		X						
Smokeless (single) brand x			X					X						
Smokeless (single) brand y								X						
Smokeless (single) brand z								X						
Tetratol				X							X			

community; however, progress is being made to aid in the intelligent selection of targets for biological as well as instrumental detectors.

7.4. Complicating the selection and necessary further research

Reliable non-hazardous training aids are needed for use in areas where live explosive aids are not practical and to provide more consistency in the odorants released in training and in order to improve canine detection performance. In the present studies, NESTT aids yielded inconsistent results with the deployed bomb dog teams tested in double-blind studies, with most dogs not alerting to these materials under field operational conditions. These results also show that NESTT aids have potentially undesirable matrix effects with a large hydrocarbon background observed for the petrolatum-based aids and dusting with the silica-based aids. There is definitely a need for improved NESTTs or alternatives that provide reliable results for use in calibrating instruments and evaluating the performance of detector dog teams.

The focus of this chapter has been on commercial and military explosives, but the detection of improvised devices must also be considered. Peroxide explosives such as TATP and HMTD should also be incorporated into training protocols. Owing to the instability of peroxide explosives, extraordinary care must be taken in handling these explosives. It has been demonstrated that inert materials, such as stainless steel bars, can be stored with peroxide explosives and later removed and used as safe training materials [42]. Additional research is needed to determine the odor mimics that might contain peroxide by-products; which can be used as effective training aids free from the hazards of having to prepare and maintain peroxide explosives.

8. Conclusions

By identifying the signature VOCs of explosives, levels of detection may be determined, and better documentation of training and deployment can be employed. In addition, identification of odor signature chemicals aids in the selection of the fewest number of target substances needed for optimal training and facilitates the development of reliable, cost-effective, non-hazardous odor mimics that can be used to enhance the capabilities of detector dogs. Reviews of electronic noses have highlighted the current limitations of instrumental methods with Yinon [16] concluding that electronic noses for detecting explosives have a long way to go before being field operational and Gopel [14] concluding that, for most applications, the performance of electronic noses containing sensor arrays is insufficient compared with established analytical instruments such as GC/MS. A recent extensive review of instrumentation for trace detection of high explosives concluded that there is still no instrument available that simultaneously solves the problems of speed, sensitivity, and selectivity required for the real-time detection of explosives [13]. For the foreseeable future, canines shall remain the gold standard in real-time, scent-to-source, standoff explosive detection. However, electronic sensors continue to improve and are important orthogonal detectors to be used in combination with biological detectors to maximize the detection of explosives in the field. As scientific studies expand our currently limited understanding of the molecular basis of olfactory chemoreception, the performance of detection canines will likely improve as will their use as models for improved electronic detectors of explosives.

References

[1] K.G. Furton and D.P. Heller, *Can. J. Police Secur. Serv.*, 3/2 (2005) 97.
[2] J. MacDonald, J.R. Lockwood, J. McFee, T. Altshuler, T. Broach, L. Carin, R. Harmon, C. Rappaport, W. Scott and R. Weaver, Appendix T, *Alternatives for Landmine Detection*, RAND Corporation, Santa Monica, CA, 2003, pp. 285–299.
[3] K.G. Furton and L.J. Myers, *Talanta*, 54 (2001) 487–500.
[4] J.M. Johnston and M. Williams, *Enhanced Canine Explosive Detection: Odor Generalization, Unclassified Final Report for Contract No. DAAD05-96-D-7019*, Office of Special Technology, 1999.
[5] U.J. Meierhentich, J. Golebiowski, X. Fernandez and D. Cabrol-Bass, *Angew. Chem. Int. Ed. Engl.*, 43 (2004) 6410–6412.
[6] R. Hudson, *Chem. Senses*, 25 (2000) 693.
[7] S. Firestein, *Nature*, 413 (2001) 211.
[8] A.B. Malnic, J. Hirono, T. Sato and L.B. Buck, *Cell*, 96 (1999) 713.
[9] P. Quignon, E. Kirkness, E. Cadieu, N. Touleimat, R. Guyon, C. Renier, C. Hitte, C. Andre, C. Fraser and F. Galibert, *Genome Biol.*, 4/12 (2003) R80.
[10] G.S. Settles and D.A. Kester, *Proc. SPIE*, 4394 (2001) 108.
[11] G.S. Settles, *J. Fluids Eng.*, 127 (2005) 189.
[12] E. Morrison, Physiology/Anatomy of the Olfactory System, Presentation at the 4th National Detector Dog Conference, Auburn, AL, 2005.
[13] D.S. Moore, *Rev. Sci. Instrum.*, 2004, 75(8) 2499.
[14] W. Gopel, *Sens. Actuators A*, 65 (2000) 70.
[15] J.M. Perr, K.G. Furton and J.R. Almirall, *J. Sep. Sci.*, 28(2) (2005) 177–183.
[16] J. Yinon, *Anal. Chem.*, (2003) 99A.
[17] C. Cummings, *Proceedings of the 1st Olfactory-Based Systems for Security Applications Meeting (OBSSA)*, London, 2004.
[18] National Research Council, *Existing and Potential Standoff Explosives Detection Techniques*, National Academies Press, Washington, DC, 2004.
[19] http://www.apopo.org/mediagallery/Limpopo-tests.doc, accessed on March 12, 2005.
[20] J. MacDonald, J.R. Lockwood, J. McFee, T. Altshuler, T. Broach, L. Carin, R. Harmon, C. Rappaport, W. Scott and R. Weaver, Appendix S, *Alternatives for Landmine Detection*, RAND Corporation, Santa Monica, CA, 2003, pp. 285–299.
[21] R.L. Fisher, J.E. Anderson, K.D. Hutchinson, J. DiBenedetto and K. Williams, *Trace Chemical Detection Through Vegetation Sentinels and Fluorescence Spectroscopy*, available at www.epa.gov/nerlesd1/land-sci/srsv/images/fischer.pdf, accessed on August 5, 2006.
[22] http://www.aresa.dk/aresa-home-english2.html, accessed on October 31, 2004.
[23] M.S. Antunes, S.-B. Ha, N.Tewari-Singh, K.J. Morey, A.M. Trofka, P. Kugrens, M. Deyholos and J.I. Medford, *Plant Biotech. J.*, 2006, pp. 605.

[24] M. Altamirano, L. Garcia-Vallida, M. Agrelo, L. Sanchez-Martin, L. Martin-Otero, A. Flores-Moya, M. Rico, V. Lopez-Rodas and E. Costas, *Biosens. Bioelectron.*, 19 (2004) 1319.
[25] L.L. Looger, M.A. Dwyer, J.J. Smith and H.W. Hellinga, *Nature*, 423 (2003) 185.
[26] A. Blake, W. Mail, *Nanodog that Can Sniff out Explosives*, available at http:/icwales.icnetwork.co.uk/0100news/1100education/tm_objectid = 174 48354&method = full&siteid = 50082&headline = scientists–nanodog-can-sniff-out-explosives-in-terror-battle-name_page.html, accessed on July 29, 2006.
[27] National Research Council, *Containing the Threat from Illegal Bombings: An Integrated National Strategy for Marking, Tagging, Rendering Inert, and Licensing Explosives and Their Precursors*, National Academy Press, Washington, DC, 1998, p. 384.
[28] F. Dubnikova, R. Kosloff, J. Almog, Y. Zeiri, R, Boese, H. Itzhaky, A. Alt and E. Keinan, *J. Am. Chem. Soc.*, 127 (2005) 1146.
[29] R. Mistafa, *K9 Explosive Detection*, Detselig Enterprises, Alberta, Canada, 1998.
[30] J. Yinon, *Modern Methods and Applications in Analysis of Explosives*, Wiley, New York, 1993.
[31] J.A. Given, *Controlled Substance Training Aids for DoD*, 3rd National Detector Dog Conference, North Miami Beach, Miami, FL, 2003.
[32] National Research Council, *Black and Smokeless Powders: Technologies for Finding Bombs and the Bomb Makers*, National Academy Press, Washington, DC, 1998.
[33] *Progress Study Report of Marking, Rendering Inert and Licensing of Explosives Materials*, US Department of the Treasury, Washington, DC, 1998.
[34] R.M. Heramb and B.R. McCord, *Forensic Sci. Commun.*, 4(2) 2002, available at http://www.fbi.gov/hg/lab/fsc/backissu/april2002/mccord.htm.
[35] K.G. Furton, R.J. Harper, J. M. Perr and J.R. Almirall, Optimization of biological and instrumental detection of explosives and ignitable liquid residues including canines, SPME/ITMS and GC/MS^n, in sensors and command, control, communications and intelligence technologies for Homeland Defense and Law Enforcement, in E.M. Carapezza (Ed.), *Proc. SPIE*, 5071 (2003) 183–192.
[36] N. Lorenzo, T. Wan, R.J. Harper, Y.-L. Hsu, M. Chow, S. Rose and K.G. Furton, *Anal. Bioanal. Chem.*, 376 (2003) 1212–1224.
[37] R.J. Harper, J.R. Almirall and K.G. Furton, *Proceedings of the 8th International Symposium on the Analysis and Detection of Explosive (ISADE)*, Ottawa, Canada, 2004.
[38] R.J. Harper, A.M. Curran, J.R. Almirall and K.G. Furton, *Proceedings of the 1st Olfactory-Based Systems for Security Applications Meeting (OBSSA)*, London, 2004.
[39] R.J. Harper, J.R. Almirall and K.G. Furton, Identification of Dominant Odor Chemicals Emanating from Explosives for use in Developing Optimal Training Aid Combinations and Mimics for Canine Detection, *Talanta*, 67 (2005) 313–327.
[40] M. Williams, J.M. Johnston, M. Cicoria, E. Paletz, L.P. Waggoner, C.C. Edge and S.F. Hallowell, in A.T. DePersia and J.J. Pennella (Eds), *Proc. SPIE*, 3575 (1998) 291–301.
[41] ATF, *Arson & Explosives*, available at http://www.atf.treas.gov/explarson/canine.htm, accessed on June 20, 2005.
[42] A. Schoon, *Detecting Illicit Substances: Explosives & Drugs*, Gordon Research Conference, Les Diablerets, Switzerland, August 28–September 2, 2005.
[43] L.R. Whalen, *Virtual Canine Anatomy*, Colorado State University, Fort Collins, Co, available at http://www.cvmbs.colostate.edu/vetneuro/index2.htm, accessed on January 13, 2006.

Index

W

X

Z

Printed and bound by CPI Group (UK) Ltd, Croydon, CR0 4YY
03/10/2024
01040328-0013